A Critical Review of the Current Approaches and Procedures of Plant Genetic Resources Conservation and Facilitating Use: Theory and Practice

A Critical Review of the Current Approaches and Procedures of Plant Genetic Resources Conservation and Facilitating Use: Theory and Practice

Editors

Johannes M. M. Engels
Andreas W. Ebert

MDPI

Basel • Beijing • Wuhan • Barcelona • Belgrade • Novi Sad • Cluj • Manchester

Editors
Johannes M. M. Engels
Alliance of Bioversity
International and CIAT
Rome
Italy

Andreas W. Ebert
Independent Researcher
Schwaebisch Gmuend
Germany

Editorial Office
MDPI
St. Alban-Anlage 66
4052 Basel, Switzerland

This is a reprint of articles from the Special Issue published online in the open access journal *Plants* (ISSN 2223-7747) (available at: https://www.mdpi.com/journal/plants/special_issues/Plant_Genetic_Resources_Conservation).

For citation purposes, cite each article independently as indicated on the article page online and as indicated below:

Lastname, A.A.; Lastname, B.B. Article Title. *Journal Name* **Year**, *Volume Number*, Page Range.

ISBN 978-3-7258-0589-1 (Hbk)
ISBN 978-3-7258-0590-7 (PDF)
doi.org/10.3390/books978-3-7258-0590-7

Cover image courtesy of Andreas W. Ebert

Contents

About the Editors

Johannes M. M. Engels

Johannes M. M. Engels is of Dutch origin, studied plant breeding, genetics, pedagogic, and didactics at the Agricultural University at Wageningen, the Netherlands, and graduated in October 1974. He completed his Ph.D. on the insights gathered on cacao genetic resources in 1986. Through his involvement in the German Agency for International Cooperation (GIZ), he implemented economic development projects in Costa Rica and Ethiopia for roughly six years each, focusing on establishing and operating regional/national genebanks. During his subsequent employment from July 1988 onwards by the International Board for Plant Genetic Resources (IBPGR), one of the CGIAR research institutes administered by FAO, he coordinated plant genetic resources conservation and use activities in South and Southeast Asia and later in Europe. The latter included coordinating the establishment and operation of the virtual European genebank AEGIS as part of the European Cooperative Programme for Plant Genetic Resources (ECPGR). He also assumed responsibilities for the coordination of IPGRI's (the IBPGR successor institute) research program, during which he closely collaborated with and contributed to the coordination of the CGIAR genebanks and was heavily involved in joint IPGRI-FAO genetic resources activities, including the legal debates and developments. This enabled him to obtain solid experience with the conservation, management, and use of plant genetic resources for food and agriculture, as well as coordination and oversight activities. After almost 50 years, he retired and operated a small farm in the hills of Umbria, Italy.

Andreas W. Ebert

Andreas W. Ebert graduated in May 1976 from the University of Hohenheim, Stuttgart, Germany, with a Diploma in Agricultural Sciences. In February 1980, he obtained his Ph.D. with distinction (magna cum laude) for his work on 'Hormonal aspects of crop regulation in apple' at the Institute for Fruit and Vegetable Production and Viticulture from the same University. Thereafter, he held a Postdoc position at the Glasshouse Crops Research Institute in Littlehampton, UK, where he researched assimilate translocation in tomatoes. At a later stage, he worked on the somatic embryogenesis of coconuts at Wye College, University of London.

He dedicated his entire professional life to agricultural research for development in the tropics and subtropics. He worked for the German Agency for International Cooperation (GIZ) from 1981 to 2002 on various development projects in Brazil, the Philippines, West Africa, Malawi, and China. From 2002 to 2008, he joined CATIE in Costa Rica as Team Leader of an interdisciplinary plant genetic resources group, directed CATIE's genebank, and served as Professor at its Graduate School. From 2008 until his retirement in December 2015, he was Genebank Manager and Global Theme Leader—Germplasm (germplasm conservation, utilization, and gene discovery) at the World Vegetable Center's headquarters in Taiwan.

He is passionate about the conservation and sustainable use of plant genetic resources, especially landraces, farmers' varieties, and wild crop relatives—critical for crop improvement and adaptation to climate change.

plants

MDPI

Editorial

How Can We Strengthen the Global Genetic Resources' Conservation and Use System?

Johannes M. M. Engels [1,*] and Andreas W. Ebert [2]

[1] Independent Researcher, 06062 Citta' della Pieve, PG, Italy
[2] Independent Researcher, 73529 Schwäbisch Gmünd, Germany; ebert.andreas6@gmail.com
* Correspondence: j.engels@cgiar.org

Abstract: Genetic resources serve as the foundation of our food supply and are building blocks for the development of new crop varieties that support sustainable crop production in the face of climate change, as well as for the delivery of healthy diets to a continuously growing global population. With the encouragement of the FAO and with technical guidance and assistance from the International Board for Plant Genetic Resources (IBPGR), almost 2000 genebanks have been established worldwide for the ex situ conservation of genetic resources since the middle of the last century. The global genetic resources' conservation and use system has evolved over several decades and presents apparent weaknesses, without a clear blueprint. Therefore, a Special Issue (SI) of *Plants* on 'A Critical Review of the Current Approaches and Procedures of Plant Genetic Resources Conservation and Facilitating Use: Theory and Practice' was initiated. This SI comprises 13 review and research papers that shed light on the history and the political dimensions of the global system; its current strengths, weaknesses, and limitations; and how the effectiveness and efficiency of the system could be improved to satisfy the germplasm users (plant breeders, researchers) and benefit consumers and society at large. This SI provides insight into new approaches and technical developments that have revolutionised ex situ conservation and the use of germplasm and related information. It also reflects on complementary conservation approaches (in situ, on-farm, home gardens) to ex situ genebanks, as well as how—through new forms of collaboration at national, regional, and global levels and through stronger links between public genebanks—synergies between the private breeding sector and botanic garden community could be achieved to strengthen the global conservation and use system. Special attention has also been given to the governance of genetic resources and access and benefit-sharing issues that increasingly hamper the needed access to a wide range of genetic resources that is essential for plant breeders to fulfil their mission.

Keywords: ex situ conservation; global germplasm conservation and use systems; International Treaty on Plant Genetic Resources; Convention on Biological Diversity; Nagoya Protocol; governance of genetic resources; access and benefit sharing; genomics; phenomics; plant breeding

Citation: Engels, J.M.M.; Ebert, A.W. How Can We Strengthen the Global Genetic Resources' Conservation and Use System? . *Plants* **2024**, *13*, 702. https://doi.org/10.3390/plants13050702

Received: 24 February 2024
Accepted: 28 February 2024
Published: 1 March 2024

1. Introduction

Targeted and considerable conservation efforts initially focused on threatened animals and subsequently on threatened ecosystems. The establishment and management of national parks and other forms of 'in situ' nature conservation were the mainstay conservation approach. It was only during the first half of the last century that crop genetic resources started to receive specific attention, often connected with plant breeding activities. A more 'systematic' ex situ conservation of plant genetic resources was possibly marked by the collection expeditions of Nikolai Iwanowitsch Vavilov, the publication of the results of his collection efforts, and the subsequent genetic diversity studies that he conducted in the 1920s/1930s [1–4]. This initial global and systematic approach was further strengthened during the so-called Green Revolution in the 1960s/1970s [5]. This resulted in more focused research on appropriate tools and methods, and a global coordination of collection and conservation efforts was undertaken by the FAO and the International Board for Plant Genetic Resources (IBPGR) [6]. Subsequently, many new institutional and national genebanks were created, in situ and on-farm conservation sites were established, and a policy and legal framework was developed and agreed upon [7,8]. Simultaneously, increasing attention was given to access and use aspects of Plant Genetic Resources for Food and Agriculture (PGRFA).

The resulting conservation and use framework was never purposely 'designed' for efficient and effective long-term conservation; it was rather the result of a spontaneous 'process' based on limited available scientific knowledge and undoubtedly influenced by political considerations [9]. Further influences resulted from the political debates and agreements such as the Convention on Biological Diversity (CBD) and, subsequently, as part of the coordinating efforts of the FAO Commission on Genetic Resources for Food and Agriculture in Rome, which were loaded with many issues, often unrelated to the subject of the effective and efficient (long-term) conservation and use of plant genetic resources, e.g., benefit sharing and property issues [10]. Since roughly the turn of the 20th century, many new technologies and scientific understandings have become available, including molecular genetics and genomics, phenomics, informatics and bioinformatics, as well as communication technologies. These technologies and the resulting scientific knowledge have revolutionised the possibilities of better understanding crop genetic diversity and improving conservation and have facilitated the use of PGRFA [11].

The developments mentioned above serve as the backdrop of this Special Issue on 'A Critical Review of the Current Approaches and Procedures of Plant Genetic Resources Conservation and Facilitating Use: Theory and Practice'. It aimed to include descriptions of the current practices/state of the art of routine conservation operations, followed by a critical review of what could or should be done (in theory) considering the newly available technologies and scientific knowledge as well as the experiences made with the current system. The scope and focus of this Special Issue were on the ex situ conservation of plant agrobiodiversity. However, due attention was also expected for the broader issues and circumstances in which the conservation efforts and the global system are embedded. Whereas short- and medium-term conservation aspects are considered necessary, especially to facilitate the use of conserved PGRFA, the primary focus of this Special Issue intended to be on the long-term conservation efforts that are expected to be rational, effective, and efficient, including the related facilitation of use efforts. It was further expected that papers contributing to this Special Issue would include a section on moving from the current scenario into a more rational, efficient, and effective long-term conservation and facilitated use approach. Social, economic, and political considerations and developments were also expected to be addressed, where relevant, to ensure a widely agreeable and supported acceptance of the proposed way forward.

In the following section, the main ideas of the 13 papers that have been published as part of the Special Issue are briefly presented. The grouping of the papers was based on a logical approach. One paper that addresses policy issues regarding the availability of PGRFA under the Plant Treaty [12], published in the section 'Plant Genetic Resources' of *Plants*, was originally also scheduled to be part of this Special Issue. In Section 3, key messages from the published papers are formulated with the primary aim of demonstrating how these contribute to strengthening the global PGRFA conservation and use system.

2. Highlights of the Papers Published in This Special Issue

There is no concept paper on which the global plant genetic resources' conservation and use system was built. The system as we know it today is the result of experiences gained with targeted ex situ conservation efforts that were triggered by massive losses of genetic resources, particularly landraces. These losses were caused by the rapid spread of newly bred varieties of the major food crops in the 1960s, especially in tropical and subtropical countries as part of the Green Revolution. The global conservation and use system was also impacted by the political debates that ensued during the late 1980s. Increasingly, the decisions on how best to proceed with the conservation in a technical manner were based on targeted research results. The history of the development of the global system in the context of the political and legal framework and the main components it eventually entailed are addressed in the paper by Engels and Ebert [5]. In a second paper, Engels and Ebert [13] described the role of active and base collections and the importance of linking germplasm conservation with use. The authors reviewed the strengths and weaknesses of the current global system and made several recommendations on how the inherent weaknesses could be overcome and how improvements could be made.

With the encouragement of the FAO Commission on Genetic Resources for Food and Agriculture to establish genebanks for the storage of collected genetic resources and with technical guidance and assistance from the IBPGR, almost 2000 genebanks have been established worldwide since the middle of the last century. Because plant genetic resources, genebanks, and the genetic resources they contain can be threatened, Herbold and Engels [14] analysed the different types of risks that undermine the safety and security of the genebanks and their collections and suggested remedies for how such risks can be reduced. Another critical aspect of the efficiency and effectiveness of genebanks is the assessment of their quality performance. This is addressed in a paper by Lusty et al. [15] in which the authors describe a genebank quality management system (QMS) for the long-term conservation of genetic resources, initially adopted by the genebanks of the CGIAR in 2014, known as the 'Genebank QMS'. The Genebank QMS is based on the FAO Genebank Standards, which defines performance targets, and adheres to international regulatory policies. It is implemented along the entire genebank operation, from germplasm acquisition to distribution, and relies on sound scientific practices that are regularly updated. The Genebank QMS provides a transparent, trusted framework for efficient operation, monitoring, auditing, and external review.

The development and application of new genomic tools and high-throughput phenotyping technologies, combined with the use of advanced information technologies to analyse the resulting vast datasets, are essential to facilitate and strengthen the effective and efficient conservation of PGRFA and their use in modern plant breeding. Volk et al. [11] provided several examples of the successful integration of genomic and phenomic approaches and demonstrated how vital access to high-quality and standardised data is for present and future PGRFA conservation and use efforts. They also indicate that advances in statistical prediction may change how germplasm characterisation data are used for further evaluation and breeding. Visioni et al. [16] provided an overview of the management and exploitation practices of barley genetic resources, predominantly illustrated with examples from the international genebank of ICARDA. They explore the relationship between genebanks and participatory plant breeding and offer insights into the diversity and utilisation of barley genetic resources. The authors highlight the importance of these genetic resources for boosting barley productivity, addressing climate change impacts, and meeting the growing food demands. They also emphasise the need for complementary genotypic and phenotypic evaluation of genebank collections to efficiently use the existing but vastly untapped biodiversity of barley genetic resources in future breeding programmes.

As is widely known, much of the world's genetic diversity of domesticated crops can still be found in farmers' fields as well as in gardens around homesteads or in garden plots in urban areas. The paper by Korpelainen [17] describes the importance of home gardens as an 'ecosystem' that harbours important and unique diversity that has sometimes developed over centuries, making a significant contribution to food and nutrition security at the local level. Home gardens have facilitated the adaptation and domestication of plants, including to extreme or specific ecological conditions, and have thus contributed to the diversification of cultivated plants. It is well known that genetic resources of public interest, not directly linked with the agricultural sector, are also conserved ex situ in genebanks. One, and possibly the most important example of such conservation efforts, is provided by botanic gardens that have increasingly established seed and field genebanks for the long-term conservation of plant genetic resources, including those of crop wild relatives and wild food plants. Unfortunately, the cooperation between the agricultural and botanic sectors is still relatively weak, and a paper by Breman and colleagues [18] from the Millennium Seed Bank at Kew Gardens, UK, intends to provide arguments and reasons to establish or strengthen such cooperation. The authors highlight the importance of networking and facilitating access to the conserved materials as well as the need to combine in situ and ex situ conservation approaches.

Whereas the initial focus of the ex situ conservation of PGRFA has been predominantly on the establishment and operation of ex situ conservation facilities, i.e., genebanks, over time, it was realised that effective and efficient conservation could only be achieved through adequate cooperation between genebanks at the national, regional, and global level. Taking into account the establishment of growing conservation activities of wild and cultivated plants, the increasing research efforts to improve conservation efforts and to facilitate use, as well as the increasing importance of coordinating the participation of national researchers in regional and global conservation programmes, the establishment of national conservation programmes and facilities was recognised. Against this backdrop, the paper of Begemann et al. [19] provides insight into the complexity of coordinating and governing such a national system in Germany, a federal state with active conservation, research, and use programmes for food and forestry plants, animals, and aquatic resources. A more specific case of collaboration between genebanks established by private plant breeding companies and publicly funded genebanks has been reviewed by Engels et al. [20]. The authors interviewed private plant breeders, assessed the published literature, and analysed specific existing cooperation arrangements to allow a more informed decision when seeking to strengthen such collaboration at the national, regional, and global level. The regional level represents yet another dimension of network coordination between the conservation of PGRFA and the facilitation of their use by genebanks and national programmes. The paper by van Hintum et al. [21] explored the establishment and operation of a decentralised regional virtual genebank in Europe, i.e., AEGIS, containing unique and important germplasm that has been designated by the respective national coordinators of the regional collaboration programme ECPGR. To further strengthen AEGIS, a system of certified genebanks with proper quality management, guaranteeing the long-term conservation of, and immediate access to, the conserved germplasm materials, is being proposed. Considering the current changes, challenges, and opportunities that impact the conservation and use of PGRFA, Lusty et al. [22] presented their views of an effective global long-term agrobiodiversity conservation system that promotes and facilitates the use of PGRFA. They argued that the rapid expansion of the applications and uses of modern genomic and phenomic technologies and approaches had a transformational impact on breeding, research, and the demand for certain genetic resources and associated data. Genebanks need to be responsive and must adapt to these changing conditions. These trends also provide important opportunities for genebanks to reorganise themselves and become more efficient individually and as a community. Ultimately, future challenges and opportunities will drive more demand for specific and well-documented genetic diversity and provide an important basis for genebanks to gear up.

Over the years, policy issues have gradually gained significant importance. At present, it is difficult to manage genebanks or breed new crop varieties without good knowledge of national, regional, and global policies and legal issues. Under the Convention on Biological Diversity (CBD) and its related Nagoya Protocol, access to PGR became increasingly restricted and cumbersome, resulting in a decrease in germplasm exchange, potentially threatening the future of plant breeding. After a critical review of current access and benefit-sharing regulations regarding PGRFA, Ebert et al. [23] recommended, among others, expanding the scope of the International Treaty on Plant Genetic Resources for Food and Agriculture (ITPGRFA) to include all PGRFA and making them and all related information accessible under a Standard Material Transfer Agreement (SMTA), combined with a subscription system or a seed sales tax, if necessary. Such a transparent, functional, and efficient system would erase legal uncertainties and minimise transaction costs for conservationists, curators, and users of genetic resources, thus aiding plant breeders in fulfilling their mission.

3. Key Messages

3.1. The Current Global Ex Situ Conservation and Use System—A Reflection on Its Inherent Weaknesses and Recommendations for Its Improvement

The current global conservation system has inherent weaknesses and limitations, partially due to its spontaneous creation out of a felt need by concerned scientists and visionaries and the subsequently required adjustments to the evolving political framework and changing realities. Because of its relatively easy and transparent access to PGRFA, the International Treaty, with its multilateral system (MLS) and standard material transfer agreement (SMTA) that stipulates benefit-sharing mechanisms for the use of germplasm, is the most significant policy instrument for PGRFA currently. As of 1 December 2023, the Treaty had 151 contracting parties, including the European Union as a member organisation [24]. However, the MLS is restricted to a list of 35 crops or crop gene pools and 29 grass and forage species that are listed in Annex 1. This is a significant limiting factor as, for example, only a small number of vegetable crops and neglected and underutilised species have been included in the list. Furthermore, commodity crops, like coffee, cacao, and tea, are excluded.

Vegetables and other minor, underutilised crops are known to play a major role in food and nutrition security [25]. Hence, their significant underrepresentation in the global system is a clear weakness. For many minor and underutilised crops and crop wild relatives (CWRs), there are still no comprehensive collections, and considerable gaps remain to be filled.

Apart from the threat to genetic resources in nature and/or farmers' fields as well as in genebanks, genebanks themselves are exposed to a series of risks that may originate from natural hazards, such as earthquakes and storms, but also from political or financial issues. Herbold and Engels [14] undertook risk analyses of 80 important national and international genebanks regarding natural hazards and political and financial risks, and they concluded that there are large differences in the risk exposure of genebanks, making a location- and institution-specific risk assessment indispensable. Such risk assessments would help create more awareness at the local or national (political) level, hopefully resulting in the implementation of measures that mitigate the impact of risks, both for the genebank structures as well as for the safety of the collections.

Well-organised and comprehensive information management in genebanks is the basis for efficient and effective conservation and use. Unfortunately, information management and the online accessibility of accession-level data remain weak in many genebanks, especially at the national level. The rationalisation of collections requires comprehensive accession-level data and is an important step toward more cost-efficient and effective PGRFA conservation and use activities. These are just some weaknesses or limitations of the current global conservation and use system. For more details, please refer to [13]. After a detailed review of the current system, Engels and Ebert [13] recommended several measures that might contribute to a more rational, effective, and efficient long-term ex situ conservation system. Some of these measures include the following:

- Targeted collection for filling genetic and geographic gaps in current ex situ collections of major and particularly minor crops to reach an adequate representation of crop gene pools in ex situ collections. This would also be important to avoid the irreversible loss of genetic diversity due to severe genetic erosion in farmers' fields and in nature.
- Rationalisation of germplasm collections through the determination of unique accessions that will form part of global base collections, and removal of the many duplicates within and among genebanks from the currently existing base collections.
- Strengthening existing and forging new collaborations among genebanks that maintain agricultural (crop) collections, including with the botanic garden community as well as with research institutes that hold collections. These are realistic mechanisms to

increase efficiency and security in genebank and germplasm management, both at the global and national level. Furthermore, establishing stronger linkages with the plant science research community at large would be another step to facilitate coordination, foster collaboration, and facilitate the sharing of responsibilities. The community of public- and private-sector breeders represents a stakeholder group of the current global conservation efforts that has a significant interest in maintaining the adequate genetic diversity of our crops. Strengthening the collaboration with the private sector, especially with the plant breeding companies that operate private genebanks, has been identified as an underdeveloped building block of the global conservation and use system [20]. Strengthening such collaborations will contribute to the more efficient and effective long-term conservation of crop gene pools. This issue is discussed in more detail in Section 3.4.

- In the fields of molecular genetics and information technologies (including artificial intelligence), the world is seeing rapid technological advances. Further (adaptive) research is needed to fully exploit and apply these new technologies to the conservation and use of genetic resources and/or collaboration with specialised or more advanced institutions. Thus, more investments in conservation research and user-oriented supportive research are needed to optimise routine genebank processes and to facilitate conserving and delivering germplasm resources of high quality and in the right form as required by the users.
- Comprehensive, reliable, and easily available information on conserved accessions is a prerequisite to facilitating targeted and sustainable use of conserved genetic resources. Consequently, there is a clear need for a better and more comprehensive accession-level description of the genetic diversity of crop collections maintained in genebanks, including genomic, phenomic, and ecological data [11,13,22]. Furthermore, users require easy access to high-quality data on conserved germplasm and associated metadata.
- Legal certainty and easy and transparent access to conserved genetic resources are possibly the most fundamental requirement to enable and facilitate their use. There are legal frameworks (ITPGRFA, CBD, Nagoya Protocol) in place that regulate germplasm access and related benefit sharing. However, due to the fact that only a limited number of crops fall under the multilateral system of the ITPGRFA and each country is free to establish its own bilateral access rules under the Nagoya Protocol, users often find it difficult to undergo such a time-consuming, bureaucratic, and also costly process, especially when genetic resources from more than one country are needed and legal certainty is not guaranteed. To strengthen and simplify the legal and policy framework, it seems unavoidable to include all PGRFA in the MLS of the International Treaty (or to create another legal system that embraces all PGRFA) to facilitate easy access to germplasm, associated information, and corresponding benefit sharing in a transparent manner.
- A model for a functional and efficient global network of base and active collections is recommended, and a lean international organisation is proposed that assumes responsibilities for the global coordination, facilitation, and oversight of the various global crop gene pool base collection networks. Such a model could build on the existing genebanks of the CGIAR, the World Vegetable Center, and ICBA, as well as on a handful of strong national genebanks that form the core of the current global system on PGRFA.
- The political oversight over the proposed global model network of base collections should remain with the FAO and the Governing Body of the International Treaty.

In summary, the proposed measures include filling genetic and geographic gaps in current ex situ collections; determining unique accessions at the global level for long-term conservation in virtual base collections; intensifying existing international collaborations among genebanks and forging collaborations with the botanic garden community; increasing investment in conservation research and user-oriented supportive research; improving the accession-level description of the genetic diversity of crop collections; improving the legal and policy framework; and overseeing the proposed network of global base collections [13].

3.2. New Approaches and Developments Regarding Ex Situ Conservation and Facilitating Use

The awareness of quality management of plant genetic resources is widely accepted as important by genebank curators. However, the daily practice in genebanks is often very different, and many circumstances do exist that 'undermine' standardised procedures, agreed (genebank) standards to be met, etc. The paper of Lusty et al. [15] makes a convincing case for how modern quality management systems can be used to improve the overall performance of genebanks and to achieve much better results in conserving materials for the long-term in base collections for efficiency and effectiveness. The Global Crop Diversity Trust (Crop Trust) served as coordinator of the CGIAR Genebanks Research Program (2012–2016) and established a monitoring system for operations across all CGIAR genebanks. The system comprises five elements: (1) performance targets, (2) online reporting, (3) a genebank quality management system (QMS), (4) a system-level Standard Operation procedure (SOP) documentation audit, and (5) external review and validation. Genebank curators are recommended to take a close look at the quality performance system that has been put in place by the CGIAR genebanks and to see which aspects could be applied to their genebank. Meanwhile, the World Vegetable Center and the Centre for Pacific Crops and Trees (CePaCT) genebanks have also started operating under the above-mentioned QMS.

In the fields of molecular genetics, phenomics, and information technologies, the world is witnessing significant advances that are gradually being utilised by genebanks to enhance the effectiveness and efficiency of genebank operations and to facilitate the use of conserved germplasm [11,13,16,22]. It is now possible to collect phenomic and genomic data for genebank accessions or entire collections, which, with the help of appropriate analytical tools, can be directly used by plant breeders to guide the selection of accessions for target traits or specific environments [11]. For traits with complex or uncertain genetic control, genomic selection is being used. For the selection of variants across the whole genome, an intelligent algorithm is required, as well as a good training population for the algorithm to learn from [26]. Genomic prediction also requires high-throughput phenotyping to develop and validate the algorithms [27].

Genebanks must catch up with these new technological developments to improve genebank operations and to accommodate the changing needs of breeders and researchers. There is a growing need for comprehensive online searchable repositories of information on genetic resources, as users increasingly require access to digital information associated with accessions, i.e., a shift to 'digital genebanks' [15,28]. An example in this direction is the AGENT project (Access to Genetic Resources and Digitisation of Plant Genetic Resources). AGENT aims to support the exploration of the untapped potential of the vast genetic resources stored in genebanks worldwide by leveraging FAIR (Findable, Accessible, Interoperable, and Reusable) international data standards and open digital infrastructure, thus facilitating germplasm use for breeding and research [16].

3.3. Other Forms of Conservation That Complement the Current Long-Term Conservation System

Home gardens may contain unique and rare, locally evolved or developed genetic diversity, as they harbour broad species and genetic diversity, including the wild relatives of our crops, and they can be found in almost any ecological condition that the inhabited world possesses. This makes this specific 'ecosystem' very interesting and relevant for the conservation of plant genetic resources as well as for the use of the frequently unique genotypes of a given crop or species that are being cultivated by smallholders. It is necessary to establish strong and effective linkages between home garden conservation efforts and the established ex situ conservation system at the local and national levels to ensure the safety of the in situ conserved materials and to facilitate their wider use [17].

Landraces still play a major role in crop cultivation and may reach up to 70% of cultivated areas, as has been shown in the case of barley [16]. Efforts by ICARDA have shown that the yield of dryland landraces can be significantly improved through seed cleaning and treatment against seed-borne diseases and through farmer-participatory selection, thus providing incentives to farmers for continued cultivation of such landraces, also as a form of on-farm conservation [29].

While ex situ conservation is a static process, on-farm conservation allows the manifestation of evolutionary processes in genetic resources and may lead to crop improvements and adaptation to changing climatic conditions over time. In this context, Evolutionary Participatory Breeding (EPB) is an exciting approach. It encompasses the planting of mixtures of diverse genotypes of the same crop in farmers' fields. This mixture may consist of early segregating generations that maximise allelic diversity for specific traits of interest. Over successive crop cycles in the same environment, the mixed population will gradually evolve and adapt to the specific environmental conditions. Genotypes that are more adapted to that environment will progressively become more dominant, ensuring resilience and long-term adaptation [30].

As stated by Engels and Ebert [5], ex situ and in situ conservation should be combined to achieve long-term security and cost-effectiveness of PGRFA conservation. In this context, the evolving concept of trans situ conservation is worth mentioning, which, in the case of crop wild relatives, dynamically integrates multiple in situ and ex situ measures, from conservation to research to education, comprising local and global scales [31].

The conservation of predominantly wild plant genetic resources in botanic gardens takes place in a largely independent evolved global conservation system from the genebank system that is more crop-genetic-resource-oriented. Considering the fact that the genetic resources that are maintained in botanic gardens include many crop wild relatives as well as locally grown or collected, edible or otherwise useful species, it appears highly relevant to establish linkages and strengthen the cooperation between crop genebanks and the community of botanic gardens. Moreover, the skill sets found within botanic gardens and agricultural genebanks complement each other and enable the development of integrated conservation approaches. The botanic garden community is highlighting the importance of networks and is willing to provide access to data and plant material [18]. Stronger linkages and cooperation between crop genebanks and the botanic garden community would be an important step toward a more effective and efficient global conservation system.

3.4. National, Regional, and Global Efforts and Strategies for the Improvement in the Current Conservation and Use System

The coordination of conservation efforts at the national level can be regarded as an essential building block of the global conservation system. Many of the current global system elements are based on (voluntary) contributions made by individual sovereign states to frameworks such as the CBD and the International Treaty. The authors have experienced the centrally coordinated German national PGRFA conservation and use system as one of the more comprehensive, efficient, and strategic approaches worldwide. It is participatory, inclusive, dynamic, and forward-looking with information management at the heart of the coordination efforts [19].

As plant genetic resources follow natural distribution patterns and not political borders, it is of crucial importance that close collaboration between neighbouring countries is a key prerequisite for the effective and efficient conservation (and use) of individual crop gene pools (or parts thereof). Thus, the regional coordination and collaboration of PGRFA (as well as of non-PGRFA) are significant to harnessing existing strengths, infrastructure, and knowledge. The European Cooperative Programme for Plant Genetic Resources (ECPGR) has tried to follow this paradigm by creating a decentralised virtual genebank, abbreviated as AEGIS (A European Genebank Integrated System). AEGIS aims to establish a European Collection of unique and important accessions maintained in various genebanks scattered over Europe that adhere to the AEGIS concept and principles, thus reducing costs due to reducing redundancy in the numerous national and institutional genebanks across Europe [21]. As AEGIS currently depends on funding from national authorities, it is far from perfect. Additional strategic funding from the European Union is required to truly set up a system of AEGIS-certified genebanks, in which the quality and continuity of conservation of, and access to, PGRFA could be guaranteed [21]. The AEGIS approach provides an excellent example of how the conservation and use facilitation of PGRFA can contribute to a more rational and efficient approach embedded in a regional governance and financial structure.

At the regional and global level, the CGIAR genebanks safeguard some of the largest and most widely used collections of crop diversity, critical to attaining the UN's Sustainable Development Goals (SDG) to end hunger and improve food and nutrition security [15]. Most CGIAR genebanks are strategically located in centres of crop diversity, therefore representing the rich genetic diversity of primary crop gene pools. Based on collection missions, donations, and genetic materials generated by breeders and researchers, the CGIAR collections have grown over the last five to six decades and harbour a rich diversity of landraces, heritage varieties, crop wild relatives, improved varieties, and breeding or research materials for specific mandate crops. The World Vegetable Center (WorldVeg), loosely aligned with the CGIAR, complements the CGIAR crop collections with a considerable diversity of vegetable genetic resources [25]). Through global and regional crop networks and collaborative projects, the CGIAR and WorldVeg have established close links with national PGRFA conservation and use programmes, thereby facilitating the efficient long-term conservation, use, and exchange of PGRFA. The efficient long-term conservation and distribution programmes established by the CGIAR and WorldVeg and their crop-specific breeding programmes serve as a model for the efficient and effective global PGRFA conservation and facilitating use efforts.

The Crop Trust served as coordinator of the CGIAR Genebanks Research Program (2012–2016) and put in place a QMS monitoring system for operations across all CGIAR genebanks. Details are provided in Section 3.2. Apart from the CGIAR genebanks, the Crop Trust also supports selected national genebanks in the Global South to safeguard the long-term conservation of critical genetic resources for food and nutrition security.

Private-sector breeders and curators of public national genebanks might have very different goals and objectives; however, genetic diversity is fundamental to both sectors. Therefore, any attempt to establish new or improve existing collaborations between the two sectors is expected to strengthen the global system. The often-existing mistrust in each other needs to be overcome through dialogue and communication to identify areas and activities of common interest.

Engels et al. [20] reported convincing examples of a close collaboration between public genebanks and private-sector breeders on the conservation of genetic resources. These examples include the Centre for Genetic Resources (CGN) in the Netherlands, the World Vegetable Center, and East-West Seed International. Over several decades, the CGN developed a fruitful collaboration with the vivid breeding industry in the Netherlands in areas that include regeneration, joint phenotyping and screening of accessions, and the funding of collection trips, including the benefit-sharing component. WorldVeg concluded agreements with private-sector companies to regenerate original accessions in order to make them available to users worldwide.

To accelerate the development and dissemination of elite vegetable crop materials, WorldVeg also concluded breeding consortia in Asia and Africa. East-West Seed International provides in-kind support to national and other domestic genebanks in the Philippines and Indonesia and also collaborates with the CGN in regenerating germplasm materials. These examples demonstrate that collaborative arrangements between (inter)national public genebanks and vegetable breeding companies, often coordinated by the respective seed associations, contribute significantly to germplasm collection, conservation, documentation, and their sustainable use, thus making a valuable contribution to the global system.

3.5. Governance and ABS Issues

Along with the evolution of the global conservation system, the 'concept' of genetic resources has evolved as well. Initially, the focus was strongly on the use of germplasm as the raw material for plant breeding, a resource that was freely available and considered a 'common heritage' of humankind. The term germplasm gradually started to also embrace the associated knowledge as well as information derived from germplasm through basic and applied research, and breeding. Simultaneously, the status and recognition of the role of cultivators/custodians of the genetic resource became part of the picture, and the concept of benefit sharing resulting from the use of the acquired resources was added to the legal arrangements to obtain these. Gradually, not only did the reproductive organ of the genetic resource become a legal 'substance', but also its genetic components as well as digital sequence information derived from germplasm. This has made arrangements to share this essential resource with others quite complex, resembling a 'legal jungle' [13,23].

The Convention on Biological Diversity (CBD), the International Treaty on Plant Genetic Resources for Food and Agriculture (ITPGRFA), and the Nagoya Protocol are relatively recent international agreements that recognise the sovereign rights of countries over their genetic resources. Under the CBD/Nagoya Protocol, countries are free to establish specific national legislations that regulate germplasm access and benefit sharing to be negotiated bilaterally. The need to negotiate bilateral agreements, often with several countries to access specific genetic resources, turned out to be a cumbersome, highly bureaucratic, time-consuming, and costly effort, resulting in a decrease in or even a cessation of germplasm exchange. The ITPGRFA attempted to ease this situation by establishing a globally harmonised multilateral system (MLS). Unfortunately, the MLS is (still) restricted to a limited number of food and forage crops, with very few vegetable crops [23].

On the other hand, crop improvement depends on access to (agro)biodiversity to source new genetic variations for breeding. Fair, transparent, and non-bureaucratic rules and regulations that provide legal certainty concerning the access to and use of germplasm in breeding and research are, therefore, a predisposition for food and nutrition security. Germplasm users need to have clarity on whether the ITPGRFA, the CBD/Nagoya Protocol, or any other ABS tool apply. Furthermore, adjustments to the current texts of national legal instruments regarding ABS regulation under the Nagoya Protocol (preferably in a common universal language) are needed to ensure legal certainty and strengthen access to genetic resources. According to Ebert et al. [23], expanding the scope of the ITPGRFA to include all PGRFA, as well as related organisms like pathogens and pests, and making these genetic resources and all related information accessible under a Standard Material Transfer Agreement (SMTA) would greatly benefit the use of new germplasm in breeding and lead to the creation of improved varieties that can cope with climate change challenges and will contribute to more sustainable forms of agriculture. To facilitate benefit-sharing arrangements, the SMTA could stipulate a subscription system or a seed sales tax. Such a transparent, functional, and efficient system would erase legal uncertainties and minimise transaction costs for conservers, curators, and users of genetic resources, thus aiding plant breeders in fulfilling their mission.

3.6. Concluding Remarks

Genebanks need to adjust and embrace new technologies in the fields of molecular genetics, phenomics, and information technologies to improve the effectiveness and efficiency of genebank operations and to meet the evolving needs of users, both in terms of genetic resources and associated data, more accurately and efficiently than they do today [22]. Given the uncertainties with climate change, the need to develop climate-smart and -resilient crop varieties for sustainable crop production, and the need to feed the still-growing population with healthy food, plant breeders depend on the easy and non-complicated availability of a wide range of genetic diversity. The form in which this diversity is needed might change and breeders might ask for larger amounts of associated data. Genebank materials will be needed for gene discovery studies and for the identification of functional variants.

With the progress in technological advancements, adjustments to the above-mentioned international agreements are mandatory to secure and facilitate germplasm exchange. ABS mechanisms need to become transparent and easy to implement and adhere to, they need to provide legal certainty, and they need to have low transaction costs, thus benefiting providers and users of germplasm.

To enhance efficiency and reduce redundancy in crop collections and costs, a model of a functional and efficient global network of base and active collections, similar to the AEGIS concept in Europe, has been recommended [13]. A lean international organisation could assume responsibilities for the global coordination, facilitation, and oversight of the various global crop gene pool base collection networks. The Guest Editors of this SI hope that the results of this review inject new ideas into the ongoing discussions at the level of the Governing Body of the ITPGRFA and other fora and contribute to the needed reform of the global genetic resources' conservation and use system.

Author Contributions: J.M.M.E. and A.W.E. conceived of, drafted, and edited this editorial equally. All authors have read and agreed to the published version of the manuscript.

Conflicts of Interest: The authors declare no conflicts of interest.

References

1. Vavilov, N.I. *Five Continents*; Rodin, L.E., Reznik, S., Stapleton, P., Eds.; International Plant Genetic Resources Institute (IPGRI): Rome, Italy, 1997.
2. Vavilov, N.I. The law of homologous series in variation. *J. Genet.* **1922**, *12*, 47–89. [CrossRef]
3. Vavilov, N.I. Centers of origin of cultivated plants. *Bull. Appl. Bot. Genet. Plant Breed.* **1926**, *16*, 248.
4. Vavilov, N.I. Linnaeus species as a system. *Bull. Appl. Bot. Genet. Plant Breed.* **1931**, *26*, 109–134.
5. Engels, J.M.M.; Ebert, A.W. A Critical Review of the Current Global Ex Situ Conservation System for Plant Agrobiodiversity. I. History of the Development of the Global System in the Context of the Political/Legal Framework and Its Major Conservation Components. *Plants* **2021**, *10*, 1557. [CrossRef] [PubMed]
6. CGIAR Technical Advisory Committee (TAC). *The Collection, Evaluation and Conservation of Plant Genetic Resources*; Report of TAC Ad Hoc Working Group: Beltsville, MD, USA, 1972. Available online: https://cgspace.cgiar.org/server/api/core/bitstreams/2189ef55-ab57-4ba6-89e4-9e7d656baacb/content (accessed on 9 February 2024).
7. Halewood, M. Moving slowly towards the light: A review of efforts to create a global system for PGRFA over the last half-century. In *Enhancing Crop Genepool Use: Capturing Wild Relative and Landrace Diversity for Crop Improvement*; Maxted, N., Dulloo, M.E., Ford-Lloyd, B.V., Eds.; CAB International: Wallingford, UK, 2016; ISBN: 978-1-78064-613-8. [CrossRef]
8. Sonnino, A. International instruments for conservation and sustainable use of plant genetic resources for food and agriculture: An historical appraisal. *Diversity* **2017**, *9*, 50. [CrossRef]
9. Pistorius, R. *Scientists, Plants and Politics—A History of the Plant Genetic Resources Movement*; IPGRI: Rome, Italy, 1997.
10. Sirakaya, A. Where access and benefit-sharing comes from: A historical overview. *Genet. Resour.* **2022**, *2022 3*, 74–88. [CrossRef]
11. Volk, G.M.; Byrne, P.F.; Coyne, C.J.; Flint-Garcia, S.; Reeves, P.A.; Richards, C. Integrating Genomic and Phenomic Approaches to Support Plant Genetic Resources Conservation and Use. *Plants* **2021**, *10*, 2260. [CrossRef] [PubMed]
12. Gullotta, G.; Engels, J.M.M.; Halewood, M. What Plant Genetic Resources for Food and Agriculture Are Available under the Plant Treaty and Where Is This Information? *Plants* **2023**, *12*, 3944. [CrossRef]
13. Engels, J.M.M.; Ebert, A.W. A Critical Review of the Current Global Ex Situ Conservation System for Plant Agrobiodiversity. II. Strengths and Weaknesses of the Current System and Recommendations for Its Improvement. *Plants* **2021**, *10*, 1904. [CrossRef]
14. Herbold, T.; Engels, J.M.M. Genebanks at Risk: Hazard Assessment and Risk Management of National and International Genebanks. *Plants* **2023**, *12*, 2874. [CrossRef]
15. Lusty, C.; van Beem, J.; Hay, F.R. A Performance Management System for Long-Term Germplasm Conservation in CGIAR Genebanks: Aiming for Quality, Efficiency and Improvement. *Plants* **2021**, *10*, 2627. [CrossRef]
16. Visioni, A.; Basile, B.; Amri, A.; Sanchez-Garcia, M.; Corrado, G. Advancing the Conservation and Utilization of Barley Genetic Resources: Insights into Germplasm Management and Breeding for Sustainable Agriculture. *Plants* **2023**, *12*, 3186. [CrossRef]
17. Korpelainen, H. The Role of Home Gardens in Promoting Biodiversity and Food Security. *Plants* **2023**, *12*, 2473. [CrossRef]
18. Breman, E.; Ballesteros, D.; Castillo-Lorenzo, E.; Cockel, C.; Dickie, J.; Faruk, A.; O'Donnell, K.; Offord, C.A.; Pironon, S.; Sharrock, S.; et al. Plant Diversity Conservation Challenges and Prospects—The Perspective of Botanic Gardens and the Millennium Seed Bank. *Plants* **2021**, *10*, 2371. [CrossRef]
19. Begemann, F.; Thormann, I.; Sensen, S.; Klein, K. Effective Coordination and Governance of PGRFA Conservation and Use at the National Level—The Example of Germany. *Plants* **2021**, *10*, 1869. [CrossRef] [PubMed]
20. Engels, J.M.M.; Ebert, A.W.; van Hintum, T. Collaboration between Private and Public Genebanks in Conserving and Using Plant Genetic Resources. *Plants* **2024**, *13*, 247. [CrossRef] [PubMed]

21. Hintum, T.V.; Engels, J.M.M.; Maggioni, L. AEGIS, the Virtual European Genebank: Why It Is Such a Good Idea, Why It Is Not Working and How It Could Be Improved. *Plants* **2021**, *10*, 2165. [CrossRef]
22. Lusty, C.; Sackville Hamilton, R.; Guarino, L.; Richards, C.; Jamora, N.; Hawtin, G. Envisaging an Effective Global Long-Term Agrobiodiversity Conservation System That Promotes and Facilitates Use. *Plants* **2021**, *10*, 2764. [CrossRef] [PubMed]
23. Ebert, A.W.; Engels, J.M.M.; Schafleitner, R.; Hintum, T.V.; Mwila, G. Critical Review of the Increasing Complexity of Access and Benefit-Sharing Policies of Genetic Resources for Genebank Curators and Plant Breeders—A Public and Private Sector Perspective. *Plants* **2023**, *12*, 2992. [CrossRef] [PubMed]
24. FAO. International Treaty on Plant Genetic Resources for Food and Agriculture—List of Contracting Parties. Available online: https://www.fao.org/plant-treaty/countries/membership/en/?page=10&ipp=20&no_cache=1&tx_dynalist_pi1[par]=YToxOntzOjE6IkwiO3M6MToiNiI7fQ== (accessed on 14 February 2024).
25. Ebert, A.W. The Role of Vegetable Genetic Resources in Nutrition Security and Vegetable Breeding. *Plants* **2020**, *9*, 736. [CrossRef]
26. Bhat, J.A.; Ali, S.; Salgotra, R.K.; Mir, Z.A.; Dutta, S.; Jadon, V.; Tyagi, A.; Mushtaq, M.; Jain, N.; Singh, P.K.; et al. Genomic Selection in the Era of Next Generation Sequencing for Complex Traits in Plant Breeding. *Front. Genet.* **2016**, *7*, 221. [CrossRef] [PubMed]
27. Akdemir, D.; Isidro-Sánchez, J. Design of training populations for selective phenotyping in genomic prediction. *Sci. Rep.* **2019**, *9*, 1446. [CrossRef] [PubMed]
28. Belzile, F.; Abed, A.; Torkamaneh, D. Time for a paradigm shift in the use of plant genetic resources. *Genome* **2020**, *63*, 189–194. [CrossRef]
29. Amri, A.; Yazbek, M.; Shehadeh, A.; Nawar, M.F.; Tsivelikas, A.; Chandrashekhar, B. ICARDA efforts to promote in situ/on-farm conservation of dryland agrobiodiversity. *Indian J. Plant Genet. Resour.* **2016**, *29*, 265–267. [CrossRef]
30. Ceccarelli, S.; Grando, S. From participatory to evolutionary plant breeding. In *Farmers and Plant Breeding: Current Approaches and Perspectives*; Westengen, O.T., Winge, T., Eds.; Routledge: London, UK, 2019; pp. 231–244.
31. Riordan, E.C.; Nabhan, G.P. Trans Situ Conservation of Crop Wild Relatives. *Crop Sci.* **2019**, *59*, 2387–2403. [CrossRef]

plants

MDPI

Review

A Critical Review of the Current Global Ex Situ Conservation System for Plant Agrobiodiversity. I. History of the Development of the Global System in the Context of the Political/Legal Framework and Its Major Conservation Components

Johannes M. M. Engels [1,*] and Andreas W. Ebert [2,†]

1 Alliance of Bioversity International and CIAT, 00153 Rome, Italy
2 World Vegetable Center, 60 Yi-Min Liao, Shanhua, Tainan 74151, Taiwan; ebert.andreas6@gmail.com
* Correspondence: j.engels@cgiar.org
† Current address: 73529 Schwäbisch Gmünd, Germany.

Abstract: The history of ex situ conservation is relatively short, not more than a century old. During the middle of last century, triggered by the realization that genetic erosion was threatening the existing landraces and wild relatives of the major food crops, global efforts to collect and conserve the genetic diversity of these threatened resources were initiated, predominantly orchestrated by FAO. National and international genebanks were established to store and maintain germplasm materials, conservation methodologies were created, standards developed, and coordinating efforts were put in place to ensure effective and efficient approaches and collaboration. In the spontaneously developing global conservation system, plant breeders played an important role, aiming at the availability of genetic diversity in their breeding work. Furthermore, long-term conservation and the safety of the collected materials were the other two overriding criteria that led to the emerging international network of ex situ base collections. The political framework for the conservation of plant genetic resources finds its roots in the International Undertaking of the FAO and became 'turbulent rapid' with the conclusion of the Convention on Biological Diversity. This paper reviews the history of the global ex situ conservation system with a focus on the international network of base collections. It assesses the major ex situ conservation approaches and methods with their strengths and weaknesses with respect to the global conservation system and highlights the importance of combining in situ and ex situ conservation.

Keywords: plant agrobiodiversity; history of the global ex situ conservation system; political and legal framework; field genebanks; in vitro collections; cryopreservation; DNA banks; pollen banks; complementary conservation approaches

Citation: Engels, J.M.M.; Ebert, A.W. A Critical Review of the Current Global Ex Situ Conservation System for Plant Agrobiodiversity. I. History of the Development of the Global System in the Context of the Political/Legal Framework and Its Major Conservation Components. *Plants* **2021**, *10*, 1557. https://doi.org/10.3390/plants10081557

Academic Editor: Ana Maria Carvalho

Received: 10 July 2021
Accepted: 26 July 2021
Published: 29 July 2021

1. Introduction

Plant genetic resources are the foundation of our food production system, thanks to the genetic diversity they contain. It is this genetic diversity, both between and within crop species and their wild relatives, that allow crops to evolve and adapt to changing conditions, either natural or human-created conditions. Since the first steps of early farmers to start the process of domesticating species from wild plants in the Near East more than 10,000 years ago, plant genetic resources and their diversity allowed humankind to develop crops according to its needs and to spread them around the world; thus, securing our plant food basis.

Since these ancient times, the number of domesticated crops has steadily increased, and the cultivated forms or varieties of most of these crops have also increased and collectively become more diverse when moving from one region to another. Human and natural selection have been the driving force behind this diversification, but this process was only possible because of the genetic diversity available within and between these crop varieties and the related wild species that collectively form the diversity gene pool [1]. Genetic mutations in the crop genome are a permanent source of genetic diversity that allowed and continue to allow human and natural selection to be successful. Human exploitation of genetic diversity drastically increased when plant breeding became established, some 150 years ago [2]. This process of purposely generating new diversity through crossing different individuals followed by subsequent selection, resulted in high(er) yielding elite varieties. The success of this human managed evolution meant a steady replacement of older and usually well-adapted cultivars and even of entire crops. The loss of genetic diversity is called genetic erosion and was the trigger for targeted conservation efforts worldwide [3].

With the steady and increasing loss of genetic diversity since the middle of the last century for many of the crops cultivated worldwide and particularly for the main food crops, the need for systematic collecting and conservation of this diversity was recognized, and global conservation activities were initiated. Gradually, the Food and Agricultural Organization of the United Nations (FAO) in Rome assumed a coordinating role, supported by the International Board for Plant Genetic Resources (IBPGR), founded in 1974, one of the CGIAR centres, whose secretariat was initially based at FAO, thus serving as a technical and advisory institute for FAO and its political bodies such as the Commission on Plant Genetic Resources and later the International Treaty on Plant Genetic Resources for Food and Agriculture (ITPGRFA). Gradually, through IBPGR's research, coordination and scientific advice and training were provided to countries worldwide, and a global network of plant genetic resources conservation centres, called genebanks, was established [4]. Political debates at FAO, and IBPGR's research efforts aimed at collecting and improving the conservation of plant genetic resources for food and agriculture (PGRFA), somehow led to a more or less spontaneous creation of a global long-term conservation system of PGRFA [5]. This system underwent an evolutionary process itself, taking advantage of new scientific and technical developments and adjusting to evolving political conditions. It is this system that we will critically assess and review, with its strengths and weaknesses, with the aim to provide a perspective on how the system can be strengthened and can be made more rational in order to enable effective and efficient long-term conservation.

1.1. Scope

Whereas natural and human made ecosystems harbour the biodiversity of plants, animals, and microbes embedded in a physical environment, the focus in this paper will be on the plant genetic resources that are used for food and agriculture, i.e., PGRFA, or plant agrobiodiversity. These PGRFA comprise landraces and primitive and obsolete cultivars, crop wild relatives and modern varieties. Sometimes, plant breeding and other research materials are also regarded as genetic resources that should be included in genebanks.

Regarding the conservation activities, the main focus of this paper will be on long-term ex situ conservation, i.e., genebanks that manage seed, field, in vitro, and cryopreserved collections as well as DNA samples. Thus, not only are seeds important organs for conservation, but entire plants, pollen, tissues, cell suspensions, and more recently, DNA are also used. As not all plant agrobiodiversity can be collected and stored in genebanks, e.g., many wild food plants, many crop wild relatives, etc., we also look at in situ or nature conservation as well as at the on-farm maintenance of landraces and other genetic resources that require keeping the population structures of the material to be protected intact and/or to ensure a continuous evolution or the maintenance through steady cultivation or management. This dynamic evolutionary conservation stands in contrast to the frozen and static conditions that genebanks practice. Whereas ex situ conservation tends to focus on genotypes, in situ and on-farm conservation aim at natural and/or human-made populations and mixtures. It might be obvious that a balanced integration of these different conservation approaches will be needed to optimize the conservation system, as these approaches are complementary.

As conservation is frequently undertaken with the aim of keeping genetic diversity available and easily accessible for use, i.e., by farmers, breeders, or researchers, availability aspects are also important to be considered when deciding on the conservation 'approach'. Therefore, due attention will be given to how to increase the use of materials conserved under long-term conservation conditions.

Detailed knowledge of the conserved genetic resources is a key requirement for rational, effective, and efficient conservation as well as to facilitate the use of the resources. Thus, research on plant genetic resources in situ or in genebanks is an essential activity to support these requirements. This aspect will be given due attention.

Besides the importance of creating new knowledge of the materials under conservation and to facilitate their use, the application of new technologies in conservation and use is critically important to achieve rational, efficient, and effective long-term conservation and to facilitate the use of plant agrobiodiversity.

Plant agrobiodiversity is distributed across the world; therefore, the sovereignty of national states is an important legal aspect that was recognized in the Convention on Biological Diversity (CBD), and the accessibility to these resources is thus determined by individual states. Moreover, genetic resources might be protected by intellectual property rights, hence the legal and policy framework for the conservation and the use of PGRFA is an important element to ensure rational, efficient, and effective conservation and use.

Other aspects that might directly or indirectly impact conservation decisions include training and capacity building, awareness creation, participatory approaches, economic considerations, and possibly others. These aspects are not the focus of this paper or of this Special Issue but can be of critical importance to achieve a rational and sustainable long-term conservation system.

1.2. Focus of This Review

In this paper, we will address the above-mentioned aspects or considerations of a long-term conservation system that might directly or indirectly impact the efficiency as well as effectiveness of the conservation and the facilitation of use of plant agrobiodiversity in all of their dimensions. A history of the (long-term) plant agrobiodiversity conservation developments will be presented to understand the 'evolution' of the system and its elements, also in the context of technical, scientific, economic, and social developments.

Brief descriptions of main conservation methods and the underlying concepts as well as of the main ex situ germplasm collection types are intended to provide a solid foundation for their critical review, as these are components of the FAO Global System for the conservation and sustainable use of plant genetic resources for food and agriculture (hereinafter called the global system) that has emerged over the past few decades. A useful definition of the 'global system' [6] refers to the worldwide community of genebanks and institutes that are working together and individually to conserve and use plant genetic resources for food and agriculture and the policy instruments and global action plans that bind them together and support their work. CGIAR genebanks, given the size and diversity of their collections, their global mandate, and the extensiveness of their partnerships form the central pillar of this system. Closely related to the previous points and possibly a conclusive statement is the need for complementary conservation approaches.

2. History of the Development of the Long-Term Conservation Practices and the Evolving Global Conservation System

Crop and related genetic diversity underpin the productivity, sustainability, resilience, and adaptive capacity of agricultural systems and, thus, their evolutionary potential [7]. This diversity, contained in the so-called plant genetic resources has played a key role in the developments of agriculture since the first steps towards the domestication of our crop plants, the subsequent diffusion of the domesticates as well as the associated weeds and wild relatives from the centres of domestication into the world and the ongoing improvement and adaptation of the crops to ever changing environments, cultural practices, and human-made and natural threats. The first farmers started to migrate out of the Fertile Crescent to new geographic areas about 10,000 years ago, carrying genetic resources with them [8]. Whereas this process will have caused bottlenecks and thus might have impacted the evolution of these crops, the introduction of new and possibly more genetic diversity, natural mutations as well as natural and human selection have resulted in an enormous diversity of crops and varieties. This traditional crop development process underwent significant changes through rediscovery, around the turn of the 20th century, of the laws of inheritance proposed by Gregor Mendel in 1865 and 1866, which formed the basis for the science of genetics and thus, the birth of scientific plant breeding [9].

One of the first persons to realize the importance and use the power of genetic diversity in crop improvement was Nicolai Vavilov, a Russian geneticist and a director of the Lenin All-Union Academy of Agricultural Sciences at Leningrad (now the Vavilov Federal Research Centre of Plant Genetic Resources—VIR) who was requested by Lenin, the head of the government of Soviet Russia and later the Soviet Union, to breed plants that could be cultivated in Siberia and thus would contribute to increased food production after the First World War [10]. Collecting about 50,000 samples of crop plants systematically and throughout the world and evaluating them to assess their traits, he realized that the collected genetic diversity was largely confined to restricted areas, the so-called centres of diversity/origin of our crops [11].

Plant introduction centres that later grew out into genebanks were established in several countries to meet the increasing demand by plant breeders for more diversity. These included the All-Union Institute for Plant Industry in St Petersburg (in 1920), the Commonwealth Potato Collection in Cambridge, UK (before the Second World War), collections for the research programmes of the Rockefeller Foundation in the USA (1943), and The National Seed Storage Laboratory (NSSL) in Fort Collins, CO, USA (1958) [12]. The latter became the long-term storage facility for valuable germplasm propagated by seeds from the four regional plant introduction stations and an inter-regional station for potato [12]. During the 1950s and 1960s, several national plant introduction centres/genebanks were established on all continents, plant quarantine regulations were initiated (such as those in West Africa), and plant exploration and collecting started (such as the initiatives in Latin American countries). During the 1940s and 1950s, advanced and well-organized global germplasm collecting missions were coordinated by the Rockefeller Foundation in the USA [12].

With the increasing successes of plant breeding and the spread of modern and frequently high-yielding varieties, especially of the major food crops, a process of variety and later, even a process of crop replacement started and resulted in significant losses of genetic diversity, a development that was called 'genetic erosion' [13]. As early as 1936, Harlan and Martini raised the issue of genetic erosion in a USDA report devoted to barley breeding [14], and Vavilov had noted the increased loss of landraces. Particularly, during the so-called 'Green Revolution', which started in the late 1950s until the early 1970s, the success of high-yielding (dwarf) varieties of wheat and rice, together with new agricultural technologies, led to drastic losses of the traditional landraces of these crops, and this triggered concern in organizations such as the European Society for Research and Plant Breeding (EUCARPIA) and FAO [12]. In 1966, the EUCARPIA delegates advised European plant breeding institutes to foster continental collaboration through the establishment of four sub-regional genebanks in what was then West Germany (FAL in Braunschweig, for NW Europe); in East Germany (Gatersleben), Poland (Radzikow), Russia (St Petersburg) and/or others for Central and Eastern Europe; in Italy (Bari, for Southern Europe); and Sweden (Lund, for the Nordic countries) [12]. Gradually, regional and global networking increased, and the contours of a global conservation system became visible.

2.1. The Role of FAO

During the 1950s and early 1960s, FAO became the major actor in the conservation of plant genetic resources. Besides the World Catalogues of Genetic Stocks for wheat, rice, maize, and barley, they started to publish the FAO Plant Introduction Newsletter and organized technical meetings/conferences (see below). Salient historical events with respect to the global conservation system are summarized in Table 1 and, where applicable, reference to the Table is made in the text. The first meeting was called the 'Technical Meeting on Plant Exploration and Introduction' and was held in 1961 (Table 1) [15]. A Panel of Experts on Plant Exploration and Introduction was established in 1965. The panel included visionary scientists such as Sir Otto Frankel (CSIRO, Australia), professor Jack Harlan (University of Illinois, Urbana, IL, USA), and Professor Jack Hawkes (University of Birmingham, Birmingham, UK); in addition, Ms. Erna Bennett, (FAO, Rome, Italy) served as one of the supporting secretarial staff members of the panel. Reports of the six panel meetings were published between 1968 and 1974 [16]. This panel also played an important role in the planning and steering of the first two International Technical Conferences that the FAO organized in collaboration with their partners [17].

- The first International Technical Conference was held in 1967 in Rome and was organized by FAO and the International Biological Programme (IBP) under the title 'Technical Conference on the Exploration, Utilization and Conservation of Plant Genetic Resources' (Table 1) [18]. Some of the major recommendations of the 1967 conference included the need to survey genetic resources in nature and in genebanks and the need for a stronger emphasis on conservation, efficient documentation, and the improved international coordination of PGR activities. It also generated important guidelines for the establishment of a global network for ex situ long-term conservation. It should also be noted that in situ conservation, especially of landraces, was a big issue, but it was given little to no importance compared to ex situ conservation [12,13].
- In 1971, the second international conference on crop genetic resources was held in Rome, and its proceedings were published in the book *Crop Genetic Resources for Today and Tomorrow*, which included a plan of action (Table 1) [19]. At this conference, the panel of experts made some major contributions with respect to global conservation plans, including the formulation of basic criteria for the conservation and the use of genetic material. These were: (i) that plant material was to be made available immediately and without restriction to all breeders requesting it and (ii) that genetic variability had to be maintained for future generations in long-term storage under conditions for maximum physical and genetic security. A third important result of the panel was a categorization of ex situ collections: base collections (for long-term conservation), active collections (for research and distribution), and working collections (usually maintained at plant breeding institutions) (for details, see [20]. They also identified regions and crops for priority collecting [3]. These collecting priorities were reformulated during the panel's last meeting in 1975, with a clear shift from crops to regions [13].
- The third international conference on crop genetic resources was held in Rome in 1981, jointly organized by FAO, UNEP, and IBPGR (Table 1) [21]. The conference addressed most of the routine genebank operational topics, including sampling, seed storage and viability monitoring, recalcitrant seeds, in vitro conservation and the genetic stability of cultures, principles of germplasm regeneration, in situ conservation, the use of back-garden and genetic reserves for regeneration, the principles and practice of germplasm distribution and exchange, the safe and rapid transfer of plant genetic resources, including a proposal to distribute only germplasm materials completely free from plant pests and pathogens, principles of characterization and evaluation, data capturing and germplasm documentation, and under-exploited and minor crops [21].
- The fourth technical conference was in the context of the FAO global system for the conservation and use of plant genetic resources and was held in Leipzig, Germany in 1996 (Table 1) [22]. The major outcome of this conference was the Global Plan of

Action (see below) and, in addition, ample information on the global conservation system [22].

Table 1. Historical events of relevance to the establishment and evolution of the global PGRFA conservation, including the international network of base collections.

Year	Event	Main Outputs and (References)	Underpinning Principles (Reference)
Since 1920	Establishment of first genebanks	VIR, St. Petersburg (1920); Commonwealth Potato Collection, Cambridge (<2nd World War); research collections by Rockefeller Foundation, USA (1943); Fort Collins, CO, USA (1958) [12]	Recognition of genetic erosion in landraces by [14]
1926	Publication *Studies on the Origin of Cultivated Plants* by N. Vavilov	Monograph in *Bulletin of Applied Botany and Plant-Breeding*; [11]	'This monograph, dedicated to the memory of De Candolle, seems to be the most substantial contribution made since his day to the history of our main cultivated plants' [23].
1960	Founding of IRRI	Jointly established by Government of the Philippines' and the Ford and Rockefeller Foundations [24]	One of the first international genebanks; focus on rice genepool.
1961	Technical Meeting on Plant Exploration and Introduction, FAO Rome	Report of the meeting [15]	Mission-driven approach: conservation and use closely linked, tied to plant breeding, dominance of ex situ collections, mainly in developed countries.
1965	Establishment of the FAO Panel of Experts on Plant Exploration and Introduction.	Six meetings and reports of same during period from 1967–1975 [16]	Formulation of criteria, standards, and procedures for the conservation and use of PGR.
1966	Formal establishment of CIMMYT	Joint Mexican—Ford Foundation breeding project in progress since 1943 [25]	Norman Borlaug awarded Nobel Peace Prize (as wheat breeder) in 1970.
1966	EUCARPIA meeting	Recommendation to foster continental collaboration through the establishment of four sub-regional genebanks in Europe [12]	First indications of establishing a (global) conservation system or network.
1967	FAO/IBP (first) Technical Conference on Plant Exploration, Utilization and Conservation of Plant Genetic Resources, Rome	Publication of *Genetic Resources in Plants—Their Exploration and Conservation* [18]	Need for surveys; concern about genetic erosion of landraces and wild relatives; long-term ex situ collections; guidelines for establishment of global network for ex situ long-term conservation; international collaboration; in situ conservation as a complementary strategy.
1969	Third Session of the FAO Panel of Experts on Plant Exploration and Introduction, Rome	Report [3]	Establishment of collecting priorities by crops (and later) by regions.

Table 1. *Cont.*

Year	Event	Main Outputs and (References)	Underpinning Principles (Reference)
1971	Second FAO Technical Conference on crop genetic resources, Rome, Italy	Book on *Crop Genetic Resources for Today and Tomorrow* [19]	Plan of action agreed; panel of experts formulated basic criteria for conservation and use of genetic material (availability; maintaining genetic variability for the long-term; categorizing ex situ collections: base, active, and working collections.
1973	FAO/IBP Technical Conference on Genetic Resources, Rome, Italy	Plan of Action [19]	Recommendation to establish in situ collections.
1974	Establishment of IBPGR	Established as secretariat for its board of trustees, administered by FAO and, technically, as one of the international centres of the CGIAR [26]	Expected to coordinate global exploration and collecting efforts and to orchestrate a global network of genebanks.
1981	Third FAO, UNEP and IBPGR Technical Conference on PGR, Rome, Italy	Report [21]	Clear focus on routine genebank operations; in vitro and in situ (CWRs) conservation; concerns about NUS.
1983	22nd Session of the FAO Conference, Rome, Italy	Adoption of the International Undertaking on Plant Genetic Resources; establishment of the Commission on Plant Genetic Resources for Food and Agriculture (CGRFA) and of the Global System on Plant Genetic Resources [27]	Shared principles; IU non-legally binding; PGRs are a common heritage of humankind; genetic stocks and breeding lines included; germplasm exchange through a network of genebanks; commission provides oversight to system.
1989	3rd Regular Session of Commission on GRFA, Rome, Italy	Call for the development of the International Network of Ex Situ Collections under the Auspices of FAO [28]	Lack of clarity regarding the legal situation of the ex situ collections.
1989	25th Session of the FAO Conference, Rome, Italy	Resolution 4/89: Adoption of an agreed interpretation of the IU; Resolution 5/89: Farmers' Rights [29]	Plant breeders' rights are not inconsistent with IU; recognition of Farmers' Rights.
1991	26th Session of the FAO Conference, Rome, Italy	Resolution 3/91 [30]	Recognition of the sovereign rights of nations over their PGRFA; agreement on development of 1st State of the World's PGRFA and Global Plan of Action on PGR.
1992	UN Conference on Environment and Development (UNCED), Rio de Janeiro, Brazil	Convention on Biological Diversity (CBD) (entered into force on 29 December 1993);	Biodiversity vs. genetic resources; national sovereignty of states over their resources.
		Chapter 14 of Agenda 21	Call for the strengthening of the FAO Global System on Plant Genetic Resources.
		Chapter 16 of Agenda 21	Biotechnology can assist in the conservation of biological resources (e.g., ex situ techniques); risk assessment of LMOs, biosafety issues.
		Adoption of Resolution 3 of the Nairobi Final Act [31]	Recognises matters not addressed by the convention: a. access to existing ex situ collections; b. questions on Farmers' Rights; requests FAO forum to address these matters.

Table 1. *Cont.*

Year	Event	Main Outputs and (References)	Underpinning Principles (Reference)
1994	1st Extraordinary Session of the CGRFA, Rome	Start of negotiations for revision of IU; 12 centres of CGIAR sign agreement with FAO, placing their collections under the Auspices of FAO [32])	CGIAR centres agree to hold the designated germplasm in trust for the benefit of the international community.
1996	4th International Technical Conference on PGR, Leipzig, Germany	Global Plan of Action for the Conservation and Sustainable Use of PGRFA [21]; First Report on the State of the World's PGRFA [33]	Recognition of in situ and ex situ approaches; fair and equitable sharing of benefits arising from the use of PGRFA.
2001	31st Session of the FAO Conference, Rome, Italy	Resolution 3/2001: adoption of the International Treaty (entered into force on 11 September 2004) [34]	A legally binding agreement; recognition of Farmers' Rights (a national responsibility); access and benefit-sharing
2004	Establishment of the Global Crop Diversity Trust	Endowment fund, the income from which will be used to support the conservation of distinct and important crop diversity in perpetuity through existing institutions [35].	Coordinates the Genebank Platform (of the CGIAR operated genebanks)
2006	First meeting of the Governing Body of the ITPGRFA, Madrid, Spain	Standard Material Transfer Agreement (SMTA); relationship between the Treaty and the Crop Trust; agreement between GB and CGIAR centres (Art. 15) [36].	SMTA is the legal instrument through which the MLS operates; recognition of the Crop Trust as an 'essential element' of the Treaty's funding strategy; ex situ genebank collections of CGIAR are put under the Treaty (replacing agreement between CG centres and FAO).
2008	Establishment of the Svalbard Global Seed Vault	Agreement [37].	Additional safety back-up for long-term ex situ collections.
2009	12th Regular Session of the CGRFA, Rome, Italy	Second Report on the State of the World's PGRFA [38]	Report developed through a participatory approach with member countries
2011	143rd Session of the FAO Council, Rome, Italy	Second Global Plan of Action for the Conservation and Sustainable Use of PGRFA [39]	Need for a roadmap on climate change and genetic resources for food and agriculture

The rising concern regarding the genetic erosion of landraces and wild relatives due to modern agriculture, and the more general, increasing need of the agro-industry for a steady flow of new germplasm convinced the members of the FAO conference to give more consideration to a generalist approach to conservation [12]. During the second conference, the availability of new cold-storage techniques was noted, thus allowing long-term ex situ storage to be undertaken, whereas advocated in situ conservation, based on genecological premises, did not materialize until much later. The focus remained on ex situ conservation, despite the arguments for in situ approaches [3].

It should be noted that during the 1960s, the discussions on PGR in general as well as within FAO were dominated by plant breeders, and this resulted in a close conceptual link between conservation and use. Moreover, germplasm was predominantly stored in industrial countries and was closely tied to plant breeding institutes. During 1967, the FAO unit of Crop Ecology and Genetic Resources was established and thus provided FAO with more in-house specialized expertise.

2.2. The Establishment of the International Board for Plant Genetic Resources (IBPGR)

During a meeting of the Technical Advisory Committee (TAC) of the CGIAR in Beltsville, USA, a group of invited external experts, including several members of the FAO panel of experts, presented an ambitious plan to establish a world network of genetic resources centres [40]. This plan consisted of four elements. The first one was to establish a coordinating centre (to become IBPGR); the second one was to stimulate the establishment of genebanks in already existing international centres in developing countries (i.e., IRRI, established in 1960; CIMMYT (1966); CIAT (1967); and IITA (1968). The third element was to establish genebanks in new international centres (WARDA, 1971; CIP, 1971; and ICRISAT, 1972). Soon thereafter, the ILCA was established in 1974, and ICARDA was established in 1976. The fourth element was the establishment of new 'regional' centres in the Vavilovian centres for crop diversity. The establishment of the International Board for Plant Genetic Resources (IBPGR) took place in 1974, as a secretariat for its board of trustees, administered by FAO and technically as one of the international institutes of the CGIAR. It was expected to coordinate global exploration and collecting efforts and to orchestrate a global network of genebanks (see also the details of this international undertaking below). Its main task was formulated as '*to promote and assist in the worldwide effort to collect and conserve the plant germplasm needed for future research and production*' [40].

The main achievements of IBPGR and its successor institute IPGRI, particularly those related to long-term conservation and the global conservation system, are updated from a list in [13] and include:

1. Organization of collecting missions, partly using consultants in addition to its own staff and through contracts with national (selected) genebanks (for details, see IBPGR Annual Reports, e.g., [41]; for an overview: [42,43].
2. Support for national and regional PGR programmes, predominantly in developing countries with the establishment of conservation facilities, documentation systems, and capacity building/training [41].
3. Establishment of regional and global PGR networks with national programmes as principal stakeholders as well as regional and global crop networks, frequently with and through CGIAR centres and their leading roles in crop specific conservation and breeding, thus trying to ensure a close link between conservation and use. The European Cooperative Program for Plant Genetic Resources (ECPGR), formerly the 'European Cooperative Programme for Crop Genetic Resources Networks'—ECP/GR), was founded in 1980 on the basis of the recommendations of the United Nations Development Programme (UNDP), the Food and Agriculture Organization of the United Nations (FAO), and the Genebank Committee of the European Association for Research on Plant Breeding (EUCARPIA); its secretariat was hosted by IBPGR [44].
4. The establishment of an international network of base collections in 52 selected genebanks located in almost 40 countries across all continents for the long-term conservation of crops or crop groups, including 80 genera and approximately 250 species [45], and the so-called Registry of Base Collections containing a total of 144,000 accessions [43].
5. Support for an international MSc course in the conservation and use of PGR at the University of Birmingham and the organization of training courses [41].
6. Establishment of a digitalized information system for genebank documentation and germplasm management.
7. Initiating, coordinating, and/or conducting plant genetic resource conservation and use research and publishing the results and procedures.
8. More recently, the successor institutes of IBPGR (IPGRI and Bioversity International), especially after their administrative separation from FAO, played an active role in developing legal and policy proposals and acted as the CGIAR representative in international meetings and activities.

2.3. The International Undertaking (IU)

The International Undertaking (IU) was established by the FAO Commission on PGR in 1983 as a non-binding intergovernmental agreement to promote the conservation, exchange, and use of plant genetic resources [27]. Its objective was to ensure that plant genetic resources of economic and/or social interest, particularly for agriculture, would be explored, preserved, evaluated, and made available for plant breeding and scientific purposes. The Undertaking was based on the universally accepted principle that plant genetic resources are a heritage of mankind and, consequently, should be available without restriction. It defined *'plant genetic resources' as the reproductive or vegetative propagating material of the following categories of plants: (i) cultivated varieties (cultivars) in current use and newly developed varieties; (ii) obsolete cultivars; (iii) primitive cultivars (landraces); (iv) wild and weedy species, near relatives of cultivated varieties; (v) special genetic stocks (including elite and current breeder lines and mutants)*. It defined *'base collection of plant genetic resources' as a collection of seed stock or vegetative propagating material (ranging from tissue cultures to whole plants) held for long-term security in order to preserve the genetic variation for scientific purposes and as a basis for plant breeding; 'active collection'* was defined as *'a collection which complements a base collection, and is a collection from which seed samples are drawn for distribution, exchange and other purposes such as multiplication and evaluation'*, and *'centre' was defined as an institution holding a base or an active collection of plant genetic resources* [46].

Furthermore, the IU foresaw the development of a global system as to ensure that (Article 7.1):

a. A well-coordinated international network of national, regional, and international genebanks, including the international network of base collections, would develop. The unrestricted availability of materials included in the active and base collections of such a network was assumed.
b. Through the progressive growth of the network, a comprehensive coverage of species and regions was aspired, and an adequate safety duplication of the germplasm was involved.
c. The exploration, collection, conservation, maintenance, rejuvenation, evaluation, and exchange of plant genetic resources should be conducted by the genebanks in accordance with scientific standards.
d. Adequate funding should be provided.
e. A global information system should be developed.
f. Genebanks should give an early warning to the FAO in the case of hazards that threaten the efficient maintenance of the collection.
g. IBPGR is expected to liaise with FAO while conducting its programme of work aiming at building institutional and human capacity within developing countries for the development and distribution of improved crop varieties.

Article 7 of the IU on International Arrangements addresses aspects of the global system and access to germplasm in the base collections. Countries are invited to notify the FAO in case their base collections are to be recognized as part of the international network of base collections. The participating genebanks are expected to make the materials in these base collections available to the participants in the IU for the purposes of scientific research, plant breeding, or conservation, free of charge and based on mutual exchange or mutually agreed terms [46].

The IU was replaced by the International Treaty on Plant Genetic Resources in 2002 (see further below).

Another component of the global system is the International Code of Conduct for Plant Germplasm Collecting and Transfer [47]. It was adopted by the FAO Conference at its 27th session in 1993. The voluntary code aims to promote the rational collecting and sustainable use of genetic resources to prevent genetic erosion and to protect the interests of both the germplasm collectors and donors. It is based on the principle of national sovereignty over PGR and is in harmony with the CBD [47].

2.4. The Convention on Biological Diversity (CBD)

The negotiation of the Convention on Biological Diversity (CBD) in the eighties and early nineties, under the auspices of the United Nations Environment Programme [48], did result in drastic changes with respect to the conservation and use of PGRFA. Besides creating a general, globally, and legally binding framework for the conservation and sustainable use of biodiversity, the CBD, which entered into force in 1993, required that access to valuable biological resources must be conducted on 'mutually agreed terms' and is subject to 'prior informed consent' of the country of origin. The national sovereignty of states over biodiversity within their borders was recognized as a key principle in the CBD, and consequently, this became the 'driving force' in the thinking and approaches to the negotiations and future developments. Besides the fact that states were expected to 'look after their own biological resources and conserve them, whenever possible in their own country', this also caused a strong incentive for countries to favour bilateral rather than multilateral arrangements for the exchange of genetic resources.

From an agricultural perspective, it should be noted that the negotiations of the CBD were strongly influenced by environmentalists and nature conservationists and, consequently, a bias towards wild (i.e., non-domesticated and non-agricultural) plant and animal species could be observed [49]. In fact, agriculturalists were hardly present in the negotiations, and it was only through a separate resolution (Resolution 3 of the Nairobi Final Act) [30] that the FAO was asked to address two important but unresolved agricultural genetic resources issues, i.e., the question of Farmers' Rights and the need to address the legal status of existing genetic resource collections established prior to 1993 [50].

The negotiation process of the CBD caused a dramatic shift concerning the overall conservation approach, i.e., from a rather technologically driven ex situ conservation approach ('putting the germplasm safely away for the future'), towards a much more people-centred conservation, with a strong emphasis on in situ and on-farm conservation and sustainable use efforts. Alongside this, due attention was being paid to participatory research (and conservation) activities to recognize the important role of local communities in the management of and their dependency on biodiversity. This also led to the recognition of traditional and indigenous knowledge to be an important component of biodiversity that needs to be collected and/or conserved. The importance of technology for the conservation and use of genetic resources should be recognized as well as the provision of access to such 'enabling' technologies. These aspects facilitated (and required) a much closer link between conservation and development and led to a greater participation of local communities and subsistence farmers in conservation and use related activities. It is against this background that the access and benefit-sharing guidelines were developed and agreed upon in 2002 within the framework of the CBD by an Ad Hoc Open Ended Working Group on Access and Benefit-Sharing [51] that eventually, in 2010, resulted in the adoption of the Nagoya Protocol on Access and Benefit Sharing (ABS), which entered into force in 2014 [52]. It is a supplementary agreement to the CBD convention of 1992 and aims at the implementation of one of the three objectives of the CBD: the fair and equitable sharing of benefits arising out of the use of genetic resources, thereby contributing to the conservation and sustainable use of biodiversity [52]. Its rather strong focus on wild species and the bureaucracy involved to apply the protocol have resulted in concerns that the added bureaucracy and legislation could be damaging to the monitoring and collecting of biodiversity, to conservation, and to research, because the protocol severely limits access to genetic resources.

The CBD recognizes the application of intellectual property rights (IPRs) on biological materials as a means of protecting inventions and stimulating innovation. This led to a further expansion of the scope and/or application of IPRs, especially patents and plant breeder rights (PBRs), in agricultural research and plant breeding. Due to concerns that the development and use of genetically modified varieties could cause a threat to the environment and its biological resources, a legal framework on biosafety aspects was demanded, and thus, the so-called Cartagena Protocol on Biosafety was developed and came into force in 2003 as a legal framework for biosafety legislation and is yet another supplementary agreement of the CBD [53].

At present, the negotiation process on the development of the post-2020 global biodiversity framework is ongoing for its adoption during the forthcoming meeting later in 2021 in Kunming, China [54].

2.5. Global Plan of Action (GPA)

The first Global Plan of Action (GPAI) for conserving and using crop diversity was adopted in 1996 by 150 countries [22]. The GPAI called for a rational global conservation system based on the principles of effectiveness, efficiency, and transparency. The Second Global Plan of Action (GPAII) reiterated that call and provided a strategic framework for the conservation and the sustainable use of plant genetic diversity. It was adopted by the FAO Council in November 2011 and reaffirmed the commitment of governments to the promotion of plant genetic resources as essential components of food security through sustainable agriculture in the face of climate change (Table 1) [39]. It is a rolling action plan and is based on the findings of the Second Report on the State of the World's PGRFA [38] and inputs from a series of regional consultations and from experts. The GPAs are a supporting component of the International Treaty on Plant Genetic Resources for Food and Agriculture [55].

The GPAII consists of four main groups of priority activities, i.e., in situ conservation and management, ex situ conservation, sustainable use, and building sustainable institutional and human capacities [39]. The in situ conservation group of four priority activities comprises: 1. surveying and inventorying PGRFA; 2. supporting on-farm management and improvement of PGRFA; 3. assisting farmers in disaster situations to restore crop systems; and 4. promoting in situ conservation and management of crop wild relatives and wild food plants. The ex situ group of priority activities includes: 5. the targeted collecting of PGRFA; 6. sustaining and expanding ex situ conservation; and 7. regenerating and multiplying ex situ accessions. The sustainable use priority activities consist of: 8. the characterization and evaluation and development of subsets of collections to facilitate use; 9. plant breeding, genetic enhancement, and base broadening; 10. promoting the diversification of crop production and broadening crop diversity; 11. the development and commercialization of varieties, primarily of farmer varieties/landraces and underutilized species; and 12. supporting seed production and distribution. The set of capacity building activities comprises: 13. building and strengthening national programmes; 14. promoting and strengthening networks for PGRFA; 15. constructing and strengthening comprehensive information systems; 16. developing and strengthening systems for monitoring and safeguarding genetic diversity and minimizing genetic erosion of PGRFA; 17. building and strengthening human resource capacity; and 18. promoting and strengthening public awareness of the importance of PGRFA [39].

The GPAII does not contain specific activities related to long-term conservation and the global system, but several comments and supporting actions are referred to throughout the text, e.g., that the network of international ex situ collections of major crops played an important role in the negotiations of the International Treaty. These collections continue to form the backbone of the global system. The Svalbard Global Seed Vault now provides an additional level of security to existing ex situ collections [37]. Furthermore, the development of a global portal of accession-level data and the imminent release of an advanced genebank information management system (recently released and called GLIS) are additional important steps towards the strengthening and more effective operation of a global system for ex situ conservation [56]. Enhancing capacity at all levels is a key strategy to implement the priority activities of the GPA, including those related to long-term conservation, sustainable use (i.e., plant breeding, genetic enhancement, and base-broadening efforts) and the global system. Whereas countries have national sovereignty over and responsibility for the PGRFA they conserve, there is nevertheless a need for the greater rationalization of the global system for ex situ collections. The fostering of partnerships and synergies among countries is a requirement to develop a more rational and cost-effective global system. Furthermore, the GPAII plays an important role in the international policy framework for world food security and as a supporting component of the International Treaty. It contributes to achieving the Millennium Development Goals and aids in the implementation of the Strategic Plan for Biodiversity [57].

2.6. International Treaty on Plant Genetic Resources for Food and Agriculture (ITPGRFA)

The International Treaty for Plant Genetic Resources for Food and Agriculture (ITPGRFA or Treaty) aims to recognize the enormous contribution of farmers to the diversity of crops that feed the world; it aims to establish a global system to provide farmers, plant breeders, and scientists with access to plant genetic materials; and it aims to ensure that recipients share the benefits they derive from the use of these genetic materials with the countries where they originated [55]. The preparations and negotiations of the revision of the IU were initiated in 1994 and were concluded in 2001 by the adoption of the International Treaty. It encompasses all PGRFA and came into force in 2004 [55].

Through the Treaty, countries agree to promote the development of national integrated approaches to the exploration, collecting, characterization, evaluation, conservation, and documentation of their PGRFA, including the development of national surveys and inventories [55]. They also agree to develop and maintain appropriate policies and legal measures to promote the sustainable use of these resources, including on-farm management, strengthening research, promoting plant-breeding efforts, broadening the genetic bases of crops, and expanding the use of locally adapted crops and varieties and underutilized species. These activities would be supported, as appropriate, by international cooperation provided in the Treaty.

The most important part of the ITPGRFA is the establishment of the so-called Multilateral System (MLS) of Access and Benefit-Sharing [58]. The MLS applies to 64 genera, including the major food crops and forages, which were agreed upon on the basis of two criteria: their importance for food security and the level of interdependence among countries. At the global level, these crops provide approximately 80% of the food that is produced by plants. Through the MLS, sovereign nations have agreed to share resources and benefits. The genetic resources included in the MLS will be made available for research, breeding, and training, and their recipients should not claim any intellectual property or other rights that limit access to these resources or their genetic parts or components in the form received from the MLS [59]. The peculiarities of PGRFA compared to biodiversity in general, e.g., the difficulty of applying the country-of-origin concept, the strong interdependency of nations on genetic diversity for crop improvement, and the critical role of these resources in traditional agriculture and in food security, formed the basis for the establishment of a multilateral rather than a bilateral system for their exchange [60]. This thinking eventually led to the establishment of the MLS, which keeps the genetic resources of the Annex 1 listed species that are formally in the public domain and under governmental control and facilitates easy access to and the use of these resources [49]. It should be noted that the diversity of the crop species or the groups of species listed in Annex I is rather limited and, for instance, the majority of the vegetable genetic resources conserved by the World Vegetable Center in Taiwan, which consist of a large proportion of indigenous vegetables that are critically important for the diversification of cropping systems, nutritional security, and livelihoods [61], are not included in Annex 1. Discussions on the extension of the Annex 1 list have been ongoing for several years, but no final decision has been reached.

The benefits arising from the use of materials from the MLS shall be shared fairly and equitably through the exchange of information, access to and transfer of technology, and capacity-building, considering the priority activity areas indicated in the above mentioned GPAII and under the guidance of the Governing Body of the Treaty. It further establishes the payment, which is in certain cases mandatory, of an equitable part of the monetary benefits that are derived from the use of PGRFA into the funding strategy of the Treaty [58]. The funding strategy aims at mobilizing funds for activities, plans, and programmes to support the implementation of the Treaty and, in particular, its implementation in developing countries while keeping in line with the priorities that have been identified in the GPA. The funding strategy includes the monetary benefits that are paid in accordance with the MLS as well as the Global Crop Diversity Trust, which is described below. The Treaty recognizes the enormous contributions that local and indigenous communities and farmers of all regions of the world have made and will continue to make for the conservation and development of PGRFA. The Treaty makes governments responsible for the realization of Farmers' Rights, including the protection of relevant traditional knowledge; provisions for farmers to participate equitably in sharing benefits; and farmer participation in national policy decision-making [55,59]. Through Article 15, the Treaty establishes its relationship with the CGIAR and other international centres: '*Ex Situ Collections of Plant Genetic Resources for Food and Agriculture held by the International Agricultural Research Centres of the Consultative Group on International Agricultural Research and other International Institutions*' and arranges that the materials listed in Annex 1 of the Treaty and that are held by the centres as well as other species than those listed in Annex 1 of this Treaty and collected before its entry into force that are held by IARCs shall be made available in accordance with the provisions of the standard material transfer agreement (SMTA) [55].

2.7. International Network of Ex Situ Collections

The international network of ex situ base collections in genebanks that are managed by national, regional, or international centres was a component of the section on the international arrangements of the International Undertaking. It was foreseen that through a steady increase of the number of genebanks participating in the network, adequate coverage in terms of species and geographical distribution would eventually be achieved. It was further foreseen in the IU to conclude agreements (four 'model agreements' were available to choose from) with countries to place their base collections within this network and/or to provide storage space for the long-term storage of base collections from elsewhere. A few countries and institutions made concrete offers to place (part of) their collections in the network. The latter would operate under the auspices and/or the jurisdiction of the FAO and a number of contracts were concluded (see below).

In 1994, the CGIAR centres expressed the wish that their designated germplasm be recognized as part of the international network of ex situ collections and signed individual agreements with FAO [62]; Chapter 3.1. in [38]. The salient features of these agreements based on one of the above-mentioned model agreements include that:

- The centre shall hold the designated germplasm in trust for the benefit of the international community.
- The centre shall not claim legal ownership over the designated germplasm, nor shall it seek any intellectual property rights over that germplasm or its related information.
- The designated germplasm shall remain in the charge of the centre.
- The FAO shall have a right of access to the premises at any time and has the right to inspect all activities performed therein.
- The centre shall undertake the management and the administration of the designated germplasm in accordance with internationally accepted standards with respect to the storage, the exchange and distribution of seeds, the international genebank standards endorsed by the Commission and that all designated germplasm is duplicated.
- The centre recognizes the intergovernmental authority of the Commission in setting policies for the International Network and shall undergo consultation with the FAO and its Commission on proposed policy changes related to the conservation of the germplasm.
- The centre shall undertake the creation of samples of the designated germplasm and will make related information available directly to users or through the FAO for the purposes of scientific research, plant breeding, or genetic resource conservation without restriction.

- The centre shall ensure that such other people or institutions and any further entity receiving samples of the designated germplasm from such a person or institution are bound by the conditions to not claim ownership over the materials or to seek any intellectual property rights over that material and, in the case of samples duplicated for safety purposes, to manage these in accordance with internationally accepted standards.

A related network, as mentioned above, is the so-called Register of Base Collections that was established by the IBPGR in the 1970s and includes genebanks that were prepared to accept a long-term commitment to conserve germplasm materials and to make these available to users. This register formed the backbone of the international network of base collections. For details, see the paper by Engels and Thormann [42].

It should be noted that further agreements have been concluded with several other international research centres (e.g., the World Vegetable Center, CATIE and CRU, and some regional organizations (e.g., South Pacific Community)). Agreements with individual countries have not been vigorously pursued. In October 2006, 11 CGIAR centres signed agreements with the Governing Body of the International Treaty to bring their in trust collections under the framework of the Treaty and to recognize the authority of the Governing Body providing policy guidance related to those collections [63,64].

With the establishment of the International Treaty and its Multilateral System, the network of ex situ collections, and the conclusion of the agreements with the centres of the CGIAR, these collections were brought under the International Treaty (Chapter 3.2 in [38]). The commitments of countries to conserve germplasm for the long-term and to make the materials available (under an SMTA) have been made by countries and genebanks through the inclusion of germplasm in the MLS.

2.8. The Institutional and Capacity Building Framework

The establishment of the IBPGR has already been mentioned above, as it was intricately linked to political debate and developments during the 1970s (see Section 2.2). Similarly, the other centres of the CGIAR that operate genebanks with the genetic resource collections of their respective mandate crops are important elements of the emerging global ex situ conservation system. Since its establishment, the IBPGR has played an active role in strengthening this global system by supporting national PGRFA programmes and facilitating the establishment of new regional genebanks as part of the global network. In 1976, the formation of regional programmes in Southeast Asia and Europe and the establishment of (regional) genebanks in Costa Rica and Ethiopia (with funding from Germany) as well as the support provided to students from developing countries to attend the MSc programme on plant genetic resources at the University of Birmingham was reported [41]. Furthermore, the annual report listed international and regional institutions that accepted the invitation to become the holders of 'world' base collections of crops of global importance. During the following years, a steady increase of arrangements for regional programmes was reported as well as the development of a computer-based information and retrieval system, support provided to establish or strengthen national programmes and training activities as well as the acceptance of recommendations on the physical and engineering design of long-term seed stores [65–67].

2.9. Global Crop Diversity Trust

The Global Crop Diversity Trust (Crop Trust) was established in October 2004 by the IPGRI, now Bioversity International, on behalf of the CGIAR and FAO to help support the global system in a sustainable way through a Crop Diversity Endowment Fund [35]. Its mission is to ensure the conservation and availability of crop diversity for food security worldwide. Among others, the Trust provides oversight of the CGIAR Genebank Platform. The 11 CGIAR genebanks safeguard a unique global resource of crop and tree diversity and respond to thousands of requests for germplasm from users in more than 100 countries worldwide every year [68]. The goal of the CGIAR Genebank Platform is to conserve these collections and to make this diversity available to breeders and researchers in a manner that meets international scientific standards and that is cost-efficient, secure, reliable, sustainable over the long-term and that is supportive of the Plant Treaty. The Crop Trust has oversight over and financial responsibility for these CGIAR genebanks [69].

2.10. Some Critical Side-Effects on the Global Conservation System

The above-described developments had some significant (perceived?) side-effects on the emerging global long-term conservation system. They included a boost to the establishment of (national) genebanks, among others, triggered by the CBD's recognition of national sovereignty. The acceptance of intellectual property rights over genetic resources resulted in a steady increase of access regulations to genetic resources. Furthermore, issues of ownership over genetic resources emerged, leading to the refusal of some countries to provide access to 'their' plant genetic resources. Against this backdrop, a rather legalistic thinking of access and benefit sharing developed and influenced the arrangements in this field of the International Treaty.

Evolving molecular and later genomic techniques allowed and facilitated the assessment of genetic diversity aspects, including the identification of duplicate accessions; a quantification of genetic diversity; the identification of alleles and genes and their functions as well as their transfer between individuals and species. A better understanding of genetic diversity also allowed for more targeted collecting, better characterization/evaluation, and greatly facilitated plant breeding. The creation of so-called GMO (genetically modified organism) varieties with the help of these new molecular and biotechnology tools became a 'hot issue', among others, due to their threat to the genetic diversity of crop germplasm collections and biodiversity hotspots [70], and this caused restrictions or even prohibition of related research or the cultivation of modified materials. The multilateral thinking became an 'alternative' to restricting ownership; more IPRs crept in and resulted in heavy debates and in more restrictive attitudes regarding sharing natural genetic resources. All of these developments and possible repercussions call for a critical review of the current global system as it has evolved in the context of the above-described developments and the mentioned side-effects to provide elements for the creation of a more efficient and rational system of global base collections of important food crops.

3. Description of Ex Situ Germplasm Conservation Methods and Their Strengths and Weaknesses

The vast majority (approx. 92%) of angiosperms comprising roughly 330,000 species of flowering plants has desiccation-tolerant and so-called orthodox seeds [71,72] that survive drying to a low moisture content, 5% or less, and subsequent rehydration without a significant loss of viability [73,74]. Orthodox seeds acquire desiccation tolerance during their late phase of development when they undergo pre-maturation drying and are later shed metabolically inactive [72]. Desiccation tolerance is lost during germination [75]. Moreover, most desiccation-tolerant species tolerate low temperature (sub-zero) storage and seed longevity increases, within certain limits, with a decrease in seed moisture content (SMC) and storage temperature [76]. Harrington [77] postulated two rules of thumb regarding seed longevity in storage that apply independently. Over the range of 14 to 4% SMC (fresh weight basis), a 1% reduction in SMC doubles the life span of the seed. Similarly, within the range of 50 to zero degrees Celsius, for each 5 °C drop in storage temperature, the life span of seed in storage would double. Therefore, the cold storage of dried seeds is a practical, efficient, and cost-effective method for the long-term storage of germplasm in genebanks. The FAO Genebank Standards recommend storage at -18 ± 3 °C and a relative humidity of 15 ± 3 percent for most original seed samples and safety duplicate samples intended for long-term storage [78]. In case seed samples are stored in hermetically sealed pouches or containers, the control of the storage room RH is not required.

In contrast to orthodox seeds, so-called 'recalcitrant' seeds are desiccation-sensitive and rapidly loose viability upon drying and do not tolerate low temperature storage [73]. Recalcitrant seeds undergo extremely limited drying during maturation and consequently, have high SMC and are metabolically active during shedding [79]. Desiccation sensitivity also seems to be linked to the non-dormant state of seeds upon shedding [71]. The SMC below which viability is lost varies between species but is generally above 20% [80]. Specifically, tree species of tropical provenance, such as avocado (*Persea americana*), cacao (*Theobroma cacao*), jackfruit (*Artocarpus heterophyllus*), breadfruit (*Artocarpus altilis*), lychee (*Litchi chinensis*), mango (*Mangifera indica*), mangosteens (*Garcinia mangostana*), etc., produce recalcitrant seeds.

There is a third category of seed storage behaviour comprising so-called 'intermediate' seeds without sharp boundaries between orthodox and recalcitrant seeds [81]. Species with intermediate seed storage behaviour can be dried to certain SMC levels but cannot be dried to a level as low as truly orthodox seeds [82] and often do not survive sub-zero storage temperatures. Moreover, seeds with intermediate storage behaviour tend to lose viability much quicker than orthodox seeds [82]. Coffee (*Coffea arabica*) seeds fall into this category of intermediate seeds [81]. Depending on the cultivar, coffee seeds tolerate drying to 5–10% SMC but viability at low or sub-zero temperatures is rapidly lost. Seeds of alpine species are also significantly shorter lived than their lowland counterparts, possibly due to abnormal seed development under the cool and wet conditions of the alpine climate [83].

As species producing seeds with intermediate or particularly recalcitrant storage behaviour have extremely limited longevity in a seed genebank, they are commonly stored in field genebanks and/or as in vitro collections for medium-term conservation and/or in liquid nitrogen for long-term conservation.

3.1. Short-, Medium- and Long-Term Ex Situ Storage of Orthodox Seeds

In general, orthodox seeds are relatively small and require little storage space for the conservation of a representative sample of the source population and further sub-samples for distribution, viability checking, and safety back-up. Crops commonly conserved in seed genebanks include cereals such as rice, wheat, barley, oats, sorghum, millet, maize, grain and forage legumes, most vegetables, and some fruit crops. True seeds of crops such as those from potato, which are commonly propagated vegetatively, can also be dried and stored at low temperature [84]. This is common practice with wild potato germplasm, and accessions are maintained as botanical seeds or true-potato seeds (TPS). A representative number of 20–50 individuals are typically collected from a wild population, and seeds are regenerated and combined to form a unique genebank accession of heterogeneous seed, which is expected to represent most alleles found in that population [85]. Seed samples of such wild potato germplasm accessions thus represent a heterogenous mix of genotypes, whereby each genotype represents a portion of the genetic make-up of the sampled population.

The core operations of a genebank conserving the seeds of orthodox species comprise cleaning, seed drying, viability and health testing, packing, storage, and distribution to users and for a safety back-up [86]. When seed stocks are running low or when seed viability drops below a minimum threshold, seed lots need to be regenerated for seed replenishment. All these genebank operational steps are documented and in many genebanks are supported by a genebank information system [87].

Most genebanks conserving PGRFA have the mandate to distribute germplasm to a range of different users and, for practical reasons, store the seeds of most collected or acquired accessions in a base and an active collection when justified. The most-original seed samples are kept in the base collection for long-term conservation, aiming at the highest level of genetic integrity of the stored sample with the original sample [78]. The active collection is oriented towards seed regeneration (triggered by low viability), characterization, evaluation, multiplication (triggered by low seed stock), and distribution and is generally kept under medium-term storage (MTS) conditions.

The base collection for any given species or a crop genepool may be distributed over several institutions, as is the case in Europe, with the implementation of a European Genebank Integrated System, abbreviated as AEGIS [88]. In contrast, the United States

Department of Agriculture (USDA-ARS) has a network of genebanks holding the active collections for different crops in 19 different locations across the country, with one main base collection held at the National Laboratory for Genetic Resources Preservation (NLGRP) in Fort Collins, Colorado, serving all of the regional genebanks. The NLGRP maintains the US system backup of more than 445,000 accessions, representing 86% of the seed collections and 15% of the clonal collections [89]. Seeds are not distributed from the base collection to the users, but rather, they are distributed from the active collections.

The active collections comprising the bulk of orthodox seeds stored in most genebanks are to be kept under medium-term storage (MTS) conditions at temperatures ranging from 5 °C to 10 °C and at a relative humidity (RH) of 15 ± 3 percent for seeds that are stored in open containers [78]. Frequently, MTS conditions have a narrower range from +2 to +5 °C [86,89,90], and RH adjustment is not required if seeds are stored in hermetically sealed pouches or containers. Refrigerated seed storage under MTS conditions is adequate for up to 30 years [78]. It should be noted that seeds stored in hermetically closed containers are to be dried in a controlled environment with a temperature range between 5 and 20 °C and a RH between 15 and 25%, depending on the species.

The base collections are stored under long-term storage (LTS) conditions at sub-zero temperatures of typically −18 to −20 °C [86,89–91], and the seeds are dried as mentioned above for MTS, maintaining high seed quality over long, species-specific periods of up to 100 years or more.

Other genebanks whose major focus is not the use plant agrobiodiversity facilitation but rather whose focus is on the long-term conservation of globally threatened species (with relatively few sample requests), store all of their seeds exclusively under LTS conditions. This applies, for example, to the Millennium Seed Bank (MSB) of the Royal Botanic Gardens Kew, where dried seeds are transferred to air-tight glass containers or aluminum foil bags and are stored in the seed vault at −20 °C [91].

Assessing 42,000 seed accessions representing 276 species in the USDA National Plant Germplasm System provided evidence that some species produce orthodox seeds of short longevity in dry storage [92]. Some plant families had typically short-lived seeds (e.g., Apiaceae and Brassicaceae) or long-lived ones (e.g., Malvaceae and Chenopodiaceae). Moreover, environmental factors seem also to determine seed longevity, as seeds from species originating from certain localities in Europe had short shelf lives, while seeds of the same species originating from localities in South Asia and Australia had much longer shelf lives. For these reasons, some genebanks additionally cryopreserve samples of those orthodox seeds that are expected to be very short-lived, even under LTS conditions [93,94].

Under short-term storage (STS) conditions, the seed quality and the viability of orthodox seeds with long shelf lives can be maintained for a minimum of eight years under ambient conditions if 25 °C is not exceeded, and the relative humidity in the storage room is kept at 10–25% [78]. At the World Vegetable Center in Taiwan, working collections of breeders and other researchers are kept in STS conditions at 15 °C and 40–45% RH [90].

3.2. Field Genebanks

Although seed desiccation sensitivity affects only about 8% of flowering plants [72], there are many field and horticultural crops as well as (agro)forestry species that cannot be conserved long-term in conventional seed storage and that require different forms of conservation, such as in field genebanks, in in vitro collections, and/or in liquid nitrogen [93]. Among those are species that only produce recalcitrant or intermediate seeds with a short storage life span. Moreover, some species take several years to produce seeds, such as yucca (*Yucca* sp.) and bamboo (a species of the Poaceae subfamily Bambusoideae), while other crop species hardly produce seeds and are only vegetatively propagated, such as edible banana and plantain (*Musa* sp.) [95].

Major food crops that are commonly clonally propagated and therefore conserved in field genebanks include herbs, shrubs, vines, and trees, and these food crops belong to about 34 families [96]. Among those are sub-tropical and tropical shrub and tree species, such as coffee (*Coffea* sp.), cacao (*Theobroma cacao*), rubber (*Hevea brasiliensis*), coconut (*Cocus nucifera*), peach palm (*Bactris gasipaes*), breadfruit (*Artocarpus altilis*), mango (*Mangifera indica*), citrus (*Citrus* sp.), avocado (*Persea americana*) many temperate fruit trees, root and tuber crops such as potato (*Solanum tuberosum*), cassava (*Manihot esculenta*), yams (*Dioscorea* sp.), sweet potato (*Ipomoea batatas*), taro (*Colocasia esculenta*), other aroids, bananas, garlic (*Allium sativum*), shallot (*Allium cepa* var. *aggregatum*), grasses such as sugarcane (*Saccharum officinarum*), and forages. Additionally, temperate and sub-tropical fruit trees like peach (*Prunus persica*) and apricot (*P. armeniaca*) are typically clonally propagated to maintain the genetic constitution of the variety. As their seeds are non-orthodox, i.e., they cannot be dried to low seed moisture content and thus cannot be stored for longer periods at low temperatures, they are maintained in field genebanks and increasingly as in vitro materials (see Section 3.3) or cryopreserved (see Section 3.4). Although some of those crops are sexually fertile, they do not breed true to type, hence, the preferred method is vegetative propagation which enables the maintenance of genotypes as clones.

In field genebanks, the plant genetic resources are kept as live plants that undergo continuous growth and require regular care and maintenance. Accessions maintained in field genebanks need considerable space, especially tree species, and require much more attention in their day-to-day management than seed or in vitro collections, as the plants are continuously exposed to biotic and abiotic stresses. Integrated pest and disease measures are essential to ensure that plants are free of pathogens [97].

Given the exposure of plants in field genebanks to biotic and abiotic stresses and physical security threats (invading animals, theft), these do not present the most secure methods of germplasm conservation; however, they are often the only practical and cost-effective choice to conserve the germplasm of clonal crops, especially when resources and skills for alternative conservation approaches, such as in vitro conservation or cryopreservation, are out of reach.

When field genebank conservation is the only viable alternative, careful planning of site selection and appropriate field management can help to mitigate those risks. The revised and updated Genebank Standards of the FAO [78] indicate the best practices for the safe establishment and management of field genebanks, including the choice of location, the acquisition of germplasm, the establishment of field collections, appropriate field management, the regeneration and propagation of plant material, characterization, evaluation, documentation, distribution, and security and safety duplication.

3.2.1. Risks Associated with Field Genebanks

Adaptation of accessions. If environmental and soil conditions as well as the elevation of the field genebank are quite different from the site where plant material was collected, some poorly adapted accessions may fail to develop properly or may grow much more slowly than better adapted accessions. Moreover, poorly adapted accessions are also more prone to pest and disease infestations, hence losses of individual plants or entire accessions might occur over time. To mitigate such risks, a decentralized field genebank approach might work better, if it is feasible, i.e., the establishment of poorly adapted accessions at sites with agro-ecological conditions that are more like the original collection site [78]. The natural environment of the original site can be simulated to some degree, as is practiced at the international coffee field collection maintained by CATIE in Turrialba, Costa Rica [98]. Dense and almost permanent shade is provided for the wild genotypes from Ethiopia, while the cultivated accessions from East Africa are exposed to full sunshine. Cultivated accessions are grown under light shade, as is the case in commercial coffee production. Curational staff must always pay special attention to the growth and performance of the accessions of wild species to avoid plant losses. Poorly adapted accessions should also be duplicated at alternative sites or grown in greenhouses to avoid the loss of entire accessions. A safe alternative backup option is the cultivation of valuable, irreplaceable accessions in in vitro conditions or their preservation in liquid nitrogen. The latter has been shown to be an interesting long-term conservation approach for coffee germplasm, as cryopreservation costs (in perpetuity per accession) were lower than conservation in field genebanks [99].

Physical safety and plant health considerations. The absence of major threats from natural calamities, such as earthquakes, volcanoes, hurricanes, typhoons, and floods is important when deciding on the location of a field genebank [78]. A safe distance of at least 10 km radius from active volcanoes should be maintained to avoid damage from lava flow and rocks. Areas that are frequently in the path of hurricanes, typhoons, or snow avalanches should be avoided. Firebreaks can be established if bushfires are a known risk. Fencing and security guards will help to avoid vandalism, theft, and damage by large animals. It is good practice to choose a location where the target crop has not been grown previously to avoid the heavy infestation of major pathogenic diseases or insect pests that might cause plant losses or make disease and pest management very costly [97]. Soils might harbor fungal, plasmodiophorid, oomycete, and bacterial pathogens as well as viruses and plant parasitic nematodes, and termites that are detrimental to plant growth and that may lead to plant death. Many of the soil borne diseases are difficult if not impossible to manage and to eradicate with conventional means. The spread of soil-borne fungi (e.g., *Rosellinia* sp.) led to the death of numerous cacao trees and the entire loss of accessions, making it necessary to relocate the international cacao collection conserved by CATIE in Costa Rica to two new alternative sites [100]. Fire blight caused by the bacterium *Erwinia amylovora* is one of the most devastating apple diseases worldwide, and it can severely damage or even eradicate susceptible apple accessions in field genebanks [101]. Given all of the above-mentioned physical safety and plant health challenges with clonally propagated materials, the only safe long-term conservation option for such crops is cryopreservation, which is described further below.

Genetic integrity. Outcrossing species that are used to produce seeds for distribution requires a safe isolation distance to avoid the potential impact of geneflow and contamination from nearby commercial crop stands or from wild populations of the same species [78]. Many forage grasses are out-breeding, and it is recommended to use an isolation distance of at least 100 m between accessions [102]. Larger isolation distances are required for peach palm, as pollination is mainly conducted by insects, particularly small beetles, over distances between 100 and 500 m; wind and gravity can also function as pollen vectors [103]. The maintenance of such large isolation distances is important to preserve rare agronomic traits such as spineless peach palm varieties, e.g., 'Putumayo' and 'tanque de San Carlos' [104] and make such germplasm with highly sought-after characteristics available for distribution to users. As shown with this specific example, the maintenance of genetic integrity is critical for the facilitation of the use aspect of the PGRFA for direct cultivation or breeding and less so for the long-term conservation of such rare alleles within a population.

Spread of systemic pathogens. While most systemic pathogens are not transmitted via seeds, clonal propagules are often associated with the spread of such pathogens [96]. Therefore, field genebanks as a source of materials for distribution present serious problems for germplasm exchange. Many national or regional disease outbreaks have been associated with the transfer of vegetative propagules, e.g., the spread of banana bunchy top virus (BBTV) to Africa and within the continent, aggressive strains of potato late blight (*Phytophthora infestans*) in Africa and Asia, and potato cyst nematode (*Globodera palladi*) in East Africa, among others [105]. To avoid the spread of dangerous pathogens through the exchange of clonally propagated germplasm, the Germplasm Health Units of the CGIAR recommend the generation of virus-free in vitro plants for germplasm exchange as per the FAO-International Board for Plant Genetic Resources (IBPGR) technical guidelines for the conservation and safe distribution of these crops [106]. All germplasm material exported and imported by CGIAR centers are tested for viruses and other pests as per guidance provided by the National Plant Protection Organizations (NPPO), and only material that is free of viruses and other pests are released to clients. Procedures for germplasm health testing, phytosanitation, and safe international transfers for clonally propagated crops as well as seed crops have been thoroughly reviewed by Kumar et al. [105].

Rejuvenation. Low plant vigour, loss of plants within accessions due to pest and disease pressure, and the high age of plants are major reasons for the rejuvenation of accessions in a field collection. The loss of a single individual plant usually could entail genetic erosion within the accession because there are normally only very few plants representing each accession, sometimes only one individual, especially in the case of woody species. According to Reed et al. [97], the number of replicates is often limited to between 5 and 10 for cassava, 10 and 12 for sweet potato, 1 and 3 for trees and shrubs, 6 and 10 for herbaceous plants, and between 3 and 20 for bananas. In the case of the USDA-ARS apple field collection, for example, trees are grafted in the nursery on M7 dwarfing rootstocks and then planted as duplicates in the fields [107]. Once the primary tree is established, the second tree is removed, thus leaving one grafted tree per accession; hence, there is a clear need to back up a collection to avoid genetic erosion. Regeneration and propagation have species-specific requirements and are very costly management interventions that need to be carefully planned. Rejuvenation might also require relocation to another site to avoid diseases, pests, and soil infestation caused by devastating pathogens. Even handling the entire process of raising rootstocks, vegetative propagation, and replanting to the field with the utmost care, human errors can easily happen, and the accessions can be mixed up [95].

To avoid genetic erosion and the loss of entire accessions, a cryopreservation back-up system is mandatory to safeguard the long-term conservation of important clonal material. Furthermore, safety duplication of field genebank accessions is an essential activity for the security of the conserved genetic diversity.

3.2.2. Advantages of Field Genebanks

Field genebanks provide ready and easy access to the conserved material for characterization, evaluation [108], and research. Phenotypic characterization of accessions in field genebanks is relatively easy to perform, as the plants are readily and permanently available in the field and do not need to be grown out, which is the case for orthodox seed collections. Because of the permanent availability of the plants in the field collection, the scoring of characterization traits can be done at the appropriate time and repeated over the years if necessary [78]. Reference accessions planted in the same field facilitate the correct scoring of specific traits and the interpretation of the results that are obtained. Herbarium specimens and high-quality voucher images will guide true-to-type identification of accessions in a field genebank.

Germplasm users can visit the collections and inspect the plants during the vegetative or reproductive stages to have a first visual impression, which will help in making an informed decision on which germplasm to select and order. Fruits and/or vegetative material are readily available for germplasm distribution. The exposure of vegetatively propagated plants in the field genebank to changing environmental conditions allows for a gradual adaptation process of the plants [96] in contrast to the seeds kept in seed storage in a frozen state over several decades. This may present a major advantage to germplasm users. In combination with the cryopreservation techniques developed for long-term conservation of clonal germplasm, field genebanks facilitate the visual germplasm selection process, while in vitro collections support the safe exchange of clonal plant germplasm.

3.3. *In Vitro Collections*

Alternative conservation strategies for vegetatively propagated crops and species with recalcitrant seeds are in vitro cultures for short- to medium term conservation (MTS) comprising a couple of months up to a few years—the so-called in vitro active genebank (IVAG) [109–111]. In the IVAGs, plant material is maintained under slow-growth conditions with species-specific successive subculturing and renewal, readily available for multiplication and distribution to germplasm users. Cryopreservation in liquid nitrogen is the technology available for long-term conservation, denominated in an in vitro base genebank (IVBG). Technical guidelines providing guidance to researchers and genebank and botanic garden managers for the establishment and management of in vitro germplasm collections have been published [95,108], and genebank standards for maintenance of PGRFA in vitro have been developed [78].

Slow-growth culture conditions are applied to in vitro collections to reduce the frequency of subculturing, which is labor-intensive and is a source of contamination of the cultures. Entire accessions might be lost due to handling errors (mixing, mislabeling, misidentification) and genetic instability (somaclonal variation) induced by the tissue culture environment [95]. Under optimal growth conditions, subculture frequencies range from one to three months, whereas under slow growth conditions, the subculture period can vary from one to two years, depending on the crop, the environmental conditions in the culture room, and the media composition. Slow-growth conditions aimed at reducing the metabolic activity of the in vitro plantlets can be achieved by applying physical, chemical, or nutrient growth limitations, either individually or in combination [110,112]. Physical growth limitations are achieved, within limits, by lowering the temperature of the growth room, often in combination with low light intensities and restricted photoperiods. Other measures comprise minimal containment in small culture vessels resulting in conditions that minimize the growth and development of plants by restricting space, gaseous exchange, and nutrient supply [112].

Species from temperate climates are, in general, more cold-tolerant than species from the tropics and subtropics. A low temperature regime of 2 °C and 10 °C is used for the MTS of in vitro grown *Allium* species at the Leibniz Institute of Plant Genetics and Crop Plant Research (IPK), Gatersleben, Germany, extending the culture cycle to 12 months [113]. For MTS storage of potatoes, the IPK applies a multi-step, sequential approach [114]. After the establishment of virus-free potato material, tissue cultures are initially kept at relatively high temperatures of 20 °C under long-day conditions for 2–3 months, followed by a microtuber-induction phase with short-day conditions at 9 °C for 2–4 months and, finally, a cold storage period with microtuber storage at 4 °C for 16–18 months.

Cold-sensitive species from the tropics and subtropics require higher storage temperatures of at least 15 °C for sweet potato [115], 16 °C for *Musa* [116], 21 °C for pineapple [117], and 25 °C for yam, with subculturing intervals of two months [113]. The higher the storage temperature, the shorter the subculture intervals.

A chemical growth limitation involves the application of osmotically active agents such as mannitol, sorbitol, and polyethylene glycol (PEG), resulting in water stress for the tissues or the addition of growth retardants such as abscisic acid (ABA), paclobutrazol, ancymidol, and hydrazides to the culture media [95,112]. A nutritional growth limitation is based on low levels of macro- and micronutrients in the culture medium, resulting in the slow growth of tissue cultures [110].

Combining physical (temperature of 6 °C; 16-h photoperiod), chemical (20 gL^{-1} mannitol inclusion in culture media), and nutritional (40 gL^{-1} sucrose) growth limitations, Sarkar and Naik [118] were able to extend viability of potato microplants in vitro for up to 30 months without subculturing. However, the only long-term conservation option is cryopreservation, which is described in Section 3.4.

3.3.1. Risks Associated with In Vitro Collections

Freedom from contamination. Tissue culture is central to the safe movement of clonal plant germplasm; hence, it is important to assure the purity of the cultures. During the germplasm acquisition process a health test is conducted, and viruses, if present, are eradicated, followed by disease indexing before entering the in vitro genebank [110]. However, it remains possible that covert, systemic endophytes go undetected and continue residing in germplasm tissues after the disease eradication and sterilization procedures. These organisms may become opportunistic pathogens and pandemic agents if spread by vectors such as mites and thrips in the culture room. Furthermore, mites, thrips and other small arthropods may cause the proliferation of fungal contaminations in the tissue cultures, and these are quite difficult to eradicate [97].

Correct identity of accessions. The identity of cultures may be compromised because of human errors, such as the physical mixing of the accession samples and documentation errors due to mislabeling or misidentification [115]. The CGIAR genebanks adopted a rigid authentication process as part of their quality management process that starts with the verification of the documentation that is associated with germplasm acquisition (passport information) followed by testing the incoming accessions with standard markers and descriptors and the application of informatics tools [110]. A wide range of molecular techniques is available to authenticate germplasm [119]. Moreover, DNA barcoding is evolving as a robust technology that allows routine checks for genetic authenticity and ensures that a mistaken identity is not perpetuated [110,120,121]. Electronic barcoding is also an important quality assurance tool that allows instant traceability and provides current information on the status of each accession in the genebank at any point in time. This information needs to be linked to an electronic inventory system to support the retention of authenticated status and to prevent errors arising from transcribing hand-written records.

Somaclonal variation. A problem that is often associated with micro-propagated plants are somaclonal variations, i.e., genetic aberrations that are caused by mutations or epigenetic effects [122]. This is especially the case when tissue is exposed to minimal (slow or sub-optimal) growth conditions over long periods of time and may be due to the accumulation of ethylene, which restricts growth and may exacerbate other stresses induced during slow growth in in vitro storage [110]. In general, the species or crop and the genotype within the same crop, the propagation methods and the nature of the tissue used as the starting material, the type and concentration of growth regulators added to the culture medium as well as the number and the duration of subcultures are some of the factors that determine the frequency of occurrence of somaclonal variation in vitro. Disorganized growth phases in tissue cultures, especially in callus and cell suspension cultures, increase the chances of mutations [122]. Banana is a crop which is frequently affected by somaclonal variation, and with increasing subculture events, the proportion of variants can reach levels of up to 72% [123]. Plant growth regulators present in the culture media seem to indirectly affect somaclonal variation by increasing the multiplication rate of the cultures. To minimize problems with somaclonal variation in micropropagated plants, it is recommended to use organized tissue systems, such as shoot cultures, upon culture initiation, rather than callus and suspension cells, and to culture the plantlets on hormone-free media for medium-term in vitro storage [95].

Cellular ageing and senescence. Cellular ageing leading to a loss of biochemical and physiological functions in cells and senescence are observed during prolonged cultivation in vitro. In eight-year-old peach palm (*Bactris gasipaes*) cultures, initiated through direct morphogenesis of adventitious buds without callus formation, Graner et al. [124] observed generalized senescence and probable ageing in clones.

Safety duplication. To avoid the aforementioned risks, it is mandatory to duplicate the collection, either in vitro or in cryopreservation, and preferably in another distant location to ensure that the duplicate collection is properly secured [95]. For example, the Bioversity International *Musa* Germplasm Transit Centre (ITC) hosted at the Katholieke Universiteit Leuven, Belgium and home to the world's largest collection of banana diversity, maintains 70% of its in vitro clones in a cryopreserved base collection. A cryopreserved sample of each in vitro clone is safely duplicated at the Institut de Recherche pour le Développement—National Research Institute for Sustainable Development (IRD), Montpellier, France [95].

3.3.2. Advantages of In Vitro Collections

In vitro conservation has several compelling advantages when compared to field genebanks, as accessions are not subjected to the risks of climate variability and pest and disease outbreaks, which are frequent in the latter. The availability of germplasm samples from a field genebank is restricted by the season, and the development stage of the plant and the international movement of vegetative propagules carries inherent risks of transmitting pernicious plant pathogens [95]. In contrast, tissue culture samples are available year-round [112], have a low space requirement, and are characterized by a high multiplication rate [95]. Moreover, tissue cultures are an internationally recognized means of the safe germplasm movement of disease-free material under aseptical conditions [105,106].

By limiting the international movement of vegetatively propagated plants to sterile in vitro plants, intracellular obligate pathogens, such as viruses, viroids, and phytoplasmas, are the only remaining concern [105]. These pathogens can be eliminated through meristem culture, thermotherapy, chemotherapy, electrotherapy, and cryotherapy [110,125–127]. In grapevine, electrotherapy consisting of the electric stimulation of grapevine herbaceous cuttings with an electric current of 40–100 mA for 5–20 min in an electrophoresis tank followed by the in vitro regeneration of new plants has been successfully used for the complete and/or partial elimination of viruses [126]. Cryotherapy is an option for pathogen eradication in those crops for which cryopreservation protocols are available, and it has been successfully applied to several crops, such as potato (*Solanum tuberosum*), sweet potato (*Ipomoea batatas*), grapevine (*Vitis vinifera*), citrus (*Citrus* sp.), raspberry (*Rubus idaeus*), banana (*Musa* sp.), apple (*Malus domestica*), kiwifruit (*Actinidia chinensis*), and gentian (*Gentiana triflora*) [127,128]. Detailed protocols for pathogen (virus, viroids) elimination and plant health status verification as they have been applied for banana, cassava (*Manihot esculenta*), potato, sweet potato, and yam (*Dioscorea* sp.) by the CGIAR genebanks have recently been summarized by Kumar et al. [105].

In summary, slow growth in an in vitro culture system is a successful method of securing plant germplasm under medium-term storage conditions, similar to field genebanks. In vitro genebanks that cultivate clonally propagated crops can apply various methods for pathogen elimination, enabling the safe distribution of clonal plant germplasm to users. Apart from field genebanks, it is the only method to conserve crops with recalcitrant seed that cannot (yet) be cryopreserved due to the lack of successful cryopreservation protocols. It is also an essential element for the recovery of cryopreserved plant germplasm and, therefore, an essential link to the long-term conservation of a crop germplasm that does not produce orthodox seed.

3.4. Cryopreservation

Given the limitations and problems associated with field genebanks and in vitro collections described above, cryopreservation, i.e., the storage of biological material at an ultra-low temperature, usually in liquid nitrogen (−196 °C) or its vapor phase (between −140 and −180 °C), is the only method that is currently available to ensure the safe and cost-effective long-term conservation of the PGRFA of species that have intermediate or recalcitrant seeds, that hardly produce seeds at all, or that are vegetatively propagated [129]. Cryopreservation can be applied to both in vivo materials, such as seed and dormant buds, as well as to in vitro materials comprising cell suspension and callus cultures, shoot tips, somatic and zygotic embryos, and embryonic axes [130].

Plant cryopreservation studies started about 60 years ago, when Sakai [131] was able to show that cold-hardened tissue sections of mulberry twigs were able to survive exposure to liquid nitrogen when first pre-frozen at −20 °C, a step that led to the dehydration of the freezable water in the cells. He clearly demonstrated that the hardening of the cells through exposure to low winter temperatures and the dehydration of their tissues were essential elements of tissue survival. The formation of ice crystals within the cells of cryopreserved material leads to cell death. Effective dehydration removes all of the freezable water from the cells and leads to the vitrification of the highly concentrated cytoplasm [132]. Vitrification means the transition of water directly from the liquid phase into an amorphous phase or glass, avoiding the formation of lethal ice crystals in the cells [95]. Cryopreservation procedures comprise slow and controlled rate cooling techniques as well as different dehydration techniques prior to direct immersion in liquid nitrogen. The latter include dehydration, vitrification, encapsulation-dehydration, encapsulation-vitrification, pre-growth, pre-growth dehydration, and droplet-vitrification [130,132].

3.4.1. Dehydration

The dehydration of explants intended for cryopreservation is mainly used with seeds, zygotic embryos, or embryonic axes extracted from seeds followed by direct immersion in liquid nitrogen for rapid cooling, except for oily seeds (e.g., *Arachis hypogea*), which require a slow pre-cooling phase in a programmable cooler before cryopreservation [130] and a slow seed imbibition phase over water [133]. The natural cold acclimatization of twigs in combination with dehydration is also a key element for dormant bud cryopreservation, which usually requires controlled-rate cooling [134,135]. At the Millennium Seed Bank of the Royal Botanic Gardens, Kew, desiccation-tolerant, orthodox seeds of wild species with short lifespans under standard long-term conservation conditions (−20 °C) are dried at about 32 ± 3% RH at 18 °C and are then stored in the vapor phase of liquid nitrogen [94]. In general, cryogenic storage extends seed longevity compared to conventional freezer storage [133]. However, the extension of seed longevity seems to be species-specific, and, above all, a high initial seed quality is critical to maximize the benefits of cryostorage [136].

Apart from orthodox seeds, dehydration has also been applied to seeds, embryos, and embryonic axes of a wide range of recalcitrant and intermediate tropical species [137]. Such species are usually dried to a SMC (fresh weight basis) ranging from 10 to 20% [130].

3.4.2. Controlled-Rate Cooling

Controlled-rate cooling is commonly employed for temperate and subtropical species, including dormant buds, apices of cold-tolerant species, and undifferentiated cell cultures, such as callus and cell suspension cultures [130,132], as well as for oily seed species [133]. The use of dormant buds for cryopreservation is a relatively easy and cost-effective cryopreservation method, as it does not involve aseptic cultures and the excision of shoots. An effective protocol for the cryopreservation of dormant apple buds (*Malus* sp.) was developed at the USDA National Center for Genetic Resources Preservation (NCGRP) in Fort Collins Collins, Colorado, USA [134], and more than 2300 apple clones have been cryopreserved and are currently being maintained in liquid nitrogen vapor conditions [107].

Volk et al. [138] provide a detailed description of the apple dormant bud cryopreservation protocol consisting of the following steps: (i) collecting dormant budwood twigs in mid-winter and cutting them into single node segments; (ii) air-dehydrating the twigs at −5 °C and 35% RH to a 25–30% moisture content (fresh weight basis—fwb); (iii) placing the dehydrated twigs in tubes that are heat-sealed, labeled, and placed in cryoboxes for slow freezing in a programmable cooler at 1 °C per hour from −5 °C to −30 °C and holding this temperature for 24 h; (iv) transferring pre-frozen boxes to the vapor phase of liquid nitrogen for long-term storage; (v) allowing the cryopreserved nodal sections to rehydrate at 2–4 °C for 14–21 days on moist peat moss for recovery; and finally, (vi) the rehydrated buds are budded onto potted seedling rootstocks (2 buds per rootstock).

Apart from apples, the described dormant bud cryopreservation has also been successfully developed for pear (*Pyrus* sp.) [139] and sour cherry (*Prunus cerasus*) [140]. Recent studies [141] confirmed that the air drying of dormant budwood to ~30% moisture content followed by slow cooling before liquid nitrogen storage was the most critical pre-storage treatment for increasing freezing resistance and cryosurvival. The fruit crops that were covered in these studies included apple, pear, sweet cherry (*Prunus avium*), apricot (*Prunus armeniaca*), and peach (*Prunus persica*). For peach, the best pre-storage moisture level was slightly higher at 35% (fwb), an indication that desiccation sensitivity may contribute to low cryosurvival. Similar protocols for the cryopreservation of dormant blueberry (*Vaccinium* sp.) are also under development, and it has been shown that the pre-harvest temperature of the twigs (which should remain below 11.2 °C for a 10-day period) is a critical factor for the successful post-cryopreservation viability of blueberry dormant buds [142]. In the case of mulberry (*Morus* sp.) [143] and blackcurrant (*Ribes nigrum*) [144], cryopreserved buds are recovered in vitro before being transferred to the field.

3.4.3. Vitrification-Based Cryopreservation Protocols

Apart from the conventional dehydration of the tissues to be cryopreserved, several protocols make use of the addition of cryoprotectants to increase the viscosity and to achieve suitable cellular dehydration, while avoiding ice formation [145]. A total of seven vitrification-based cryopreservation protocols can be distinguished [129,130], which consist of (i) encapsulation dehydration; (ii) vitrification; (iii) encapsulation-vitrification; (iv) dehydration; (v) pre-growth; (vi) pre-growth dehydration, and (vii) droplet vitrification.

Droplet vitrification is now the most common and most widely used cryopreservation protocol for hydrated tissues, such as in vitro cultures [95]. This method exposes meristem tips to plant vitrification solution (PVS), leading to a more concentrated, vitrifiable cell solution, which can then be exposed to liquid nitrogen for long-term cryostorage [95,132]. Recent modifications to the droplet vitrification method comprise the use of aluminum cryoplates with encapsulation dehydration or encapsulation vitrification [146–148]. With these more recent protocols, meristems are enclosed in tiny drops of calcium alginate and placed on the aluminum plate before being dehydrated and subsequently exposed to liquid nitrogen. Cryopreservation by droplet vitrification has been successfully tested in grapevine (*Vitis vinifera*), gentian (*Gentiana triflora*), potato (*Solanum tuberosum*), kiwifruit (*Actinidia chinensis*), and raspberry (*Rubus idaeus*) in New Zealand. This technology is also being applied for the pathogen eradication of viruses and bacteria infecting those crops, thus ensuring the long-term conservation of healthy clonal plant material [127].

Unfortunately, there is no 'generic cryopreservation protocol' that can easily be adopted and adapted to a wide range of species. The science and methodology of cryopreservation, i.e., protocol development, is still a major challenge for many crop species. A further difficulty is the successful implementation of available cryopreservation protocols to an entire crop collection, as some genotypes within the same species might not respond favorably to a specific protocol requiring further modifications [95,132,149].

Major cryopreserved collections of temperate, subtropical, and tropical plant species include apple (*Malus* sp.), pear (*Pyrus communis*), *Citrus* sp., mulberry (*Morus* sp.), potato (*Solanum tuberosum*), grape (*Vitis vinifera*), coffee (*Coffea arabica*), *Musa*, sweet potato (*Ipomoea batatas*), cassava (*Manihot esculenta*), yam (*Dioscorea* sp.), strawberries (*Fragaria* × *ananassa*), hops (*Humulus lupulus*), garlic (*Allium sativum*), chives (*Allium schoenoprasum*), mint (*Mentha* sp.), and medicinal plants (for an overview of conservation institutes, crops conserved, and cryopreservation methods used, please refer to Panis [150]). Recently, O'Brien et al. [148] reported on the successful cryopreservation of the somatic embryos and shoot tips of avocado (*Persea* sp.). It has been estimated that about 100,000 unique accessions of vegetatively propagated and recalcitrant seed crops require long-term conservation through cryopreservation, while currently, about 18,500 accessions are conserved by this method [151], up from the approximately 10,000 accessions reported by Acker et al. [149]. Most cryopreserved accessions belong to five crops: potato (38%), cassava (22%), bananas and plantains (11%), mulberry (12%), and garlic (5%) [149].

3.4.4. Advantages and Limitations of Cryopreservation for Long-Term Conservation

The major benefit of cryopreservation protocols is the fact that this technology is the only available method that allows the safe and long-term conservation of many species that are vegetatively propagated or that have recalcitrant seeds (mostly from the tropics and subtropics), which otherwise can only be conserved in field genebanks or in in vitro collections. The inherent risks and the short- to medium-term nature of these conservation methods have been described above. In general, introducing an accession into cryopreserved storage is more expensive than establishing an accession in in vitro culture or in the field. However, the costs of maintaining an accession in cryopreserved storage for the long-term (above 20 years) are considerably lower than those of maintaining an accession in the field or in vitro, particularly when dealing with a large number of accessions [99,130,132,149]. Moreover, cryopreservation is a conservation method that ensures genetic stability over long periods of time. In addition, cryotherapy offers additional benefits for the removal of viruses from a wide range of vegetatively propagated crops [127,128].

At present, over one million seed samples from national and international genebanks are being conserved at the Svalbard Global Seed Vault (SGSV), Norway, a global security back-up system for long-term seed conservation, at −18 °C [152]. Although clonal crop collections can be duplicated for safety reasons at other locations, either in the field or in vitro, the safest backup approach would be the cryopreservation of a safety duplicate. A recent feasibility study concluded that a safety backup facility similar to the one in Svalbard, Norway is required to accommodate a duplicate of the approximately 10,000 unique accessions currently cryopreserved at the global level and to offer space for additional safety duplicates arising from on-going cryopreservation activities [149]. Unfortunately, the implementation of this important proposal has not yet started.

3.5. DNA Banks

DNA storage is regarded as an emerging complementary ex situ technique for safeguarding the genetic diversity of a crop's genepool, especially for species that are difficult to conserve by conventional means in the form of seeds or vegetatively in field genebanks, in in vitro collections, or via cryopreservation and that are highly threatened in the wild [153]. The transfer of genetic material in the form of DNA samples rather than seed would be especially meaningful for programmes that focus mainly on genetic and genomic studies and not on agronomic performance. It is a lot easier and safer to exchange DNA samples than seed or vegetative propagules, as the latter require seed/planting material inspection, phytosanitary certificates, and post-entry quarantine testing to ensure that the requested genetic plant resources are free from undesirable diseases and pests [154]. Moreover, shipping costs of DNA samples are considerably lower than those of seed or vegetative material.

DNA banks can also serve as backup or safety duplicates of the physical seed, field, or in vitro collections in case of catastrophic losses [154]. Although it is not (yet) possible to recover a plant from a DNA sample, the storage of entire genomes (total DNA) or genome fragments (genomic libraries) would permit the preservation of its valuable genetic information thus contributing to the objective of gene or genome conservation [155,156]. With the impressive advances in molecular genetics, these preserved genes or genomes might be of high relevance in the future. Genome conservation could play a major role for species that are currently under threat of extinction or that are already extinct [156]. While DNA banks are considered as a common genetic biodiversity repository [157], Datlof et al. [158] were able to demonstrate that the tissues of target species stored in DNA banks also harbour their corresponding microbial symbionts, many of which are yet to be discovered.

In anticipation of the emerging role of genomics in the conservation of PGRFA, the International Plant Genetic Resources Institute (IPGRI, now Bioversity International) conducted a global survey on the feasibility of DNA storage and use in 2004 and published its findings in a book on "DNA banks—providing novel options for genebanks" [159]. Guidelines for the management of DNA banks have been reviewed and described by Hodkinson et al. [160]. DNA resources can be maintained at −20 °C for short- and medium-term storage, i.e., up to 2 years, and at −70 °C or in liquid nitrogen for much longer periods, comparable to long-term seed storage [161]. Several factors, such as space, containers, frequency of access, cost, stability (temperature fluctuations), and security (breakdown of equipment) impact decisions about using conventional freezers (−18 to −20 °C), −80 °C freezers, or liquid nitrogen storage [162]. Liquid nitrogen (LN_2) freezers are the most secure option, as they do not require mechanical compressors; hence, the equipment does not fail in the event of power outages. However, this option is more costly and is primarily used for the long-term storage of hydrated samples. Preservation stresses, such as the drying of tissue to be stored, freezing, or the factor time, may inflict some damage to the DNA, but most chromosomal aberrations are repaired in the surviving cells after a few cell divisions [162].

Purified DNA dissolved in buffer may be safely stored for 1–2 years at 4 °C for 4–7 years at −18 °C and for more than 4 years at −80 °C, however, the overall fragment size and, consequently, the DNA quality decreases with storage time [162]. Long-term DNA conservation can also be achieved using a solid medium, such as cellulose-based cards, instead of DNA dissolved in buffer [154]. The paper conservation method is also an efficient means of inactivating pathogens and protecting plant DNA from degradation. DNA can either be stored within the tissue after transfer to the paper or as extracted DNA after submitting the plant tissue to an extraction protocol and transferring the nucleic acid to the paper. The DNA that is conserved on paper can be safely stored at room temperature and 30% relative humidity, at least for medium-term storage [162]. DNA samples on paper can be easily exchanged among institutions, and identification is facilitated by bar-coded tags that allow for a complete recovery of the sample information.

An interesting further development of the use of paper for DNA storage and exchange is the development of DNA books. DNA clones or PCR products are printed directly onto the pages of books and are delivered to users along with the relevant scientific information [163]. The DNA sheets are not damaged by high temperatures and humidity, conditions that might be imposed on the sheets during bookbinding and delivery to the recipients. Recipients can extract the DNA fragments from the DNA sheets and can amplify them using a polymerase chain reaction (PCR). In this context, we can refer to the Rice Full-Length cDNA Encyclopedia DNABook™, which contains 32,000 clones printed on special paper and is bound as a book [153]. A DNA book allows the efficient maintenance of tens of thousands DNA materials in a small space and under ambient conditions. It is an approach that is much less costly than DNA storage in a freezer and allows distribution using ordinary mail.

However, a study conducted by Colotte et al. [164] clearly demonstrated the necessity of protecting DNA from the air (humidity, ozone) to preserve its integrity at room temperature. Such conditions can be created by DNA encapsulation in laser-sealed capsules and accelerated ageing studies at a high temperature (76 °C) and at 50% RH for 30 h did not show any detectable DNA degradation [165]. Storing DNA samples for longer periods under these accelerated aging conditions required the addition of trehalose, which provides a protective matrix to the encapsulated DNA. By extrapolation, this could correspond to 100 years of storage at 25 °C, according to the Arrhenius model [165]. Therefore, DNA encapsulation seems to be a safe method for long-term DNA storage at room temperature, guaranteeing durable DNA stability and facilitating the international movement of DNA samples for molecular biology research.

Within living organisms, DNA is physically and chemically isolated from the environment, and this barrier can keep DNA intact, sometimes for hundreds of thousands of years, as seen in DNA extracted from frozen environments [166], but it can also protect it from arid, hot environments [167]. Lake sediments have also been suggested for ancient DNA studies [168]. Given the fact that DNA can be degraded during extraction and storage, most DNA banks store cells or tissues and extract DNA upon request [158,162]. Seeds are an efficient and inexpensive means of storing the DNA of individual genotypes. As long as seeds are viable, the supply of DNA is guaranteed. However, even seeds that have lost viability can still be used for DNA extraction and amplification, as has been shown in the case of 70- and 135-year-old seeds that were stored under ambient conditions [169]. This is of special relevance for accessions collected from wild populations, i.e., crop wild relatives, which might be threatened in situ.

As DNA can withstand significant variations in temperature as well as modest variations in moisture and offers tremendous information density, DNA storage is currently being explored beyond biological systems for the safe, long-term preservation of important information, such as a global seed vault [170]. Koch et al. [171] developed a storage architecture, called the DNA of things (DoT), for storing DNA-encoded information in

3D-printed objects. To protect the DNA from degradation at the elevated temperatures of 3D-printing, the DNA is encapsulated in nanometer silica beads and is then fused into the raw materials used for 3D-printing. Through this approach, the molecular memory can be concealed in the object and recovered at any time, even after the object has been damaged [172]. The encoded information can be retrieved by sequencing the DNA that has been extracted from a tiny portion of the object.

3.6. Pollen Banks

Pollen conservation is a complementary tool for the management and exchange of plant genetic resources, as it helps to conserve the alleles of a genotype or a population [173]. Pollen storage also facilitates crosses in breeding programmes, for example, for wide crosses, when natural pollen production is low or to overcome flowering asynchrony between parents [174]. Other uses of pollen storage include fertility research and studies in basic physiology, biochemistry, and biotechnology for gene expression, transformation, and in vitro fertilization [175]. Pollen should be harvested at peak anthesis, preferably in the morning hours [176]. To save collecting and processing time, it is often recommended to collect anthers in the field and then to separate the pollen grains from the anthers in the laboratory [173]. Pollen is quite sensitive and deteriorates quickly when kept at room temperature and at high relative humidity (75%) [177].

Cytological studies undertaken with 265 plant families by Brewbaker [178] revealed that about 68% release pollen in the bicellular state at anthesis (e.g., Rosaceae), 20% in the tricellular state (e.g., Compositae), and the remaining 12% releases in both types. The nuclear state of pollen grains at anthesis is a major determining factor for pollen viability during storage. While tricellular pollen has high moisture levels at anthesis (approx. 40–60%) and is desiccation-sensitive, bicellular pollen usually is drier at anthesis and can be safely dried to moisture levels below 10%, and its storage behaviour is similar to that of a desiccation-tolerant seed [175]. Longevity is increased by storing pollen at lower temperatures and at a lower moisture content. Apart from storage conditions (temperature and moisture content), the storage atmosphere can also affect longevity. Freeze-dried and vacuum-dried pollen showed greater longevities when stored in a vacuum compared to storage in air [179]. Similarly, pollen viability was enhanced when stored in nitrogen [180]. The beneficial effects of vacuum and nitrogen atmospheres on pollen viability are especially evident at temperature ranges from −5 °C to ambient conditions [175]. The pollen of several species can be successfully stored at temperatures ranging from 4 °C to −20 °C for the short-term [173].

Long-term storage is required if pollen is intended for germplasm conservation and exchange. In this case, pollen should remain viable and functional for about 10 years [181], and the safest way to achieve this is storage in freezers at −80 °C or in cryogenic storage [174,175,182]. Pollen cryopreservation has been successfully demonstrated for a vast range of horticultural crops as well as for staple food crops, forage grasses, ornamental and medicinal plants, and forestry species [182]. As shown by Ren et al. [183], the longevity of cryopreserved pollen seems to be species-specific. The pollen of 102 ornamental plant species/cultivars affiliated to 32 genera of 14 families showed the following changes in pollen viability after cryogenic storage for about 10 years: 11.7% (12 species/cultivars) had increased viability, 16.7% (17 species/cultivars) had stable viability, and the viability of 71.6% (73 species/cultivars) showed a decreasing trend.

Pollen with high moisture levels does not survive exposure to freezing temperatures, most likely due to intracellular ice formation [175]. Therefore, pollen grains are dehydrated before their immersion in liquid nitrogen using silica gel, saturated salt solutions, or drying in an airflow cabinet or oven at 35 °C [174,184].

Desiccation-sensitive pollen such as maize can also be cryopreserved by partially dehydrating pollen to safe moisture levels where no freezable water exists [185]. The highest maize seed set occurred with pollen grains that were dried to about a 12–20% moisture content. Rapid air-drying using pollen dryers that expose the pollen to air at

20–40% RH and at 20 °C has been shown to be beneficial for desiccation-sensitive species of the Poaceae, extending the tolerance of the pollen to freezing temperatures and their longevity [181].

After cryopreservation, a quick thawing protocol is mostly completed by placing samples in a water bath (37–45 °C) or holding them under running water, as reviewed by Dinato et al. [174]. Dried pollen is susceptible to imbibitional injury during rehydration, and this may significantly reduce their viability [186]. Slow rehydration, which can be achieved by placing the pollen in an environment with high RH for a couple of hours at room temperature, minimizes imbibitional damage to pollen grains [187].

Pollen viability can be assessed through the vital staining of pollen grains with fluorescein diacetate (FDA) or tetrazolium-based stains, through in vitro germination, or through effective in vivo fertilization and subsequent seed production [173].

Despite several limitations, such as low the pollen production of some species, the high labour requirement for collecting pollen, the lack of standardized protocols for pollen processing and viability testing, and difficulties in replenishing pollen supplies when quantities are depleted or when the pollen has deteriorated, pollen remains a valuable genetic resource for long-term conservation in cryogenic storage. Moreover, from a biosecurity point of view, pollen is a relatively safe means of germplasm exchange, as pests and diseases are rarely transferred through pollen [108].

In summary, pollen conservation is an additional tool for the maintenance of plant genetic resources and can assist plant breeders to overcome problems such as flowering asynchrony between different parent genotypes and the production of insufficient pollen in nature. Similar to orthodox seeds, the exchange of pollen is a safe means of germplasm exchange, as harmful pathogens are hardly transferred through pollen. For long-term conservation, pollen needs to be cryopreserved, and protocols have already been established for many species. As in other plant structures, the freezable water content needs to be removed from pollen for cryogenic storage in order to safeguard pollen viability during long-term storage at ultra-low temperatures.

4. Need for Complementary Conservation Approaches

The ex situ conservation of crop genetic resources largely takes place in genebanks and, to a lesser extent, in botanic gardens. In the case of wild species, such as the relatives of our crops, they are either conserved in their natural habitat or are collected and stored in genebanks or botanic gardens [188]. A special category of crop genetic resources are primitive varieties and landraces of our crop plants. Many of these are still found on farms as part of traditional production systems, and consequently, such materials are maintained 'on-farm' or, when collected for the purpose of PGR conservation, are placed in a genebank. In the case of species that grow in natural habitats but that are used by humans for food or medicine, these are mostly left in nature [189].

The use of in situ and on-farm conservation for the routine conservation of PGRFA had a difficult start and was fiercely debated at the FAO [13]. The strong influence of plant breeders and of those that had food production in mind as the most important objective to counter the strongly increasing genetic erosion in the 1960s resulted in a clear preference for ex situ conservation. However, with the increasing interest to widen the conservation to all cultivated plant species and more difficult crops, such as those producing recalcitrant seed or being vegetatively propagated, have become a target for collecting and conservation.

Driven by the strong push for in situ and on-farm conservation by the CBD during the early 1990s and the realization of the importance to also conserve the 'difficult crops', a stronger focus on the use of in situ and on-farm approaches became apparent, which is also true for agricultural crops [13]. This development is based on the fact that in situ conservation allows the conserved materials (typically populations in equilibrium with the ecosystem they occur in or traditional varieties and landraces to be cultivated on farm) to remain part of the natural or agricultural environment, in which evolutionary processes continue to manifest themselves. Thus, adaptation to changing conditions can happen, with or without human intervention. Furthermore, as wild plant species and crops are typically widely growing or being cultivated, respectively, much more genetic diversity within and between species can be conserved. The targeted conservation taxa develop naturally under 'local conditions'; thus, some of the political and managerial issues that apply to ex situ conservation can be avoided. An additional advantage is that the cost of conservation can be limited, which is predominantly confined to monitoring the genetic and species diversity. In the case of on-farm conservation, a close link between people and crops or species is maintained and allows adaptation to changing environmental, cultural, and economic conditions. The on-farm conservation approach is very suitable for 'crowdfunded, conducted and orientated' projects and programmes [190]. Possible disadvantages of this conservation method are the limited access to specific subsets of the resources conserved; the lack of adequate characterization and evaluation of the material; and the potential and continuous danger that farmers abandon the cultivation of traditional landraces because of their frequently disadvantaged competitive status. To conserve a given set of genetic diversity on-farm, it will be required that the traditional agro-ecosystem continues to play a livelihood role for the farmers. Due to the dynamic economic, social, and environmental nature of in situ and on-farm conservation, there is a need for careful monitoring practices [191].

The advantages of ex situ seed conservation are the capability of storing large numbers of accessions in a collection, which is relatively cost-efficient; the reproducibility of the results due to the availability of standardized procedures for all major food crops [78]; the possibility to maintain specific genotypes over time; the ready access of the germplasm for characterization, evaluation, research, and distribution; the perceived secure conservation conditions; the generally better health conditions of conserved material and thus the lower risk of spreading diseases; and possibly more specific aspects [191,192]. It should also be noted that within the ex situ approach, complementarity of specific methods do exist, e.g., maintenance in a field genebank can be complemented by in vitro or even cryopreservation storage, as mentioned in the previous section.

The drawback of ex situ conservation is that the germplasm materials are under static and artificial genebank conditions during storage; thus, these accessions are 'only' exposed to the selection pressures that are caused by these artificial environmental conditions and not by the (dynamic) natural environmental or cultivation conditions under which the conserved materials could evolve and adapt to the changing conditions.

To facilitate decision-making regarding which conservation method(s) to apply, it is important to know the strengths and weaknesses of both in situ and ex situ methods. The reproductive biology of the species is certainly the most critical one [193]. Genetic erosion and other threat considerations will certainly impact the urgency and the coverage of the genetic diversity that one must address through conservation efforts. Furthermore, it is also important to realize that some of the decision criteria will depend on other factors, such as available infrastructure, trained staff, budget, and the prevailing legal and policy framework as well as collaboration with other institutions inside and outside the country. Furthermore, when planning complementary conservation strategies, the following additional points might also be relevant to consider: the extent of the gene pool coverage and the distribution of the genetic diversity, both within the gene pool as well as geographically [7]. The reproductive biology of a species is critically important to decide which methods are applicable. The extent of genetic erosion and other threats need to be considered [194] as well as non-biological aspects, including the socio-economic feasibility, possible support from governmental agencies, and the availability of markets (in the case of on-farm conservation of traditional crops) are other aspects to take into consideration when deciding on the combination of available conservation methods [191,195]. At the end of the day, it will have to be practical, long-term, and sustainable aspects that should prevail.

The CBD explicitly states that in situ conservation should be given the highest priority but also states that ex situ conservation has an important role to play. Considering the pros and cons of the various conservation approaches, the prevailing conclusions and recommendations are that in situ and ex situ conservation should be combined to achieve more sustainability, long-term security, efficiency, and cost-effectiveness of PGRFA conservation [191,192,196]. Specifically for the efficient protection of crop wild relatives, the concept of so-called trans situ conservation has been introduced, which dynamically integrates multiple in situ and ex situ measures, from conservation to research to education, spanning local to global scales [197]. The conservation of wild chili (*Capsicum annuum* L. var. *glabriusculum*) in southern Arizona is demonstrating this evolving concept.

5. Concluding Remarks

The history of the creation and growth of the global conservation system, particularly of the international network of ex situ collections, provides a useful foundation for the critical review of this global system. This foundation is further strengthened by a detailed analysis of the routine genebank operations and of the importance to aim for an integration of in situ and ex situ conservation approaches. In part two of this paper, we will critically review routine germplasm conservation activities, including the active and base collection concept, evaluate new developments that facilitate germplasm conservation and use, assess factors that facilitate or limit the participation of genebanks in the global system, and provide a concluding long-term perspective for an efficient and effective global conservation system.

Author Contributions: J.M.M.E. and A.W.E. conceived of, drafted, and edited this paper equally. All authors have read and agreed to the published version of the manuscript.

Funding: This research received no external funding.

Conflicts of Interest: The authors declare no conflict of interest.

References

1. Harlan, J.R. *Crops and Man*; American Society of Agronomy, Madison, Wisconsin, and Crop Science Society of America: Madison, WI, USA, 1975; p. 295. ISBN 0-89118-032-X.
2. Schlegel, R.H.J. *History of Plant. Breeding*; CRC Press: Boca Raton, FL, USA, 2021; p. 330. ISBN 9781032095813.
3. FAO. *Report of the Third Session of the FAO Panel of Experts on Plant. Exploration and Introduction*; Plant Production and Protection Division, FAO: Rome, Italy, 1969; pp. 25–28.
4. Plucknett, D.L.; Smith, N.J.H.; Williams, J.T.; Anishetty, N.M. Crop germplasm conservation and developing countries. *Science* **1983**, *220*, 163–169. [CrossRef] [PubMed]
5. Khoury, C.K.; Laliberte, B.; Guarino, L. Trends in ex situ conservation of plant genetic resources: A review of global crop and regional conservation strategies. *Genet. Resour. Crop. Evol.* **2010**, *57*, 625–639. [CrossRef]
6. Anonymous. CGIAR-Crop Trust System Level Review of Genebank Costs and Operations (GCO). *Report October 2020.* Available online: https://www.genebanks.org/wp-content/uploads/2020/11/GCO-Report_261020.pdf (accessed on 30 June 2021).
7. Hoban, S.; Bruford, M.; Funk, W.C.; Galbusera, P.; Griffith, M.P.; Grueber, C.E.; Heuertz, M.; Hunter, M.E.; Hvilsom, C.; Kalamujic, S.B.; et al. Global commitments to conserving and monitoring genetic diversity are now necessary and feasible. *BioScience* **2021**, 13. [CrossRef]
8. Prance, S.G.; Nesbitt, M. *The Cultural History of Plants*, 1st ed.; Routledge: New York, NY, USA, 2005; 460p. [CrossRef]
9. Hallauer, A.R. Evolution of plant breeding. *Crop Breed. Appl. Biotechnol.* **2011**, *11*, 197–206. [CrossRef]
10. Vavilov, N.I. *Five Continents*; International Plant Genetic Resources Institute: Rome, Italy, 1997; 198p, ISBN 92-9043-302-7.
11. Vavilov, N. Studies on the origin of cultivated plants. *Bull. Appl. Bot. Plant. Breed.* **1926**, *26*, 1–248.
12. Scarascia-Mugnozza, G.T.; Perrino, P. The history of ex situ conservation and use of plant genetic resources. In *Managing Plant Genetic Diversity*; Engels, J.M.M., Ramanatha Rao, V., Brown, A., Jackson, M.T., Eds.; CABI Publishing: New York, NY, USA, 2002; pp. 1–22.
13. Pistorius, R. *Scientists, Plants and Politics—A History of The Plant. Genetic Resources Movement*; International Plant Genetic Resources Institute: Rome, Italy, 1997; 197p. Available online: https://www.bioversityinternational.org/fileadmin/_migrated/uploads/tx_news/Scientists__plants_and_politics_240_01.pdf (accessed on 30 June 2021).
14. Harlan, H.V.; Martini, M.L. Problems and Results in Barley Breeding. In *Yearbook of Agriculture*; USDA: Washington, DC, USA, 1936; pp. 303–346. Available online: https://naldc.nal.usda.gov/download/IND43893522/PDF (accessed on 28 July 2021).
15. FAO. Report on the Technical Meeting on Plant Exploration and Introduction. In *FAO Plant Introduction Newsletter No. 10*; FAO: Rome, Italy, 1962. Available online: https://www.bioversityinternational.org/fileadmin/user_upload/online_library/publications/pdfs/Plant_Introduction_Newsletter_No_10.pdf (accessed on 29 June 2021).

16. Hawkes, J.G. *The Diversity of Crop Plants*; Harvard University Press: Cambridge, MA, USA, 1983; 184p.
17. Sonnino, A. International instruments for conservation and sustainable use of plant genetic resources for food and agriculture: An historical appraisal. *Diversity* **2017**, *9*, 50. [CrossRef]
18. Frankel, O.H.; Bennett, E. *Genetic Resources in Plants—Their Exploration and Conservation*; IBP Handbook No. 11; Blackwell Scientific Publications: Oxford, UK, 1970; 554p.
19. Frankel, O.H.; Hawkes, J.G. *Crop. Genetic Resources for Today and Tomorrow*; Cambridge University Press: Cambridge, UK; New York, NY, USA, 1975; p. 514.
20. Frankel, O.H.; Soulé, M.E. *Conservation and Evolution*; Cambridge University Press: Cambridge, UK, 1981.
21. FAO. *Report of the FAO/UNEP/IBPGR International Conference on Crop Genetic Resources*; AGP: 1981/M/6; FAO: Rome, Italy, 1981; 80p.
22. FAO. *Global Plan of Action for the Conservation and Sustainable Use of Plant Genetic Resources for Food and Agriculture*; FAO: Rome, Italy, 1996.
23. Lowie, R.H. Reviewed work: Studies on the origin of cultivated plants by N. Vavilov. *Am. Anthropol.* **1928**, *30*, 716–719. Available online: www.jstor.org/stable/661137 (accessed on 29 June 2021). [CrossRef]
24. Chandler, R.F., Jr. *An Adventure in Applied Science: A History of the International Rice Research Institute*; International Rice Research Institute (IRRI): Manila, Philippines, 1992; p. 252.
25. CIMMYT. Our History. 2021. Available online: https://www.cimmyt.org/about/our-history/ (accessed on 7 July 2021).
26. CTA. *IBPGR: International Board for Plant Genetic Resources*; Spore 38; CTA: Wageningen, The Netherlands, 1992; Available online: https://hdl.handle.net/10568/45742 (accessed on 26 June 2021).
27. Rose, G.L. The International Undertaking on Plant Genetic Resources for Food and Agriculture: Will the Paper Be Worth the Trees? Available online: https://ro.uow.edu.au/lawpapers/32 (accessed on 29 June 2021).
28. Moore, G.; Tymowski, W. *Explanatory Guide to the International Treaty on Plant Genetic Resources for Food and Agriculture*; IUCN: Gland, Switzerland; Cambridge, UK, 2005; 212p.
29. FAO. Interpretation of the International Undertaking on Plant. Genetic Resources. Available online: http://www.fao.org/3/z4968en/z4968en.pdf (accessed on 29 June 2021).
30. FAO. Resolution 3/91. Annex 3 to the International Undertaking on Plant Genetic Resources. Available online: http://www.fao.org/3/x5587E/x5587e06.htm#Resolution3 (accessed on 7 July 2021).
31. CBD. Section IX. Nairobi Final Act of the Conference for the Adoption of the Agreed Text of the Convention on Biological Diversity. Available online: https://www.cbd.int/doc/handbook/cbd-hb-09-en.pdf (accessed on 7 July 2021).
32. FAO. *Report of the Commission on Plant Genetic Resources*; First Extraordinary Session: Rome, Italy, 1994. Available online: http://www.fao.org/3/aj681e/aj681e.pdf (accessed on 7 July 2021).
33. FAO. *Report on the State of the World's Plant Genetic Resources*; FAO: Rome, Italy, 1998.
34. FAO. *Multilateral Treaties Deposited with the Director General*; Thirty-First Session: Rome, Italy, 2001. Available online: http://www.fao.org/3/Y1938e/Y1938e.htm (accessed on 7 July 2021).
35. Crop Trust. Our History. 2021. Available online: https://www.croptrust.org/about-us/our-history/ (accessed on 7 July 2021).
36. FAO. Commission on Genetic Resources for Food and Agriculture. In Proceedings of the First Session of the Governing Body of the International Treaty on Plant Genetic Resources for Food and Agriculture, Madrid, Spain, 12–16 June 2006. Available online: http://www.fao.org/cgrfa/meetings/detail/en/c/1112354/ (accessed on 7 July 2021).
37. Crop Trust. Svalbard Global Seed Vault. 2021. Available online: https://www.croptrust.org/our-work/svalbard-global-seed-vault/ (accessed on 1 July 2021).
38. FAO. *The Second Report on the State of the World's Plant Genetic Resources*; FAO: Rome, Italy, 2010.
39. FAO. *Second Global Plan of Action for Plant Genetic Resources for Food and Agriculture*; FAO: Rome, Italy, 2011.
40. CGIAR Technical Advisory Committee (TAC). *The Collection, Evaluation and Conservation of Plant Genetic Resources*; Report of TAC Ad Hoc Working Group: Beltsville, MD, USA, 1972. Available online: https://cgspace.cgiar.org/bitstream/handle/10947/1524/tc7204d.pdf?sequence=1&isAllowed=y (accessed on 1 July 2021).
41. IBPGR. *Annual Report 1976*; AGPE:IBPGR/76/27; International Board for Plant Genetic Resources: Rome, Italy, 1977; 31p.
42. Engels, J.M.M.; Thormann, I. *IBPGR/IPGRI Register of Base Collections*; Bioversity International: Rome, Italy, 2000; 33p. Available online: https://www.researchgate.net/publication/251880833_IBPGRIPGRI_REGISTER_OF_BASE_COLLECTIONS (accessed on 24 July 2021).
43. Thormann, I.; Engels, J.M.M.; Halewood, M. Are the old international board for plant genetic resources (IBPGR) base collections available through the plant treaty's multilateral system of access and benefit sharing? A review. *Genet. Resour. Crop Evol.* **2019**, *66*, 291–310. [CrossRef]
44. ECPGR. Overview. 2021. Available online: https://www.ecpgr.cgiar.org/about/overview (accessed on 25 July 2021).
45. Thormann, I.; Fiorino, E.; Halewood, M.; Engels, J.M.M. Plant genetic resources collections and associated information as a baseline resource for genetic diversity studies: An assessment of the IBPGR-supported collections. *Genet. Resour. Crop Evol.* **2015**, *62*, 1279–1293. [CrossRef]
46. FAO. *Resolution 8/83 International Undertaking on Plant. Genetic Resources*; FAO: Rome, Italy, 1983; 8p. Available online: http://www.fao.org/wiews-archive/docs/Resolution_8_83.pdf (accessed on 24 April 2021).

47. FAO. XIV. Appendix E—International Code of Conduct for Plant Germplasm Collecting and Transfer. 2021. Available online: http://www.fao.org/3/x5586E/x5586e0k.htm (accessed on 7 July 2021).
48. United Nations. Convention on Biological Diversity. 1992. Available online: https://www.cbd.int/doc/legal/cbd-en.pdf (accessed on 7 July 2021).
49. Engels, J.M.M. Plant genetic resources management and conservation strategies: Problems and progress. *Acta Hortic.* **2003**, *623*, 179–192. [CrossRef]
50. Bragdon, S. *International Law of Relevance to Plant Genetic Resources: A Practical Review for Scientists and Other Professionals Working with Plant Genetic Resources*; Issues in Genetic Resources No. 10; International Plant Genetic Resources Institute: Rome, Italy, 2004.
51. WIPO. *Report of the ad hoc Open-Ended Working Group on Access and Benefit-Sharing of the Convention on Biological Diversity (Submitted by the Secretariat of the Convention on Biological Diversity)*; WIPO: Geneva, Switzerland, 2001; Available online: https://www.wipo.int/meetings/en/doc_details.jsp?doc_id=1996 (accessed on 6 July 2021).
52. CBD. Text of the Nagoya Protocol. 2019. Available online: https://www.cbd.int/abs/text/ (accessed on 1 July 2021).
53. CBD. The Cartagena Protocol on Biosafety. 2021. Available online: http://bch.cbd.int/protocol (accessed on 1 July 2021).
54. CBD. UN Biodiversity Convention Confirms Dates for Subsidiary Body Meetings Critical to Development of Post-2020 Global Biodiversity Framework. 2021. Available online: https://www.cbd.int/doc/press/2021/pr-2021-04-09-sbstta-sbi-en.pdf (accessed on 1 July 2021).
55. FAO. International Treaty on Plant Genetic Resources for Food and Agriculture. 2021. Available online: http://www.fao.org/plant-treaty/overview/en/ (accessed on 1 July 2021).
56. FAO. The Global Information System for PGRFA. 2021. Available online: https://ssl.fao.org/glis/ (accessed on 1 July 2021).
57. FAO. Commission on Genetic Resources for Food and Agriculture. 2021. Available online: http://www.fao.org/cgrfa/policies/global-instruments/gpa/en (accessed on 1 July 2021).
58. Manzella, D. The design and mechanics of the multilateral system of access and benefit sharing. In *Crop Genetic Resources as a Global Commons. Challenges in International Law and Governance*; Bioversity International: Routledge, UK, 2013; pp. 150–163.
59. Esquinas-Alcázar, J. Protecting crop genetic diversity for food security: Political, ethical and technical challenges. *Nat. Rev. Genet.* **2005**, *6*, 946–953. [CrossRef] [PubMed]
60. Cooper, D.; Engels, J.M.M.; Frison, E.A. *A Multilateral System for Plant Genetic Resources: Imperatives, Achievements and Challenges*; Issues in Genetic Resources No. 2; International Plant Genetic Resources Institute: Rome, Italy, 1994; p. 41. ISBN 92-9043-238-1.
61. Ebert, A.W. Potential of underutilized traditional vegetables and legume crops to contribute to food and nutritional security, income and more sustainable production systems. *Sustainability* **2014**, *6*, 319–335. [CrossRef]
62. CGIAR Secretariat. CGIAR Mid Term Meeting, New Delhi, 23–27 May 1994, Summary of Proceedings and Decisions. July 1994. Available online: https://hdl.handle.net/10947/306 (accessed on 6 July 2021).
63. FAO. Draft Agreements Between the Governing Body and the IARCs of the CGIAR and Other Relevant International Institutions. March 2006. Available online: http://www.fao.org/3/be221e/be221e.pdf (accessed on 6 July 2021).
64. López Noriega, I.; Halewood, M.; Abberton, M.; Amri, A.; Angarawai, I.I.; Anglin, N.; Blümmel, M.; Bouman, B.; Campos, H.; Costich, D.; et al. CGIAR operations under the plant treaty framework. *Crop Sci.* **2019**, *59*, 819–832. [CrossRef]
65. IBPGR. *Annual Report 1977*; AGPE:IBPGR/78/8; International Board for Plant Genetic Resources: Rome, Italy, 1978; 80p.
66. IBPGR. *Annual Report 1978*; AGPE:IBPGR/79/8; International Board for Plant Genetic Resources: Rome, Italy, 1979; 102p.
67. IBPGR. *Annual Report 1979*; AGPE:IBPGR/80/5; International Board for Plant Genetic Resources: Rome, Italy, 1980; 98p.
68. Halewood, M.; Jamora, N.; Noriega, I.L.; Anglin, N.L.; Wenzl, P.; Payne, T.; Ndjiondjop, M.N.; Guarino, L.; Kumar, P.; Yazbek, M.; et al. Germplasm acquisition and distribution by CGIAR genebanks. *Plants* **2020**, *9*, 1296. [CrossRef] [PubMed]
69. Crop Trust. The Genebank Platform. 2021. Available online: https://www.croptrust.org/project/the-genebank-platform/ (accessed on 1 July 2021).
70. Engels, J.M.M.; Ebert, A.W.; Thormann, I.; De Vicente, M.C. Centres of crop diversity and/or origin, genetically modified crops and implications for plant genetic resources conservation. *Genet. Resour. Crop Evol.* **2006**, *53*, 1675–1688. [CrossRef]
71. Tweddle, J.C.; Dickie, J.B.; Baskin, C.C.; Baskin, J.M. Ecological aspects of seed desiccation sensitivity. *J. Ecol.* **2003**, *91*, 294–304. [CrossRef]
72. Wyse, S.V.; Dickie, J.B. Predicting the global incidence of seed desiccation sensitivity. *J. Ecol.* **2017**, *105*, 1082–1093. [CrossRef]
73. Roberts, E.H. Predicting the storage life of seeds. *Seed Sci. Technol.* **1973**, *1*, 499–514.
74. Hong, T.D.; Ellis, R.H. *A Protocol to Determine Seed Storage Behaviour*; IPGRI Technical Bulletin No. 1; International Plant Genetic Resources Institute IPGRI: Rome, Italy, 1996; 62p.
75. Costa, M.C.D.; Cooper, K.; Hilhorst, H.W.; Farrant, J.M. Orthodox seeds and resurrection plants: Two of a kind? *Plant Physiol.* **2017**, *175*, 589–599. [CrossRef]
76. Roberts, E.H. The influence of storage conditions on seed viability. In *Genetic Resources of Mediterranean Pasture and Forest Legumes. Current Plant Science and Biotechnology in Agriculture*; Bennett, S.J., Cocks, P.S., Eds.; Springer: Dordrecht, The Netherlands, 1999; Volume 33, pp. 132–140. [CrossRef]
77. Harrington, J.F. Seed and pollen storage for conservation of plant gene resources. In *Genetic Resources in Plants: Their Exploration and Conservation*; Frankel, O.H., Bennet, E., Eds.; Blackwell: Oxford, UK, 1970; pp. 501–521.
78. FAO. *Genebank Standards for Plant. Genetic Resources for Food and Agriculture, Revised ed.*; FAO: Rome, Italy, 2014; Available online: http://www.fao.org/3/i3704e/i3704e.pdf (accessed on 29 April 2021).

79. Berjak, P.; Pammenter, N.W. Seed recalcitrance—Current perspectives. *S. Afr. J. Bot.* **2001**, *67*, 79–89. [CrossRef]
80. Hong, T.D.; Linnington, S.; Ellis, R.H. *Compendium of Information on Seed Storage Behaviour*; Royals Botanic Gardens, Kew: Richmond, UK, 1998; Volume 1–2.
81. Ellis, R.H.; Hong, T.D.; Roberts, E.H. An intermediate category of seed storage behaviour? I. Coffee. *J. Exp. Bot.* **1990**, *41*, 1167–1174. [CrossRef]
82. Walters, C. Orthodoxy, recalcitrance and in-between: Describing variation in seed storage characteristics using threshold responses to water loss. *Planta* **2015**, *242*, 397–406. [CrossRef] [PubMed]
83. Mondoni, A.; Probert, R.J.; Rossi, G.; Vegini, E., Hay, F.R. Seeds of alpine plants are short lived: Implications for long-term conservation. *Ann. Bot.* **2011**, *107*, 171–179. [CrossRef] [PubMed]
84. Howard, H.W. Storage of true seeds of potatoes for 25 years. *Potato Res.* **1980**, *23*, 241–242. [CrossRef]
85. Ellis, D.; Salas, A.; Chavez, O.; Gomez, R.; Anglin, N. Ex situ conservation of potato [*Solanum* section *Petota* (Solanaceae)] genetic resources in genebanks. In *The Potato Crop*; Campos, H., Ortiz, O., Eds.; Springer: Cham, Switzerland, 2020; pp. 109–138. [CrossRef]
86. Hay, F.R.; Sershen, N. *New Technologies to Improve the Ex Situ Conservation of Plant Genetic Resources*; Burleigh Dodds Series in Agricultural Science; Burleigh Dodds Science Publishing Limited: Cambridge, UK, 2021; pp. 1–32. [CrossRef]
87. Weise, S.; Lohwasser, U.; Oppermann, M. Document or Lose It—On the Importance of Information Management for Genetic Resources Conservation in Genebanks. *Plants* **2020**, *9*, 1050. [CrossRef] [PubMed]
88. Engels, J.M.; Maggioni, L. Managing germplasm in a virtual European genebank (AEGIS) through networking. In *Theorien der Lebendsammlung: Pflanzen, Mikroben und Tiere als Biofakte in Genbanken. (Lebenswissenschaften im Dialog, Band 25)*; Karafyllis, N.C., Ed.; Verlag Karl Alber: Freiburg, Germany, 2018; pp. 169–197.
89. Byrne, P.F.; Volk, G.M.; Gardner, C.; Gore, M.A.; Simon, P.W.; Smith, S. Sustaining the future of plant breeding: The critical role of the USDA-ARS National Plant Germplasm System. *Crop Sci.* **2018**, *58*, 451–468. [CrossRef]
90. Schreinemachers, P.; Ebert, A.W.; Wu, M.H. Costing the ex situ conservation of plant genetic resources at AVRDC–The World Vegetable Center. *Genet. Resour. Crop Evol.* **2014**, *61*, 757–773. [CrossRef]
91. Liu, U.; Cossu, T.A.; Davies, R.M.; Forest, F.; Dickie, J.B.; Breman, E. Conserving orthodox seeds of globally threatened plants ex situ in the Millennium Seed Bank, Royal Botanic Gardens, Kew, UK: The status of seed collections. *Biodivers. Conserv.* **2020**, *29*, 2901–2949. [CrossRef]
92. Walters, C.; Wheeler, L.M.; Grotenhuis, J.M. Longevity of seeds stored in a genebank: Species. *Seed Sci. Res.* **2005**, *15*, 1–20. [CrossRef]
93. Davies, R.M.; Dickie, J.B.; Ballesteros, D. Evaluation of short-lived seeds' cryopreservation as alternative to conventional seed banking. *Cryobiology* **2018**, *85*, 140–141. [CrossRef]
94. Ballesteros, D.; Fanega-Sleziak, N.; Davies, R. Cryopreservation of seeds and seed embyros in orthodox-, intermediate, and recalcitrant-seeded species. In *Cryopreservation and Freeze-Drying Protocols, Methods in Molecular Biology*; Wolkers, W.F., Oldenhof, H., Eds.; Humana: New York, NY, USA, 2021; Volume 2180. [CrossRef]
95. Panis, B.; Nagel, M.; Van den Houwe, I. Challenges and prospects for the conservation of crop genetic resources in field genebanks, in in vitro collection and/or in liquid nitrogen. *Plants* **2020**, *9*, 1634. [CrossRef] [PubMed]
96. McKey, D.; Elias, M.; Pujol, B.; Duputié, A. The evolutionary ecology of clonally propagated domesticated plants. *New Phytol.* **2010**, *186*, 318–332. [CrossRef]
97. Reed, B.M.; Engelmann, F.; Dulloo, M.E.; Engels, J.M.M. *Technical Guidelines for the Management of Field and In Vitro Germplasm Collections*; IPGRI Handbooks for Genebanks No. 7; International Plant Genetic Resources Institute: Rome, Italy, 2004.
98. Vega, F.E.; Ebert, A.W.; Ming, R. Coffee Germplasm Resources, Genomics and Breeding. *Plant Breed. Rev.* **2008**, *30*, 415–447.
99. Dulloo, M.E.; Ebert, A.W.; Dussert, S.; Gotor, E.; Astorga, C.; Vasquez, N.; Rakotomalala, J.J.; Rabemiafara, A.; Eira, M.; Bellachew, B.; et al. Cost efficiency of cryopreservation as a long-term conservation method for coffee genetic resources. *Crop Sci.* **2009**, *49*, 2123–2138. [CrossRef]
100. Phillips-Mora, W.; Mora, A.; Johnson, E.S.; Astorga, C. Recent efforts to improve the genetic and physical conditions of the international cacao collection at CATIE. In Proceedings of the International Cocoa Producer's Conference, San Jose, Costa Rica, 9–14 October 2006; Volume I, pp. 611–623.
101. Peil, A.; Emeriewen, O.F.; Khan, A.; Kostick, S.; Malnoy, M. Status of fire blight resistance breeding in Malus. *J. Plant Pathol.* **2020**, *103*, 3–12. [CrossRef]
102. Crop Genebank Knowledge Base. Regeneration Guidelines for Forage Grass Genetic Resources. Available online: https://cropgenebank.sgrp.cgiar.org/index.php/crops-mainmenu-367/forage-grasses-mainmenu-271/regeneration-mainmenu-286 (accessed on 30 April 2021).
103. Graefe, S.; Dufour, D.; Van Zonneveld, M.; Rodriguez, F.; Gonzalez, A. Peach palm (*Bactris gasipaes*) in tropical Latin America: Implications for biodiversity conservation, natural resource management and human nutrition. *Biodivers. Conserv.* **2013**, *22*, 269–300. [CrossRef]
104. Ebert, A.W.; Astorga, C.; Ebert, I.C.M.; Mora, A.; Umaña, C. *Securing Our Future: CATIE's Germplasm Collections—Asegurando Nuestro Futuro: Colecciones de Germoplasma del CATIE*; CATIE Technical Series, Boletín Técnico no. 26; CATIE: Turrialba, Costa Rica, 2007; p. 204.

105. Kumar, P.L.; Cuvero, M.; Kreuze, J.F.; Muller, G.; Kulkarni, G.; Kumar, S.G.; Massart, S.; Mezzelema, M.; Alakonya, A.; Muchugi, A.; et al. Phytosanitary Interventions for Safe Global Germplasm Exchange and Prevention of Transboundary Pest Spread: The Role of CGIAR Germplasm Health Units. *Plants* **2021**, *10*, 328. [CrossRef] [PubMed]
106. Frison, E.A.; Putter, C.A.J. FAO/IBPGR Technical Guidelines for the Safe Movement of Germplasm. In Proceedings of the 5th International Congress of Plant Pathology, Kyoto, Japan, 20–27 August 1988; p. 279.
107. Volk, G.M.; Chao, C.T.; Norelli, J.; Brown, S.K.; Fazio, G.; Peace, C.; McFerson, J.; Zhong, G.Y.; Bretting, P. The vulnerability of US apple (*Malus*) genetic resources. *Genet. Resour. Crop Evol.* **2015**, *62*, 765–794. [CrossRef]
108. Engels, J.M.M.; Visser, L. *A Guide to Effective Management of Germplasm Collections*; International Plant Genetic Resources Institute: Rome, Italy, 2003.
109. Engelmann, F. Use of biotechnologies for the conservation of plant biodiversity. *Cell. Dev. Biol. Plant* **2011**, *47*, 5–16. [CrossRef]
110. Benson, E.E.; Harding, K.; Debouck, D.; Dumet, D.; Escobar, R.; Mafla, G.; Panis, B.; Panta, A.; Tay, D.; Van den Houwe, I.; et al. *Refinement and Standardization of Storage Procedures for Clonal Crops—Global Public Goods Phase 2: Part. 1. Project Landscape and General Status of Clonal Crop. In Vitro Conservation Technologies*; System-Wide Genetic Resources Programme: Rome, Italy, 2011.
111. Chauhan, R.; Singh, V.; Quraishi, A. In Vitro Conservation Through Slow-Growth Storage. In *Synthetic Seeds*; Faisal, M., Alatar, A., Eds.; Springer: Cham, Switzerland, 2019; pp. 397–416. [CrossRef]
112. Cha-um, S.; Kirdmanee, C. Minimal growth in vitro culture for preservation of plant species. *Fruit Veg. Cereal Sci. Biotechnol.* **2007**, *1*, 13–25.
113. Keller, E.R.J.; Senula, A.; Leunufna, S.; Grube, M. Slow growth storage and cryopreservation—Tools to facilitate germplasm maintenance of vegetatively propagated crops in living plant collections. *Int. J. Refrig.* **2006**, *29*, 411–417. [CrossRef]
114. Thieme, R. An in-vitro potato cultivar collection: Microtuberization and storage of microtubers. *Plant Genet. Resour. Newsl.* **1992**, *88–89*, 17–19.
115. Jarret, R.L.; Florkowski, W.J. In vitro active vs. field genebank maintenance of sweet potato germplasm: Major costs and considerations. *HortScience* **1990**, *25*, 141–146. [CrossRef]
116. Van den Houwe, I.; Desmet, K.; Dumontcel, H.T.; Swennen, R. Variability in storage potential of banana shoot cultures under medium-term storage-conditions. *Plant Cell Tissue Organ. Cult.* **1995**, *42*, 269–274. [CrossRef]
117. Souza, F.V.D.; Soares, T.L.; Cabral, J.R.S.; Reinhardt, D.H.; Cardoso, J.L.; Benjamin, D.A. Slow-growth conditions for the in vitro conservation of pineapple germplasm. *Acta Hortic.* **2006**, *702*, 41–45. [CrossRef]
118. Sarkar, D.; Naik, P.S. Factors affecting minimal growth conservation of potato microplants in vitro. *Euphytica* **1998**, *102*, 275–280. [CrossRef]
119. De Vincente, M.C.; Metz, T.; Alercia, A. *Descriptors for Genetic Markers Technologies*; International Plant Genetic Resources Institute: Rome, Italy, 2004.
120. Kane, N.; Sveinsson, S.; Dempewolf, H.; Yang, J.; Zhang, D.; Engels, J.M.M.; Cronk, Q. Ultra-barcoding in cacao (*Theobroma* spp.; Malvaceae) using whole chloroplast genomes and nuclear ribosomal DNA. *Am. J. Bot.* **2012**, *99*, 320–329. [CrossRef] [PubMed]
121. Abbas, B.; Tjolli, I. Genetic diversity of sago palm (*Metroxylon sagu*) accessions based on plastid cpDNA matK gene as DNA barcoding. *Biodiversitas J. Biol. Divers.* **2020**, *21*, 219–225. [CrossRef]
122. Bairu, M.W.; Aremu, A.O.; van Staden, J. Somaclonal variation in plants: Causes and detection methods. *Plant Growth Regul.* **2011**, *63*, 147–173. [CrossRef]
123. Bairu, M.W.; Fennell, C.W.; van Staden, J. The effect of plant growth regulators on somaclonal variation in Cavendish banana (*Musa* AAA cv. 'Zelig'). *Sci. Hortic.* **2006**, *108*, 347–351. [CrossRef]
124. Graner, E.M.; Brondani, G.E.; de Almeida, C.V.; Batagin-Piotto, K.D.; de Almeida, M. Study of senescence in old cultures of the *Bactris gasipaes* Kunth in vitro. *Plant Cell Tissue Organ. Cult.* **2015**, *120*, 1169–1189. [CrossRef]
125. Bamberg, J.B.; Martin, M.W.; Abad, J.; Jenderek, M.M.; Tanner, J.; Donnelly, D.J.; Nassar, A.M.K.; Veilleux, R.E.; Novy, R.G. In vitro technology at the US Potato Genebank. *Cell. Dev. Biol. Plant* **2016**, *52*, 213–225. [CrossRef]
126. Guta, I.C.; Buciumenau, E.C.; Tataru, L.D.; Oprescu, B.; Topala, C.M. New approach of electrotherapy for grapevine virus elimination. *Acta Hortic.* **2019**, *1242*, 697–702. [CrossRef]
127. Pathirana, R.; Mathew, L.; Jibran, R.; Hunter, D.A.; Morgan, E.R. Cryopreservation and cryotherapy research on horticultural crops in New Zealand. *Acta Hortic.* **2019**, *1234*, 29–36. [CrossRef]
128. Wang, Q.; Valkonen, J.P. Cryotherapy of shoot tips: Novel pathogen eradication method. *Trends Plant Sci.* **2009**, *14*, 119–122. [CrossRef]
129. Engelmann, F. Importance of cryopreservation for the conservation of plant genetic resources. In *Cryopreservation of Tropical Plant Germplasm: Current Research Progress and Application, Proceedings of the An International Workshop, Tsukuba, Japan, October 1998*; International Plant Genetic Resources Institute (IPGRI): Rome, Italy, 2000; pp. 8–20.
130. Engelmann, F.; Dussert, S. Cryopreservation. In *Conservation of Tropical Plant Species*; Normah, M.N., Chin, H.F., Reed, B.M., Eds.; Springer Science & Business Media: New York, NY, USA, 2013; pp. 107–119. [CrossRef]
131. Sakai, A. Survival of plat tissue at super-low temperatures. IV. Cell survival with rapid cooling and rewarming. *Plant Physiol.* **1966**, *41*, 1050–1054. [CrossRef] [PubMed]
132. Reed, B.M. Plant cryopreservation: A continuing requirement for food and ecosystem security. *Cell. Dev. Biol. Plant* **2017**, *53*, 285–288. [CrossRef]

133. Pritchard, H.W.; Nadarajan, J. Cryopreservation of orthodox (desiccation tolerant) seeds. In *Plant Cryopreservation: A Practical Guide*; Reed, B.M., Ed.; Springer: New York, NY, USA, 2008; pp. 485–501. [CrossRef]
134. Towill, L.E.; Ellis, D.D. Cryopreservation of dormant buds. In *Plant Cryopreservation—A Practical Guide*; Reed, B.M., Ed.; Springer Science + Business Media: New York, NY, USA, 2008; pp. 421–442.
135. Jenderek, M.M.; Ambruzs, B.; Tanner, J.; Holman, G.; Ledbetter, C.; Postman, J.; Ellis, D.; Leslie, C. Extending the dormant bud cryopreservation method to new tree species. *Acta Hortic.* **2014**, *1039*, 133–136. [CrossRef]
136. Ballesteros, D.; Pence, V.C. Survival and death of seeds during LN storage: A case study on seeds with short lifespans. *CryoLetters* **2017**, *38*, 278–289.
137. Engelmann, F. Importance of desiccation for the cryopreservation of recalcitrant seed and vegetatively propagated species. *Plant Genet. Res. Newsl.* **1997**, *112*, 9–18.
138. Volk, G.M.; Jenderek, M.; Chen, K. Cryopreservation of Dormant Apple Buds. In *Training in Plant Genetic Resources: Cryopreservation of Clonal Propagules*; Volk, G.M., Ed.; Colorado State University: Fort Collins, CO, USA, 2020; Available online: https://colostate.pressbooks.pub/clonalcryopreservation/chapter/apple-dormant-bud-cryopreservation/ (accessed on 20 May 2021).
139. Reed, B.M.; Denoma, J.; Luo, J.; Chang, Y.; Towill, L. Cryopreservation and long-term storage of pear germplasm. *Cell. Dev. Biol. Plant* **1998**, *34*, 256. [CrossRef]
140. Towill, L.E.; Forsline, P.L. Cryopreservation of sour cherry (*Prunus cerasus* L.) using a dormant vegetative bud method. *Cryoletters* **1999**, *20*, 215–222.
141. Tanner, J.D. Cryopreservation and Recovery of Temperate Fruit Germplasm Using Dormant Bud Technology. Ph.D. Thesis, Colorado State University, Fort Collins, CO, USA, 2020.
142. Jenderek, M.M.; Tanner, J.D.; Ambruzs, B.D.; West, M.; Postman, J.D.; Hummer, K.E. Twig pre-harvest temperature significantly influences effective cryopreservation of *Vaccinium* dormant buds. *Cryobiology* **2017**, *74*, 154–159. [CrossRef]
143. Fukui, K.; Shirata, K.; Niino, T.; Kashif, I.M. Cryopreservation of Mulberry Winter Buds in Japan. *Acta Hortic.* **2011**, *908*, 483–488. [CrossRef]
144. Rantala, S.; Kaseva, J.; Karhu, S.; Veteläinen, M.; Uosukainen, M.; Häggman, H. Cryopreservation of *Ribes nigrum* (L.) dormant buds: Recovery via in vitro culture to the field. *Plant Cell Tissue Organ. Cult.* **2019**, *138*, 109–119. [CrossRef]
145. Volk, G.M.; Walters, C. Preservation of genetic resources in the national plant germplasm clonal collections. *Plant Breed. Rev.* **2010**, *23*, 291–344.
146. Yamamoto, S.I.; Rafique, T.; Priyantha, W.S.; Fukui, K.; Matsumoto, T.; Niino, T. Development of a cryopreservation procedure using aluminium cryo-plates. *CryoLetters* **2011**, *32*, 256–265. [PubMed]
147. Yamamoto, S.I.; Rafique, T.; Fukui, K.; Sekizawa, K.; Niino, T. V-cryo-plate procedure as an effective protocol for cryobanks: Case study of mint cryopreservation. *CryoLetters* **2012**, *33*, 12–23. [PubMed]
148. O'Brien, C.; Hiti-Bandaralage, J.; Folgado, R.; Hayward, A.; Lahmeyer, S.; Folsom, J.; Mitter, N. Cryopreservation of Woody Crops: The Avocado Case. *Plants* **2021**, *10*, 934. [CrossRef]
149. Acker, J.P.; Adkins, S.; Alves, A.; Horna, D.; Toll, J. *Feasibility Study for a Safety Back-Up Cropreservation Facility*; Independent Expert Report: July 2017; Bioversity International: Rome, Italy, 2017; p. 100.
150. Panis, B. Sixty years of plant cryopreservation: From freezing hardy mulberry twigs to establishing reference crop collections for future generations. *Acta Hortic.* **2019**, *1234*, 1–8. [CrossRef]
151. CGIAR Genebank Platform. Key Performance Indicators of CGIAR Genebanks. 2021. Available online: https://www.genebanks.org/resources/genebanks-in-numbers/key-performance/ (accessed on 2 July 2021).
152. Asdal, A.; Guarino, L. The Svalbard Global Seed Vault: 10 Years—1 million samples. *Biopreservation Biobanking* **2018**, *16*, 391–392. [CrossRef] [PubMed]
153. Dulloo, E.; Nagamura, Y.; Ryder, O., III. DNA storage as a complementary conservation strategy. In *DNA Banks—Providing Novel Options for Genebanks? Topical Reviews in Agricultural Biodiversity*; De Vicente, M.C., Andersson, M.S., Eds.; International Plant Genetic Resources Institute: Rome, Italy, 2006; pp. 11–24.
154. Ebert, A.W.; Karihaloo, J.L.; Ferreira, M.E. Opportunities, limitations and needs for DNA banks. In *DNA Banks—Providing Novel Options for Genebanks? Topical Reviews in Agricultural Biodiversity*; De Vicente, M.C., Andersson, M.S., Eds.; International Plant Genetic Resources Institute: Rome, Italy, 2006; pp. 61–68.
155. Andersson, M.S. *Feasibility Study for DNA Storage and Use: DNA Banks—Providing Novel Options for Genebanks? A Worldwide Survey*; IPGRI Report No. 6900/D10A; International Plant Genetic Resources Institute (IPGRI): Rome, Italy, 2004; 38p.
156. Spooner, D.M.; Ruess, H.M. Curating DNA specimens. In *Curating Biocultural Collections: A Handbook*; Salick, J., Konchar, K., Nesbitt, M., Eds.; University of Chicago Press: Chicago, IL, USA, 2014; pp. 97–106.
157. Seberg, O.; Droege, G.; Barker, K.; Coddington, J.A.; Funk, V.; Gostel, M.; Petersen, G.; Smith, P.P. Global Genome Biodiversity Network: Saving a blueprint of the Tree of Life–a botanical perspective. *Ann. Bot.* **2016**, *118*, 393–399. [CrossRef] [PubMed]
158. Datlof, E.M.; Amend, A.S.; Earl, K.; Hayward, J.; Morden, C.W.; Wade, R.; Zahn, G.; Hynson, N.A. Uncovering unseen fungal diversity from plant DNA banks. *PeerJ* **2017**, *5*, e3730. [CrossRef] [PubMed]
159. De Vicente, M.C. *DNA Banks—Providing Novel Options for Genebanks?* Topical Reviews in Agricultural Biodiversity; Andersson, M.S., Ed.; International Plant Genetic Resources Institute: Rome, Italy, 2006; p. 93.

160. Hodkinson, T.R.; Waldren, S.; Parnell, J.A.; Kelleher, C.T.; Salamin, K.; Salamin, N. DNA banking for plant breeding, biotechnology and biodiversity evaluation. *J. Plant Res.* **2007**, *120*, 17–29. [CrossRef]
161. De Vicente, M.C.; Andersson, M.S.; Engels, J.M.M. Introduction and overview. In *DNA Banks—Providing Novel Options for Genebanks? Topical Reviews in Agricultural Biodiversity*; De Vicente, M.C., Andersson, M.S., Eds.; International Plant Genetic Resources Institute: Rome, Italy, 2006; pp. 1–5.
162. Walters, C.; Hanner, R. Platforms for DNA banking. In *DNA Banks—Providing Novel Options for Genebanks? Topical Reviews in Agricultural Biodiversity*; De Vicente, M.C., Andersson, M.S., Eds.; International Plant Genetic Resources Institute: Rome, Italy, 2006; pp. 25–35.
163. Kawai, J.; Hayashizaki, Y. DNA book. *Genome Res.* **2003**, *13*, 1488–1495. [CrossRef] [PubMed]
164. Colotte, M.; Coudy, D.; Tuffet, S.; Bonnet, J. Adverse effect of air exposure on the stability of DNA stored at room temperature. *Biopreservation Biobanking* **2011**, *9*, 47–50. [CrossRef]
165. Clermont, D.; Santoni, S.; Saker, S.; Gomard, M.; Gardais, E.; Bizet, C. Assessment of DNA encapsulation, a new room-temperature DNA storage method. *Biopreservation Biobanking* **2014**, *12*, 176–183. [CrossRef] [PubMed]
166. Willerslev, E.; Cappellini, E.; Boomsma, W.; Nielsen, R.; Hebsgaard, M.B.; Brand, T.B.; Hofreiter, M.; Bunce, M.; Poinar, H.N.; Johnsen, S.; et al. Ancient biomolecules from deep ice cores reveal a forested southern Greenland. *Science* **2007**, *317*, 111–114. [CrossRef]
167. Mascher, M.; Schuenemann, V.J.; Davidovich, U.; Marom, N.; Himmelbach, A.; Hübner, S.; Korol, A.; David, M.; Reiter, E.; Riehl, S.; et al. Genomic analysis of 6000-year-old cultivated grain illuminates the domestication history of barley. *Nat. Genet.* **2016**, *48*, 1089–1093. [CrossRef]
168. Parducci, L.; Bennett, K.D.; Ficetola, G.F.; Alsos, I.G.; Suyama, Y.; Wood, J.R.; Pedersen, M.W. Ancient plant DNA in lake sediments. *New Phytol.* **2017**, *214*, 924–942. [CrossRef]
169. Walters, C.; Reilley, A.A.; Reeves, P.A.; Baszczak, J.; Richards, C.M. The utility of aged seeds in DNA banks. *Seed Sci. Res.* **2006**, *16*, 169–178. [CrossRef]
170. Greengard, S. Cracking the code on DNA storage. *Commun. ACM* **2017**, *60*, 16–18. [CrossRef]
171. Koch, J.; Gantenbein, S.; Masania, K.; Stark, W.J.; Erlich, Y.; Grass, R.N. A DNA-of-things storage architecture to create materials with embedded memory. *Nat. Biotechnol.* **2020**, *38*, 39–43. [CrossRef]
172. Farzadfard, F. DNA storage in everyday objects. *Nat. Biotechnol.* **2020**, *38*, 31–32. [CrossRef]
173. Volk, G.M. Chapter 25: Collecting pollen for genetic resources conservation. In *Collecting Plant Genetic Diversity: Technical Guidelines—2011 Update*; Guarino, L., Rao, V.R., Reid, R., Eds.; CAB International: Wallingford, UK, 2011.
174. Dinato, N.B.; Imaculada Santos, I.R.; Zanotto Vigna, B.B.; de Paula, A.F.; Fávero, A.P. Pollen cryopreservation for plant breeding and genetic resources conservation. *Cryoletters* **2020**, *41*, 115–127.
175. Towill, L.E.; Walters, C. Cryopreservation of pollen. In *Cryopreservation of Tropical Plant Germplasm*; International Plant Genetic Resources Institute (IPGRI): Rome, Italy, 2000; pp. 115–129.
176. Ganeshan, S.; Rajasekharan, P.E.; Shashikumar, S.; Decruze, W. Cryopreservation of pollen. In *Plant Cryopreservation: A Practical Guide*; Reed, B.M., Ed.; Springer: New York, NY, USA, 2008; pp. 443–464.
177. Van Bilsen, D.G.; Hoekstra, F.A.; Crowe, L.M.; Crowe, J.H. Altered phase behavior in membranes of aging dry pollen may cause imbibitional leakage. *Plant Physiol.* **1994**, *104*, 1193–1199. [CrossRef] [PubMed]
178. Brewbaker, J.L. The distribution and phylogenetic significance of binucleate and trinucleate pollen grains in the angiosperms. *Am. J. Bot.* **1967**, *54*, 1069–1083. [CrossRef]
179. Jensen, C.J. Pollen Storage under Vacuum. In *Yearbook 1964*; Royal Veterinary and Agricultural College: Amsterdam, The Netherlands, 1964; pp. 133–146.
180. Hoekstra, F.A. Stress Effects on the Male Gametophyte. In *Sexual Plant Reproduction*; Cresti, M., Tiezzi, A., Eds.; Springer: Berlin/Heidelberg, Germany, 1992; pp. 193–201. [CrossRef]
181. Barnabás, B.; Kovacs, G. Storage of pollen. In *Pollen Biotechnology for Crop Production and Improvement*; Shivanna, K.R., Sawhney, V.K., Eds.; Cambridge University Press: Cambridge, UK, 1997; pp. 293–314.
182. Rajasekharan, P.E.; Ganeshan, S. Current perspectives on pollen cryopreservation in horticultural species. *Acta Hortic.* **2019**, *1234*, 47–56. [CrossRef]
183. Ren, R.; Li, Z.; Li, B.; Xu, J.; Jiang, X.; Liu, Y.; Zhang, K. Changes of pollen viability of ornamental plants after long-term preservation in a cryopreservation pollen bank. *Cryobiology* **2019**, *89*, 14–20. [CrossRef]
184. Yates, I.E.; Sparks, D.; Connor, K.; Towill, L. Reducing pollen moisture simplifies long-term storage of pecan pollen. *J. Am. Soc. Hortic. Sci.* **1991**, *116*, 430–434. [CrossRef]
185. Barnabás, B. Preservation of Maize Pollen. In *Maize. Biotechnology in Agriculture and Forestry*; Bajaj., Y.P.S., Ed.; Springer: Berlin/Heidelberg, Germany, 1994; Volume 25, pp. 607–618. [CrossRef]
186. Hoekstra, F.A.; van der Wal, E.W. Initial moisture content and temperature of imbibition determine extent of imbibitional injury in pollen. *J. Plant Physiol.* **1988**, *133*, 257–262. [CrossRef]
187. Hanna, W.W.; Towill, L.E. Long-term pollen storage. *Plant Breed. Rev.* **1995**, *13*, 179–207.
188. Engels, J.M.; Thormann, I. Main Challenges and Actions Needed to Improve Conservation and Sustainable Use of Our Crop Wild Relatives. *Plants* **2020**, *9*, 968. [CrossRef] [PubMed]

189. Borelli, T.; Hunter, D.; Powell, B.; Ulian, T.; Mattana, E.; Termote, C.; Pawera, L.; Beltrame, D.; Penafiel, D.; Tan, A.; et al. Born to eat wild: An integrated conservation approach to secure wild food plants for food security and nutrition. *Plants* **2020**, *9*, 1299. [CrossRef] [PubMed]
190. Gallo-Cajiao, E.; Archibald, C.; Friedman, R.; Steven, R.; Fuller, R.A.; Game, E.T.; Morrison, T.H.; Ritchie, E.G. Crowdfunding biodiversity conservation. *Conserv. Biol.* **2018**, *32*, 1426–1435. [CrossRef]
191. Engels, J.M.M. Complementary strategies for improved conservation and use of plant genetic resources. In Proceedings of the International Workshop. Towards Sustainable National Plant Genetic Resources Programmes—Policy, Planning and Conservation Issues, Zschortau, Germany, 2000. Available online: https://www.researchgate.net/publication/270450607_Complementary_strategies_for_improved_conservation_and_use_of_plant_genetic_resources (accessed on 29 April 2021).
192. Gepts, P. Plant Genetic Resources Conservation and Utilization: The Accomplishments and Future of a Societal Insurance Policy. *Crop Sci.* **2006**, *46*, 2278–2292. [CrossRef]
193. Van den Hurk, A.; (Plantum NL, Randstad, The Netherlands). Personal communication, 2008.
194. Hoban, S.; Arntzen, J.A.; Bruford, M.W.; Godoy, J.A.; Hoelzel, A.R.; Segelbacher, G.; Vilà, C.; Bertorelle, G. Comparative evaluation of potential indicators and temporal sampling protocols for monitoring genetic erosion. *Evol. Appl.* **2014**, *7*, 984–998. [CrossRef]
195. Zegeye, H. In situ and ex situ conservation: Complementary approaches for maintaining biodiversity. *Int. J. Res. Environ. Stud.* **2017**, *4*, 1–12.
196. Maxted, N.; Ford-Lloyd, B.V.; Hawkes, J.G. *Plant Genetic Conservation: The In Situ Approach*; Chapman and Hall: London, UK, 1997.
197. Riordan, E.C.; Nabhan, G.P. Trans Situ Conservation of Crop Wild Relatives. *Crop Sci.* **2019**, *59*, 2387–2403. [CrossRef]

plants

MDPI

Review

A Critical Review of the Current Global Ex Situ Conservation System for Plant Agrobiodiversity. II. Strengths and Weaknesses of the Current System and Recommendations for Its Improvement

Johannes M. M. Engels [1,*] and Andreas W. Ebert [2,†]

[1] Alliance of Bioversity International and CIAT, 00153 Rome, Italy
[2] World Vegetable Center, 60 Yi-Min Liao, Tainan 74151, Taiwan; ebert.andreas6@gmail.com
* Correspondence: j.engels@cgiar.org
† Current address: 73529 Schwäbisch Gmünd, Germany.

Abstract: In this paper, we review gene bank operations that have an influence on the global conservation system, with the intention to identify critical aspects that should be improved for optimum performance. We describe the role of active and base collections and the importance of linking germplasm conservation and use, also in view of new developments in genomics and phenomics that facilitate more effective and efficient conservation and use of plant agrobiodiversity. Strengths, limitations, and opportunities of the existing global ex situ conservation system are discussed, and measures are proposed to achieve a rational, more effective, and efficient global system for germplasm conservation and sustainable use. The proposed measures include filling genetic and geographic gaps in current ex situ collections; determining unique accessions at the global level for long-term conservation in virtual base collections; intensifying existing international collaborations among gene banks and forging collaborations with the botanic gardens community; increasing investment in conservation research and user-oriented supportive research; improved accession-level description of the genetic diversity of crop collections; improvements of the legal and policy framework; and oversight of the proposed network of global base collections.

Keywords: plant agrobiodiversity; routine gene bank operations; active collection; base collection; linking conservation and use; genomics; phenomics; conservation strategies; global conservation network

Citation: Engels, J.M.M.; Ebert, A.W. A Critical Review of the Current Global Ex Situ Conservation System for Plant Agrobiodiversity. II. Strengths and Weaknesses of the Current System and Recommendations for Its Improvement. *Plants* **2021**, *10*, 1904. https://doi.org/10.3390/plants10091904

Academic Editor: Ana Butrón

Received: 8 August 2021
Accepted: 9 September 2021
Published: 14 September 2021

1. Introduction

Plant agrobiodiversity, i.e., the diversity of plants that is used or has the potential to be used in agriculture and horticulture, has been the foundation for human subsistence for millennia and will continue to play a decisive role in securing global food and nutrition security for a growing population, especially under the current threat of climate change. While 7039 edible plant species are known to science and 417 are considered food crops [1], today, only 12 plant and five animal species are used to supply 75% of human food [2]. This over-reliance of global food production on a very small number of crops and animals with a largely very limited number of genetically uniform, high-yielding varieties of crop plants and breeds of animals presents a major challenge for both the conservation of agrobiodiversity and for human nutrition and health.

While we (still) enjoy such an enormous agrobiodiversity, we are being made aware that two out of five plant species are threatened with extinction according to current estimates [1]. The major threats of genetic erosion of plant species are of anthropogenic nature and include agriculture and modern plant breeding; overexploitation of biological resources in the wild; modification, fragmentation, and destruction of natural ecosystems; rapidly expanding residential and commercial developments; pollution; and climate change. Conscious of the threat of modern agriculture and mankind to (plant) agrobiodiversity, plant introduction centers evolved since the early 20th century in several countries that later grew out into gene banks. These efforts were made to meet the growing demand of plant breeders for broad genetic diversity, essential for the development of well-adapted, high-yielding varieties with resistance to biotic threats and tolerance to abiotic stresses. The history and the major players in the development of global long-term conservation practices and the evolving global (ex situ) conservation system have been described in a previous paper [3].

While aiming at an accurate description of the "global conservation system" that gradually emerged under the auspices of FAO, it has become clear that the individual components of that system evolved somewhat "spontaneously" and that no precise goal of that system existed. Consequently, not all components are logically embedded in the system and suffered adjustments to the "in parallel" evolving political framework and changing realities. Some components "disappeared", and others were announced but did not materialize. Thus, the authors felt it was necessary to aim at a "working definition" of the global conservation system as follows: "A long-term global plant agrobiodiversity conservation system of well-defined national and international ex situ seed, tissue and plant collections that is managed under agreed genebank quality management standards and in harmony with the prevailing political framework regarding access and benefit-sharing, and that aims at safe, effective, efficient and rational long-term conservation and facilitating use by making high-quality accession-level information available".

In this paper, we focus on the major routine gene bank activities and assess several constraints that might affect long-term ex situ conservation activities. A specific aspect of the current long-term conservation and facilitation of use "system" is the concept of base and active collections, as this concept has been designed in the past to address and resolve the difficult issue of linking conservation and use.

Having looked at the different conservation approaches as well as at the major ex situ gene bank management activities, it will be indispensable to describe the major components of the current global (long-term) conservation system and to identify and describe its strengths and weaknesses.

Molecular techniques, genomics, and bioinformatics have developed since the current global conservation system emerged and have only been applied to a limited extent in managing germplasm in gene banks and facilitating their use. In addition, communication technologies, digitalization capacities, bioinformatics, and possibly other recent developments could well provide opportunities for strengthening the current system and possibly help to rationalize the long-term conservation and the facilitation of the use of conserved materials.

This paper will focus on: (i) the description of the main conservation approaches and activities; (ii) their strengths and weaknesses; (iii) the advancements made in molecular genetics, genomics, bioinformatics, and the considerably increased knowledge regarding genetic diversity aspects; (iv) considerations regarding other developments that have an impact on long-term conservation and offer opportunities for possible improvements of the ex situ long-term conservation system, including the policy framework. Furthermore, suggestions and recommendations will be formulated on how the current global long-term conservation system can be made more rational and effective to allow more efficient conservation of plant agrobiodiversity.

2. Brief Description and Critical Review of Key Routine Germplasm Conservation Activities

In this section, we will focus on the major routine germplasm conservation activities, typically being part of gene bank operations across the world (see Figure 1). We critically review these operations with the intention to identify weak and/or critical aspects that should be considered to ensure optimum performance and thus contribute to an effective and efficient global conservation and sustainable use system. It is understood that other aspects such as sustainable funding of gene bank operations are important if not essential prerequisites for effective and efficient conservation and for facilitating the use of conserved germplasm materials. Funding, in particular, is such a complex, diverse and circumstantial aspect that would require specialized knowledge and expertise to be treated comprehensively, and to do this topic justice would blow the remit and scope of this paper. For details on the risk of decreasing funding and staffing of conservation and breeding programs, see the recent paper of Coe et al. [4].

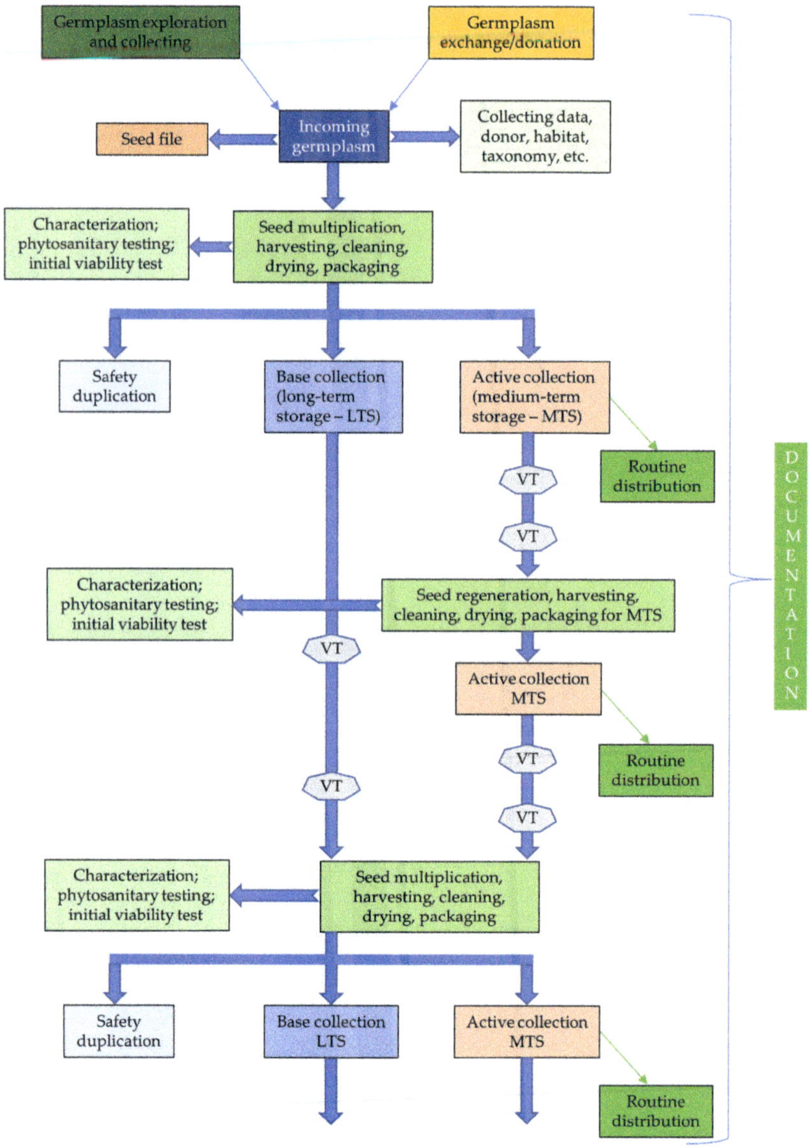

Figure 1. Overview of standard operations in a seed gene bank, based on the FAO Genebank Standards [5] and adapted from Hay and Sershen [6]. Germplasm arriving at a gene bank through a collecting mission or exchange/donation from

other collection owners is assessed for its uniqueness, documented with all available information (passport data), and if seed quantity allows a sample of 20–30 seeds is separated to create a seed file for future reference. As the seed amount received is usually insufficient for storage as a base and active collection, the incoming seed lot needs to undergo a seed multiplication phase, followed by careful seed processing. During the multiplication phase, plants and seeds are usually characterized according to crop descriptor lists based on heritable morpho-agronomic traits. Seed subsamples will be taken for phytosanitary testing and for an initial viability test. Each subsequent regeneration event offers the opportunity for additional/complementary characterization and requires phytosanitary testing and an initial viability test. After equilibrium drying, usually at 5–20 °C and 10–25% RH, seeds will be hermetically packed for long-term storage (base collection) at 18 ± 3 °C and medium-term storage at 5–10 °C (active collection). Safety duplicate samples will be sent to another gene bank for long-term conservation under a "black box" agreement and/or to the Svalbard Global Seed Vault. Due to the higher storage temperature under medium-term storage conditions, viability testing (VT) must be conducted more frequently compared to long-term storage conditions. Should the viability of the active collection fall below 85% of the initial viability or seed quantity become insufficient, a seed regeneration cycle must be programmed for replenishment/replacement of the active collection. Up to four regeneration cycles can be conducted with seeds from the active collection before seeds from the base collection should be used. At the time of replacement of the seeds in the active collection, the viability of the seeds in the base collection needs to be monitored independently as seed lots are now derived from different regeneration cycles. Once seed viability of the accessions in the base collection falls below the threshold value, a new regeneration cycle with seeds from the base collection is required, following the same steps as described for the first seed multiplication cycle. This includes a replacement of the safety duplicate samples as their viability might also have fallen below the critical threshold value. Across all gene bank operations, a huge amount of data is generated, which needs to be captured and adequately managed in the gene bank information system for in-house use as well as for the benefit of germplasm users (passport, characterization, and evaluation data).

2.1. Exploration and Collecting

Exploration of plant genetic resources is the first step after the decision has been made to conduct a collecting mission. One critically important aspect to check before making this decision is to see if the gene bank has sufficient capacity, adequate expertise, and financial resources for timely and effective regeneration and sustainable long-term storage or that maintenance of the collected materials is secured [7]. Exploration can be defined as "the act of searching or traveling around a terrain for the purpose of discovery of resources or information" [8]. The process of gathering information starts already well before a trip/mission is undertaken, through literature searches regarding the "where" (i.e., the areas and places where a given species naturally occurs, or from where it has been reported and/or is cultivated), "what" kind of material (i.e., species and populations or varieties to be collected) can be found, etc.

The decision to implement a mission should be based on the following actions:

- Contacting scientists in the areas concerned, asking for "on the ground information" about aspects such as species distribution, genetic diversity, time of seed or material maturity, genetic erosion, and threat status;
- Whether or not there have been collecting missions conducted in the past, where the collected materials are being conserved (in situ and/or ex situ), and if the materials would be readily available;
- Which sampling strategy had been used to obtain a better idea if "re-collecting" of wanted genetic diversity would be justified, also from a long-term conservation perspective;
- Targeted gap analysis on geographical coverage of the collection as well as genetic diversity presence in the collection; and
- Possibly other specific considerations.

Such a well-informed decision also includes aspects on which geographic area(s) to concentrate on, if a more refined search for specific information is needed through literature searches, local contacts, etc. This refined search should include the collecting of all available information on the population structure of the target species prior to the collecting mission [9,10].

The next step will be the search for collaborators and possibly specialists on the target species and, whenever possible, local plant collectors that would be prepared to join the exploration mission, to define the best period in the year to conduct the exploration/collection, to define what kind of outfit and equipment would be needed, etc. Finally, a formal request will have to be made to the local and national authorities to be allowed to explore (and possibly collect) plant genetic resources and thus, to be able to conclude a formal agreement for traveling to the identified sites and eventually to be allowed to take collected resources out of the country, when applicable (for details, see the work of [11]).

Frequently, exploration is combined with the actual collecting of the genetic resources taxa that had been prioritized by the conservation program to be added to the collection, for whatever reasons. Engels [12] distinguishes different reasons to collect as well as different types of collecting missions, i.e., (a) rescue collecting; (b) collecting for immediate use; (c) gap-filling for future use; (d) research; and (e) opportunistic collecting. Each of these types of collecting missions has different objectives, might require different sampling strategies, and might not all lead to an adequate representation of the genetic diversity for the target taxa in the collected samples. Furthermore, he also distinguishes several types of collecting missions, including multi-species vs. species-specific collecting, wild species vs. crop collecting, etc. These types require different preparations, different sampling strategies, etc. Another important aspect to ascertain before embarking on a collecting mission is to establish what the possible precise collecting sites are, i.e., natural habitats, disturbed areas, farming fields, marketplaces, home gardens, etc. in order to decide on the required (transport) equipment, accommodation, possible need to prepare meals, etc. and to determine what to collect where [12].

When in the field, taking careful notes on the collecting site (for a definition, see the work of [12]); deciding on the most effective sampling strategy (that might well vary with: a. the biology of the target taxon or taxa, e.g., annual/perennial, self- or outbreeder, seeds or vegetative parts to be collected; b. the precise purpose of collecting (see above); c. here we focus on collecting the maximum amount of genetically useful variability in the target species while keeping the number of samples within the practical limits for long-term conservation. Possibly the most important criterium is the frequency of alleles in the population, i.e., the conceptual class of the alleles: (1) common, geographically widespread, (2) rare, widespread, (3) common, localized, and (4) rare, localized. For a comprehensive treatment on this subject, please refer to the work of [9,10]. Besides genetic considerations, the number of individuals per population or variety to be collected also depends on the "viability" of the materials collected and on how many gene banks or collections are expecting subsamples of the collected materials (each with the same genetic diversity). Materials collected in the field need to be treated with great care to avoid that the viability of the collected organs (seeds, cuttings, tubers, etc.) would drop during the travel and/or shipment to the home base before the adequate processing for storage in the gene bank (i.e., an important time factor). Observations on possible selection pressure parameters that the collected material has been or could have been exposed to should always be noted.

From the above, it can be deduced that for a species to be conserved long-term, sampling is critically important. Therefore, due consideration should be given not only to the number of individuals of a population or taxon per site to be collected but also how many populations or sites in the area should be collected and how these should be distributed over the area of a given species in a country or even region to obtain an adequate representation of the "total diversity" present in that area or region. As for most species, detailed information on the distribution of the genetic variation required to decide on the best sampling strategy is missing. One has to extrapolate the information from those species that have been studied to those species where basic information is lacking [9].

A generally accepted benchmark criterion for collecting germplasm is to ensure that at least one copy of 95% of the alleles with a frequency greater than 0.05 is included in the collected sample. Random and unrelated gametes from a population of a target species will meet this criterion. This would be assured by collecting and bulking seeds or vegetative material from 30 randomly chosen individuals in a fully outbreeding sexual species or from 30 random genotypes in an apomictic species or from 59 random individuals in a self-fertilizing species. A sample of 50 individuals from each population and about 50 populations in an ecogeographic area is considered as a benchmark [9,13]. In a later analysis, Hoban concluded that for a metapopulation of wild species, one should collect a minimum of 1000 individuals per metapopulation or an ecogeographic range of the species to compensate for migration between populations and the loss of plants through germination failure, disease, and active use, to preserve enough allele copies to account for various degrees of collection attrition [14].

However, in many instances, the collector is not able to collect such a high number of individuals from a population, as the maturity time of the species might not be optimal, the number of growing plants is limited, etc. and thus the samples collected do not represent the variation of the target species in the area collected. In the case of cultivated crops (which require a different and, in general, more simple collecting strategy), many of the collecting has been carried out for the sake of convenience in marketplaces with all possible implications this might have for the diversity collected, especially for heterogeneous landraces and traditionally mixed materials where sampling the variation in the target materials is not possible and its representation in the market sample likely inadequate. Another limiting factor is that the records taken at the collecting site are limited or sometimes completely lacking, but for the sake of the possibly threatened diversity, such materials are added to the collection [15].

From a gene bank management perspective, a practice of gene bank curators is to collect sufficient seeds and to use the collected material directly for long-term storage and thus to avoid the initial seed multiplication/regeneration. This practice can only work if sufficient seeds of high quality can be collected, a pre-condition that often does not apply [16].

As has been mentioned in the section on the history of conservation, with the entrance into force of the CBD in 1993 in which the sovereignty of states over the (plant) genetic resources in their territories is recognized, a greater hesitation of readily sharing genetic resources with other countries can be observed [17]. In the case of the CGIAR centers, a significant decrease in collecting activities has been observed as several countries and organizations have difficulties providing permission to access genetic diversity for inclusion into the in-trust collections. Some of the reasons for this are uncertainties regarding institutional ownership over genetic resources and unresolved tensions concerning benefit sharing [18].

It should be noted, however, that not all germplasm samples entering gene banks are a result of a collecting mission. In the more recent history of gene banks, many accessions are also obtained as a donation, upon request, from other public or private collection owners (see Figure 1).

2.2. Processing

Processing of collected, harvested, regenerated/multiplied, or donated germplasm materials refers to activities during which the materials are being prepared for (long-term) storage or maintenance (in the case of perennial crops kept in field gene banks, in in vitro collections, or tissues and plant propagules cryopreserved). The focus here will be on seed materials as these are the bulk of conserved germplasm. Such preparation steps include the threshing of the seeds from the collected culms; the removal of the seeds from the fruits and where necessary their washing; cleaning or winnowing; the removal of broken, diseased seeds or seeds from different taxa (e.g., weeds), seed drying, packaging, and storage. For details on these various steps, especially on seed drying and other factors that might impact the quality of seed for long-term conservation, see the work of [19] and the Crop Genebank Knowledge Base [20].

Whereas most of the processing steps are straightforward, some aspects that might impact the seed quality for long-term conservation are briefly treated here. In summary, seeds of high quality can be obtained by planting for regeneration/multiplication in suitable areas/fields and at appropriate times; applying suitable crop management practices; adoption of proper harvesting and drying techniques; careful handling and processing to minimize mechanical injuries and unwanted seed mixing with other accessions; and ensuring minimum deterioration before reaching the designated storage, in particular, fast processing and no exposure of seeds to high humidity and temperature. However, seed production and post-harvest handling are highly dependent on the biology and agronomy of the species [19].

While the FAO Genebank Standards [5] recommend seed drying to equilibrium in a controlled environment of 5–20 °C and 10–25% RH and many gene banks operate their drying room at 15 °C and 15% RH, such a drying regime has been found to be not suitable for rice [21,22]. If, due to weather conditions, rice seeds have a high moisture content at the time of harvest (>16.5%), initial drying at 40–45 °C is recommended, followed by final equilibrium drying at 15 °C and 15% RH. This practice has been shown to significantly improve seed longevity during storage compared with standard drying at 15 °C and 15% RH [22]. A similar response was observed with accessions of wild rice [23]. Based on this research, the IRRI gene bank is now routinely using a two-stage drying process for the entire rice collection [6]. Freshly harvested seeds are first dried for three days in a drying room set at 40 °C and 30% RH, followed by equilibrium drying in a drying room set at 15 °C and 15% RH. Some other species might also respond favorably to an initial drying at higher temperatures than 15 °C in terms of seed longevity [22].

One important seed processing step relates to the creation of subsamples of the materials that belong to the same accession with the aim to facilitate "easy access" when material is needed for viability testing, regeneration, or distribution. The idea is that one subsample represents the diversity of the accession adequately and that the number of seeds is meeting the requirements for viability testing (typically 100 seeds, allowing for four replicates of 25 seeds each), for regeneration (typically to ensure that the entire genetic variation is represented in one subsample, i.e., not less than 50 seeds) and/or distribution (typically very limited numbers of seeds). Thus, correct sub-sampling and including sufficient seeds to represent the diversity adequately, especially for long-term conservation, is critical. Of the same importance and nature is the number of individuals used in the regeneration of an accession. Genetic drift is likely to happen when the number of plants is below the effective population size, and thus, genetic erosion might happen in the gene bank.

Another important germplasm processing activity with a possible direct and significant impact on the longevity of the stored seeds is the (timely) drying of the collected/harvested seeds. Whereas one can hardly influence the quality of collected seeds in the field during a collecting mission, many factors can be influenced and optimized to produce high-quality seeds during a regeneration cycle. These factors include cultivation and harvest practices, but also the proper drying of seeds in the gene bank before storage [19]. As the optimum drying is, among others, depending on the species and as the possibility of over-drying has been reported, possibly decreasing the longevity of seeds, it is advisable to conduct straightforward tests to define the optimum seed moisture content for long-term storage [24–26].

2.3. Seed Longevity

Knowledge of expected seed longevity in storage is important for determining viability monitoring intervals. Seed viability can be predicted with the help of a viability equation developed by Ellis and Roberts [27] for orthodox seeds, using parameters derived from seed storage experiments under different temperature regimes and moisture contents. The Ellis and Roberts viability equation $v = K_i - p/\sigma$ shows the relationship between viability and storage period, where v is the viability after p years in storage, whereby σ represents the slope of the curve and K_i the initial viability of the seeds. Meanwhile, improved equations have been developed by Hay et al. [28] and Probert et al. [29]. For a limited number of about 70 species, the seed viability constants can be found in the Seed Information Database (SID) of the Royal Botanic Gardens Kew [30].

Detailed knowledge of crop-specific seed longevity is important for the determination of seed viability testing intervals. The Genebank Standards [5] recommend setting monitoring intervals at "one-third of the time predicted for viability to fall to 85% of initial viability or lower depending on the species or specific accessions, but no longer than 40 years. If this deterioration period cannot be estimated and accessions are being held in long-term storage at −18 °C . . . , the interval should be 10 years for species expected to be long-lived and five years or less for species expected to be short-lived". To arrive at reliable seed longevity estimates, analyses of regular viability monitoring data over the entire storage period are essential. However, even among CGIAR gene banks, there is a lack of robust, reliable historical data on the long-term viability of seed lots [31]. Reliable seed longevity estimates would enable gene bank curators to forecast more reliably regeneration requirements, to estimate the size of seed lots required for long-term storage, and to adapt accession monitoring intervals.

USDA seed longevity research has shown that some plant families are characterized by predominantly short-lived seeds (e.g., Apiaceae and Brassicaceae), while others (e.g., Malvaceae and Chenopodiaceae) have long-lived seeds [32]. A meta-analysis of seed longevity studies indicated that seed of maize (*Zea mays*), oat (*Avena sativa*), barley (*Hordeum vulgare*), sorghum (*Sorghum bicolor*), many grain legumes (*Cicer arietinum*, *Vicia* sp., *Vigna radiata*, *Lens culinaris*, *Phaseolus vulgaris*, *Pisum sativum*, *Trifolium repens*, *Melilotus alba*), and vegetable crops (*Raphanus sativum*, *Abelmoschus esculentus*, *Cucumis melo*, *Cucumis sativus*, *Solanum melogena*, *Solanum lycopersicum*, *Spinacea oleracea*) are long-lived, while seed of rye (*Secale cereale*), groundnut (*Arachis hypgea*), sunflower (*Helianthus annuus*), rapeseed (*Brassica napus*), and some vegetables (*Allium cepa*, *Allium ampeloprasum*, *Lactuca sativa*, *Capsicum annuum*, *Apium graveolens*, *Daucus carota*, *Pastinaca sativa*) and some forage grasses tend to be short-lived [32,33]. Surprisingly, the comparison of seed longevity between wild and cultivated species under the same storage conditions did not reveal significant differences. Across all species, the meta-analysis carried out by Solberg et al. [33] indicated a viability loss in the range of 0.2–0.3% per year if seeds were stored under the recommended conditions according to the Genebank Standards [5]. These viability losses are much higher than would be expected by the published viability equations. The multi-faceted aspects and approaches to understanding the inter- and intra-specific differences in seed longevity have been discussed by the global research community during a workshop organized by the International Society for Seed Science (ISSS) in July/August 2018 in Fort Collins, CO, USA, and synthesized by Pritchard [34].

There are many factors that determine initial seed quality and viability and, consequently, have an impact on seed longevity in storage. Among these are crop management practices, climatic factors, stage of seed development at harvest time, and post-harvest seed processing [19]. Differences in geographic origins and, hence, climatic and environmental factors appear to contribute to the variation of P_{50} values within genera and families [32]. Seeds from *Brassica* and *Lolium* species that originated from Europe had characteristically shorter shelf lives than seeds from the same species originating from South Asia and Australia. Ellis [35] stressed that the interaction of genotype with environmental factors determines when maximum seed quality is first attained and for how long it is maintained during the seed development and maturation phase. The period of maximum seed quality may be brief or could be extended depending on several factors. Regarding seed processing factors, research at IRRI has shown that a two-stage drying process significantly improves seed longevity during storage compared to standard drying at 15 °C and 15%, and this modified drying approach has now been adopted for all rice accessions at IRRI [6]. Other species might also respond favorably to an initial drying at higher temperatures than 15 °C in terms of seed longevity [22].

Storage conditions clearly affect seed longevity. Experiments conducted at the CGN gene bank in the Netherlands with seeds of wheat (initial germination rate of 95%) and barley (initial germination rate of 94%), stored at either +4 °C or −20 °C retained high viability of 94% for wheat and 90% for barley after 23–33 years of storage at −20 °C [36]. In contrast, the viability of seeds stored at +5 °C for the same period declined to 62% for wheat and 75% for barley with concomitant losses of seed vigor. A subset of the wheat accessions tested only seven years later showed a further drastic decline in mean germination to 35% when kept at 4 °C, while the samples conserved at −20 °C remained stable at 95%. Similarly, seed longevity studies in maize accessions stored for an average of 48 years at the CIMMYT gene bank in Mexico revealed a significantly lower and more variable seed germination rate of 81.4% for seed lots conserved as an active collection for distribution (at −3 °C), as compared with a high and more stable germination rate of 92.1% of the seed lots conserved as a base collection in a chamber maintained at −15 °C [37]. Based on these long-term storage results of maize accessions, it has been suggested to apply base collection storage conditions (−15 °C) to both the active and base collection to improve seed longevity and reduce the need for costly regeneration events [37].

In a relatively short storage experiment of five years only, no loss in seed viability was detected in any of five species tested during this period when seeds were stored at −20 °C with either low (5.5–6.8%) or ultra-low (2.0–3.7%) seed moisture content [37]. However, significant viability losses were measured after a 5-year storage period at +20 °C, and losses occurred faster at low SMC compared to ultra-low SMC [38].

Molecular approaches to understanding and predicting seed longevity in storage are briefly discussed in Section 4.1.

2.4. Seed Regeneration

As gene bank accessions are often collected from a wide range of geographical locations, there is a high probability that original phenotypic variance is lost during ex situ conservation and seed regenerations. This applies more to crop wild relatives than to landraces and commercial cultivars and seems to be caused by selection or gene flow [39]. Multispectral image analysis of seed, i.e., seed phenotyping, has shown to be an effective method for identifying different seed types within a sample of seeds and for verifying whether incoming seeds from a regeneration cycle match the original seeds [40]. While DNA fingerprinting is an effective method to verify the genetic integrity of regenerated seed materials, a complete phenotypic assessment of accessions through high-throughput phenotyping (HTP) during periodic seed regenerations constitutes an alternative option to ensure that original phenotypic features are preserved [41]. The creation of a digital seed file could be the basis for high-speed authentication [42].

HTP tools such as hypospectral imaging have also been successfully used for seed quality, purity, viability, vigor testing, and variety identification in commercial seed lots of various crop species [43,44]. However, these tools can also be used as objective methods for managing gene bank accessions, starting from acquisition to seed regeneration, avoiding physical contamination, and maintaining genetic integrity [40].

2.5. Germplasm Exchange

Given the history of crop domestication and global dispersal of crops for food and agriculture, all countries are highly dependent upon plant genetic resources originating from beyond their borders. This dependency has increased over the past 50 years in connection with economic and agricultural development, the globalization of food systems [45], population growth, and climate change. The increasing challenges of crop adaptation to biotic and abiotic stresses exacerbated by climate change and the need to satisfy food and nutrition security of a still-growing global population, reshuffling alleles within a subset of well-performing breeding lines is no longer sufficient to address the global challenges. Plant breeders, therefore, need to broaden the genetic base and introduce specific traits into their breeding populations, and this can be done by resorting to diverse landraces or crop wild relatives that harbor genetic diversity, which was lost during the domestication bottleneck [46].

The international germplasm collections hosted by 11 CGIAR centers include over 760,000 accessions of crops, forages, and trees [47] and constitute a major proportion of the international germplasm exchange. Over the last 10 years (2010–2019), the CGIAR gene banks distributed over 1.1 million PGRFA samples to recipients in 163 countries. During the period from 2017 to 2019, landraces were the most frequently requested materials (50%), followed by breeding materials (24%) and crop wild relatives (13%). Most samples were distributed to advanced research institutes and universities (42%), followed by National Agricultural Research Systems (NARS; 38%), Non-Governmental Organizations (NGOs) and farmers (85), the commercial sector (7%), and others (5%) [47].

Despite major efforts by gene banks to facilitate and enhance the use of the genetic materials conserved in gene banks, these resources are far from being used exhaustively by breeding programs and/or farmers [48]. This is possibly attributable to the scarcity of descriptive information related to accessions conserved in gene banks, the limited use of genomic, phenomic, and information technologies, and, finally, obstacles in implementing national and international policies for benefit sharing [49].

In developing countries, public breeding programs are faced with financial, technical, and policy-related challenges that are limiting the more widespread use of landraces and crop wild relatives [50]. Apart from technical and financial issues, it was especially the lack of a supportive policy environment that was perceived by public sector breeders in developing countries as a major bottleneck restricting their sourcing and use of more diverse genetic resources.

The multilateral system (MLS) established under the ITPGRFA governs the access to the genetic resources of a pool of 64 food and forage crops (referred to as Annex 1 crops to the treaty) under a standard material transfer agreement (SMTA) and the benefit-sharing arising from their use [5]. Many European gene banks also adopted the use of the SMTA for non-Annex 1 crops (for further details on the ITPGRFA, please refer to the work of [3]). Apart from the SMTA, other material transfer agreements (MTAs) are also in use. The Nagoya Protocol regulates access and benefit-sharing under the CBD [5]. In contrast to the SMTA used under the MLS of the ITPGRFA, this is a bilateral agreement between the provider country and germplasm user.

Prior to the shipment of PGRFA (seed, clonal propagules, DNA), the beneficiary needs to sign the SMTA or other MTA. The MTAs regulate the intellectual property rights (IPR) of the requested material and related information, the conditions of its use and distribution to third parties, as well as benefit-sharing arrangements [51]. The SMTA of the International Treaty and most other MTAs only regulate the exchange of physical germplasm materials and do not refer to the exchange of digital sequence information (DSI) or DNA samples extracted from the genetic resources. According to Andersson [52], the following institutions make explicit reference to the exchange of DNA in their MTA: CATIE, Costa Rica; the National Institute of Agrobiological Sciences (NIAS), Japan, the Missouri Botanical Garden, USA; and the Royal Botanic Gardens, Kew, U.K. However, even though some MTAs cover the exchange of DNA samples, there are still different interpretations regarding the question of whether this precludes the patenting of specific genes. In recent years, the governance of digital genomic sequence information has become a contentious issue, and this, in turn, is leading to international disagreement over access and benefit-sharing regulations and blocks the intended expansion of the list of Annex 1 crops [47,53]. The political dimension of DSI is extensively covered in Section 4.4.

Seeds and especially vegetative propagules used for germplasm exchange are known to potentially harbor harmful pathogens, which may lead to transboundary disease spread along with the international movement of germplasm. Quarantine and phytosanitary measures have been adopted by most countries around the globe to minimize the threat of disease spread by screening export and import consignments of germplasm. The effectiveness of these measures depends on seed phytosanitation treatments, the actual knowledge of pathogen distribution and associated risks, the development, adaptation, and availability of diagnostic tools and protocols for seed health testing, qualified operators, procedures for inspection, and post-entry quarantine facilities [54]. Within the CGIAR gene banks, germplasm health units (GHUs) are responsible for germplasm phytosanitation and testing of the health status to guarantee safe global germplasm movement and exchange and the prevention of the transboundary spread of pests and diseases [54]. In their recent review, Kumar et al. [54] describe in detail current procedures for germplasm health testing and pathogen elimination for the major CGIAR mandate crops. As GHUs are widely distributed in developing countries and are known for their high-level expertise and technical capability, they could evolve into a global network of phytosanitary hubs for the research, diagnoses, control of established and emerging pests and their elimination from germplasm propagules, thus, guaranteeing the safe international movement of germplasm.

2.6. Documentation

All routine gene bank operations produce data that need to be captured and documented for internal use and, in many instances, for sharing with germplasm users (passport, characterization, and evaluation data). Adequate information management is important for the safe operation of a gene bank. This includes data on the acquisition, registration, storage conditions and collection type (base, active collection, safety duplicate), monitoring of the viability of accessions prior to storage and in storage, regeneration, characterization, and evaluation, germplasm health testing, distribution, and number of sub- samples and seed quantity of each accession kept in the gene bank [5,55]. In addition, among CGIAR gene banks, weaknesses in effective and consistent documentation of routine gene bank operations have recently been revealed [31]. Accession-level data, which are of high relevance for germplasm users, are passport, characterization, evaluation, and lately also omics data [56]. Internationally accepted multi-crop passport descriptors (MCPD) [57] have been adopted by most gene banks as standards for documenting passport, characterization, and evaluation data. These descriptors allow the exchange of accession-level data between gene banks and the operation of international information data portals on PGRFA, such as Genesys, the FAO-led World Information and Early Warning System on Plant Genetic Resources for Food and Agriculture (WIEWS), the European Search Catalog for Plant Genetic Resources (EURISCO), and GRIN-Global [55,58]. As noted by CGIAR gene bank managers, one of the most important factors affecting demand for PGRFA is the quality, comprehensiveness, and relevance of the accession-level information that gene banks compile about the materials in their collections and make accessible online [47]. Although gene banks are encouraged to also record molecular data being generated through genomics, proteomics, metabolomics, phenomics, and bioinformatics [5,56], these are mostly not generated by gene bank staff themselves but through collaborative projects with other research teams or by specialized research institutes. Given their complexity and exponential increase in volume over time, omics data are stored and made accessible in specific public databases or dedicated to other systems [56]. The GenBank platform [59] hosted by the National Center for Biotechnology Information (NCBI) is such a public database for genomics data.

In line with the Plant Treaty's global information system (GLIS), digital object identifiers (DOIs) have been introduced for gene bank material [60], and the CGIAR gene banks, as well as the Dutch gene bank CGN, have assigned DOIs to almost all their gene bank accessions [47,56]. DOIs provide a globally unique and permanent mechanism for identifying germplasm [61] and derived products such as DNA and related publications, thus assisting the user community as well as facilitating access- and benefit-sharing mechanisms. However, gene banks still face technical challenges, such as how many different DOIs should be assigned to the same accession. As regeneration in the gene bank environment might result in changes of the original material, often collected in far-away places, should; therefore, accessions of each regeneration have their own (new) DOI [62]? Another question relates to different forms of conservation (seed, in vitro culture, cryopreservation, herbarium) and storage conditions (short, medium, and long term). Should the same material conserved under different forms and conditions receive a different DOI? Should purified lines from heterogenous accessions receive a new DOI? There are also challenges with including DOIs in publications to be automatically discovered as this would require that all DOIs must be listed in the reference section of a paper, and this is not the standard practice of journals.

2.7. Research

Long-term conservation of plant genetic resources in gene banks aims at maintaining the genetic diversity of individual accessions as authentic and as close as possible to the original genetic composition. In the critical assessment of current practices and the underpinning theories, we have identified possible research topics that gene banks could (or possibly even should) undertake, where relevant or necessary in collaboration with specialists, to contribute to more effective and efficient long-term conservation and/or to rationalize the current global long-term conservation system. Another reason for making such suggestions is to demonstrate that gene banks are not "dead conservation morgues" but lively, dynamic, and essential institutions that use advances in science to continuously improve the knowledge on the conserved materials and on the applied procedures and to make these as cost-efficient and effective to reduce the burden on future generations. A third reason to promote research on the conservation procedures and on the materials conserved is to involve and widen the participation of researchers in the actual conservation of PGRFA efforts, as that is being seen as an essential responsibility by society. Any contribution to make this happening is important and should be duly recognized.

Examples of how (routine) research activities can underpin optimum management of germplasm accessions and collections, and thus, e.g., contribute to extending seed longevity of accessions are summarized in Table 1.

Table 1. Examples of how (routine) research activities can underpin optimum management of germplasm accessions and collections.

Research Activities	Description
Determine genetic diversity of an accession	In-depth characterization of accessions using internationally agreed descriptor lists, making use of advanced molecular, genomic, and phenotyping tools.
Optimal management of gene bank accessions	Determine adequate, minimum numbers of individuals per accession for viability testing, regeneration, characterization, evaluation, and other gene bank activities with the aim of preserving the original genetic diversity.
Elucidate flower biology of crop species, when not known	Full understanding of the flower biology of a given species helps to avoid cross-pollination during regeneration and to maximize high-quality seed production for subsequent long-term storage.
Optimize seed production procedures	Genebanks conserving a wide range of different species (e.g., The World Vegetable Center) may need to conduct research to gain insight in crop-specific knowledge for optimizing seed production procedures to improve initial seed quality and, consequently, seed longevity.
Optimize seed drying procedures	The FAO Genebank Standards include clear and specific recommendations on all routine gene bank operations, including seed drying. However, rice accessions, for example, require modified drying procedures to enhance seed longevity (see Section 2.2).
Determine optimum seed moisture content (SMC)	Research has shown that SMC levels aiming at maximizing seed longevity differ among species. Thus, gene banks should consider conducting their own research to determine the species-specific optimum.
Optimize species- and accession-specific seed viability monitoring	Optimizing the schedule and procedures of routine viability monitoring of long-term stored accessions provides an early warning for deteriorating accessions and helps to rationalize the number of seeds used per test. Comparing the physiological response and storage behavior of seed lots produced in different crop seasons and/or environments improves our understanding of seed longevity. Weekly scoring of germination during a viability monitoring test helps to obtain information on seed vigor and how vigor declines as seeds age.
Predicting seed longevity	Conducting studies on the integrity of DNA and RNA in seeds under long-term storage helps to predict species- and accession-specific seed longevity.
Optimizing the genetic diversity representation of populations	Extending the knowledge of genetic parameters that allow optimizing the genetic diversity representation of populations of a given species will improve germplasm collecting and the establishment of truly representative collections.
Publishing research results	Publishing research results on the above-described topics in scientific journals would benefit staff from other gene banks and would boost the reputation of the gene bank itself.

During the 1980s, IBPGR, later denominated IPGRI and now Bioversity International, initiated research on ultra-dry seed storage, based on the assumption that seeds dried to levels well below 5% SMC could be stored at room temperature for extended periods [63].

However, concerns had been raised regarding a possible over-drying of seeds, potentially leading to a loss of longevity [25]. This issue led to a scientific debate and a global seed project to resolve the controversial points [64]. Results of experiments with *Lactuca* seeds showed that crop species have an optimum seed moisture content and that drying seeds below the optimum water content does neither benefit nor damage seed longevity. Furthermore, it was observed that there is a temperature x water content interaction affecting seed longevity and that for storage of seeds at room temperature (about 25 °C), the appropriate RH of the storage room is about 14%. Conducting research on groundnut (*Arachis hypogaea*) seeds, Sastry et al. [65] were able to demonstrate the potential benefit of ultra-dry storage. When groundnut seeds were dried to 1.7% SMC and stored at 50 °C in aluminum foil bags, seeds retained viability up to 192 weeks (3.7 years) under vacuum storage in incubators, compared to only 144 weeks (2.7 years) when stored in normal atmosphere in an incubator. Seed storage at a temperature of 35 °C and at SMC levels of 1.7% or 3.4% retained seed viability even after more than 5 years.

3. Strengths and Weaknesses of the Current Active and Base Collection Concept

The term germplasm collection requires some attention as it is a term that clearly refers to the ex situ conservation scenario and that can encompass all the accessions of a given species or crop, a subset of selected accessions (e.g., a core collection), or to all accessions of the various species that make up the gene pool of a given crop. However, it can also refer to all the accessions stored in a gene bank. Thus, some care in using this term is required, also as it is frequently used for the activity of collecting germplasm. Thus, regarding the latter, it is proposed to use the term "collecting".

In many instances, gene bank collections have grown out of collections established by plant breeders that were eventually "converted" into a germplasm collection. In addition, research collections established at universities often formed the starting point of subsequent gene bank collections. It should be noted that in both cases, the collections were likely not built for long-term conservation of genetic diversity per se, and thus, their origin might less reflect the genetic diversity of collecting areas, as would germplasm collections that have been formed with samples that were collected for that purpose. Traditionally, such breeding or research collections were conserved at cool temperatures (possibly +15 °C or less) and in paper or cloth bags under controlled relative humidity. It was only during the late 1950s that research on storage temperatures started and that a two-tiered conservation strategy was defined: 1. long-term conservation of "base collections" of adequately dried seeds, usually stored at −18 °C in hermetically closed containers; and 2. medium-term storage of "active collections" under less stringent conditions, e.g., +5 °C and controlled relative air humidity of approximately 35%; both collection types were maintained in insulated storage rooms [66].

In the following, we will take a critical look at the various collection types, along with their proposed storage conditions, but with a clear focus on their long-term conservation strategic aspects. The latter comprises the base and the security backup collections that jointly provide the conditions for long-term conservation. The active collection is meant for the storage of samples for characterization and evaluation, multiplication/regeneration, research, and distribution purposes, usually at a higher temperature than the base collection. A fourth collection type is the archive collection that consists of accessions that are meant to be disposed of but that are kept without any management at −18 °C for security purposes only [67].

3.1. Active Collection

Accessions that are being conserved for their use in research activities, i.e., characterization and evaluation, molecular studies, and/or for distribution, are kept under conditions that would provide for short- or medium-term duration, i.e., up to 30 years, depending on the species and seed quality. Thus, a well-cooled storage room at plus 5 °C and with controlled humidity at +/−35% RH would provide these conditions and allow storage in paper or net bags, or any other "open" storage forms, and would allow rather easy access (no cold room protection suites, etc.), whereas also the use of hermetically sealed containers can be used. The agreed FAO Genebank Standards define an active accession as "A germplasm accession that is used for regeneration, multiplication, distribution, characterization and evaluation. Active collections are maintained in short to medium term storage and are usually duplicated from a base collection maintained in medium to long term storage.' Active collection samples for medium-term conditions should be stored under refrigeration at 5–10 °C and relative humidity of 15% ± 3%" [5]. In this early conservation concept, it was assumed that the material would be turned over within the medium term and restocked with newly regenerated germplasm samples from the base collection. This "rough" concept worked well in cases where both the active and base collections of the same materials were stored at the same gene bank, or in cases where a well-developed networked system between base and active collections had been established, e.g., as at NPGRS and the regional centers of the USDA. However, in many other gene banks where the two collection types were split, this concept failed, for a range of reasons, including inadequate refrigeration, the lack of adequate viability testing, and a too strong focus (if any) on the active collection.

Because of varying local conditions with respect to the precise objectives of the gene banks, the availability of adequate infrastructure, as well as of human and financial resources, refinements of this concept should be undertaken to optimize the conservation [67]. Besides the "local fine-tuning" of the traditional concept, there are several specific reasons that would justify a more critical assessment of this traditional concept and possibly a revision of the accepted practices. These reasons have been updated and expanded from Sackville Hamilton et al. [66] (see Table 2).

Table 2. Reasons for adjustments and refinements of the concept of active collections to optimize long-term conservation.

Topic	Description
Refrigeration issues	The cost of refrigeration, the unreliable electricity supply, and/or lack of adequate maintenance and repair opportunities of the cooling equipment may hamper adequate long-term conservation.
Overly large active collections	Many gene banks try to maintain overly large active collections (e.g., too many samples and subsamples of the same accession that stem from different regeneration cycles); the anticipated use of the materials is frequently over-estimated, and due to "sub-optimal" storage conditions for such accessions, avoidable higher regeneration frequencies are the consequence of keeping seed viability at the desired level.
Rationalizing collections	Improved germplasm management technologies, such as the use of barcodes, molecular tools, and digitalized information management, including early warning systems, can facilitate more effective and efficient gene bank management and allow to rationalize collections, e.g., sorting out genetic duplicates, removing accessions from the active collection that are never requested or used but are included in the base collection.
Accession management	Regenerated materials of a given accession are kept in the active collection under medium-term storage conditions. In case the regenerated subsamples do not suffice for further distribution or use, one could continue to use a regenerated subsample for a maximum of four regeneration cycles (and possibly consider regenerating more materials in case of high demands) before returning to the primary MOS from the base collection for the next regeneration cycle.
Optimizing seed management and storage procedures	Genebanks should carefully consider under which conditions to store the active collection. Lower seed moisture content and lower storage temperatures would result in much-prolonged storage periods with less total operational costs and increased (genetic) security due to reduced regeneration frequencies. Certainly, accessions that have low distribution numbers due to a lack of accession-level data could best be maintained under the same storage conditions as the base collection to avoid more frequent regeneration cycles triggered by a drop in seed viability. Maintaining those materials under long-term storage (base collection) conditions, materials would still be available for distribution, and costly regeneration cycles could be reduced [36].
New conservation technologies	Because of the rapid development of in vitro gene bank conservation techniques and the wider availability of cryopreservation protocols [68], germplasm materials previously conserved under more threatening conditions in field gene banks can now also be maintained as tissue in in vitro collections, thus increasing the security of the material, and/or be cryopreserved, and thus adding a long-term cryopreservation perspective.
Complementary conservation approaches	The increased availability and use of complementary conservation approaches and methods increase the overall security of accessions and allow to opt for the most effective combination of methods, both from a management as well as an economic perspective.

3.2. Base Collection

The objective of the base collection of a given species is to maintain accessions that are distinct, with respect to the genetic integrity as close as possible to the original sample, conserved for the long-term and not intended for distribution [66]. Furthermore, the base collection should contain as much as possible genetic diversity in a rational (i.e., as few as possible accessions), effective and efficient manner under controlled, secured, and safe conditions for the longest possible time. This collection type is strongly focused on long-term conservation but should also consider the facilitation of the use of the conserved materials whenever possible and without compromising the objective of the conservation. According to the FAO Genebank Standards, the agreed conditions are as follows: "Most-original-samples and safety duplicate samples should be stored under long-term conditions (base collections) at a temperature of −18 ± 3 °C and a relative humidity of 15% ± 3%" [5]. If samples conserved under LTS and MTS conditions are kept in hermetically sealed containers (as recommended), then RH control of the storage room would not be required and is common practice in many gene banks [6,55,69].

The most original sample (MOS) is being defined as: A sample of seeds that have undergone the lowest number of regenerations since the material was acquired by the gene bank, as recommended for storage as a base collection. It may be a subsample of the original seed lot or a seed sample from the first regeneration cycle if the original seed lot required regeneration before storage [5]. Furthermore, the MOS should be prepared and stored under the best possible conditions for safe long-term survival; the seed from the MOS should never be distributed for use. The number of seeds in the "primary MOS" (the sample stored at the gene bank for its conservation) should be sufficient to: (a) allow for the optimum regeneration of the MOS (at least the minimum amount to represent the genetic diversity of the original sample and/or the minimum amount needed to reproduce sufficient seeds for the next generation plus a safety margin). The seeds of the primary MOS should not be touched until the viability begins to drop; (b) conduct routine and smart viability tests to determine when the MOS must be regenerated; and (c) supply the seed that is required for regenerating materials for distribution as part of the active collection. It should be considered to allow stock for several regeneration cycles to avoid that the materials of the primary MOS are depleted before it starts losing viability [66]. The storage of the primary MOS samples can be performed in one or in several containers, but all under the same optimal storage conditions.

Besides the primary MOS samples that make up the base collection, a subsample (the "secondary MOS") should be stored for security reasons under the same or better conditions than the base collection at another distant gene bank to protect the base collection material against accidental loss [20]. It is called the security backup or safety duplication collection and will be maintained under black-box conditions. The latter means that the recipient gene bank has no responsibilities for viability testing and should never use, regenerate, or distribute these safety duplicates without instructions from the duplicating gene bank. The secondary MOS should only be recalled in case of loss of the primary MOS and should contain sufficient seeds for one regeneration cycle. A viability monitoring routine of the primary MOS needs to be established and should be performed in the most efficient way, i.e., to use as few as possible seeds. In case regenerated subsamples for distribution are stored separately in the active collection, they could serve as an indicator for the viability of the primary MOS if stored under the same or less strict conditions (Figure 1) [66].

In another conceptual scenario, the term "base collection" is used to define a set of accessions that are designated to form a base collection of a given crop. These designated accessions can be stored in the respective gene banks that maintain part of the unique diversity of a given species (i.e., each gene bank conserves a fraction of the global genetic diversity of that species), and collectively all gene banks maintain the "global base collection" for that species. Through its role to stimulate and facilitate the collecting of threatened germplasm, IBPGR established a network of gene banks that had formally agreed to maintain germplasm materials collected with the help of IBPGR/IPGRI for a given crop gene pool in their respective base collections for the long term (the Register of Base Collections) [70] and to make this material readily available to bona fide users. The gene banks of the CGIAR are an important part of this register as they hold global collections of their mandate crops and aim to cover an adequate representation of the total diversity in the respective base collections. In many instances, also supportive crop networks have been established to coordinate and implement these efforts with the collaborating national programs. This network concept was developed by IBPGR/IPGRI, also at the regional level, to network national and institutional gene bank programs to strengthen the collaboration between active and base collections. Furthermore, it was felt that not every genebank had to have a base collection and that the collaboration between active and base collections could be organized at the regional level through regional networks. Europe initiated since approximately 2009 a virtual gene bank collection of unique and important accessions, spread across the gene banks of the continent and collectively recorded in the European germplasm database EURISCO [71]. However, possibly except for Europe, these regional PGRFA networks have disappeared or are dysfunctional.

Weaknesses of existing base collections and options to overcome those weaknesses to arrive at a more secure and rational long-term conservation system of PGRFA are summarized in Table 3.

Table 3. Weaknesses of existing base collections and options to overcome those weaknesses to arrive at a more secure and rational long-term conservation system of PGRFA.

Topic	Description
Poor representation of genetic diversity of natural populations and landraces in gene bank accessions	Original collecting often resulted in accessions that represented only a fraction of the prevailing genetic diversity of a population or landrace. In case sampling was performed of natural populations, one could consider lumping samples/accessions of the same population into one. In the case of landraces, one could consider lumping the samples collected from the same field.
Accession duplicates	In case one or more genetic duplicates of a given accession are identified in the collection, duplicates could be lumped into one accession. However, if an identified duplicate accession is phenotypical of special interest and has substantial research/evaluation data, it might be justified to keep that one separate.
Rationalizing base collections	When two or more subsamples of the same accession are found in the base collection (possibly from different regeneration years), the gene bank may opt to identify the most original sample (MOS) among them and to proceed with that subsample while moving the remaining subsamples to the active or the archive collection (see Section 3.3).
Reducing seed viability testing	A practice that is being recommended at the Center for Genetic Resources, the Netherlands (CGN) to rationalize routine gene bank operations is the decision to delay the first germination monitoring tests to 25 years after regeneration [72] or to the time when the samples of the active collection are undergoing the first regeneration cycle, and the origin of seed lots start to diverge between active and base collection (see Figure 1).
Base collection concept for vegetatively propagated materials	Whereas we have focused in the above completely on orthodox seed-producing species, it is understood that the concept of base collections might not apply to vegetatively propagated materials directly, commonly maintained in field gene banks due to the lack of available in vitro and cryopreservation options. In cases where in vitro techniques and cryopreservation protocols are available, the concept of base collections might apply as well. In all other cases, suitable maintenance in the field and adequate safety duplication might be the only option.

3.3. Other Collection Types

a. Backup collections. Besides the base and active collections, we have referred in the above also to safety duplicate or security backup collections that are arranged based on black-box agreements between different gene banks. These backup collections can consist of subsamples of accessions from the base as well as the active collection. It is important to stress the importance that the storage conditions of the safety duplicates at the recipient gene bank should be the same or better than those at the "conservation" gene bank. Apart from these bilateral arrangements, the Svalbard

Global Seed Vault serves as a global long-term seed storage facility to provide an additional security backup to germplasm stored in gene banks around the world. It is built into a permafrost mountain at Svalbard, and the storage temperature is maintained at −18 °C through an additional solar energy-based cooling system that counters the global rise of the earth temperature caused by climate change, which is also witnessed at Svalbard. The seed vault will only agree to receive seeds that are shared under the multilateral system (see above) or under Article 15 of the International Treaty or seeds that have originated in the country of the depositor. The black-box system entails that the depositor is the only one that can withdraw the seeds and open the boxes [73].

b. Archive collections. The archive collection consists of germplasm accessions that are stored under optimal conditions at relatively low cost but that are not actively maintained. The gene bank does not have (anymore) the responsibility for conserving or distributing these accessions. The type of accessions or materials stored in the archive collection could include the following: a. black-box conservation of experimental materials that could have an IPR protection; b. in case a collection has to be disbanded and yet no other gene bank could be identified to accept that collection, the accessions should be temporarily stored in the archive; c. as per some examples mentioned before (e.g., possible duplicates; extra subsamples), in cases when the curator decides to discard accessions they could be archived instead; d. in case of a "forced" rationalization of the collection selected accessions might be removed from the collection, and they should be considered for archiving until a solution is found [66].

c. Research collections. Research collections contain materials that stem from past research activities and have been kept by researchers or their institutes, sometimes over long periods of time. In addition, collections of plant breeding materials could have been stored or maintained by individual breeders, researchers, or institutes. Depending on the type of activities, some collections might contain very specific materials that are difficult to keep and/or to regenerate and might require specialized knowledge. Since the advent of molecular research, the increased importance of DNA materials can be observed but also of single-seed descendent collections of very uniform quality. Whereas the latter might not be important from a genetic diversity perspective, they might well contain important material for molecular and genomic research as they contain the diversity in a suitable form for such research.

d. Structured collections. Another type of collection that stems from research activities on germplasm materials by structuring the collections on the basis of a specific characteristic or trait (e.g., core, mini-core, trait collections) or also to select the genetically unique accessions maintained by gene banks in a regional context (e.g., the AEGIS initiative in Europe) to form a virtual collection are examples of structured collections. They could be virtual or physical collections.

e. Reference collections. Another type of collection that possibly falls somewhat outside the "direct conservation" related objective of a gene bank is the seed reference collection. The concept stems from the botanic garden world as part of the herbarium "system" that was adapted by gene banks to increase the security of the accessions by allowing the detection of possible mistakes, for instance, during regeneration by comparing the phenotypic features of the seeds of a given harvested accession with those stored in the reference collection and comparing the accession number(s).

f. Non-seed collections. In ex situ conservation, other collection types have been created to maintain specific forms of plants, e.g., field gene bank collections in which accessions are being maintained of entire plants for practical reasons such as the need to maintain the genetic constitution of a vegetatively propagated crop (e.g., potato and many other root and tuber crops), or that the seeds are recalcitrant and cannot be dried without killing the seeds (e.g., avocado, cacao, many other especially tropical crops or species); when tissue cultures of plants are maintained in specially equipped rooms the term in vitro collection is used; in case such materials have been cryopreserved by placing them in liquid nitrogen the term cryo-collection is used [3].

3.4. Linking Conservation and Use

In the long-term conservation concept, the base collection is at the center of the strategy and comprises accessions that include the most original sample as well as a representative smaller sample deposited at a distant gene bank for safety reasons, all stored under optimum conditions. In addition, regenerated subsamples of the MOS are maintained in the active collection and are intended for research and distribution. Whereas the storage conditions can be less stringent to facilitate access and as the turn-over of materials is assumed to be faster as the loss of viability, this collection type is the joint between conservation and use.

Possibly the biggest hurdle between (long-term) conservation and use is the strong focus on genetic diversity integrity of accessions and their representation of the sampled diversity in a population or landrace, whereas users are predominantly interested in materials that can be easily used and that have ample data on their genetic makeup and agronomic performance. In addition, the ease of use, both in terms of time and of preparatory steps needed before the material can be used, is an important aspect, and this frequently is related to the degree of uniformity of accession or sample. It might be noteworthy, as already mentioned before, that many of the current base collections stem from past breeding collections and, thus, it can be assumed that many of the accessions are relatively uniform. In addition, during the 1980s and 1990s, it was observed that some genebanks had adopted a strategy to remove "off-types" from (landrace) accessions of self-breeding crops, e.g., ICRISAT in its sorghum collection [74]. Thus, we present ideas and examples of how long-term conservation and the facilitation of use can be improved:

- The creation of core and mini-core collections facilitates germplasm screening and selection for breeding purposes (see Section 4.2). Similarly, the inclusion of specific breeding or discovery populations created through introgressions from the wild into cultivated backgrounds into the active collection will enhance germplasm use in plant breeding (see Section 4.2);
- Whenever possible, more uniform materials created from a diverse accession could (or possibly should) be kept as separate subsamples in the active collection for distribution purposes. Examples of such "more uniform" subsamples could be pure lines selected and created from genotypes of self-breeding landraces and single-seed descendent lines, prepared for sequence studies. However, such subsamples will always remain part of the active collection and will not become part of the base collection;
- Possibly the most critical factor that triggers the use of accessions by plant breeders is the availability of comprehensive data on the performance of the accession, on specific traits or characteristics obtained through characterization or evaluation activities, molecular and genomics data, as well as data from genotyping and phenotyping efforts. Thus, a gene bank should generate or facilitate the generation of such information and make the data readily available online as well as through publications. In addition, the availability of comprehensive passport data will be of relevance, for example, for the FIGS approach (see Section 4.2) to see if certain traits or characteristics that are environment-related have a specific and well-defined origin;
- One aspect that could reduce the "tension" between conservation and use is the economic rationalization of the conservation operation through optimizing the storage conditions for the active collection whenever possible. As already mentioned before, keeping the active collection under suboptimal conditions, i.e., triggering regeneration by loss of viability and not by depletion of the stock, is a real cost factor as well as a potential threat to the integrity of the genetic diversity of an accession.

4. New Developments That Facilitate More Effective and Efficient Conservation and Use of PGRFA

Major technological advances, particularly in DNA sequencing, molecular biology, and omics technologies, phenomics (including sensors, imaging, robotics,) computation, information science, and the management of big data enable a transformation of the way in which plant genetic resources are managed and used. These advances are attracting a considerable number of new clients to gene banks, such as molecular biologists and geneticists alongside molecular and traditional plant breeders and may affect the operations of gene banks and aspects of their future role [75,76]. It is a major challenge for gene banks to satisfy the needs of this wide range of users, each group with a different set of expectations.

4.1. Role of Molecular Biology and Genomics in Promoting Long-Term Conservation

New applications of modern molecular biology tools and techniques such as next-generation sequencing (NGS) and genotyping-by-sequencing (GBS) enable scientists to enhance the quality, efficiency, and cost-effectiveness of gene bank operations, as well as the depth of scientific knowledge of gene bank holdings, thereby also guiding conservation management [77–79].

4.1.1. Redundancy in Crop Collections

Molecular tools are certainly helpful for making informed decisions on reducing redundancy in crop collections, thus contributing to efficient long-term conservation [80–82]. However, this approach is not straightforward. Given the fact that even in self-pollinating species within accession variation may be considerable, there is a need to statistically quantify variation within and among accessions to decide whether they are sufficiently different to consider them distinct accessions [80]. However, from a user perspective, the functional diversity of a single trait, such as disease resistance, is often of high relevance. Given the uniformity of modern cultivars, a certain accession/cultivar may appear redundant based on molecular data but nevertheless differ in a single important trait that is highly relevant for a breeder.

Verification of passport data is the first element that might provide a clue on potential duplicates in a crop collection. This suspicion needs to be verified by morphological comparison of accessions followed by molecular marker techniques that can detect genetic differences between and within accessions [83]. Combining phenotyping and genotyping with single sequence repeats (SSR) markers allowed the identification of duplicate accessions in lettuce (*Lactuca sativa*) and the determination of the most appropriate accessions (MAA) for inclusion into AEGIS [84]. Similarly, the combination of morphological characterization with genotyping using an SNP (single nucleotide polymorphism) array, originally developed for *Brassica napus,* allowed the identification of duplicate accessions in *Brassica oleracea,* and a subset of 500 SNP markers have been suggested for genotyping *Brassica oleracea* accessions [85].

However, the correct identification of accession duplicates within a given crop collection across different institutes is challenging as collaborating institutes have to agree on a common, crop-specific set of markers and, subsequently, there might be problems to reliably reproduce DNA marker data between different laboratories [78]. Such difficulties could be overcome by using next-generation sequencing platforms to tackle the issue of redundancy within and between crop collections of different holding institutes [77,79,86,87]. In a case study comprising three gene banks (CIMMYT; the Wheat Genetics Resource Center (WGRC) at Kansas State University in Manhattan, KS, USA; the Punjab Agricultural University (PAU), Ludhiana, India) and focusing on *Aegilops tauschii,* a wild crop relative of wheat and source of genetic diversity for wheat improvement, Singh et al. [87] identified and characterized over 50% duplicated accessions on average within gene banks. With increasingly more powerful tools to compare genetic information between individuals/accessions, the likelihood of finding two absolute duplicates will decrease. Therefore, in deciding whether two accessions are duplicates, it is advisable to use phenotypic and genetic data for the comparison and to take a decision thereafter.

Genebank scientists of the International Rice Research Institute (IRRI) have pioneered the incorporation of genomics-based research into gene bank activities. IRRI scientists use such data to classify the degree of genetic similarity between accessions and the diversity within accessions to shed light on population structure and admixture, to classify potential duplicates, and to identify genetic novelty [77]. Genomic information will also provide a rationale for avoiding redundancies, thus limiting the size of collections, as well as facilitate genetic gap analyses to guide future collecting and acquisitions. Before incorporating new accessions into a gene bank collection, sequencing data can also be used to make an informed decision whether the new material possesses genuine genetic novelty to deserve inclusion into the base and active collections of the gene bank [77].

4.1.2. Inferring Missing Passport Data

Often, vital metadata, such as geographical data or taxonomic information on the species of accessions conserved in gene banks, is missing or incorrect. Curating such data is important as accessions with incomplete passport data and missing associated metadata are rarely requested by users [88]. A combination of existing genomic tools and statistical analyses can be used to infer missing pieces such as geographical region of origin as shown by Singh et al. [87] with accessions of *Aegilops tauschii*, a wild relative of wheat.

4.1.3. Predicting Seed Longevity

DNA protection and repair are important for maintaining genome integrity and seed longevity in plants. DNA damage in stored seeds results in faulty transcription and replication, thus, affecting key processes that are activated during the imbibition stage of seed germination [89,90]. Telomere lengthening has been proposed as a tool to distinguish between short- and long-lived species [90]. Telomere lengthening occurs after seed imbibition when metabolic activities resume, whereas telomere degradation is associated with seed aging. Reduction in seed longevity is often associated with the oxidation of cellular constituents such as nucleic acids, proteins, and lipids [91].

Seeds possess protective mechanisms to prevent damage to their cellular constituents through the formation of glassy cytoplasm that reduces cellular metabolism and the production of antioxidants that prevent the accumulation of oxidized macromolecules during seed storage [91]. Moreover, seeds also have repair mechanisms that remove damage in DNA, RNA, and proteins that accumulate during seed storage. This repair mechanism sets in during seed imbibition through the activation of enzymes such as DNA glycosylase and methionine sulfoxide reductase [89,91]. Through genome-wide association (GWA) analysis in diverse Indica rice varieties, eight major loci associated with seed longevity parameters were identified [92]. Based on their research, Lee et al. [92] concluded that high seed longevity in rice might be related to DNA repair and transcription mechanisms, sugar metabolism, reactive oxygen species scavenging, and embryonic/root development.

A complex network of putative longevity-related genes has been reported by Righetti et al. [93] that links seed longevity to biotic defense-related pathways. Genotypic variation of seed longevity in storage might be determined by two sets of genes [34]. A major set of genes evolved specifically for storability, while the other set is linked to seed dormancy. Metabolomics is a complementary approach to dissect the complexity of seed longevity. A shorter-lived rice cultivar (IIT998) showed a 2- to 6-fold increase in the change of sugar-related metabolites and glutathione-related proteins during natural seed aging compared with another cultivar (BY998) with extended seed longevity [34]. The rapidly increasing availability of reference genomes and pan-genomes constitutes another approach to dissecting and understanding the complexity of seed longevity [34].

Experiments with seeds of several vegetable crops have shown that RNA integrity declines with storage time in dry seeds [94]. As a decrease in RNA integrity was usually observed before viability loss, this assessment can be used to predict the onset of viability decline. Observing DNA and RNA integrity loss and understanding repair pathways in stored seed could help predict seed longevity and determine seed viability testing intervals. This information could also be used to develop crop varieties with improved seed storability and enhanced germination performance [89,95].

4.2. Role of Functional Genomics and Phenomics in Facilitating the Use of Plant Genetic Resources Conserved in Genebanks

Different strategies have been developed to select and prioritize potentially useful accessions from gene banks that can be used for crop improvement. Among those is the development of core or mini-core collections. Frankel [96] coined the term "core collection" meant to "represent with a minimum of repetitiveness, the genetic diversity of a crop species and its relatives". A core collection is a subset of a large collection, consisting of about 10% of the accessions and capturing most of the genetic diversity available in the entire collection [97]. As core collections of large crop collections such as those maintained by CGIAR gene banks are still too large for use by breeders, Upadhyaya and Ortiz [98] developed a mini-core collection concept, which is based on the evaluation and selection of a further subset of about 10% accessions from an existing core. ICRISAT has sent mini-core collections of chickpea, groundnut, pigeonpea, sorghum, pearl millet, foxtail millet, and finger millet to different research groups in 14 countries [99]. The World Vegetable Center has developed mungbean core and mini-core collections [100] and even a core collection of the wild tomato species *Solanum pimpinellifolium* [101] and offers these special collections together with the respective accession-level phenotypic and genotypic data for distribution to interested researchers and breeders to enhance the access to biodiverse vegetable germplasm for breeding and research.

Another strategy of selecting accessions based on their phenotype and associated passport data is the focused identification of germplasm strategy (FIGS) that is based on the assumption that the adaptive traits expressed by accessions are the direct result of environmental conditions of their respective place of origin and that the genetic diversity of the specific traits of interest can be maximized by sampling accessions based on their diverse contrasting geographic regions [102,103]. However, accessions conserved in gene banks around the globe are often lacking phenotypic data, and passport data might also be incomplete or have incorrect location data, which limits the application of FIGS. Hence other methods are required to facilitate the use of germplasm for breeding.

With the advances in molecular biology and genomics, DNA extracted from nuclei, mitochondria and chloroplasts are increasingly being used to evaluate patterns of genetic variation within and among species, map and characterize desirable traits and underlying genes of interest for breeding, for taxonomic studies telling species apart [104] and to infer the evolution of genome structure in plants [105,106]. Genotyping-by-sequencing (GBS) combines genotyping and genome-wide molecular marker discovery in one and the same process [79]. It facilitates the exploration of new germplasm sets or species that have not yet been characterized without the need to first discover and characterize polymorphisms through molecular marker studies. Moreover, with the advancements of bioinformatics, the development of new reference genomes, and the availability of an increased volume of sequence data, GBS data sets can later be reanalyzed to uncover further information, such as new polymorphisms or annotated genes.

Large crop collections cannot be sequenced in one go. There is a need to develop or use existing core collections and to transform them into genetic stocks for reference purposes and comparative and integrative genomic studies [107]. Heterogenous accessions must be purified through single-seed descent (SSD) before DNA extraction and initiating systematic molecular characterization. The purity of accessions can be assessed through genotyping with various molecular marker types such as inter-simple sequence repeats (ISSRs) or amplified fragment length polymorphism (AFLPs) [78].

These purified accessions or lines will also serve as the source material for phenotyping and ensure that phenotypic information can be properly linked with the sequence information in a meaningful way [77]. As has been demonstrated for rice, the demand for core subsets may substantially increase if all the accessions in the subset have been sequenced, and this information is made easily accessible to breeders or other researchers [42]. This information allows users to perform their own genome-wide association studies to elucidate the genetic control of multiple traits of interest. Therefore, this concept should be applied to all major crop collections conserved by a gene bank. DNA genotyping and sequencing results in combination with precise phenotyping are perfect assets for trait mapping, gene analysis, and allele mining in support of modern plant breeding.

An important objective of functional genomics in agricultural species is the use of sequence polymorphisms for phenotypic predictions and the selection of improved plant types. Prediction models are built by correlating phenotype and genotype in a breeding population of interest, and these models allow the identification of individuals with superior breeding values [108,109]. The suitability of GBS markers in developing genomic selection models has been verified in the complex wheat genome with the prediction for yield and other agronomic traits [108]. In rice inbred lines, genomic prediction models outperformed prediction based on pedigree records alone for three traits, i.e., grain yield, plant height, and flowering time [110]. Meanwhile, genomic selection has been recognized as an excellent tool to estimate genomic breeding values and is widely used in crop breeding [111].

Even if genetic stocks required for GWS and genomic prediction do not contain unique genetic novelty and, therefore, do not merit long-term conservation, they constitute important assets for genomic research. They could be kept together with specific breeding populations in the active collection under medium-term storage conditions to support future research and breeding needs [77]. Such specific breeding or discovery populations created through introgressiomics [112] would include collections of recombinant inbred lines (RILs), backcross introgression lines (BILs), chromosome segment substitution lines (CSSLs), multiparent advanced generation intercross (MAGIC) lines, nested association mapping (NAM) populations [113] and other training populations developed to represent different breeding pools [77]. It should be noted that many of these genetic stocks are very difficult to regenerate and that it might require the involvement of the breeders concerned to assist in such efforts.

With the current advances in NGS and GBS, the growing volume of fully annotated genomes, and knowledge of candidate genes, genotypic accession-level data is no longer of major concern [76]. The lack of high-quality phenotypic data is currently the major bottleneck for functional genomics and the efficient exploitation and use of germplasm accessions in modern breeding. With the advances in high-throughput phenotyping (HTP), there should now be a major focus on phenomics in crop collections to complement genotyping data. Phenotyping is an expensive yet indispensable component of plant research and crop improvement programs that helps to understand the genetic basis of traits and the interaction between genotypes and the environment [41]. Phenomics aims at bridging the gap between genomics, plant function, and agricultural traits [114]. While "forward phenomics" uses phenotyping tools to "sieve" collections of germplasm for valuable traits, such as yield components, biotic or abiotic stresses, "reverse phenomics" dissects those traits to reveal underlying mechanisms, such as biochemical or biophysical processes and ultimately the gene(s) regulating those processes [114].

Genebank phenomics is a novel approach in modern gene banking, and Nguyen and Norton [41] shed light on new HTP methods that enable capturing traits during seed regeneration events. One of the valuable features of HTP is that multiple sensors can be deployed at the same time to simultaneously and non-destructively capture several independent observations, which would not be possible through manual observations and measurements as practiced until recently in most gene banks. HTP will allow for more targeted prioritization of accessions from large crop collections for further downstream studies and identification of traits of interest for breeding. Seed phenomics has also been shown to aid genomic prediction for seed traits in barley breeding lines [115]. A convention on the minimum information about plant phenotyping experiments (MIAPPE) has been recommended by the plant phenomics community to ensure easy and correct interpretation, assessment, review, and reproducibility of published data [116].

However, the majority of gene banks might not be able to afford the investment needed for setting up and operating a phenotyping platform. Imaging costs include the imaging hardware, the cost of the vector (e.g., manual measurements, drones, hand-held or automated/robotized ground vehicles), and associated software/pipelines for data capture, storage, organization, and analysis [117]. The latter may represent 30–200% of the cost of image capture, a considerable cost factor of phenotyping, and might best be achieved through research consortia.

Nguyen and Norton [41] proposed a strategic phenomics approach to benefit the management of gene bank collections and to enhance the value and use of PGRFA. If possible, seed regeneration blocks should be replicated with a reasonable number of individuals to facilitate statistical analysis and ensure that the sample size is sufficient to maintain the genetic diversity and integrity of accessions. Using HTP from routine seed regeneration events over subsequent years, an enormous volume of morphological, agronomic, physiological, and environmental data [118] can be collected simultaneously. As most quantitative traits, such as grain yield, cannot be assessed in small regeneration plots, the measurement of secondary correlated traits such as early vigor, height, canopy properties, and biomass during the growth phase may serve as indirect indicators of grain yield and can be used together with GBS data for phenomic and genomic selection from diverse landrace accessions [119–121]. The described strategic gene bank phenomics approach has been implemented, for example, in the Australian grains gene bank (AGG) using different HTP platforms [41].

With the advances in biotechnology, the term "synthetic biology" has been created, which may be described as combining functional elements in novel configurations to modify existing properties or to create new ones [122]. Synthetic biology comprises a variety of techniques ranging from systems biology, metabolic engineering ("Golden Rice"), protein engineering, and genetic engineering and is a topic of regulatory concern regarding the biosafety of new products that could potentially fall into the category of living modified organisms (LMO). In addition, the use of digital sequence information (DSI) derived from germplasm is of concern regarding the third objective of the CBD on access to genetic resources and benefit-sharing and the recently concluded Nagoya Protocol [123]. These regulatory concerns are discussed further below in Section 4.4.

Synthetic biology encompasses genome editing and the respective enabling tools such as the "CRISPR" (clustered regularly interspaced short palindromic repeats) technology [124], and applications of CRISPR such as organisms containing engineered gene drives, a genetic strategy to control populations of disease-vectoring insects [125]. Furthermore, synthetic biology also allows de novo domestication of species, such as wild *Solanum pimpinellifolium* with enhanced fruit size, number, and nutritional value of the fruits [126], and multiplex editing, i.e., the simultaneous targeting of several genes with a single molecular construct [126,127].

The discovery of sequence-specific nucleases, including zinc-finger nucleases (ZFNs), transcription activator-like effector nucleases (TALEN), and the CRISPR-Cas system [128], enabled targeted genome editing in a precise and predictable manner in transformable plants by inducing a DNA double-strand break (DSB) at a target site [129]. Thereafter, either the non-homologous end joining (NHEJ) pathway or the donor-dependent homology-directed repair (HDR) pathway repairs the DSB, thereby introducing genetic changes [128,130], which might take the form of gene knockouts or gene replacements.

Gene knockouts are useful to eliminate genes that are detrimental to food quality or that confer susceptibility to plant pathogens [128]. For example, CRISPR/Cas-9 targeted mutagenesis has been used to knock out the powdery mildew susceptibility gene *PMR4* in tomatoes, resulting in enhanced resistance against this pathogen [131]. Wang et al. [132] used both TALEN and CRISPR/Cas9 technologies to target the genes of the mildew-resistance locus (MLO) in hexaploid bread wheat and successfully knocked out all three

MLO homoeo-alleles resulting in heritable, broad-spectrum resistance to powdery mildew. The CRISPR–Cas9 system has also shown the potential to directly target plant-infecting gemini viruses by inhibiting virus replication and, thus, enhancing plant resistance to those virus diseases [127,133,134].

While the NHEJ pathway for DSB repair is error-prone and the HDR pathway has a low editing efficiency, newly developed precise CRISPR-Cas technologies rely on deaminase-mediated base editing and reverse transcriptase-mediated prime editing that do not induce DSB formation and do not require donor DNA [130]. These newly developed technologies allow precise nucleotide sequence editing, are more efficient than HDR in plant genome modifications, and show great promise for rapid plant improvement. CRISPR-Cas technologies can work alone or can be combined with conventional breeding methods, thus, accelerating the breeding progress. This has been demonstrated for haploid induction in wheat, maize, and rice, for generating male sterile lines (wheat, tomato), inducing apomixis for fixation of hybrid vigor, for overcoming and restoring incompatibility, and crosses among distant gene pools [130].

The value of underutilized species and wild food plants for food and nutrition security and crop diversification aiming at more sustainable production systems has been demonstrated (see, for example, the work of [135,136]). CRISPR-Cas technologies, with their capacity for precise genome editing, could be used to accelerate the domestication and breeding process of such underutilized crops. For example, the wild tomato species *Solanum pimpinellifolium* shows salt tolerance as well as resistance to fungal and bacterial pathogens [101,137]. Using a multiplex CRISPR–Cas9 strategy to edit genes related to unsatisfactory traits in the wild form such as day-length sensitivity, shoot architecture, flower and fruit production, and nutrient content, Li et al. [138] were able to accelerate the domestication process of *S. pimpinellifolium* without compromising its abiotic and biotic stress tolerance traits. Similarly, the orphan solanaceous crop ground cherry, also called husk tomato (*Physalis pruinosa*), was partially domesticated by mutating orthologous domestication genes of tomato, resulting in plants that were shorter and had more flowers and larger fruits [139]. This is clear evidence that knowledge from model crops such as tomatoes can be used to edit genes to improve agronomic traits of distantly related underutilized crops.

According to Gao [140], the use of CRISPR technology in plant breeding could simply be considered as "a new breeding method that can produce identical results to conventional methods in a much more predictable, faster and even cheaper manner". Seeing genome editing with CRISPR technology in such a way could eventually help overcome the current prohibition of using this technology in the European Union, where the resulting products are still defined as "GMOs" even if no foreign DNA is introduced.

4.3. Specialized Databases, Portals and Networks for Genomics and Phenomics Data Related to Plant Agrobiodiversity

DNA or digital sequences in themselves are of no real value in the absence of information about the samples they were derived from [141]. It is crucial that a collection of plant DNA extracts is intricately linked to the original plant material (and associated information) from which the genomic DNA was derived. To manage the huge amounts of genetic and phenotypic data and make this and other valuable information pertaining to germplasm accessions available for users around the globe, a cooperative platform for data collection, analysis and sharing is required [41,42,142]. Information networks need to be unified, globally accessible, and updated as new research results become available [143]. For this to succeed, the gene bank community will need to link with other information specialists to build a truly global information system with a searchable interface [42]. With the Plant Treaty's global information system (GLIS), introducing digital object identifiers (DOIs) for gene bank material [60], a first step has been made in that direction. The DOIs assigned to germplasm accessions allow the storage of omics data in specialized databases without losing the link to the original accession from where it was derived from.

Several components and international initiatives already exist to develop different aspects of the required information infrastructure and are listed below. The first three are widely used in the gene bank and germplasm user community, while other listed resources try to better connect the user community with gene banks or with phenomics and genomics data portals.

- Genesys, a global platform on PGRFA with free online search engines, provides access to passport and characterization data on accessions conserved in gene banks worldwide [144];
- GRIN-Global [145], the global germplasm resource information network, provides a scalable version of the USDA-ARS Germplasm Resource Information Network (GRIN) that is suitable for use by any interested gene bank around the world;
- EURISCO [146], the European plant genetic resources search catalog, receives data from the European National Inventories (NI) and provides accession-level information of PGR conserved in European gene banks or other collections;
- The DivSeek network, founded in 2012, aims at catalyzing the advanced conservation, management, and traceability of PGRFA through a collaborative network of gene banks, breeders, plant and crop scientists, and database and computational experts [49]. To achieve the goal of value addition to germplasm conserved in gene banks, DivSeek has assembled three working groups that are focusing on genomics, phenomics, and policy;
- The Breeding API (BrAPI), an interface for exchanging plant phenotype and genotype data between crop breeding applications [147];
- The Research Data Alliance (RDA), which aims at enabling data sharing, exchange, and interoperability [148];
- The Global Open Data for Agriculture and Nutrition (GODAN) with the objective of making agricultural and nutritional data available, accessible, usable, and unrestricted [149];
- The Global Biodiversity Information Facility (GBIF) [150] was established in 2001 in Copenhagen, Denmark, based on a recommendation from the Organization for Economic Cooperation and Development (OECD) Global Science Forum as an international mechanism to promote standardization and aggregation of biodiversity data and (updated) information and make it accessible worldwide. As of 8 May 2021, GBIF had registered an amazing 1.7 billion occurrence records. However, despite this impressive number, only about 21% of preserved collections are digitally accessible via GBIF [151];
- The Global Genome Biodiversity Network (GGBN) [152], created in 2011 [153], links through its data portal globally distributed biodiversity databases of genomic samples, ensures easy access to DNA and/or tissue samples, and bridges the gap between biodiversity repositories, sequence databases and research results [154]. Within GGBN, a pilot project called GGI-Gardens focuses on the approximately A total of 460 vascular plant families [153]. Under CGBN, access is governed through standard material transfer agreements in compliance with regulations of the Convention on Biological Diversity [155], the Nagoya Protocol on Access and Benefit Sharing [156], and the Convention on International Trade in Endangered Species of Wild Flora and Fauna (CITES) [157];
- The International Nucleotide Sequence Database Collaboration (INSDC) [158], comprising the DNA DataBank of Japan (DDBJ), the European Nucleotide Archive (ENA), and GenBank at the National Center for Biotechnology Information (NCBI), an annotated collection of all publicly available DNA sequences. It is an archival database that rarely provides updates about specimen or locality data;
- The International Plant Phenotyping Network (IPPN) [159] represents plant phenotyping centers globally, runs multidisciplinary working groups, and facilitates the sharing of up-to-date information about new HTP infrastructures and methodolo-

gies for various crop phenotypes [41,160]. Several regional and national partners are associated with IPPN, two of which are mentioned below;

- The North American Plant Phenotyping Network (NAPPN) [161] brings together scientists in the evolving area of plant phenomics and is a regional partner of IPPN;
- EMPHASIS, also a regional partner of IPPN, enables researchers to use facilities, resources, and services for plant phenotyping across Europe [162].

There are also crop-specific consortia such as the International Rice Informatics Consortium (IRIC) [163], the International Wheat Improvement Network (IWIN) [164], the APSA/WorldVeg Vegetable Breeding Consortium [165], and many others.

Taxonomic and evolutionary studies make increasingly use of information provided through data portals or virtual collections where specimen data and images are made available through the Internet. To annotate virtual specimens, online annotation tools are required. A generic online annotation system called *AnnoSys* [166] has recently been developed to access collection data from both conventional web resources or the Biological Collection Access Service (BioCASe) and accepts XML-based data standards such as Access to Biological Collection Data (ABCD) [167] or DarwinCore [168] for data exchange [169]. GBIF and other biodiversity portals are already integrating AnnoSys. Filter types for specimen records in queries and for notifications include: family, genus, species, collector name, collector's number, country, institution code, collection code, catalog number, identified by, and annotator.

4.4. The Political Dimension of Digital Sequence Information (DSI) Sharing

Progress in life sciences, including health, biodiversity protection, and working toward reaching the sustainable development goals is relying on open access to sequence data provided by public sequence databases. A clear example is the current pandemic caused by SARS-CoV-2. Without the rapid sharing of pathogen genetic resources and digital sequence information (DSI) (called genetic sequence data (GSD)) by the World Health Organization (WHO), it would not have been possible to create effective vaccines within a truly short timeframe of under a year. While rapid progress by the scientific community relies on openness and public availability of genetic sequences, fears have been expressed regarding the increasing ease with which genetic material can be transformed into digital information, transmitted, reproduced, and manipulated through advances in sequencing technologies, genome editing and synthetic biology [170,171]. Progress in synthetic biology might soon enable de novo biological design [170], thus, confirming fears of the dematerialization of genetic resources, i.e., making physical access superfluous and in that way threatening the principles of ABS as established under the International Treaty and the Nagoya Protocol of the CBD [171].

Free-for-all access to sequence information (and associated germplasm) is considered by biodiverse countries to be mainly beneficial for user countries and the biotechnology industry and is seen as counterproductive for provider countries, their local communities, and indigenous people who are the custodians of plant agrobiodiversity and who will not be able to benefit if access to DSI is not subject to ABS regulations of prior informed consent (PIC) and mutually agreed terms (MAT) under the Nagoya Protocol of the CBD [172]. Discussions on the current and future access to digital sequence information are currently ongoing under the Plant Treaty, CBD, Nagoya Protocol, and the multilateral Prepared Influenza Preparedness (PIP) Framework as reviewed by Lawson et al. [173]. The United Nations Convention on the Law of the Sea (UNCLOS), which governs international waters and the deep sea, is also developing a new multilateral treaty with the aim to enhance the conservation and sustainable use of marine biological diversity in areas beyond national jurisdiction [174]. Best practices have been outlined to regulate access to marine genetic resources and sequencing data while sharing the benefits derived from such access, a process that is meant to support science and society.

For the time being, DSI is still used as a placeholder, lacking a precise and generally agreed definition [175]. According to Houssen et al. [176], the term DSI may comprise four groups of information:

(i) Narrow, covering DNA and RNA only;
(ii) Intermediate, including DNA, RNA, and proteins;
(iii) Intermediate, including DNA, RNA, proteins, and metabolites;
(iv) Broad, including DNA, RNA, protein, metabolites, and traditional knowledge, ecological interactions, etc.

The unsolved definition of DSI is of major concern for gene banks as they share germplasm with associated accession-level information. Should a broad definition of DSI (group iv of Houssen et al. [176]) be adopted, this would seriously affect the disclosure of passport and descriptive data of accessions, and in consequence, would have a major impact on the distribution of germplasm to users [15]. A narrower definition of DSI with restricted access to that information would be less harmful. Should more countries include DSI in the national ABS legislation, gene banks may no longer wish to conserve material from those countries due to increased complexity of germplasm handling and distribution. As has been shown by the current use of an SMTA under the MLS of the ITPGRFA for germplasm sharing, a multilateral approach to the DSI issue would be the best solution; hence, the incorporation of DSI into the SMTA of the ITPGRFA has been proposed to facilitate the smooth operation of gene banks including germplasm distribution in the future [15]. A further open question is whether ABS regulation would only apply to DSI acquired after entry into force of this regulation or would be applied retroactively.

While the next UN Biodiversity Conference of the Parties (CBD COP15) to be held in October 2021 in Kunming, China will discuss the DSI topic, some countries went ahead and have already included DSI interpretations in their national ABS legislation, thus creating legal uncertainties for scientists accessing publicly available DSI. In fact, 15 countries have already adopted legislation on DSI, and 18 more are planning to do so [171]. If a country decides to include DSI in its national ABS regulation, access to and use of DSI derived from their genetic resources may not be free anymore. This means that PIC and MAT would apply to the use of DSI as well in case access to genetic resources is governed by the Nagoya Protocol.

To solve this contentious issue, Lawson et al. [173] proposed two options: (i) a risk framework matrix for valuing information as part of the ABS transaction by attributing an estimated worth to a particular kind of information; (ii) a charge, tax, or levy that would allow externalizing the costs so that information would remain available to be disclosed and exchanged in support of the scientific community. Based on the matrix, passport data on accessions would be considered as of low value, without restrictions (public domain data), while descriptive (phenotypic) data would be treated as restricted public access data, and sequence data would have a time limit restriction with the requirement for reporting the results obtained to the germplasm/DNA provider. The tax or levy might need to be paid by the party accessing the resources (similar to the PIP Framework) or as a levy on contracting parties, similar to the Norway seed sales tax under the MLS of the International Treaty. For 12 consecutive years, Norway has made an annual contribution to the benefit-sharing fund of the International Treaty, an amount that is equivalent to 0.1 per cent of the value of seed and plant material traded in agriculture in Norway every year [177]. The tax or levy option avoids the high ABS transaction costs required to negotiate the value of information in every single transaction and allows the scientific community to disclose and share the generated information freely [173].

The International Nucleotide Sequence Database Collaboration (INSDC) is the central foundation for global sequence information as it connects over 1700 scientific databases and platforms [178]. The INSDC provides the free core infrastructure for DSI deposition, preservation, and global dissemination as part of a scientific collaboration between the European Molecular Biology Laboratory (EMBL; inter-governmental treaty organization), the National Center for Biotechnology Information (NCBI); USA), which is hosting GenBank, and the DNA Databank of Japan (DDBJ) [179]. Rhoden and Scholz [171] noted that tracking and tracing the movement of nucleotide sequence data (NSD) in GenBank, the largest public database platform, is challenging and that scientists in every country of the world are accessing this platform for their work. Therefore, any financial or administrative burden for accessing NSD will affect all scientists worldwide and limit their ability to undertake research and collaborate. A recently published white paper formulated five different policy options under which DSI could be governed, generate revenues for benefit sharing while preserving the current open access system for scientific discovery and publication [178]. Similar options have been proposed by Oldham [141].

Scholz et al. [178] noted historical parallels between CGIAR gene banks that became part of the MLS of the International Treaty about 25 years after the creation of the CGIAR institutes and the INSDC, established in the early 1980s, which now has assembled DSI from every country, continent, ocean, and region in the world and its databases are accessed by users in every country in the world. There is hope that a multilateral perspective, or even a multilateral mechanism that covers multiple international organizations/agreements, such as the CBD, the International Treaty with its MLS, WHO with its PIP Framework, and UNICLOS, could be a practical way forward to solve the contentious DSI issue. From the foregoing, it becomes apparent that the scientific community requires a multilateral, universal framework for accessing DSI if it is to thrive and to contribute to solving current and future global challenges, similar to the situation of PGRFA.

Despite the growing concern of reducing germplasm collections to dematerialized and digitized genomic sequences, the conservation of physical specimens (seed, tissue, living plants) will retain value for future research beyond the DNA code, although we may not be able to anticipate and specify those values right now [180].

5. Strengths, Limitations and Opportunities of the Existing Regulatory Framework, the International Network of Ex Situ Base Collections and the Routine Genebank Operations for Genebanks to Effectively Participate in and Contribute to Regional or Global Long-Term Conservation Efforts

Having described the global long-term conservation and exchange/use system above, in this section, we will take a closer look at the existing policy framework in which the global system is embedded (for a detailed description of the policy framework, please see the previous paper [3]). We provide an assessment of the strengths and weaknesses as well as of possible limitations of this framework that exist from the perspective of national, regional, and global gene banks regarding their ability to participate in the regional/global system. We will conduct a similar assessment of possible strengths, weaknesses, and limitations that characterize the current international network of ex situ base collections with a view on the possible participation of gene banks in that network. In the second part of this section, we will critically assess the strengths, weaknesses, and limitations of routine gene bank operations that have an impact on the effective and efficient participation of national, regional, and global gene banks in the global system.

5.1. Regulatory Strategic Framework

The Global System for the Conservation and Sustainable Use of Plant Genetic Resources is a system that evolved in and is being managed and coordinated by FAO, under the oversight of the commission, with the aim to ensure safe conservation and to promote the availability and sustainable use of PGRFA. The global system was also an element of the International Undertaking on Plant Genetic Resources. The latter was adopted in 1983 and entirely devoted to the conservation and facilitation of use [181]. More details on this global system are provided in Engels and Ebert [3]. This global system is largely based on the national PGRFA programs around the world, the botanic gardens maintaining PGRFA, the gene banks of the regional and international research centers of the CGIAR and AIRCA (Association of International Research and Development Centers for Agriculture, among which the World Vegetable Center and ICBA [International Center for Biosaline Agriculture] maintain considerable germplasm resources), the regional PGRFA networks, the global crop networks and other loosely related institutions that are concerned with the conservation and research of biodiversity.

As in any global "project", it is important to have a well-defined strategy that determines the scope (the "what"), procedures (the "how"), timeframe (the "when"), the participants (the "whom") and the rules (the "commitment", including operative principles, financial, infrastructural, and human resources). The most important elements of the policy framework for the international network of ex situ base collections address: a. the strategy; b. the global plan of action; c. monitoring and reporting; d. the international network of ex situ base collections as the implementing agencies; e. the International Treaty and its MLS as an oversight body; f. the CGRFA that provides oversight, in close consultation with the treaty.

Components of the global system listed below contribute in their respective capacity to such a strategy, including the listed "political" elements that underpin the actions.

5.1.1. The Global Strategy for Plant Conservation (GSPC)

The GSPC was adopted by the Convention on Biological Diversity (CBD) in 2002. It provides a framework for the policies and actions required to prevent the loss of plant diversity and promote plant conservation. Target 9 of the GSPC is closely linked to Aichi Target 13 [182], which addresses the key objective relating to genetic diversity and provides a clear entry point to the work carried out in the framework of FAO, its Global Plan of Action for Plant Genetic Resources for Food and Agriculture (see below) and the work of the CBD. While crop diversity is well represented in crop gene banks, crop wild relatives (CWRs) and other socio-economically important species are significantly underrepresented. In this respect, botanic gardens and other plant conservation organizations play an important role. The FAO Commission reports to the CBD on progress achieved with the implementation of the GSPC and participates in the post-2020 strategy. The GSPC addresses PGRFA through the implementation of the strategic plan of the FAO Commission. As such, it is at a rather high level for actual gene bank operations and impacts mainly at setting priorities for actions through the GPA, as mentioned below. The FAO agreed on a strategic plan for the Commission on Genetic Resources for Food and Agriculture (2019–2027), providing high-level strategic guidance to the commission and its member states [183]. There are no specific points to mention that directly impact our assessment as this strategic plan provides merely a framework of its components.

5.1.2. Global Crop Conservation Strategies

For 26 food crops, global crop conservation strategies have been developed over the past 20 years. These global crop conservation strategies provide a very useful framework for prioritizing and planning collecting activities for the most important food crops, based on existing collections, on gaps in terms of geographical areas that are underrepresented in collections, for conservation approaches followed, including research and what needs to be performed based on latest knowledge and information to help plan and prioritize actions to ensure the long-term conservation and availability of plant genetic resources for food and agriculture. The evolution toward a rational global system necessitates the identification of the location and status of unique genetic diversity and the outlining of the processes by which this diversity can be most efficiently and effectively conserved and used for the benefit of the global community [184]. The strategies cover not only the crop species but also the related crop wild relatives and treat the different types of germplasm to some extent and make mention of in situ and on-farm conservation activities. They do set priorities for recommended actions and thus, are a particularly useful and important starting point for gene banks to consult when planning new activities that would include one or more of the food crops covered by these strategies.

The most common constraint in developing such crop strategies is the lack of sufficient accession- and collection-level information to categorize collections by importance, to identify duplicates, and to fulfill other tasks needed for the ideal strategy. In addition, the use of different taxonomic systems for the same crop and misidentification of accessions are reported problems to generate well-informed crop strategies [184]. Another constraint is obviously that only a limited number of crops are covered, that only information has been covered that was readily available to the crop experts, and thus, that the holdings of many (smaller) gene banks in several countries are not covered. At the same time, the strategies do provide ample information on gene banks holding major collections, existing networks, and other useful aspects. The global crop strategies represent a major undertaking in the field of plant genetic resources conservation, mobilizing experts to collaboratively design plans for more efficient and effective conservation and use of crop diversity ex situ. Supporting the development and updating of crop strategies allows moving forward toward a more efficient and useful global system of (long-term) conservation and use of these invaluable plant genetic resources [184]. The Crop Trust provides access to the existing 26 strategies and is in the process of revising and updating several of the existing strategies and developing new ones for 10 more crops/crop groups [185].

5.2. Global Plan of Action (GPA) for the Conservation and Sustainable Use of PGRFA (GPAII)

The GPA is one of the important instruments for countries and gene banks to obtain relevant and useful information about general conservation and use aspects, identified global priorities, reported needs and opportunities, etc., for 18 global priority activities. The GPA is the most important reference document for national, regional, and global efforts to conserve and use PGRFA sustainably and to share the benefits that derive from their use in an equitable and fair way. The strength of the GPA is that it squarely addresses all PGRFA, in a general sense, in situ and ex situ conservation, sustainable use, and building institutional and human capacity. It also provides guidance in linking conservation and use in achieving more sustainable production through breeding and broadening crop diversity. Strengthening of national programs, the promotion, and strengthening of PGRFA networks, constructing comprehensive information systems, and strengthening public awareness of the importance of PGRFA are important topics addressed. Furthermore, it was a very consultative process that had led to the formulation and adoption of the plan, involving experts and specialists for the main activity areas. The updating of GPAI was based on the first state of the world's PGRFA (SOW) report [186]. The GPA also anticipates developments and trends in agriculture that might impact the conservation and use of PGRFA, e.g., the industrialization of agriculture, low input agriculture, the globalization of markets, including the seed sector, the increasing use of genetically modified varieties, the strategic use of PGRFA to better cope with climate change, major advances in science and technology, the advances in molecular and genomic methods as well as policy developments [187]. For more details, see also Section 2.5 of [3].

A possible weakness of SOWs to be guiding documents to gene banks and countries is that each (updated) GPA covers a relatively long period of 10 and more years, that it is a complex and rather bureaucratic process, and that its implementation is "voluntary", with limited opportunities for monitoring its implementation. The lack of financial resources is one of the constraints to implement the GPA to its full extent.

The opportunities the GPA offers to governments, research, and conservation communities, as well as to the users of PGRFA, are directly related to the extent of its implementation. The more the GPA is used as a priority-setting mechanism, the better it will be implemented, and the more efficient and effective the conservation and sustainable use efforts become worldwide.

Reports on the State of the World's Plant Genetic Resources for Food and Agriculture (SOWs) [186,188] present the outcome of periodic assessments by FAO of the state of the world's PGRFA, thereby facilitating analyses of changing gaps and needs and, thus, contributing to the process of updating the "rolling" GPA. To facilitate this monitoring process, and thus the preparation of the SOW report, an online reporting tool of WIEWS has been made available [189] for the preparation of SOW III, scheduled for 2023. The monitoring for SOW II was conducted through a participatory process with the appointed national focal points in the member states that were guided to formulate a country report through a nationwide consultation and to use tools to guide this process. Ninety-one country reports have been prepared between 2014 and 2017 [190]. These country reports are particularly useful and important documents as they contain ample and typically detailed information on the state of PGRFA conservation and use in the respective country. The preparation process of the SOW report is possibly the most comprehensive global assessment on the state of PGRFA, in countries, in regions (in the past through regional consultation meetings), and globally and thus, the publications this process generates are indispensable resources for planning activities, for prioritizing efforts at the various levels and to obtain a better overview of the current situation of individual crops, crop gene pools as well as of countries. It should be noted that the individual country reports vary greatly in quality, but all follow a defined structure and can be (relatively) easily consulted. A major drawback might be that almost half of the countries did not produce a country report, and thus, the coverage of the country reports collectively might well be incomplete on aspects such as distribution of species, their conservation, and use status, etc.

5.3. International Treaty on Plant Genetic Resources for Food and Agriculture (ITPGRFA) and Its Multilateral System (MLS)

As the successor of the International Undertaking, the International Treaty with its MLS [191] is the most significant policy instrument that we have for PGRFA. For details on this legal agreement, operational since 2006, see the work of [3]. In November 2020, the treaty had 148 contracting parties, including the European Union. The treaty aims at establishing a global system to provide farmers, plant breeders, and scientists with access to plant genetic materials. Its main provision is the establishment and operation of the MLS, which puts 64 of the most important crops into an easily accessible global pool of genetic resources that are freely available to potential users in the treaty's ratifying nations for specified uses. With the creation of the MLS and placing genetic resources of the above-mentioned 64 crops and species into the public domain, it was intended to make access to PGRFA easy and simple.

However, it should be mentioned that this access is somewhat "bureaucratic" as it is regulated by a relatively complex standard material transfer agreement (SMTA) that defines several conditions and restrictions, including for the use of such materials. Furthermore, despite the impressive number of contracting party states, many of them did not contribute materials to the MLS, thus limiting the total genetic diversity in the pool. Until mid-2019, a total of 58 notification letters designating PGRFA to the MLS were submitted to the secretariat of the treaty by 44 countries and six organizations. Furthermore, 18 international and regional centers have included their ex situ collections in the MLS [192]. In addition, the rather restricted list of 35 crops and 29 grass and forage species that have been included in Annex 1 of the treaty is a significant limiting factor. For instance, only a small number of vegetables and neglected and underutilized species has been included in the list, no commodity crops, etc. Consequently, for many countries, the current MLS is of limited interest. Yet, another limiting factor of the MLS is the hesitation of the private sector to use materials from the MLS, especially triggered by the benefit-sharing conditions. Because of the difficulty to ease the access conditions through negotiations at the Governing Body meetings, many of the bigger breeding companies have established their own private collections, partly officially through the MLS and partly through their own initiatives with gene banks and countries [193].

The strong link between the Global Crop Diversity Trust and the treaty is a major strength of the MLS. The Crop Trust has fully embraced supporting the major food crops and their wild relatives that are included in Annex 1 as their priority crops, but also realizes that other important and minor food crops would further strengthen the MLS and started a project to develop global crop strategies for gene pools that have not yet been covered so far. Since the Crop Trust is a major partner in the international network of ex situ base collections and as Genesys (initiated and operated by the Crop Trust) is one of the main information resources on global germplasm collections at the accession level, there is a direct link between effective and efficient long-term conservation of major food crops' gene pools and the MLS.

5.4. The FAO Commission on Genetic Resources for Food and Agriculture (CGRFA)

The CGRFA provides a platform for all the PGRFA as well as for animal, aquatic, and forest genetic resources [194]. For details on the historical achievements of the Commission, see Engels and Ebert [3]. With the establishment of the International Treaty, replacing the International Undertaking, and the inclusion of other sectors of genetic resources, in particular forest, animal, and aquatic resources for food and agriculture, the role of the commission changed. However, as most of the food and agricultural crop species are not covered by the treaty, the commission continues to be of direct relevance as an oversight body for most of the PGRFA. The relevant provisions of the commission for the international network of base collections and global conservation system have been mentioned by Engels and Ebert [3], in particular, the Genebank Standards, the rolling GPA, WIEWS, and the SOW and country reports. In terms of global coverage of countries, in July 2014, 178 countries and the European Union were members of the commission [195].

The strength of the commission is that it has a broad scope and covers all genetic resources of relevance to food and agriculture, with a clear focus on food security and sustainability issues. Another strength is that most of the countries are members of the commission. A possible weakness is a fact that many member states do not place the conservation and sustainable use of PGRFA high on their agenda and that the commission as a forum for discussion has no possibilities for sanctioning members that do not implement agreed activities or do not adhere to agreed standards. Furthermore, the commission has to share the oversight over the international network of base collections with the treaty, and thus, a clear focus is somewhat lost. When looking at the global conservation system, it should also be noted that other elements, organizations, operations, and activities that directly or indirectly contribute to international collaboration on conservation and use of PGRFA are currently outside the global system, or their contributions are not fully appreciated, e.g., many NGOs, farmers' organizations, botanic garden networks, and others. The global system does not provide many opportunities to strengthen linkages and connections between the existing components of the system. The global dimension of in situ conservation is not well developed. In addition, the financial arrangements that support the system are underdeveloped [196].

The following websites provide relevant information regarding the different elements of the international network of ex situ base collections:

- WIEWS—World Information and Early Warning System on Plant Genetic Resources for Food and Agriculture (http://www.fao.org/wiews/en/) [197];
- CGIAR Genebank Platform (https://www.genebanks.org/the-platform/) [198];
- Multilateral System of the International Treaty (http://www.fao.org/plant-treaty/areas-of-work/the-multilateral-system/overview/en/) [191];
- Svalbard Global Seed Vault (https://www.seedvault.no/) [199];
- Global Safety Backup Cryopreservation Facility, still being developed (http://www.fao.org/3/CA1371EN/ca1371en.pdf) [200]. The Global Safety Backup Cryopreservation Facility is an initiative that has been developed by many key players in the field of cryopreservation and coordinated by the CGIAR gene banks, as a parallel to the SGSV in Norway, to provide a global security backup for cryopreserved accessions. This mechanism is awaiting political acceptance;
- The European Cooperative Program on Genetic Resources (ECPGR; https://www.ecpgr.cgiar.org/about/overview) [201];
- GRIN-Global (https://www.nal.usda.gov/grin-global) [145] is based on the plant database of the Germplasm Resources Information Network (GRIN).

5.5. International Network of Ex Situ Base Collections

In the following, we will focus on the base collections maintained by the CGIAR centers, CATIE, and the Pacific community, as they have concluded agreements with the secretariat of the International Treaty and thus, formally included their designated germplasm in the MLS. In addition to these international centers, also the national collections in countries such as the USA, South Korea, Japan, Germany, The Netherlands, and some others are important partners in the MLS. The 5th largest international gene bank in the world, dedicated to the conservation of global and indigenous vegetable germplasm and maintained by the World Vegetable Center in Taiwan [69], is distributing germplasm globally under the SMTA, thus applying the guidelines of the International Treaty and the MLS but is yet to officially sign an agreement with the Treaty Secretariat.

The CGIAR collections, as well as those maintained by the World Vegetable Center and CATIE, are to a large extent collections with global coverage, have a strong focus on (major) crop gene pools, are relatively large, well-managed, and form the backbone of the MLS of the International Treaty. Furthermore, most of the centers that included germplasm in the MLS also have active breeding programs with a global focus and actively distribute breeding materials as well as newly bred varieties worldwide. The integration of the gene bank and its collections and the breeding cum research activities into a research institute has many advantages and has proven to work well. It should be noted that materials of other (minor) crops that are maintained by international and regional centers (Article 15 species) are also included in the MLS and thus available to all users through a standard material transfer agreement (SMTA).

The maintenance of global germplasm collections, usually with a suitable representation of the total genetic diversity, and targeted (mandate) crop breeding activities "under one roof" are important strengths that the international centers offer. Furthermore, the global coordinating role that these centers play with respect to the conservation of their respective mandate crops, in many cases following the elaborated activities in the respective global crop strategies, and the coordinated management of these collections with respect to gene bank standards, policies on access and distribution, information and research, through the Genebank Platform should be mentioned as another significant strength. In addition, the assurance of stable funding over the long-term, thanks to arrangements through the Endowment Fund of the Crop Trust, combined with the provision of training and advice to collaborating scientists from developing countries, this network of international centers is a true pillar of the international network [188].

The current international network of ex situ collections also has some obvious weaknesses. Possibly the most important one is the strong focus on the major food crops, i.e., those listed in Annex 1, and the relative neglect of minor (including global and regional) crops, such as vegetables and the locally adapted NUS or orphan crops. Another weakness is the, in general, less active involvement of smaller developing countries with collections consisting of mostly local minor crops, frequently enriched with breeding materials and varieties of the international research centers. As many of the centers of diversity are situated in developing countries, and since these are usually less well collected (possibly except for the mandate crops of the international research centers), there is likely also a bias toward the more accessible and bigger developing countries. Yet another factor that relates to the previous points is the lack of functional breeding programs, possibly also other research programs worldwide for (most of) these marginal crops, and thus, the countries that are "left out of the system" do hardly benefit from the genetic resources in their territories. However, more recently, some donors have recognized this aspect and initiated regional or continental programs or projects to address, for instance, the genetic diversity of African orphan crops [202] or to build molecular research capacity as part of an agricultural transformation program [203].

The Governing Body of the treaty is well aware of the limitations of the restricted coverage of Annex 1 crops and species causes, and for a number of years, negotiations have been ongoing to overcome this shortcoming. An "Ad hoc open-ended working group to enhance the functioning of the multilateral system" has been meeting nine times, but no results could yet be achieved [204].

For the CGIAR centers, optimal functioning of the MLS is critically important to fulfill their mission. Unfortunately, some challenges have been faced since its inception. Since 2018, only one payment has been made to the Plant Treaty's benefit-sharing fund, in line with the mandatory monetary benefit-sharing conditions included in the SMTA that are triggered when breeding products derived from MLS germplasm materials are commercialized. Furthermore, many potential providers are demonstrating reluctance [16] to proactively provide access to plant genetic resources in the multilateral system until more money from commercial users is contributed to the benefit-sharing fund of the treaty. In addition, the treaty's relatively low profile in many countries makes it difficult to obtain political support to implement national measures. Moreover, some companies and universities have declined to take germplasm materials from the MLS because of difficulties with the SMTA. Finally, very few requests for CGIAR germplasm come from farmers or farmers' organizations, civil society organizations, or countries with small or no plant breeding programs for direct use [16]. In this context, it should be noted that direct use of materials from the MLS is not explicitly foreseen by the SMTA. In addition, other aspects that limit the transfer of germplasm from the MLS to users have been noted, including the uncertainties whether SMTAs are used when passing germplasm on to third party users, the lack of reporting evaluation results by recipients of the germplasm to the CGIAR centers, despite reminders. The same is the case with respect to feedback regarding improved lines that are being released by countries as cultivars and thus, no, or only limited value adding on the germplasm in the MLS takes place through information sharing [16].

Whereas through the establishment of the MLS, a global system is, in principle. Available for placing and keeping (including for the long-term) PGRFA important for food security, nutrition, and sustainable production practices in the public domain, thus making those freely available, there is a need to further strengthen the system. One important issue is the restrictive and very limited list of species included in Annex 1; it should rather be all PGRFA. Furthermore, the benefit-sharing arrangements are too bureaucratic and complex, and they should be replaced by a system that altruistically supports the financially poorer but genetic diversity-rich countries with their conservation efforts. A global system could play a key role in this, especially through the establishment of close linkages between conservation and use. The international network of ex situ base collections could be the central conservation component of such a system, ensuring the effective and efficient long-term conservation of unique, carefully selected, and representative samples that are being made readily available to users worldwide. The efficient management of the related information is another key aspect of such a rational global PGRFA system.

5.6. Observations on Strengths and Weaknesses of Routine Genebank Operations, and Opportunities to Facilitate Cooperation between Genebanks at Large to Contribute to the Global System

In this subsection, we will assess strengths and weaknesses for each of the key routine gene bank operations vis-à-vis the relevant aspects of the global conservation system and, where opportune, identify and describe possibilities for more active participation of individual gene banks in the global system. In order to provide a more concrete context, global data are reported, sometimes repeating earlier facts that might help to understand this context better; they stem largely from the comprehensive but somewhat outdated SOWII report [188], from the Genebank Platform [198], and have been updated with personal observations and experiences, whenever possible.

5.6.1. Exploration and Collecting; Prioritization

- Whereas for most of the major food crops, a large part of the genetic diversity is represented in collections, for many other crops, especially many neglected and underutilized species, and CWR, comprehensive collections still do not exist, and considerable gaps (genetically as well as geographically) remain to be filled [188];

- Crop conservation strategies are possibly the most powerful tools to set priorities where to collect what as they typically provide a global account of existing collections, their diversity coverage (as far as known), reported threat status, etc. For details, see Section 5.1.2. For understandable reasons, there has been a strong focus on major food crops, in particular orthodox seed-producing crops. However, of the 10 new crop gene pools, several vegetables, and neglected and underutilized species have been included;
- About 55% of accessions that are being conserved in gene banks globally, and for which the country of origin is known, has originated in the country where the collection is maintained [188]. The lack of information on the other 45% of accessions is a serious limiting factor in planning as well as developing convincing crop strategies to guide collecting priority setting;
- Since the mid-1980s, the average number of accessions collected annually decreased [188]. The number of newly acquired materials by the centers to be included in the international collections dropped down to the lower levels that characterized the mid-1990s to 2009 [47];
- Largely triggered by particularly the legal access and benefit-sharing issues, a significant shift from internationally organized toward national collecting missions has been observed since the mid-1990s. This has resulted in slower growth of new accessions that are globally being made available. Thus, increased technical and financial support to national and local programs will be indispensable to ensure that timely collecting will be undertaken;
- Priorities for collecting at the local and national level should be on local and threatened genetic resources, whenever possible, with support from national or international crop gene pool specialists to ensure the highest possible quality sampling and treatment of the genetic resources;
- Collaboration with partners from outside a given country can be critically important to obtain specific expertise (e.g., taxonomic knowledge of targeted taxa for collecting) in preparing and conducting this demanding and complex activity;
- Sharing collected materials with other gene banks can be important to ensure effective and efficient long-term conservation if specific expertise is required for the conservation, e.g., cryopreservation, need for specialized infrastructure, and possibly other aspects.

5.6.2. Conservation, General

- Of the total 7.4 million accessions, national government gene banks conserve about 6.6 million, 45% of which are held in only seven countries [188];
- Of the analyzed 6,998,760 accessions, 10% were CWR, 24% landraces, 11% breeding materials, 9% advanced cultivars and 46% others and/or no information available [188]. A more balanced overall representation of some germplasm types is desirable;
- Total number of reported accessions in approximately 1750 gene banks or germplasm collections is about 7.4 million; of these, approximately 2 million are unique [205], and about 4.6 million are Annex I crops [188];
- Percentages of duplication within and between collections have been estimated to be high; only about 25%–30% (or between 1.9 and 2.2 million) are distinct [188];
- Many crops and important collections remain inadequately safely duplicated; for vegetatively propagated species and species with recalcitrant seeds, the situation is worse [188];
- Total number of accessions notified to be included in the MLS through a notification letter is 532,545, and 1,256,680 have been reported on the treaty website [192]. The number of CGIAR accessions reported in Genesys having been included in the MLS as of May 2021 is 766,680 [192];
- With respect to the security of stored materials, there are two main areas of concern, i.e., the extent of safety duplication and regeneration backlogs [188];

- Whenever possible, locally occurring or cultivated and thus, likely adapted germplasm should be locally/nationally conserved, either in situ and/or on farms, and preferably a copy of the material be included in an international collection as well. It would be ideal if such locally adapted and ex situ conserved materials could be regenerated/multiplied under local conditions that would allow a high-quality performance with limited risks of losing genetic diversity due to constraining conditions in the field. Where possible and/or necessary, collaboration with national or regional, and ideally with international experts and gene banks, should be sought;
- Priority setting at the local and national level will logically be based on criteria that are best known at the national level, especially regarding the threat status of taxa occurring in nature or on-farm;
- Whereas local limitations for conservation can be manifold, reasons and needs for participation in global (and regional) conservation efforts might be important and justified in case-specific infrastructure is missing, specialized expertise is not available, or local capacity constraints for storage and/or management exist;
- Participation of countries, including their relevant local gene banks, in the global system through the respective national PGRFA program is strongly encouraged and recommended in order to add value to and increase the safety of the nationally conserved genetic resources. This can be achieved through international collaboration but also through more sustainable "bottom-up" approaches;
- As for exploration and collecting, global crop conservation strategies do provide an excellent tool for planning and participation in global (and regional) conservation activities.

5.6.3. Conservation—Linking in Situ and Ex Situ

- As already implied in a recent publication by Engels and Ebert [3], linking in situ (including on-farm) conservation with ex situ conservation might well be indispensable to combine the strengths of each of these approaches and to complement the efforts at the local level with those at the national and international levels. This would ensure that a maximum amount of genetic diversity is conserved in the most appropriate and effective way and that biological and cultural information is not lost inadvertently [188];
- As in situ and on-farm conservation is per definition an activity that can only take place "locally", it is a given that the coordinating national PGRFA program should provide the link between such in situ conservation efforts and complementing national, regional, or global ex situ activities;
- The important aspect of enabling effective and efficient (long-term) conservation applies predominantly to the local situation, i.e., there where the genetic materials occur in nature or on-farm and where also ex situ conservation should be applied. However, such complementary effort could also be applicable to situations where materials are being conserved "only" ex situ (and possibly under very constraining conditions for the germplasm itself) but where, for instance, a permanent evolution/adaptation of this material to changing conditions would be important;
- There is one more reason to actively promote collaboration between in situ and ex situ conservation efforts, and this concerns the facilitation of germplasm use. While direct access to populations or landraces conserved in situ is difficult, after collecting such material and storing it in a gene bank, access to those accessions by users can be targeted and routinely provided;
- Whenever genetic diversity occurring locally is being collected for conservation and use elsewhere, adequate arrangements should be made to ensure that benefits deriving from the use of such resources will be shared, in whatever form, with the local communities;
- There could be strong justification to combine the two conservation approaches in case materials are predominantly conserved in nature or on-farm, and a "strategic

representation" of the genetic diversity in the gene bank would be important to ensure easy and targeted access to materials;

- Direct and effective links between germplasm conserved ex situ using techniques such as in vitro storage or cryopreservation and the corresponding materials maintained on-farm (and/or in field gene banks) will be important in case genetic instability is critical and regular "refreshment" of the ex situ conserved materials is required;
- The establishment and operation of local "community gene banks" might well be considered to strengthen local conservation capacity and to facilitate collaboration with the national (and eventually international) level and thus, to empower/strengthen the ownership over the local resources by the local communities [206].

5.6.4. Germplasm Management, including Processing

- An important prerequisite for effective and efficient germplasm and gene bank management is the formulation of clear conservation and use objectives. Knowing why and what to conserve allows better planning and priority setting [14];
- Strong germplasm (and gene bank) management capacity is a prerequisite for efficient and effective long-term conservation. Many of the routine operations provide opportunities for increasing the longevity of conserved germplasm, of improving the quality of the management and thus of the materials, facilitating the use of the materials by adding value through characterization and evaluation, etc. Such improvements will make the gene banks more attractive to play a more prominent role in regional and global efforts, to access more easily additional funding through projects, etc., and thus, to become increasingly more attractive as a partner in collaborative research activities locally, nationally, and internationally;
- A well-coordinated national conservation effort [207] enables the country to be cost-efficient and effective with its national conservation activities, including in situ and ex situ, to make strategic decisions on assigning conservation responsibilities, strengthening the link between conservation and use, participating actively and effectively in regional and global initiatives, etc. [207];
- The CGIAR gene banks have a well-developed germplasm quality management system [208], basically following the Genebank Standards [5], the division of active and base collections with the most original sample concept [5,67] and other aspects such as risk management and user satisfaction [209] and thus, could well provide guidance to national and local management efforts.

5.6.5. Storage—Seed; Field; In Vitro; Others

- Of the reported more than 1750 gene banks worldwide, about 130 maintain more than 10,000 accessions each [188];
- Of the reported 3.6 million accessions (about half of the global total), maintained by 488 gene banks and germplasm collections, about 6% is conserved in field gene banks; almost 1% is conserved in in vitro collections, and 0.2% accessions are cryopreserved. Another 1.458 accessions are "conserved" as DNA [209]. For the CGIAR gene banks, most accessions are held and distributed as seed; just 23,862 (3.1%) are conserved as tissue in in vitro and 29,122 (3.8%) in field gene bank collections [16];
- About 96% of all conserved accessions in gene banks are maintained as seeds [188,198];
- In CGIAR gene banks, about 18,500 or 2.5% of the accessions are cryopreserved or safety duplicated as in vitro samples [198];
- Germplasm of crops listed in Annex 1 of the ITPGRFA is conserved in more than 1240 gene banks worldwide, and they add up to a total of about 4.6 million samples. Of these, about 51% is conserved in more than 800 gene banks of the Contracting Parties of the ITPGRFA, and 13% is stored in the collections of the CGIAR centers [188]. The fact that 96% of accessions are being kept as seeds in cold storage, it can be expected that the "global system" is being dominated by the conceptual thinking that relates

to this type of genetic resources, possibly at the detriment of the non-seed conserved genetic resources or those with recalcitrant seeds;
- However, many of the gene banks do maintain germplasm falling in two or more categories and not seldom managing big numbers of different species, as for instance, botanic gardens do;
- A useful tool for planning and/or rationalizing gene bank operations and germplasm management practices is the published guide to effective management of germplasm collections [14],
- In cases that gene banks do not meet the aspired thresholds in conservation, such as the number of seeds per accession, seed viability, etc., it might be advantageous to seek advice and possibly assistance from experts, either in the national network or in the regional or certainly the global network. CGIAR centers are well geared toward providing such assistance, including capacity building;
- As human and infrastructural capacities for the more demanding conservation methodologies such as in vitro and cryopreservation are often missing in small or national gene banks, collaboration with other gene banks/countries that do possess such facilities will enable access to required knowledge and resources.

5.6.6. Safety Duplication

- The Crop Genebank Knowledge Base is a slightly outdated but still a very resourceful tool for most of the routine gene bank operations, including on the safety duplication for seed, clonal, and in vitro materials [210];
- It is estimated that more than one-third of the globally distinct accessions of 156 crop genera stored in gene banks as orthodox seeds are conserved in the Svalbard Global Seed Vault, with high coverage of Annex 1 crops and of those crops for which there is a CGIAR mandate. Cereals and food legumes together constitute 87% of the accessions in the Seed Vault [211];
- As per the CGIAR criterion for mitigation of risk of loss, seed accessions in long-term storage should be safely duplicated in two external locations, one of which is the SGSV; on this basis, 73% of the seed accessions have been adequately secured against risks of loss [16];
- The number of accessions stored at SGSV is 1,081,026 seed samples from 87 gene banks and 66 countries [199];
- The safety duplication of accessions is an essential step to increase the security of base collection materials as this germplasm is unique and forms part of a global inheritance. However, this should also apply to local and national germplasm collections to avoid possible losses;
- Safety duplication does not have to be in all instances under long-term storage conditions, in particular when germplasm is maintained in field gene bank collections. The duplication of the collection (or part thereof) at another location or another gene bank might be sufficient;
- Arrangements for safety duplication are best made with and through the respective national PGRFA program to ensure adequate coordination, conclude proper agreements and use the existing network of gene bank contacts.

5.6.7. Germplasm Health

- The Crop Genebank Knowledge Base provides very useful information on plant health testing [212];
- In 2018 and 2019, the germplasm health units (GHUs) of the CGIAR facilitated 3900 events of international germplasm transfers from gene banks and breeding programs, reaching >100 countries per year. In this process, GHUs tested 453,972 samples and eliminated 6% of those that were pest-affected [213];
- The CGIAR gene banks reported that during 2019, out of the 717,693 total accessions, 76,766 accessions were tested on their health status [213];

- Germplasm conserved as part of the international network of base collections, in fact, stored and distributed from the related active collections, should be free of pests and diseases when being distributed, in particular of quarantine pests and diseases [54];
- In addition, during regeneration/multiplication activities, regular inspection of the cultivated accessions in the field/greenhouse will be important to ensure the health status of materials that are conserved for long-term storage or maintained for distribution. For the latter, IBPGR created a series of technical guidelines for the safe movement of crops [214];
- Even for well-equipped and staffed gene banks, this specific routine operation on germplasm health continues to be a challenge, and collaboration with specialized institutions nationally or internationally might be required to handle germplasm in conformity with the rules and regulations of the International Plant Protection Convention (IPPC) and National Plant Protection Organizations (NPPOs) to avoid distributing harmful pests and diseases with germplasm samples;
- In the case of local conservation efforts, e.g., nature conservation, on-farm management, and community gene banks, proper links to the national PGRFA system that can provide germplasm health assistance will be important. Such contacts can also be useful to introduce, for instance, germplasm with resistance to locally occurring devastating diseases from elsewhere.

5.6.8. Distribution and Exchange

- The Crop Genebank Knowledge Base provides a section on germplasm distribution, including related biological as well as legal aspects [215];
- Relevant information on the transfer of germplasm of 18 crops/crop groups with respect to import and export requirements, guidelines for the detection and treatment of relevant pests and diseases, and best practices for seed and clonal germplasm materials can provide useful assistance [216];
- The total germplasm distribution remained steady over the period from 1996 to 2007 at about 100,000 accessions each year, and it peaked in 2004 [203];
- Over the last 10 years, the CGIAR gene banks have distributed more than 1.1 million PGRFA samples to recipients in 163 countries, or 23% of all PGRFA samples transferred following the rules of the MLS [47,217]. Over the first 10 years of operation under the MLS of the treaty, the CGIAR centers distributed almost 4 million samples of PGRFA with over 47,000 SMTAs. This represents 93% of the reported global distribution of germplasm under the multilateral system [16];
- The CGIAR breeding programs were the source of an additional 66% (approximately 3.3 million samples) of the PGRFA transferred through the MLS in addition to the above-mentioned 23%. The remaining 11% of materials exchanged were transferred by organizations and individuals outside the CGIAR [47];
- The CG gene banks reported 1238 external germplasm requests during 2019, and a total of 45,941 germplasm samples were distributed outside the CG, belonging to 38,099 accessions [218];
- Germplasm exchange and distribution have been important activities of gene banks and possibly the main source of acquisition of germplasm for many gene banks around the world (also recorded as donations). This has certainly been triggered by the fact that the first (oldest) gene banks were all outside the centers of crop diversity, and thus, germplasm collecting missions were typical to foreign countries, costly, and demanding. Consequently, interesting germplasm from the main crops was in many cases of interest to many other gene banks, and over time, such samples were globally distributed, resulting in a significant duplication, certainly from a global perspective;
- As described above, careful and detailed planning of new collecting missions, also with a clear understanding of what had already been collected in the past (also by other gene banks), is an important step to avoid unnecessary duplication. This also

applies to the acquisition of germplasm from other gene banks. Consequently, this check should possibly be best performed at the national level;

- Since there are many steps involved in germplasm management that might jeopardize the genetic constitution or authenticity of the sampled populations or landraces, there is a growing recognition to a stronger focus on genetic diversity-related aspects of individual samples/accessions and to avoid steps that put this at risks;
- The above also requires better and comprehensive information to be provided along with the exchanged or distributed samples or accessions;
- As for some other routine gene bank operations, also the exchange and distribution of germplasm from a given country, especially when numerous gene banks and collections exist in the country, is best carried out in a coordinated manner by and through the national PGRFA programs, as they know the rules and procedures, know the collaborating institutions, etc.

5.6.9. Regeneration and Multiplication

- The Crop Genebank Knowledge Base provides a section on germplasm regeneration, why, and how it is being performed [219];
- Guidelines for the regeneration of 16 predominantly minor crop plants are an important tool and can be found at the Crop Genebank Knowledge Base [220];
- With respect to the security of stored materials at the global level, FAO identified two main areas of concern, i.e., the low extent of safety duplication and the large backlogs with respect to regeneration [188];
- Regeneration (and multiplication) is a complex, demanding, costly, and rather risky gene bank activity, certainly from a genetic diversity point of view. Consequently, significant backlogs have built up in many gene banks, with all the risks this might have of losing materials due to loss of viability. In particular, CWRs present many problems and challenges for their regeneration, as many are cross-pollinating, produce little seeds, and are highly shattering [184];
- One of the major constraints is that germplasm, in particular crop wild relatives, obtained from other gene banks outside the country might not be well-adapted (e.g., photoperiod and ecological requirements) to the growing conditions of the gene bank in question, and this can result in losses of genetic information or even of entire accessions and may need international collaboration [184];
- The CGIAR gene banks reported that during 2020 of the 721,574 accessions maintained, 11,414 were regenerated and 68,616 multiplied [221];
- Information reported and gathered on almost 900,000 accessions for the period 2012–2014 showed that 18% had been regenerated, whereas 38% needed regeneration. For about 40% of those that were due for regeneration, an adequate budget was reported not to be available [209];
- As regeneration requires solid knowledge of cultivating a given species, it might well have advantages to seek cooperation with institutions that possess such knowledge and experience. For instance, the Dutch gene bank CGN developed such collaboration with interested plant breeders in the country to regenerate accessions for the gene bank according to the gene bank standards. The breeder concerned was entitled to keep a subsample of the regenerated materials and expected to return the harvested produce along with any pertinent information to the gene bank. The World Vegetable Center followed the same procedure with Asian breeding companies [222];
- The Crop Trust offered countries assistance with the regeneration of germplasm accessions in the country, during the operation of a big global regeneration project, and expected materials of these accessions to be included in the collection of the respective center for long-term conservation with the intention that this material would be made available to users under an SMTA;

- Regeneration of germplasm materials provides several opportunities to complement gene bank functions such as gathering characterization data, increasing seed stock, and eliminating diseased plants in accessions [184].

5.6.10. Characterization and Evaluation

- The Crop Genebank Knowledge Base provides very useful and practical information on the characterization of germplasm materials [223];
- Characterization is regarded as a routine gene bank activity, typically using characteristics that are highly heritable. It is an essential activity that facilitates proper management of the materials and provides a solid information basis for its use. In contrast, evaluation is a step that requires specific knowledge of the conserved species, usually advanced facilities, and it needs proper experimental design to eliminate the generally high dependence of the evaluated traits on the environmental circumstances;
- Characterization produces valuable agronomic and breeding data and allows the identification of unwanted duplicates in the collection and thus, provides the basis for rationalizing the collection(s) by eliminating unnecessary duplicates [184];
- Collaboration of the gene bank with, for instance, plant breeders or specialized researchers is a common approach to get accessions evaluated. However, it seems advisable for the gene banks to make clear agreements with the plant breeders as their willingness to return evaluation results seems to be limited;
- The extent of characterization of collections held by CGIAR centers and World Vegetable Center (a total of 585,193 accessions were analyzed) is 77%, varying from 17% to 88%, depending on the crop [188]. Please note that these percentages vary greatly from one species or crop group to another;
- The average extent of characterization and evaluation of national collections in 40 reporting countries for the main crop groups is 64% of almost 320,000 accessions, 63% of the 410,000 cereal accessions were morphologically characterized, and 65% of the 48,000 vegetable accessions [188];
- When characterizing local germplasm materials, it is important to use characters and traits that are of interest to the local communities and, whenever possible, to use descriptors agreed among gene banks and to engage the local community in this activity.

5.6.11. Documentation

- Proper gene bank documentation and information management are essential preconditions for any gene bank to be effectively linked with any larger conservation system, to share germplasm and related information, and to be able to effectively and efficiently conserve PGRFA;
- Many attempts have been made since the 1970s to develop gene bank information management systems to ensure connectivity and the easy exchange of information between gene banks and users. The advances of the Internet, the recent development of management systems such as GRIN-Global [145] have greatly improved the myriad of approaches by individual gene banks or national PGRFA programs;
- Greater standardization of data and information management systems is needed [188] to facilitate their exchange, analysis, and thus their use in setting priorities, facilitating monitoring, and allowing the creation of a more effective and efficient global conservation system;
- Unfortunately, information management in many gene banks remains weak, and thus, this situation does not support effective and efficient participation in coordinated national or international conservation efforts;
- Of the accessions maintained in CGIAR gene banks, 87% have passport or characterization data accessible online [16]. This figure is much higher than the average percentage of passport and characterization data of gene banks at large;

- Well-organized and comprehensive information management in a gene bank is the basis for efficient and effective conservation; for orchestrating the active and base collections efficiently; to monitor the viability of stored accessions timely and efficiently (and thus, among others, avoiding unnecessary regeneration efforts); providing a sound and solid foundation for decision-making; enables targeted rationalization of collections and conservation practices; is a prerequisite for local, national, regional and global collaboration and cooperation; facilitates targeted use of conserved germplasm material, etc.;
- As for conservation activities, especially those that are integrated into a bigger "system" or network, the availability and curation of high quality and comprehensive data at all levels is an indispensable prerequisite for effective and efficient collaboration, as has been mentioned already in Section 4.3.

5.6.12. Research

- Whereas research can be regarded to steadily improve and strengthen gene bank operations, it is not an essential requirement per se to conserve germplasm efficiently and effectively. However, as already mentioned above, several routine conservation operations require at least some applied research activities to identify, for instance, the optimum SMC of seeds of less well-known species, to adjust conservation procedures to locally prevailing conditions, etc.;
- In general, a gene bank that fosters the culture of research, of interacting with researchers to understand local genetic resources better, to seek solutions of farmers to improve sustainable production, etc. will play a more important role in local and national PGRFA conservation and sustainable use and thus contribute to sustainable development, to better incomes for farmers and to protect the environment;
- Active researchers in gene banks are also more attractive to others that seek collaboration, increase the willingness of governments to support routine and new activities, etc.;
- Whereas applied research might not require collaboration, more advanced research activities certainly do, as the infrastructure might be lacking, the right expertise has not yet been built, or for other reasons. In addition, research on important contextual aspects of the germplasm in question might further add value to the germplasm. Examples could be on gathering specific traditional uses of collected landraces, agronomic information obtained from local farmers, and others.

5.6.13. Collaboration and Networking

- As conservation of PGRFA is still a relatively new science, as conservation operations are frequently localized, not widely recognized as critical, and in general rather complex by nature, collaboration offers solutions for problems that can hardly be resolved in isolation. Such need for collaboration applies at all levels and offers opportunities to learn new aspects, to share strengths and overcome weaknesses, etc. For aspects related to international collaboration, see also Section 6;
- Greater efforts are needed to build a truly rational global system of ex situ collections. This requires particularly strengthened regional and international trust and cooperation [188], and, as observed by the authors, this has significantly decreased in most of the regions over the past 10 years or so (among others triggered by the disengagement of regional and/or crop networks coordinated for several decennia by Bioversity International and others);
- To improve the management of collections and to facilitate increased use of the germplasm collections as part of a network, documentation, characterization, and evaluation need to be strengthened and harmonized, and the data need to be made more accessible [188];
- As plant genetic resources for food and agriculture are global for many of the cultivated crop species and, to a much lower extent for the related wild species, it is

obvious that some sort of coordination from a global perspective is indispensable to achieve effective and efficient rational conservation of PGRFA and to facilitate their sustainable use. Thus, it is advantageous for all involved in conservation activities to seek collaboration with others, certainly at the national, depending on the scope and objectives of the conservation and use activities at the regional level, and whenever possible through the national program with the international efforts;
- Participation in regional or global PGRFA networks has proven to be advantageous to all involved from a contributing, receiving, and capacity-building perspective;
- Rationalization of collections is best performed at the regional/international level, requires comprehensive data and information about the local accessions/collections, and is an important step toward more cost-efficient and effective PGRFA activities;
- Active engagement of the lower in the next higher level in PGRFA activities through networking is a key prerequisite to enable and strengthen collaboration. One or more focal points for key areas might be one way to organize such participation effectively;
- Active and inclusive engagement of individual partners and participatory approaches seem to be pre-conditions to facilitate open, transparent, and motivating participation in conservation activities.

5.6.14. Human Resources, Infrastructure, and Financial Resources

- Many countries do not have the adequate human capacity, funds, or facilities to carry out the necessary work to the required standards. Many valuable collections are in jeopardy as their storage and management are suboptimal [188];
- In view of the afore-mentioned point, also, in this case, cooperation with other gene banks and institutions involved in the conservation and sustainable use of PGRFA locally or nationally is a possible solution to strengthen human capacity, to share facilities, etc.
- Many countries lack the resources needed to maintain adequate levels of the viability of the materials conserved [188], thus resulting in frequent regeneration cycles with the related difficulties of geneflow, loss of materials due to weather, and other conditions, at a high cost, and others. Thus, networking at the different levels is an important step in overcoming problems, constraints and in building capacity;
- Capacity building is at the core of the CGIAR centers' work. CGIAR research programs support about 1000 students in their BS, MSc, and Ph.D. degrees annually [16];
- Training and education at schools, high school, and university in PGRFA conservation and use are important premises to build a sufficiently strong force of human capacity, as well as to create a broad awareness;
- A comprehensive overview of strengthening institutions and organizations and building capacity for the conservation and use of agricultural biodiversity was compiled in 2017 [224];
- In order to participate effectively in local, national, or international networking, a country needs a minimum level of conservation expertise;
- Additional resources for ex situ conservation need to be mobilized. Greater efforts are required to raise awareness among policymakers and the general public on the importance of PGRFA and the need to conserve it [188];
- In case farmers and other members of a community actively engage in conservation (and use) activities, it seems to be beneficial to place the local activities in a broader context and to explain the basic principles of evolution, genetic diversity, breeding and improvement, genetic erosion and of conservation activities in general.

Without a minimum of well-trained and knowledgeable human resources, effective and efficient conservation activities are not possible. Thus, reaching such a minimum "threshold" is an important step to achieve this. Seeking the best possible means and ways to achieve this would be to consult and possibly collaborate with the next higher level and certainly with the national-level PGRFA program.

5.6.15. Linking Conservation and Use

- Stronger links are needed between the managers of collections and those whose primary interest lies in using the resources, especially for plant breeding [188];
- An aspect of strengthening the link between conservation and use is the provision of adequate and comprehensive information on the germplasm materials, either conserved in situ, on-farm, and/or ex situ in genebanks;
- Typically, access to the materials is easier when conserved ex situ involving healthy and vigorous seeds or propagules and in adequate sample size for the user (albeit that this size is always small, such as research materials but hopefully sufficiently large to represent the genetic diversity the accession entails);
- Yet, at another level, users also expect to receive materials that can be easily used. In case the requesting user would like to have an adequate representation of the diversity present in the collection, the gene bank might want to create a core or mini-core collection to facilitate the use. Such subsets of a collection might also be "constructed" of accessions that possess a specific trait. An alternative approach for heterogeneous accessions that possess a defined trait at the accession level could be to split such accessions into either pure lines or in single-seed descents and to keep such "pure" lines separate, evaluate such materials for individual traits, and offer those with the wanted trait to users. This practice could work for self-pollinating species and be a real service to users;
- Constraints in using conserved germplasm in gene banks include accession-level data existence and accessibility, quality and status of the material, policy and legal obstacles, and awareness/education/ outreach [184];
- Priorities outlined in the crop strategies to enhance user relationships with collections include refinement of collections for breeders (use of marker-assisted selection technologies, creation of advanced core and mini-core collections, pre-breeding, further work on identifying diseases and resistance), and strengthening or creating new relationships with other users [184]. Further aspects have been mentioned in Section 3.4.

5.6.16. Genebank Standards

- Another important operational element of this international network of base collections is the Genebank Standards, a formally endorsed set of standards for routine gene bank operations for the conservation of orthodox seeds, non-orthodox seeds, and vegetatively propagated plants by the FAO Commission [5]. They do provide the foundation for worldwide collaboration between gene banks, as in fact is being demonstrated by the gene banks of the CGIAR using the CGIAR Quality Management System [208]) and the European gene banks using AQUAS, the Quality Management for AEGIS [225].

Whereas the use of standards has been widely accepted, it is very difficult to provide actual data on percentages of accessions that are conserved under "standardized" conditions. Only a handful of gene banks have implemented a gene bank-level quality management certification system (e.g., ISO 9001:2000), e.g., CGN (The Netherlands), IPK (Germany), CRI (Czech Republic), and the CGIAR centers CIP and CIMMYT, whereas the other CGIAR gene banks have a strict and active quality management system in place. Unfortunately, many of the gene bank collections are maintained without clear adherence to the Genebank Standards for a range of reasons. However, to effectively engage in a region of global conservation effort, following agreed standards is a prerequisite that cannot be avoided. The importance of quality management is demonstrated in the context of operating a regional virtual collection, i.e., AEGIS, in Europe [225]. Further details on the quality management of gene banks are provided by the Crop Genebank Knowledge Base [226].

5.6.17. International Information Sources on Conserved PGRFA

A third important operational element of the global system is an information management system that includes relevant information on the materials included in the international network of base collections. An inventory that was started by IBPGR in the late 1970s eventually provided the foundation for WIEWS, a global information system on PGRFA facilitating information exchange. It consists of a global network of national focal points, a registry of more than 17,000 national, regional and international institutes and organizations dealing with the conservation and use of PGRFA. The WIEWS database is populated with information from direct contributions made to the 2020 data assemblage by countries on the implementation of the GPA and, as of July 2021, has registered over 5.7 million accessions from 420 genera and 54,306 species, conserved under medium- (30.4%) or long-term conditions (61.6%) in 831 gene banks from 114 countries and 19 international/regional centers [227].

CGIAR centers, and IRRI in particular, are contributing to the creation of the Global Information System for PGRFA (GLIS) [228] under the framework of the Plant Treaty. Work on the GLIS has focused on the development of digital object identifiers (DOIs) as permanent, unique identifiers for PGRFA accessions. Through the CGIAR Genebank Platform, the global version of the Germplasm Resource Information Network (GRIN-Global) and Genesys [145] have been enhanced to accommodate DOIs and link with the GLIS server. The CGIAR gene banks have already assigned DOIs to 73% of their accessions as of 1 April 2018 [16]. Other important international PGRFA information sources are EURISCO, the European Search Catalog for Plant Genetic Resources [229], linked to Genesys; the Svalbard Global Seed Vault's Seed Portal [199], also linked to Genesys; the Kew Millennium Seed Bank List [230]; and the Global Biodiversity Information Facility (GBIF) [150].

6. Suggested Measures to Achieve a More Rational, Effective, and Efficient Conservation System

In the foregoing, we have addressed the weaknesses of the current long-term PGRFA conservation and facilitating use system. Here we identify measures that should be undertaken to overcome those weaknesses with the aim to obtain a more rational, effective, and efficient global system. In this process, we have been guided by the following statement: "The ideal future genebank will be part of a rational, efficient, and effective system in which genebanks work in close partnerships with each other, and in harmony with the scientific and the policy dimensions of research-for-development, ensuring that the benefits of innovation reach those who most need them" [42].

The weaknesses, limitations, or deficiencies of the current system include:

1. Many crop gene pools are not adequately represented in gene banks, especially those of neglected and underutilized species as well as of crop wild relatives, and/or the genetic diversity of crops is inadequately represented in the gene bank collections. Given the ongoing and accelerating threat of genetic erosion, there is a need to systematically collect henceforth neglected genetic resources, following priorities yet to be established;
2. In general, there is a lack of long-term storage and supporting facilities. Gene bank management practices are weak, and severe regeneration backlogs have been reported;
3. Many gene banks have weak information management systems in place with concomitant poor coverage of basic information on the germplasm conserved;
4. Many gene banks do not have or use advanced genotyping and phenotyping technologies and do not collaborate with other gene banks having such facilities;
5. Many gene banks have only weak or no linkages with the breeding community and other germplasm users;
6. There are many unwanted duplicates within and between collections that do not need to be included in a long-term base collection system;
7. Over the past two or three decennia, a decrease in regional and global coordination, as well as cooperation between gene banks, has been observed;

8. The international distribution of germplasm materials included in the MLS as well as of breeding materials is primarily performed by centers of the CGIAR and AIRCA who have the technical expertise and resources to conduct adequate germplasm health testing and to eliminate pathogens that fall under quarantine restrictions. Unfortunately, most national gene banks do not have the facilities and expertise to fulfill the increasingly demanding requirements for the safe distribution of germplasm;
9. The current legal and policy framework restricts rather than facilitates collecting, conservation, and distribution/exchange of threatened plant agrobiodiversity;
10. The oversight and management of the global conservation system are weak, and there is no strong, active participation of many countries in the system.

From the above list, it becomes clear that the current arrangements and mechanisms need significant improvement and expansion to achieve efficient, effective, and rational global PGRFA long-term conservation and facilitated germplasm use. While the current global system has many of the required components in place, one must conclude that it is not functioning well, possibly due to a high level of bureaucracy as well as mistrust between diversity-rich, developing countries, and advanced countries with a stronger economy. This can be deduced from the current lack of progress in negotiations on the expansion of the list of Annex 1 species and the ongoing debate on how to solve the issue of digital sequence information sharing (see Section 4.4). More visible and tangible benefits, either monetary or otherwise, should become available to biodiversity-rich countries and communities that are willing to participate in such a system for the sharing of genetic resources and traditional knowledge.

The following measures might help to overcome some of the identified weaknesses and contribute to a more rational, effective, and efficient long-term conservation system with use facilitation.

a. Filling genetic and geographic gaps in current collections. While genetic diversity of major food crops is generally well represented in collections, underutilized crops and wild food plants, as well as CWR, are clearly underrepresented, and considerable gaps (both genetically and geographically) do exist. Many of the underutilized crops, wild food plants, as well as CWR, still play a critical role in achieving food security of the people in developing countries, especially in rural areas. Existing crop conservation strategies are powerful tools to set priorities where and what to collect as they typically provide a global account of existing collections, their diversity coverage (as far as known), reported threat status, etc. (for details, see Section 5.1.2). However, those crop conservation strategies are currently only existing for major crop gene pools and for only a few underutilized food crops.
b. Determining unique accessions of germplasm collections at the global level. The total number of PGRFA accessions conserved in approximately 1750 gene banks or germplasm collections is about 7.4 million [188], yet only about two million of those are estimated to be unique [205]. With the current advances in genomics and phenomics, we envision genetic curation across international, regional, and national gene banks around the world to identify unique accessions across all crop collections or crop gene pools (see also Singh et al. [87]) and the European AEGIS approach [231]. Each unique accession would receive a globally unique ID, and duplicate accessions within and between collections could be removed, based on the decision of the holding gene bank, from long-term conservation in the respective base collections and could serve for germplasm exchange and user-oriented research. With such global curation, ex situ long-term conservation of crop collections and their wild relatives will become more effective and efficient.
c. International collaborations. Intensifying existing international collaborations among institutes maintaining agricultural crop collections and forging new collaborations with the botanic gardens' community offer important opportunities for a more rational, effective, and efficient long-term conservation system of plant genetic resources, their sustainable use, including the exploration of useful traits to strengthen breeding

programs around the globe, thus, enhancing food and nutrition security of a growing population under the challenges of climate change. Intensification of collaboration with the private plant breeding sector would be yet another step to achieve this.

Based on several case studies, Pearce et al. [232] highlight a number of key benefits of such collaborations, such as (i) synergy by bringing together collaborative teams with complementary skills and expertise; (ii) cost efficiency by sharing technologies and gaining access to local knowledge and other local resources; (iii) enhanced confidence for sharing and aggregating of resources such es accessions and specimens and associated information kept in globally dispersed collections; (iv) long-term positive change that brings about impact and leverage beyond the specific objectives of the collaboration; and (v) transfer of knowledge and technologies lead to building local expertise and strengthens national research capacity for plant diversity studies and conservation efforts. Furthermore, setting and applying standards, potentially extending plant breeding to "orphan crops", and facilitating better coordination of activities would be other significant benefits that enable the emergence of a more rational and effective global system.

d. Investment in research. Conserving and delivering germplasm resources "in the right way" requires investment in conservation research and user-oriented supportive research and optimization of routine processes, which may include automation. Research investment should be directed toward seed longevity studies and viability testing intervals of orthodox, intermediate, and recalcitrant seed species and of crop wild relatives as well as cryopreservation, as both areas are of critical importance to improving efficiency and effectiveness of ex situ conservation protocols and gene bank management [42]. Other examples of (applied) research activities are listed in Section 5.6.12.

e. With a focus on facilitating germplasm use, targeted investments in characterization (highly heritable descriptors that are also used by breeders), evaluation (including "smart" phenotyping and genotyping) will be required, and collaborations among gene banks need to be forged as investments in corresponding equipment and expertise are significant and should be shared for effective use. To enable gene discovery studies, it is essential to phenotype the pure lines (derived from single-seed descent) that have been used for GBS. The resulting data needs to be professionally managed and packaged for easy use by breeders and other researchers [42].

f. Digital quality information on accessions. A better accession-level description of the genetic diversity of crop collections maintained in gene banks (including genomic, phenomic, and ecological data) and easy access for users to quality data on conserved germplasm and associated metadata is critical to meet evolving challenges in crop diversity conservation and to facilitate more efficient use of those resources in breeding and research [42,76,233,234].

Access to detailed digital information associated with each accession is increasingly becoming as important for breeders and researchers as to the physical material itself, although this is currently proving to be a very contentious issue [53]; see also respective references in Section 4.4).

According to Sackville Hamilton [42], forward-looking gene banks should envision "a digital catalogue of the functional genetic variants existing in each accession, linked to the corresponding information on genomes outside the gene bank. This will enable DNA-based decisions on conservation and, in conjunction with knowledge of gene function and associated phenotypic data, on use".

g. Legal and policy framework. The current legal and policy framework has evolved "spontaneously" along with the global conservation system. Since the entrance into force of the CBD, and the recognition of national sovereignty with respect to biodiversity, its conservation, and sustainable use, a strong focus was given to issues related to a property right, as well as access and benefit-sharing. With the establishment of the International Treaty, the MLS was intended to play a central role with respect to ABS arrangements of germplasm that had been placed into the global system, restricted to materials under governmental control and in the public domain, thus largely confined to the species listed in Annex 1. From a PGRFA perspective, it would make more sense to aim at the coverage of all PGRFA and to avoid any exceptions. Whereas many countries are members of the FAO Commission on PGRFA as well as the International Treaty, the active participation of many countries in the debates in the Commission and Governing Body is limited, as is the level of implementation of the agreed actions. Furthermore, the lack of trust (possibly caused by Western dominance), impressive bureaucracy, the lack of clear assignments of responsibilities, and the lack of shared benefits deriving from the use of the genetic resources provided by the diversity-rich countries are some of the main impediments for a conducive legal and policy framework that need to be addressed to achieve an inclusive and effectively functioning global system.

Oversight and management. The co-existence of the FAO Commission on PGRFA and the treaty as oversight and policy- and legislation-setting bodies are sometimes difficult to understand and lead easily to non-transparent operations. With the understandable focus on member states, many actively engaged entities in the management, conservation, and use of PGRFA feel not to be represented at "the table", and this leads to distrust and tensions, both at the national as well as international level. Furthermore, the perceived non-committal role of the private sector makes it difficult for countries to freely share their resources and to arrive at an effective, inclusive, and rational oversight of any global PGRFA initiative and the global conservation and use system.

The following "model" for a functional global network or system of base and active collections is proposed:

1. For each crop gene pool, or where meaningful a set of related gene pools, a virtual global base collection is created and coordinated by a carefully selected lead gene bank to ensure that the identified unique base collection accessions (whenever possible also the "most original samples (MOS)" that for any accession could be identified) are stored under optimal conditions for the long-term. Such a virtual global base collection facilitates cost-efficient conservation operations through strategic coordination; it requires the coordination of a global crop gene pool network, that among other responsibilities, develops and keeps the global crop conservation strategy up to date; the coordinating gene bank should have the facilities and personnel to manage the virtual base collection under agreed quality standards, including the management of the MOS accessions, where necessary through the support to the gene banks that physically hold base collection accessions; each gene bank conserving accessions of the global base collection will also manage the associated active collection samples/subsamples for regeneration and arrangements for safety duplication, and backup storage at the SGSV as well as for distribution; all activities will be conducted in close consultation and agreement with the lead coordination gene bank;
2. It would make logical and practical sense that current CGIAR gene banks that do hold global base collections for their "mandate crops" would continue these "assignment(s)", including the operation of their active collections;
3. Regeneration of the base collection accessions will be carried out by the gene banks that physically curate the agreed "base collection accessions" when the viability has dropped below the agreed viability threshold or the stock below the minimum stock size; if cultivation conditions for to-be-regenerated accessions are better at one of the other gene banks of a given accession, they could be asked to regenerate;
4. The gene bank holding base collection accessions take in principle responsibility for the regeneration of those accessions and use the samples of a given accession from the active collection for at most four regeneration cycles before turning to the MOS subsample from the corresponding base collection accession; the gene bank holding base and active collection accessions are also responsible for regenerating

the MOS subsamples, in close consultation with the coordinating lead base collection gene bank;

5. Characterization and evaluation activities are coordinated by the lead base collection gene bank and in principle implemented by those gene banks holding physical base collection accessions; obviously, characterization will be combined with the regeneration of the accessions, whenever possible; each gene bank holding base collection accessions will coordinate research activities on their accessions, whenever relevant with additional accessions, in close agreement with the coordination gene bank;
6. The coordinating lead base collection gene bank, in close consultation with all the operational gene banks holding physical base collection accessions, looks after human resources training and capacity-building activities; the lead base collection gene bank represents the base collection for a given crop gene pool (or as a "general" responsibility, with other lead crop base collection gene banks) in the network, and triggers policy and legal framework related activities;
7. The base and active collection gene banks will conduct the distribution of those accessions they hold physically, if applicable upon request from the coordinating lead base collection gene bank, as the latter manages the global conservation as well as the global user databases for that crop and is the proposed recipient of such requests from users. The base and active collection gene banks are responsible for the germplasm health aspects of those accessions they are physically managing;
8. It is foreseen that the "services" that gene banks provide for the management of the assigned base collection accessions will be paid for through the coordinating lead base collection gene bank from a to-be-formed global base collection fund, similar to the arrangements the Global Crop Diversity Trust has made with the CGIAR and a few additional gene banks;
9. Countries/gene banks that participate in the global crop conservation network, e.g., through the inclusion and management of unique accessions of the base collection, and/or agree to conserve materials of the crop gene pool in situ or on-farm, or provide other "services" to the network, are entitled to participate in the sharing of benefits (e.g., monetary, non-monetary, membership in global and regional networks) that are derived from the network activities;
10. All germplasm materials that are included in the global base collection network are "automatically" part of the MLS of the International Treaty and thus, in the public domain and freely available to all users without restrictions, as set out in the yet to be adjusted standard material transfer agreement with the stakeholder community of the group of global base collection networks. The International Treaty will be recognized as the global policy setting mechanism, but for all PGRFA; and
11. A lean international organization, e.g., the Governing Body of the International Treaty, should assume responsibility for the global coordination, facilitation, and oversight over the various global base collection networks.

7. Conclusions and Recommendations

The current global conservation system has inherent weaknesses and limitations, partially due to its spontaneous creation out of a felt need by concerned scientists and visionaries and the subsequently required adjustments to the evolving political framework and changing realities. Here we recommend some measures that might contribute to a more rational, effective, and efficient long-term ex situ conservation system. These measures include:

- Filling genetic and geographic gaps in current ex situ collections through continued collecting of threatened genetic diversity with the aim of reaching an adequate representation of crop gene pools in ex situ collections, especially of neglected, currently underutilized species and crop wild relatives;
- Determining unique accessions of germplasm collections in base collections at the global level and removing the many duplicates within and among gene banks from

the base collections. Identified duplicates could be used for research and germplasm distribution, rather than resorting to the base collection accessions;

- Intensifying existing international collaborations among gene banks maintaining agricultural crop collections, and forging new collaborations with the botanic gardens' community offer important opportunities as well as stronger linkages with the plant science research community at large;
- The private sector, in particular the plant breeding community, is highly dependent on genetic resources and genetic diversity, and thus, their involvement in the conservation and sustainable use activities is desirable. Furthermore, plant breeders have specific knowledge of "their" crops and have the required expertise for regenerating and evaluating gene bank materials; hence, it could serve as a key link between long-term conservation and sustainable use. At the same time, it is obvious that adequate benefit-sharing arrangements, in the broadest possible sense, will be required to allow and strengthen this collaboration;
- More investment in conservation research and user-oriented supportive research, as well as optimization of routine gene bank processes, will help with conserving and delivering germplasm resources of high quality and in the right form as required by the users;
- A better accession-level description of the genetic diversity of crop collections maintained in gene banks (including genomic, phenomic, and ecological data) and easy access for users to quality data on conserved germplasm and associated metadata is critical to meet evolving challenges in crop diversity conservation and to facilitate more efficient use of those resources in breeding and research;
- The legal and policy framework clearly requires improvements, ideally by including all PGRFA in the MLS of the International Treaty and reaching a more equitable sharing of benefits derived from germplasm use, benefiting those who most need them and most often are the custodians of the rich agrobiodiversity on which people around the globe rely to satisfy food and nutrition security;
- A model for a functional and efficient global network of base and active collections and a lean international organization is proposed that assumes responsibilities for the global coordination, facilitation, and oversight over the various global crop gene pool base collection networks. This model should build on the existing gene banks of the CGIAR, the World Vegetable Center, and ICBA, as well as on a handful of strong national gene banks that form the core of the current global system on plant genetic resources for food and agriculture (PGRFA);
- The political oversight over the proposed global model network of base collections should remain with FAO and the Governing Body of the International Treaty.

Author Contributions: J.M.M.E. and A.W.E. conceived of, drafted, and edited this paper equally. Both authors have read and agreed to the published version of the manuscript.

Funding: This research received no external funding.

Institutional Review Board Statement: Not applicable.

Informed Consent Statement: Not applicable.

Data Availability Statement: Not applicable.

Conflicts of Interest: The authors declare no conflict of interest.

References

1. Antonelli, A.; Fry, C.; Smith, R.J.; Simmonds, M.S.J.; Kersey, P.J.; Pritchard, H.W.; Abbo, M.S.; Acedo, C.; Acedo, C.; Adams, J.; et al. *State of the World's Plants and Fungi*; Research Report; Sfumato Foundation; Royal Botanic Gardens (Kew): Richmond, UK, 2020; p. hal-02957519.
2. CBD. UN Biodiversity Convention Partners with Slow Food International in Celebrating the International Day for Biological Diversity. 22 May 2019. Available online: https://www.cbd.int/doc/press/2019/pr-2019--05--22-idb-en.pdf (accessed on 5 July 2021).

3. Engels, J.M.M.; Ebert, A.W. A Critical Review of the Current Global Ex Situ Conservation System for Plant Agrobiodiversity. I. History of the Development of the Global System in the Context of the Political/Legal Framework and its Major Conservation Components. *Plants* **2021**, *10*, 1557. [CrossRef]
4. Coe, M.T.; Evans, K.M.; Gasic, K.; Main, D. Plant breeding capacity in US public institutions. *Crop Sci.* **2020**, *60*, 2373–2385. [CrossRef]
5. FAO. *Genebank Standards for Plant Genetic Resources for Food and Agriculture*; Revised edition; FAO: Rome, Italy, 2014; 168p. Available online: http://www.fao.org/3/i3704e/i3704e.pdf (accessed on 29 April 2021).
6. Hay, F.R.; Sershen, N. *New Technologies to Improve the Ex Situ Conservation of Plant Genetic Resources*; Burleigh Dodds Series in Agricultural Science; Burleigh Dodds Science Publishing Limited: Cambridge, UK, 2021; pp. 1–32. [CrossRef]
7. Visser, L.; Engels, J.M.M. Context of genebank management. In *A Guide to Effective Management of Germplasm Collections*; IPGRI Handbooks of Genebanks No. 6; Engels, J.M.M., Visser, L., Eds.; IPGRI: Rome, Italy, 2003; pp. 8–19.
8. Definitions. Exploration. Available online: https://www.definitions.net/definition/exploration (accessed on 27 July 2021).
9. Marshall, D.R.; Brown, A.H.D. Optimum sampling strategies in genetic conservation. In *Crop Genetic Resources for Today and Tomorrow*; Frankel, O.H., Hawkes, J.G., Eds.; Cambridge University Press: Cambridge, UK, 1975; pp. 53–80.
10. Brown, A.H.D.; Marshall, D.R. A basic sampling strategy: Theory and practice. In *Collecting Plant Genetic Diversity: Technical Guidelines*; Guarino, L., Rao, V.R., Reid, R., Eds.; CAB International: Wallingford, UK, 1995; pp. 75–91. Available online: https://cropgenebank.sgrp.cgiar.org/images/file/procedures/collecting1995/Chapter5.pdf (accessed on 10 September 2021).
11. Guarino, L.; Ramanatha Rao, V.; Goldberg, E. (Eds.) *Collecting Plant Genetic Diversity: Technical Guidelines. 2011 Update*; Bioversity International: Rome, Italy, 2011; ISBN 978-92-9043-922-6. Available online: https://www.bioversityinternational.org/e-library/publications/detail/collecting-plant-genetic-diversity-technical-guidelines-2011-update/ (accessed on 10 September 2021).
12. Engels, J.M.M. Chapter 3: An introduction to plant germplasm exploration and collecting: Planning, methods and procedures, follow-up. In *Collecting Plant Genetic Diversity: Technical Guidelines. 2011 Update*; Guarino, L., Ramanatha Rao, V., Goldberg, E., Eds.; Bioversity International: Rome, Italy, 2011; p. 6, ISBN 978-92-9043-922-6. Available online: https://www.bioversityinternational.org/e-library/publications/detail/collecting-plant-genetic-diversity-technical-guidelines-2011-update/ (accessed on 10 September 2021).
13. Hoban, S.; Volk, G.; Routson, K.J.; Walters, C.; Richards, C. Sampling Wild Species to Conserve Genetic Diversity. In *North American Crop Wild Relatives*; Greene, S., Williams, K., Khoury, C., Kantar, M., Marek, L., Eds.; Springer Nature Switzerland AG: Cham, Switzerland, 2018; Volume 1, pp. 209–228. [CrossRef]
14. Hoban, S. New guidance for ex situ gene conservation: Sampling realistic population systems and accounting for collection attrition. *Biol. Conserv.* **2019**, *235*, 199–208. [CrossRef]
15. Engels, J.M.M.; (Bioversity International, Rome, Italy). Personal communication, 2021.
16. Engels, J.M.M.; Visser, L. Genebank management procedures. In *A Guide to Effective Management of Germplasm Collections*; IPGRI Handbooks for Genebanks No. 6; Engels, J.M.M., Visser, L., Eds.; IPGRI: Rome, Italy, 2003; pp. 60–79, ISBN 92-9043-582-8. Available online: https://www.researchgate.net/publication/235929868 (accessed on 29 July 2021).
17. Brink, M.; Van Hintum, T. Genebank operation in the arena of access and benefit-sharing policies. *Front. Plant Sci.* **2020**, *10*, 1712. [CrossRef] [PubMed]
18. López Noriega, I.; Halewood, M.; Abberton, M.; Amri, A.; Angarawai, I.I.; Anglin, N.; Blümmel, M.; Bouman, B.; Campos, H.; Costich, D.; et al. CGIAR operations under the plant treaty framework. *Crop Sci.* **2019**, *59*, 819–832. [CrossRef]
19. Kameswara Rao, N.; Dulloo, M.E.; Engels, J.M.M. A review of factors that influence the production of quality seed for long-term conservation in genebanks. *Genet. Resour. Crop Evol.* **2017**, *64*, 1061–1074. [CrossRef]
20. Crop Genebank Knowledge Base. Sample Processing. 2021. Available online: https://cropgenebank.sgrp.cgiar.org/index.php/procedures-mainmenu-242/sample-processing-mainmenu-196 (accessed on 15 July 2021).
21. Whitehouse, K.J.; Hay, F.R.; Ellis, R.H. Increases in the longevity of desiccation-phase developing rice seeds: Response to high temperature drying depends on harvest moisture content. *Ann. Bot.* **2015**, *116*, 247–259. [CrossRef]
22. Whitehouse, K.J.; Hay, F.R.; Ellis, R.H. Improvement in rice seed storage longevity from high-temperature drying is a consistent positive function of harvest moisture content above a critical value. *Seed Sci. Res.* **2018**, *28*, 332–339. [CrossRef]
23. Timple, S.E.; Hay, F.R. High-temperature drying of seeds of wild *Oryza* species intended for long-term storage. *Seed Sci. Technol.* **2018**, *46*, 107–112. [CrossRef]
24. Walters, C.; Kameswara Rao, N.; Hu, X. Optimizing seed water content to improve longevity in ex situ genebanks. *Seed Sci. Res.* **1998**, *8*, 15–22.
25. Walters, C.; Engels, J.M.M. The effects of storing seeds under extremely dry conditions. *Seed Sci. Res.* **1998**, *8*, 3–8.
26. Walters, C. Orthodoxy, recalcitrance and in-between: Describing variation in seed storage characteristics using threshold responses to water loss. *Planta* **2015**, *242*, 397–406. [CrossRef]
27. Ellis, R.H.; Roberts, E.H. Improved equations for the prediction of seed longevity. *Ann. Bot.* **1980**, *45*, 13–30. [CrossRef]
28. Hay, F.R.; Mead, A.; Manger, K.; Wilson, F.J. One-step analysis of seed storage data and the longevity of *Arabidopsis thaliana* seeds. *J. Exp. Bot.* **2003**, *54*, 993–1011. [CrossRef]
29. Probert, R.J.; Daws, M.I.; Hay, F.R. Ecological correlates of ex situ seed longevity: A comparative study on 195 species. *Ann. Bot.* **2009**, *104*, 57–69. [CrossRef] [PubMed]

30. Royal Botanic Gardens Kew. Seed Information Database—SID. 2021. Available online: https://data.kew.org/sid/ (accessed on 15 July 2021).
31. Hay, F.R.; Whitehouse, K.J.; Ellis, R.H.; Hamilton, N.R.S.; Lusty, C.; Ndjiondjop, M.N.; Tia, D.; Wenzl, P.; Santos, L.G.; Yazbek, M.; et al. CGIAR genebank viability data reveal inconsistencies in seed collection management. *Glob. Food Secur.* **2021**, *30*, 100557. [CrossRef]
32. Walters, C.; Wheeler, L.M.; Grotenhuis, J.M. Longevity of seeds stored in a genebank: Species characteristics. *Seed Sci. Res.* **2005**, *15*, 1–20. [CrossRef]
33. Solberg, S.Ø.; Yndgaard, F.; Andreasen, C.; Von Bothmer, R.; Loskutov, I.G.; Asdal, Å. Long-term storage and longevity of orthodox seeds: A systematic review. *Front. Plant Sci.* **2020**, *11*, 1007. [CrossRef]
34. Pritchard, H.W. Diversity in seed longevity amongst biodiverse seeds. *Seed Sci. Res.* **2020**, *30*, 75–80. [CrossRef]
35. Ellis, R.H. Temporal patterns of seed quality development, decline, and timing of maximum quality during seed development and maturation. *Seed Sci. Res.* **2019**, *29*, 135–142. [CrossRef]
36. Van Treuren, R.; Bas, N.; Kodde, J.; Groot, S.P.; Kik, C. Rapid loss of seed viability in ex situ conserved wheat and barley at 4 C as compared to −20 C storage. *Conserv. Physiol.* **2018**, *6*, coy033. [CrossRef] [PubMed]
37. Guzzon, F.; Gianella, M.; Velazquez Juarez, J.A.; Sanchez Cano, C.; Costich, D.E. Seed longevity of maize conserved under germplasm bank conditions for up to 60 years. *Ann. Bot.* **2021**, *127*, 775–785. [CrossRef] [PubMed]
38. Ellis, R.H.; Hong, T.D.; Astley, D.; Pinnegar, A.E.; Kraak, H.L. Survival of dry and ultra-dry seeds of carrot, groundnut, lettuce, oilseed rape, and onion during five years' hermetic storage at low temperatures. *Seed Sci. Technol.* **1996**, *24*, 347–358.
39. Solberg, S.O.; Yndgaard, F.; Palmè, A. Morphological and phenological consequences of ex situ conservation of natural populations of red clover (*Trifolium pratense* L.). *Plant Genet. Resour.* **2015**, *15*, 97–108. [CrossRef]
40. Hansen, M.A.E.; Hay, F.R.; Carstensen, J.M. A virtual seed file: The use of multispectral image analysis in the management of genebank seed accessions. *Plant Genet. Resour.* **2016**, *14*, 238–241. [CrossRef]
41. Nguyen, G.N.; Norton, S.L. Genebank phenomics: A strategic approach to enhance value and utilization of crop germplasm. *Plants* **2020**, *9*, 817. [CrossRef]
42. Sackville Hamilton, R. *Scope and Roles of the CGIAR Genebanks: 2030 Vision. Background Paper No. 4 Prepared for the 2020 System Level Review of Genebank Costs and Operations (GCO)*; CGIAR and The Global Crop Diversity Trust: Bonn, Germany, 2020; 22p. Available online: https://www.genebanks.org/wp-content/uploads/2020/09/Session-2_CGIAR-genebanks-2030-vision.pdf (accessed on 10 June 2021).
43. ElMasry, G.; Mandour, N.; Al-Rejaie, S.; Belin, E.; Rousseau, D. Recent applications of multispectral imaging in seed phenotyping and quality monitoring—An overview. *Sensors* **2019**, *19*, 1090. [CrossRef]
44. Feng, L.; Zhu, S.; Liu, F.; He, Y.; Bao, Y.; Zhang, C. Hyperspectral imaging for seed quality and safety inspection: A review. *Plant Methods* **2019**, *15*, 91. [CrossRef]
45. Khoury, C.K.; Achicanoy, H.A.; Bjorkman, A.D.; Navarro-Racines, C.; Guarino, L.; FloresPalacios, X.; Engels, J.M.M.; Wiersema, J.H.; Dempewolf, H.; Ramírez-Villegas, J.; et al. *Where Our Food Crops Come from: A New Estimation of Countries' Interdependence in Plant Genetic Resources*; CIAT Policy Brief No. 25; Centro Internacional de Agricultura Tropical (CIAT): Cali, Colombia, 2015; 4p.
46. Louafi, S.; Welch, E. Improving the global exchange of germplasm for crop breeding. In *Plant Genetic Resources*; Dulloo, M.E., Ed.; Burleigh Dodds Science Publishing: Cambridge, UK, 2021; pp. 81–100.
47. Halewood, M.; Jamora, N.; Noriega, I.L.; Anglin, N.L.; Wenzl, P.; Payne, T.; Ndjiondjop, M.N.; Guarino, L.; Kumar, P.; Yazbek, M.; et al. Germplasm acquisition and distribution by CGIAR genebanks. *Plants* **2020**, *9*, 1296. [CrossRef]
48. Louafi, S.; Manzella, D. International treaty: Heterogeneity and equity in global resources management. In *The Commons, Plant Breeding and Agricultural Research: Challenges for Food Security and Agrobiodiversity*; Girard, F., Frison, C., Eds.; EarthScan: Routledge: London, UK, 2018; pp. 255–271.
49. DivSeek International Network Inc. Strategic Plan. Available online: https://divseekintl.org/strategic-plan/ (accessed on 4 June 2021).
50. Galluzzi, G.; Seyoum, A.; Halewood, M.; López Noriega, I.; Welch, E.W. The Role of Genetic Resources in Breeding for Climate Change: The Case of Public Breeding Programmes in Eighteen Developing Countries. *Plants* **2020**, *9*, 1129. [CrossRef]
51. Ebert, A.W.; Karihaloo, J.L.; Ferreira, M.E., IX. Opportunities, limitations and needs for DNA banks. In *DNA Banks—Providing Novel Options for Genebanks? Topical Reviews in Agricultural Biodiversity*; De Vicente, M.C., Andersson, M.S., Eds.; International Plant Genetic Resources Institute: Rome, Italy, 2006; pp. 61–68.
52. Andersson, M.S. *Feasibility Study for DNA Storage and Use: DNA Banks—Providing Novel Options for Genebanks? A Worldwide Survey*; IPGRI Report No. 6900/D10A; International Plant Genetic Resources Institute (IPGRI): Rome, Italy, 2004; 38p.
53. Halewood, M.; Chiurugwi, T.; Hamilton, R.S.; Kurtz, B.; Marden, E.; Welch, E.W.; Michiels, F.; Mozafari, J.; Sabran, M.; Patron, N.J.; et al. Plant genetic resources for food and agriculture: Opportunities and challenges emerging from the science and information technology revolution. *New Phytol.* **2018**, *217*, 1407–1419. [CrossRef]
54. Kumar, P.L.; Cuvero, M.; Kreuze, J.F.; Muller, G.; Kulkarni, G.; Kumar, S.G.; Massart, S.; Mezzelema, M.; Alakonya, A.; Muchugi, A.; et al. Phytosanitary Interventions for Safe Global Germplasm Exchange and Prevention of Transboundary Pest Spread: The Role of CGIAR Germplasm Health Units. *Plants* **2021**, *10*, 328. [CrossRef] [PubMed]
55. Weise, S.; Lohwasser, U.; Oppermann, M. Document or Lose It—On the Importance of Information Management for Genetic Resources Conservation in Genebanks. *Plants* **2020**, *9*, 1050. [CrossRef]

56. Brink, M.; van Hintum, T. Practical consequences of digital sequence information (DSI) definitions and access and benefit-sharing scenarios from a plant genebank's perspective. *Plants People Planet* **2021**, 1–10. [CrossRef]
57. Alercia, A.; Diulgheroff, S.; Mackay, M. *FAO/Bioversity Multi-Crop Passport Descriptors, V.2.1 [MCPD V.2.1]—December 2015*; Bioversity International: Rome, Italy, 2015; 11p. Available online: https://hdl.handle.net/10568/69166 (accessed on 10 September 2021).
58. Byrne, P.F.; Volk, G.M.; Gardner, C.; Gore, M.A.; Simon, P.W.; Smith, S. Sustaining the future of plant breeding: The critical role of the USDA-ARS National Plant Germplasm System. *Crop Sci.* **2018**, *58*, 451–468. [CrossRef]
59. NCBI GeneBank. GenBank Overview. 2021. Available online: https://www.ncbi.nlm.nih.gov/genbank/ (accessed on 15 July 2021).
60. Alercia, A.; López, F.M.; Sackville Hamilton, N.R.; Marsella, M. *Digital Object Identifiers for Food Crops: Descriptors and Guidelines of the Global Information System*; FAO: Rome, Italy, 2018; 45p.
61. CGIAR Genebank Platform. DOIs: A New Era for Germplasm Labelling. 2021. Available online: https://www.genebanks.org/news-activities/news/dois/ (accessed on 20 May 2021).
62. CGIAR Genebank Platform. CGIAR Centers' Use of Digital Object Identifiers (DOIs): A Submission to the Advisory Committee on the Global Information System. 2019. Available online: https://cgspace.cgiar.org/bitstream/handle/10568/102457/CGIAR%20report%20regarding%20use%20of%20DOIs.pdf?sequence=1 (accessed on 22 June 2021).
63. Ellis, R.H.; Hong, T.D.; Roberts, E.H. Logarithmic relationship between moisture content and longevity in sesame seeds. *Ann. Bot.* **1986**, *57*, 499–503. [CrossRef]
64. Dulloo, M.E.; Engels, J.M.M.; Walters, C.; Hill, L.; Hu, X.; Lu, X.; Chen, X.; Xin, P.; Kameswara Rao, N.; Upadhyaya, H.D.; et al. Presentation on the Global Seed Experiment Regarding Ultra-Dry Seed Technology. 2015. Available online: https://www.researchgate.net/publication/279687146_Presentation_on_the_Global_seed_experiment_regarding_ultra-dry_seed_technology/stats (accessed on 26 May 2021). [CrossRef]
65. Sastry, D.V.S.S.R.; Upadhyaya, H.D.; Gowda, C.L.L. Survival of groundnut seeds under different storage conditions. *J. SAT Agric. Res.* **2007**, *5*, 1–3.
66. Sackville Hamilton, R.; Engels, J.M.M.; Van Hintum, T. Considerations for improved conservation and utilization concepts and strategies. In *A Guide to Effective Management of Germplasm Collections*; IPGRI Handbooks of Genebanks No. 6; Engels, J.M.M., Visser, L., Eds.; IPGRI: Rome, Italy, 2003; pp. 43–59. Available online: https://www.researchgate.net/publication/235929868 (accessed on 10 September 2021).
67. Engels, J.M.M. Plant genetic resources management and conservation strategies: Problems and progress. *Acta Hortic.* **2003**, *623*, 179–192. [CrossRef]
68. Reed, B.M.; Engelmann, F.; Dulloo, M.E.; Engels, J.M.M. *Technical Guidelines for the Management of Field and In Vitro Germplasm Collections*; IPGRI Handbooks for Genebanks No. 7; International Plant Genetic Resources Institute: Rome, Italy, 2004.
69. Schreinemachers, P.; Ebert, A.W.; Wu, M.H. Costing the ex situ conservation of plant genetic resources at AVRDC–The World Vegetable Center. *Genet. Resour. Crop Evol.* **2014**, *61*, 757–773. [CrossRef]
70. Thormann, I.; Engels, J.M.M.; Halewood, M. Are the old International Board for Plant Genetic Resources (IBPGR) base collections available through the Plant Treaty's Multilateral System of Access and Benefit Sharing? A Review. *Genet. Res. Crop Evol.* **2019**, *66*, 291–310. [CrossRef]
71. Engels, J.M.M.; Maggioni, L. Managing germplasm in a virtual European genebank (AEGIS) through networking. In *Theorien der Lebendsammlung: Pflanzen, Mikroben und Tiere als Biofakte in Genbanken*; (Lebenswissenschaften im Dialog, Band 25); Karafyllis, N.C., Ed.; Verlag Karl Alber: Freiburg, Germany, 2018; pp. 169–197.
72. Van Treuren, R.; de Groot, E.C.; van Hintum, T.J. Preservation of seed viability during 25 years of storage under standard genebank conditions. *Genet. Resour. Crop Evol.* **2013**, *60*, 1407–1421. [CrossRef]
73. Asdal, A.; Guarino, L. The Svalbard Global Seed Vault: 10 Years—1 million samples. *Biopreserv. Biobank.* **2018**, *16*, 391–392. [CrossRef]
74. Engels, J.M.M.; (Bioversity International, New Delhi, National Capital Territory, India). Personal communication, 1991.
75. Van Treuren, R.; van Hintum, T.J. Next-generation genebanking: Plant genetic resources management and utilization in the sequencing era. *Plant Genet. Resour.* **2014**, *12*, 298. [CrossRef]
76. Wambugu, P.W.; Ndjiondjop, M.N.; Henry, R.J. Role of genomics in promoting the utilization of plant genetic resources in genebanks. *Brief. Funct. Genom.* **2018**, *17*, 198–206. [CrossRef]
77. McCouch, S.R.; McNally, K.L.; Wang, W.; Sackville Hamilton, R. Genomics of gene banks: A case study in rice. *Am. J. Bot.* **2012**, *99*, 407–423. [CrossRef] [PubMed]
78. Kilian, B.; Graner, A. NGS technologies for analyzing germplasm diversity in genebanks. *Brief. Funct. Genom.* **2012**, *11*, 38–50. [CrossRef]
79. Poland, J.A.; Rife, T.W. Genotyping-by-sequencing for plant breeding and genetics. *Plant Genome* **2012**, *5*, 92–102. [CrossRef]
80. Van Treuren, R.; van Hintum, T.J. Marker-assisted reduction of redundancy in germplasm collections: Genetic and economic aspects. *Acta Hortic.* **2003**, *623*, 139–150. [CrossRef]
81. Mace, E.; Mathur, P.N.; Izquierdo, L.; Hunter, D.; Taylor, M.B.; Singh, D.; DeLacy, I.H.; Jackson, G.V.H.; Godwin, I.D. Rationalization of taro germplasm collections in the Pacific Island region using simple sequence repeat (SSR) markers. *Plant Genet. Resour.* **2006**, *4*, 210–220. [CrossRef]

82. Upadhyay, A.; Aher, L.B.; Shinde, M.P.; Mundankar, K.Y.; Datre, A.; Karibasappa, G.S. Microsatellite analysis to rationalize grape germplasm in India and development of a molecular database. *Plant Genet. Resour.* **2013**, *11*, 225–233. [CrossRef]
83. Van Treuren, R.; de Groot, E.C.; Boukema, I.W.; van de Wiel, C.C.M.; van Hintum, T.J.L. Marker-assisted reduction of redundancy in a genebank collection of cultivated lettuce. *Plant Genet. Resour. Charact. Util.* **2010**, *8*, 95–105. [CrossRef]
84. Sochor, M.; Jemelková, M.; Doležalová, I. Phenotyping and SSR markers as a tool for identification of duplicates in lettuce germplasm. *Czech J. Genet. Plant Breed.* **2019**, *55*, 110–119. [CrossRef]
85. Palmé, A.E.; Hagenblad, J.; Solberg, S.Ø.; Aloisi, K.; Artemyeva, A. SNP Markers and Evaluation of Duplicate Holdings of *Brassica oleracea* in Two European Genebanks. *Plants* **2020**, *9*, 925. [CrossRef]
86. Anglin, N.L.; Amri, A.; Kehel, Z.; Ellis, D. A case of need: Linking traits to genebank accessions. *Biopreservation Biobanking* **2018**, *16*, 337–349. [CrossRef] [PubMed]
87. Singh, N.; Wu, S.; Raupp, W.J.; Sehgal, S.; Arora, S.; Tiwari, V.; Vikram, P.; Singh, S.; Chhuneja, P.; Gill, B.S.; et al. Efficient curation of genebanks using next generation sequencing reveals substantial duplication of germplasm accessions. *Sci. Rep.* **2019**, *9*, 650. [CrossRef] [PubMed]
88. Meyer, R. Encouraging metadata curation in the Diversity Seek initiative. *Nat. Plants* **2015**, *1*, 15099. [CrossRef] [PubMed]
89. Waterworth, W.M.; Masnavi, G.; Bhardwaj, R.M.; Jiang, Q.; Bray, C.M.; West, C.E. A plant DNA ligase is an important determinant of seed longevity. *Plant J.* **2010**, *63*, 848–860. [CrossRef] [PubMed]
90. Dona, M.; Balestrazzi, A.; Mondoni, A.; Rossi, G.; Ventura, L.; Buttafava, A.; Macavei, A.; Sabatini, M.E.; Valassi, A.; Cabonera, D. DNA profiling, telomere analysis and antioxidant properties as tools for monitoring ex situ seed longevity. *Ann. Bot.* **2013**, *111*, 987–998. [CrossRef] [PubMed]
91. Sano, N.; Rajjou, L.; North, H.M.; Debeaujon, I.; Marion-Poll, A.; Seo, M. Staying alive: Molecular aspects of seed longevity. *Plant Cell Physiol.* **2016**, *57*, 660–674. [CrossRef]
92. Lee, J.S.; Velasco-Punzalan, M.; Pacleb, M.; Valdez, R.; Kretzschmar, T.; McNally, K.L.; Ismail, A.M.; Cruz, P.C.S.; Sackville Hamilton, N.R.; Hay, F.R. Variation in seed longevity among diverse Indica rice varieties. *Ann. Bot.* **2019**, *124*, 447–460. [CrossRef]
93. Righetti, K.; Vu, J.L.; Pelletier, S.; Vu, B.L.; Glaab, E.; Lalanne, D.; Pasha, A.; Patel, R.V.; Provart, N.J.; Verdier, J.; et al. Inference of longevity-related genes from a robust coexpression network of seed maturation identifies regulators linking seed storability to biotic defense-related pathways. *Plant Cell* **2015**, *27*, 2692–2708. [CrossRef]
94. Fleming, M.B.; Hill, L.M.; Walters, C. The kinetics of ageing in dry-stored seeds: A comparison of viability loss and RNA degradation in unique legacy seed collections. *Ann. Bot.* **2019**, *123*, 1133–1146. [CrossRef]
95. Waterworth, W.; Bray, C.; West, C. The importance of safeguarding genome integrity in germination and seed longevity. *J. Exp. Bot.* **2015**, *66*, 3549–3558. [CrossRef]
96. Frankel, O.H. Genetic perspective of germplasm conservation. In *Genetic Manipulations: Impact on Man and Society*; Arber, W., Limensee, K., Peacock, W.J., Stralinger, P., Eds.; Cambridge University Press: Cambridge, UK, 1984; pp. 161–170.
97. Van Hintum, T.h.J.L.; Brown, A.H.D.; Spillane, C.; Hodgkin, T. *Core Collections of Plant Genetic Resources*; IPGRI Technical Bulletin No. 3; International Plant Genetic Resources Institute: Rome, Italy, 2000.
98. Upadhyaya, H.D.; Ortiz, R. A mini core subset for capturing diversity and promoting utilization of chickpea genetic resources in crop improvement. *Theor. Appl. Genet.* **2001**, *102*, 1292–1298. [CrossRef]
99. Upadhyaya, H.D.; Yadav, D.; Dronavalli, N.; Gowda, C.L.L.; Singh, S. Mini core germplasm collections for infusing genetic diversity in plant breeding programs. *Electron. J. Plant Breed.* **2010**, *1*, 1294–1309.
100. Schafleitner, R.; Nair, R.M.; Rathore, A.; Wang, Y.W.; Lin, C.Y.; Chu, S.H.; Lin, P.Y.; Chang, J.C.; Ebert, A.W. The AVRDC—The World Vegetable Center mungbean (*Vigna radiata*) core and mini core collections. *BMC Genom.* **2015**, *16*, 344. [CrossRef] [PubMed]
101. Rao, E.S.; Kadirvel, P.; Symonds, R.C.; Geethanjali, S.; Ebert, A.W. Using SSR markers to map genetic diversity and population structure of *Solanum pimpinellifolium* for development of a core collection. *Plant Genet. Resour.* **2012**, *10*, 38. [CrossRef]
102. Bari, A.; Street, K.; Mackay, M.; Endresen, D.T.F.; De Pauw, E.; Amri, A. Focused identification of germplasm strategy (FIGS) detects wheat stem rust resistance linked to environmental variables. *Genet. Resour. Crop Evol.* **2012**, *59*, 1465–1481. [CrossRef]
103. Khazaei, H.; Street, K.; Bari, A.; Mackay, M.; Stoddard, F.L. The FIGS (focused identification of germplasm strategy) approach identifies traits related to drought adaptation in *Vicia faba* genetic resources. *PLoS ONE* **2013**, *8*, e63107. [CrossRef] [PubMed]
104. Hollingsworth, P.M.; Li, D.Z.; van der Bank, M.; Twyford, A.D. Telling plant species apart with DNA: From barcodes to genomes. *Philos. Trans. R. Soc. B Biol. Sci.* **2016**, *371*, 20150338. [CrossRef]
105. Gugerli, F.; Parducci, L.; Petit, R.J. Ancient plant DNA: Review and prospects. *New Phytol.* **2005**, *166*, 409–418. [CrossRef]
106. Zhang, G.J.; Dong, R.; Lan, L.N.; Li, S.F.; Gao, W.J.; Niu, H.X. Nuclear integrants of organellar DNA contribute to genome structure and evolution in plants. *Int. J. Mol. Sci.* **2020**, *21*, 707. [CrossRef]
107. Hawkins, R.D.; Hon, G.C.; Ren, B. Next-generation genomics: An integrative approach. *Nat. Rev. Genet.* **2010**, *11*, 476–486. [CrossRef]
108. Poland, J.; Endelman, J.; Dawson, J.; Rutkoski, J.; Wu, S.; Manes, Y.; Dreisigacker, S.; Crossa, J.; Sánchez-Villeda, H.; Sorrells, M.; et al. Genomic selection in wheat breeding using genotyping-by-sequencing. *Plant Genome* **2012**, *5*, 103–113. [CrossRef]
109. Spindel, J.; Iwata, H. Genomic Selection in Rice Breeding. In *Rice Genomics, Genetics and Breeding*; Sasaki, T., Ashikari, M., Eds.; Springer: Singapore, 2018; pp. 473–496. [CrossRef]

110. Spindel, J.; Begum, H.; Akdemir, D.; Virk, P.; Collard, B.; Redoña, E.; Atlin, G.; Jannink, J.-L.; McCouch, S.R. Genomic Selection and Association Mapping in Rice (*Oryza sativa*): Effect of Trait Genetic Architecture, Training Population Composition, Marker Number and Statistical Model on Accuracy of Rice Genomic Selection in Elite, Tropical Rice Breeding Lines. *PLoS Genet.* **2015**, *11*, e1004982. [CrossRef]
111. Wang, X.; Xu, Y.; Hu, Z.; Xu, C. Genomic selection methods for crop improvement: Current status and prospects. *Crop J.* **2018**, *6*, 330–340. [CrossRef]
112. Prohens, J.; Gramazio, P.; Plazas, M.; Dempewolf, H.; Kilian, B.; Díez, M.J.; Fita, A.; Herraiz, F.J.; Rodríguez-Burruezo, A.; Soler, S.; et al. Introgressiomics: A new approach for using crop wild relatives in breeding for adaptation to climate change. *Euphytica* **2017**, *213*, 158. [CrossRef]
113. Song, Q.; Yan, L.; Quigley, C.; Jordan, B.D.; Fickus, E.; Schroeder, S.; Song, B.H.; Charles, A.Y.Q.; Hyten, D.; Nelson, R.; et al. Genetic characterization of the soybean nested association mapping population. *Plant Genome* **2017**, *10*, 1–14. [CrossRef]
114. Furbank, R.T.; Tester, M. Phenomics–technologies to relieve the phenotyping bottleneck. *Trends Plant Sci.* **2011**, *16*, 635–644. [CrossRef] [PubMed]
115. Nielsen, N.H.; Jahoor, A.; Jensen, J.D.; Orabi, J.; Cericola, F.; Edriss, V.; Jensen, J. Genomic prediction of seed quality traits using advanced barley breeding lines. *PLoS ONE* **2016**, *11*, e0164494. [CrossRef]
116. Ćwiek-Kupczyńska, H.; Altmann, T.; Arend, D.; Arnaud, E.; Chen, D.; Cornut, G.; Fiorani, F.; Frohmberg, W.; Junker, A.; Klukas, C.; et al. Measures for interoperability of phenotypic data: Minimum information requirements and formatting. *Plant Methods* **2016**, *12*, 44. [CrossRef] [PubMed]
117. Reynolds, D.; Baret, F.; Welcker, C.; Bostrom, A.; Ball, J.; Cellini, F.; Lorence, A.; Chawade, A.; Khafif, M.; Noshita, K.; et al. What is cost-efficient phenotyping? Optimizing costs for different scenarios. *Plant Sci.* **2019**, *282*, 14–22. [CrossRef] [PubMed]
118. Xu, Y. Envirotyping for deciphering environmental impacts on crop plants. *Theor. Appl. Genet.* **2016**, *129*, 653–673. [CrossRef]
119. Crossa, J.; Jarquín, D.; Franco, J.; Pérez-Rodríguez, P.; Burgueño, J.; Saint-Pierre, C.; Vikram, P.; Sansaloni, C.; Petroli, C.; Akdemir, D.; et al. Genomic prediction of gene bank wheat landraces. *G3 Genes Genom. Genet.* **2016**, *6*, 1819–1834. [CrossRef] [PubMed]
120. Gorjanc, G.; Jenko, J.; Hearne, S.J.; Hickey, J.M. Initiating maize pre-breeding programs using genomic selection to harness polygenic variation from landrace populations. *BMC Genom.* **2016**, *17*, 30. [CrossRef]
121. Bustos-Korts, D.; Boer, M.P.; Malosetti, M.; Chapman, S.; Chenu, K.; Zheng, B.; Van Eeuwijk, F.A. Combining crop growth modeling and statistical genetic modeling to evaluate phenotyping strategies. *Front. Plant Sci.* **2019**, *10*. [CrossRef] [PubMed]
122. Oldham, P.; Hall, S.; Burton, G. Synthetic Biology: Mapping the Scientific Landscape. *PLoS ONE* **2012**, *7*, e34368. [CrossRef]
123. Keiper, F.; Atanassova, A. Regulation of Synthetic Biology: Developments Under the Convention on Biological Diversity and Its Protocols. *Front. Bioeng. Biotechnol.* **2020**, *8*, 310. [CrossRef] [PubMed]
124. Jiang, F.; Doudna, J.A. CRISPR–Cas9 structures and mechanisms. *Ann. Rev. Biophys.* **2017**, *46*, 505–529. [CrossRef] [PubMed]
125. Legros, M.; Xu, C.; Morrison, A.; Scott, T.W.; Lloyd, A.L.; Gould, F. Modeling the dynamics of a non-limited and a self-limited gene drive system in structured *Aedes aegypti* populations. *PLoS ONE* **2013**, *8*, e83354. [CrossRef]
126. Zsögön, A.; Čermák, T.; Naves, E.R.; Notini, M.M.; Edel, K.H.; Weinl, S.; Freschi, L.; Voytas, D.; Kudla, J.; Peres, L.E.P. De novo domestication of wild tomato using genome editing. *Nat. Biotechnol.* **2018**, *36*, 1211–1216. [CrossRef] [PubMed]
127. Borrelli, V.M.; Brambilla, V.; Rogowsky, P.; Marocco, A.; Lanubile, A. The enhancement of plant disease resistance using CRISPR/Cas9 technology. *Front. Plant Sci.* **2018**, *9*, 1245. [CrossRef] [PubMed]
128. Yin, K.; Gao, C.; Qiu, J.L. Progress and prospects in plant genome editing. *Nat. Plants* **2017**, *3*, 1–6. [CrossRef] [PubMed]
129. Bortesi, L.; Fischer, R. The CRISPR/Cas9 system for plant genome editing and beyond. *Biotechnol. Adv.* **2015**, *33*, 41–52. [CrossRef]
130. Zhu, H.; Li, C.; Gao, C. Applications of CRISPR–Cas in agriculture and plant biotechnology. *Nat. Rev. Mol. Cell Biol.* **2020**, *21*, 661–677. [CrossRef]
131. Santillán Martínez, M.I.; Bracuto, V.; Koseoglou, E.; Appiano, M.; Jacobsen, E.; Visser, R.G.; Wolters, A.M.A.; Bai, Y. CRISPR/Cas9-targeted mutagenesis of the tomato susceptibility gene *PMR4* for resistance against powdery mildew. *BMC Plant Biol.* **2020**, *20*, 284. [CrossRef]
132. Wang, Y.; Cheng, X.; Shan, Q.; Zhang, Y.; Liu, J.; Gao, C.; Qiu, J.L. Simultaneous editing of three homoeoalleles in hexaploid bread wheat confers heritable resistance to powdery mildew. *Nat. Biotechnol.* **2014**, *32*, 947–951. [CrossRef] [PubMed]
133. Baltes, N.J.; Hummel, A.W.; Konecna, E.; Cegan, R.; Bruns, A.N.; Bisaro, D.M.; Voytas, D.F. Conferring resistance to geminiviruses with the CRISPR–Cas prokaryotic immune system. *Nat. Plants* **2015**, *1*, 1–4. [CrossRef]
134. Ji, X.; Zhang, H.; Zhang, Y.; Wang, Y.; Gao, C. Establishing a CRISPR–Cas-like immune system conferring DNA virus resistance in plants. *Nat. Plants* **2015**, *1*, 15144. [CrossRef]
135. Ebert, A.W. Potential of underutilized traditional vegetables and legume crops to contribute to food and nutritional security, income and more sustainable production systems. *Sustainability* **2014**, *6*, 319–335. [CrossRef]
136. Borelli, T.; Hunter, D.; Powell, B.; Ulian, T.; Mattana, E.; Termote, C.; Pawera, L.; Beltrame, D.; Penafiel, D.; Tan, A.; et al. Born to Eat Wild: An Integrated Conservation Approach to Secure Wild Food Plants for Food Security and Nutrition. *Plants* **2020**, *9*, 1299. [CrossRef] [PubMed]
137. Rao, E.S.; Kadirvel, P.; Symonds, R.C.; Geethanjali, S.; Thontadarya, R.N.; Ebert, A.W. Variations in DREB1A and VP1.1 Genes Show Association with Salt Tolerance Traits in Wild Tomato (*Solanum pimpinellifolium*). *PLoS ONE* **2015**, *10*, e0132535. [CrossRef] [PubMed]

138. Li, T.; Yang, X.; Yu, Y.; Si, X.; Zhai, X.; Zhang, H.; Dong, W.; Gao, C.; Xu, C. 2018. Domestication of wild tomato is accelerated by genome editing. *Nat. Biotechnol.* **2018**, *36*, 1160–1163. [CrossRef] [PubMed]
139. Lemmon, Z.H.; Reem, N.T.; Dalrymple, J.; Soyk, S.; Swartwood, K.E.; Rodriguez-Leal, D.; Van Eck, J.; Lippman, Z.B. Rapid improvement of domestication traits in an orphan crop by genome editing. *Nat. Plants* **2018**, *4*, 766–770. [CrossRef] [PubMed]
140. Gao, C. The future of CRISPR technologies in agriculture. *Nat. Rev. Mol. Cell Biol.* **2018**, *19*, 275–276. [CrossRef]
141. Oldham, P. Digital Sequence Information—Technical Aspects. 2020. Available online: https://ec.europa.eu/environment/nature/biodiversity/international/abs/pdf/Final_Report_technical_aspects_of_DSI.pdf (accessed on 8 May 2021).
142. Wang, C.; Hu, S.; Gardner, C.; Lübberstedt, T. Emerging avenues for utilization of exotic germplasm. *Trends Plant Sci.* **2017**, *22*, 624–637. [CrossRef]
143. Walters, C.; Hanner, R. Platforms for DNA Banking. In *Issues in Genetic Resources DNA Banks—Providing Novel Options for Genebanks?* de Vicente, M.C., Andersson, M.S., Eds.; International Plant Genetic Resources Institute: Rome, Italy, 2006; pp. 25–35.
144. Genesys. Online platform for PGRFA. 2021. Available online: https://www.genesys-pgr.org/ (accessed on 13 July 2021).
145. USDA. GRIN Global. 2021. Available online: https://www.nal.usda.gov/grin-global (accessed on 13 July 2021).
146. ECPGR. EURISCO Catalogue. Available online: https://www.ecpgr.cgiar.org/resources/germplasm-databases/eurisco-catalogue/ (accessed on 2 August 2021).
147. BrAPI. BrAPI V1.3. Available online: https://brapi.docs.apiary.io/#introduction/url-structure (accessed on 2 August 2021).
148. Research Data Alliance (RDA). About RDA. Available online: https://www.rd-alliance.org/about-rda (accessed on 2 August 2021).
149. Global Open Data. Harnessing Open Data to Achieve Global Food Security. Available online: https://www.godan.info/ (accessed on 2 August 2021).
150. GBIF. Free and Open Access to Biodiversity Data. 2021. Available online: https://www.gbif.org/ (accessed on 13 July 2021).
151. Paton, A.; Antonelli, A.; Carine, M.; Forzza, R.C.; Davies, N.; Demissew, S.; Dröge, G.; Fulcher, T.; Grall, A.; Holstein, N.; et al. Plant and fungal collections: Current status, future perspectives. *Plants People Planet* **2020**, *2*, 499–514. [CrossRef]
152. Global Genome Biodiversity Network (GGBN). Portal. Available online: https://www.ggbn.org/ggbn_portal/ (accessed on 2 August 2020).
153. Seberg, O.; Droege, G.; Barker, K.; Coddington, J.A.; Funk, V.; Gostel, M.; Petersen, G.; Smith, P.P. Global Genome Biodiversity Network: Saving a blueprint of the Tree of Life–a botanical perspective. *Ann. Bot.* **2016**, *118*, 393–399. [CrossRef]
154. Droege, G.; Barker, K.; Astrin, J.J.; Bartels, P.; Butler, C.; Cantrill, D.; Coddington, J.; Forest, F.; Gemeinholzer, B.; Hobern, D.; et al. The global genome biodiversity network (GGBN) data portal. *Nucleic Acids Res.* **2014**, *42*, D607–D612. [CrossRef]
155. CBD. Information about the Secretariat. Available online: https://www.cbd.int/ (accessed on 2 August 2021).
156. CBD. The Nagoya Protocol on Access and Benefit-Sharing. Available online: https://www.cbd.int/abs/ (accessed on 2 August 2021).
157. CITES. What Is CITES? Available online: https://cites.org/eng/disc/what.php (accessed on 2 August 2021).
158. International Nucleotide Sequence Database Collaboration (INSDC). About INSDC. Available online: https://www.insdc.org/ (accessed on 2 August 2021).
159. IPPN. Plant Phenotyping for a Sustainable Future. Available online: https://www.plant-phenotyping.org/IPPN_home (accessed on 2 August 2021).
160. Pieruschka, R.; Schurr, U. Plant phenotyping: Past, present, and future. *Plant Phenomics* **2019**, *2019*, 7507131. [CrossRef]
161. NAPPN. North American Plant Phenotyping Network. Available online: https://nappn.plant-phenotyping.org/# (accessed on 2 August 2021).
162. EMPHASIS. About. Available online: https://emphasis.plant-phenotyping.eu/ (accessed on 2 August 2021).
163. IRIC. International Rice Informatics Network. Available online: http://iric.irri.org/ (accessed on 2 August 2021).
164. CIMMYT. International Wheat Improvement Network (IWIN). Available online: https://www.cimmyt.org/funder_partner/international-wheat-improvement-network-iwin/ (accessed on 2 August 2021).
165. World Vegetable Center. APSA-WorldVeg Vegetable Breeding Consortium. Available online: https://avrdc.org/apsa-worldveg-vegetable-breeding-consortium/ (accessed on 2 August 2021).
166. AnnoSys. AnnoSys—Online Annotation of Biodiversity Data. Available online: https://annosys.bgbm.fu-berlin.de/AnnoSys/AnnoSys (accessed on 2 August 2021).
167. Holetschek, J.; Dröge, G.; Güntsch, A.; Berendsohn, W.G. The ABCD of primary biodiversity data access. *Plant Biosyst. Int. J. Deal. All Aspects Plant Biol.* **2012**, *146*, 771–779. [CrossRef]
168. TDWG. Darwin Core. Available online: https://dwc.tdwg.org/ (accessed on 2 August 2021).
169. Suhrbier, L.; Kusber, W.H.; Tschöpe, O.; Güntsch, A.; Berendsohn, W.G. AnnoSys—implementation of a generic annotation system for schema-based data using the example of biodiversity collection data. *Database* **2017**, *2017*, 1–20. [CrossRef]
170. Gilliot, P.A.; Gorochowski, T.E. Sequencing enabling design and learning in synthetic biology. *Curr. Opin. Chem. Biol.* **2020**, *58*, 54–62. [CrossRef]
171. Rhoden, F.; Scholz, A.H. The international political process around Digital Sequence Information under the Convention on Biological Diversity and the 2018–2020 intersessional period. *Plants People Planet* **2021**, *00*, 1–10. [CrossRef]
172. Kharb, D. The legal conundrum over regulation of access and benefit sharing obligations in digital sequence information over genetic resources-assessing Indian position. *J. World Intellect. Prop.* **2021**, *24*, 152–166. [CrossRef]

173. Lawson, C.; Humphries, F.; Rourke, M. The future of information under the CBD, Nagoya Protocol, Plant Treaty, and PIP Framework. *J. World Intellect. Prop.* **2019**, *22*, 103–119. [CrossRef]
174. Rabone, M.; Harden-Davies, H.; Collins, J.E.; Zajderman, S.; Appeltans, W.; Droege, G.; Brandt, A.; Pardo-Lopez, L.; Dahlgren, T.G.; Glover, A.G.; et al. Access to marine genetic resources (MGR): Raising awareness of best-practice through a new agreement for biodiversity beyond national jurisdiction (BBNJ). *Front. Mar. Sci.* **2019**, *6*, 520. [CrossRef]
175. Laird, S.; Wynberg, R.; Rourke, M.; Humphries, F.; Muller, M.R.; Lawson, C. Rethink the expansion of access and benefit sharing. *Science* **2020**, *367*, 1200–1202. [CrossRef]
176. Houssen, W.; Sara, R.; Jaspers, M. Digital sequence information on genetic resources: Concept, scope and current use. In Proceedings of the Convention on Biological Diversity, Montreal, QC, Canada, 17–20 March 2020. Available online: https://www.cbd.int/doc/c/fef9/2f90/70f037ccc5da885dfb293e88/dsi-ahteg-2020--01--03-en.pdf (accessed on 31 May 2021).
177. FAO. *Norway Contributes to International Treaty*; FAO: Rome, Italy, 2020. Available online: http://www.fao.org/plant-treaty/news/news-detail/en/c/1296008/ (accessed on 1 June 2021).
178. Scholz, A.; Upneet, H.; Freitag, J.; Cancio, I.; dos Santos Ribeiro, C.; Haringhuizen, G.; Oldham, P.; Saxena, D.; Seitz, C.; Thiele, T.; et al. *Finding Compromise on ABS & DSI in the CBD: Requirements & Policy Ideas from a Scientific Perspective*; Leibniz Institute DSMZ-German Collection of Microorganisms and Cell Cultures GmbH: Braunschweig, Germany, 2020. [CrossRef]
179. Cochrane, G.; Karsch-Mizrachi, I.; Takagi, T. International Nucleotide Sequence Database Collaboration. The international nucleotide sequence database collaboration. *Nucleic Acids Res.* **2016**, *44*, D48–D50. [CrossRef]
180. Gollin, D. Conserving genetic resources for agriculture: Economic implications of emerging science. *Food Secur.* **2020**, *12*, 919–927. [CrossRef]
181. FAO. *Resolution 8/83 International Undertaking on Plant Genetic Resources*; FAO: Rome, Italy, 1983; 8p. Available online: http://www.fao.org/wiews-archive/docs/Resolution_8_83.pdf (accessed on 24 April 2021).
182. CBD. Aichi Biodiversity Targets. 18 September 2020. Available online: https://www.cbd.int/sp/targets/ (accessed on 12 July 2021).
183. FAO. Strategic Plan for the Commission on Genetic Resources for Food and Agriculture (2019–2027). 2021. Available online: http://www.fao.org/3/ca8345en/ca8345en.pdf (accessed on 12 July 2021).
184. Khoury, C.; Laliberté, B.; Guarino, L. Trends in ex situ conservation of plant genetic resources: A review of global crop and regional conservation strategies. *Genet. Res. Crop Evol.* **2010**, *57*, 625–639. [CrossRef]
185. Crop Trust. Global Crop Conservation Strategies. *Breathing New Life into Global Crop Conservation Strategies*. 2021. Available online: https://www.croptrust.org/project/conservation-strategies/ (accessed on 12 July 2021).
186. FAO. *The State of the World's Plant Genetic Resources for Food and Agriculture*; FAO: Rome, Italy, 1997. Available online: http://www.fao.org/3/w7324e/w7324e.pdf (accessed on 12 July 2021).
187. FAO. *Second Global Plan of Action for Plant Genetic Resources for Food and Agriculture*; FAO: Rome, Italy, 2011.
188. FAO. *The Second Report on the State of the World's Plant Genetic Resources*; FAO: Rome, Italy, 2010. Available online: http://www.fao.org/docrep/013/i1500e/i1500e.pdf (accessed on 12 July 2021).
189. FAO. WIEWS Reporting Tool for the Second Global Plan of Action for PGRFA. 2021. Available online: http://www.fao.org/pgrfa/ (accessed on 12 July 2021).
190. FAO. Commission on Genetic Resources for Food and Agriculture. Country Reports. 2021. Available online: http://www.fao.org/cgrfa/topics/biodiversity/countryreports/en/ (accessed on 12 July 2021).
191. FAO. International Treaty for Plant Genetic Resources. The Multilateral System. 2021. Available online: http://www.fao.org/plant-treaty/areas-of-work/the-multilateral-system/overview/en/ (accessed on 12 July 2021).
192. Gullotta, G.; (Bioversity International, Rome, Italy). Personal communication, 2021.
193. Díez, M.J.; De la Rosa, L.; Martín, I.; Guasch, L.; Cartea, M.E.; Mallor, C.; Casals, J.; Simó, J.; Rivera, A.; Anastasio, G.; et al. Plant genebanks: Present situation and proposals for their improvement. The case of the Spanish Network. *Front. Plant Sci.* **2018**, *9*, 1794. [CrossRef]
194. FAO; Commission on Genetic Resources for Food and Agriculture. Conserving Biodiversity for Food and Agriculture and Promoting Its Use in Support of Global Food Security and Sustainable Development, for Present and Future Generations. 2021. Available online: http://www.fao.org/cgrfa/en/ (accessed on 13 July 2021).
195. FAO. Members of the Commission on Genetic Resources for Food and Agriculture. 2021. Available online: http://www.fao.org/3/I8638EN/i8638en.pdf (accessed on 13 July 2021).
196. Hodgkin, T.; Demers, N.; Frison, E. The evolving global system of conservation and use of plant genetic resources for food and agriculture. What is it and where does the Treaty fit in? In *Crop Genetic Resources as a Global Commons*, 1st ed.; Halewood, M., Lopez Noriega, I., Louafi, S., Eds.; Routledge: London, UK, 2012; 24p. [CrossRef]
197. FAO. WIEWS—World Information and Early Warning System on Plant Genetic Resources for Food and Agriculture. 2021. Available online: http://www.fao.org/wiews/en/ (accessed on 13 July 2021).
198. CGIAR Genebank Platform. The Genebank Platform. 2021. Available online: https://www.genebanks.org/the-platform/ (accessed on 13 July 2021).
199. Svalbard Global Seed Vault. Safeguarding Seeds for the Future. 2021. Available online: https://www.seedvault.no/ (accessed on 3 August 2021).

200. FAO. Global Safety Back-up Cryopreservation Facility. 31 August 2018. Available online: http://www.fao.org/3/CA1371EN/ca1371en.pdf (accessed on 13 July 2021).
201. ECPGR. Overview. 2021. Available online: https://www.ecpgr.cgiar.org/about/overview (accessed on 13 July 2021).
202. African Orphan Crops Consortium. Healthy and Prosperous Africa through Nutritious Local African Crops and Trees. 2021. Available online: http://africanorphancrops.org/ (accessed on 13 July 2021).
203. AGRA. Our Story. 2021. Available online: https://agra.org/our-story/ (accessed on 13 July 2021).
204. FAO. Ninth Meeting of the Ad Hoc Open-Ended Working Group to Enhance the Functioning of the Multilateral System. October 2019. Available online: http://www.fao.org/3/nb653en/nb653en.pdf (accessed on 13 July 2021).
205. Fu, Y.-B. The vulnerability of plant genetic resources conserved ex situ. *Crop Sci.* **2017**, *57*, 2314–2328. [CrossRef]
206. Vernooy, R.; Akhter, F.; Alonzo, S.; Mokoena, L.M.; Mushita, A.; Otieno, G.; Shrestha, P. Strengthening Smallholder Farmers' Capacity to Adapt to Climate Change: Roles of Community Seedbanks. In *Handbook of Climate Change Management*; Leal Filho, W., Ed.; Springer Nature Switzerland AG: Cham, Switzerland, 2021; 22p. [CrossRef]
207. Spillane, C.; Engels, J.M.M.; Fassil, H.; Withers, L.; Cooper, D. *Strengthening National Programmes for Plant Genetic Resources for Food and Agriculture: Planning and Coordination*; Issues in Genetic Resources No. 8; IPGRI: Rome, Italy, 1999; 51p.
208. CGIAR Genebank Platform. Quality Management. 2021. Available online: https://www.genebanks.org/the-platform/quality-management/ (accessed on 13 July 2021).
209. FAO. *Assessment of the Implementation of the Second Global Plan of Action for Plant Genetic Resources for Food and Agriculture 2012–2014*; CGRFA-16/17/Inf.17.2; FAO: Rome, Italy, 2016. Available online: http://www.fao.org/3/a-mr796e.pdf (accessed on 3 August 2021).
210. Crop Genebank Knowledge Base. Safety Duplication. 2021. Available online: https://cropgenebank.sgrp.cgiar.org/index.php/procedures-mainmenu-242/safety-duplication-mainmenu-207 (accessed on 13 July 2021).
211. Westengen, O.T.; Jeppson, S.; Guarino, L. Global ex-situ crop diversity conservation and the Svalbard Global Seed Vault: Assessing the current status. *PLoS ONE* **2013**, *8*, e64146. [CrossRef]
212. Crop Genebank Knowledge Base. Health Diagnosis. 2021. Available online: https://cropgenebank.sgrp.cgiar.org/index.php/procedures-mainmenu-243/germplasm-testing-mainmenu-197/health-diagnosis-mainmenu-183 (accessed on 13 July 2021).
213. CGIAR Genebank Platform. Germplasm Health. 2021. Available online: https://www.genebanks.org/the-platform/germplasm-health/ (accessed on 13 July 2021).
214. Frison, E.A.; Putter, C.A.J. FAO/IBPGR Technical Guidelines for the Safe Movement of Germplasm. In Proceedings of the 5th International Congress of Plant Pathology, Kyoto, Japan, 20–27 August 1988; p. 279.
215. Crop Genebank Knowledge Base. Distribution. 2021. Available online: https://cropgenebank.sgrp.cgiar.org/index.php/procedures-mainmenu-243/distribution-mainmenu-208 (accessed on 13 July 2021).
216. Crop Genebank Knowledge Base. Safe Transfer of Germplasm (STOG). 2021. Available online: https://cropgenebank.sgrp.cgiar.org/index.php/management-mainmenu-433/stogs-mainmenu-238 (accessed on 13 July 2021).
217. Crop Trust. Welcome to the Genebank Online Reporting Tool. 2021. Available online: https://grants.croptrust.org/ (accessed on 13 July 2021).
218. CGIAR Genebank Platform. Distribution of Samples Within and Outside the CGIAR. 2021. Available online: https://www.genebanks.org/resources/genebanks-in-numbers/distribution/ (accessed on 13 July 2021).
219. Crop Genebank Knowledge Base. Regeneration. 2021. Available online: https://cropgenebank.sgrp.cgiar.org/index.php/procedures-mainmenu-243/regeneration-mainmenu-206 (accessed on 13 July 2021).
220. Crop Genebank Knowledge Base. Introduction for the Regeneration Guidelines of Crops. 2021. Available online: https://cropgenebank.sgrp.cgiar.org/index.php/crops-mainmenu-367/other-crops-regeneration-guidelines-mainmenu-290 (accessed on 13 July 2021).
221. CGIAR Genebank Platform. Genebank Operations. 2021. Available online: https://www.genebanks.org/resources/genebanks-in-numbers/genebank-operations-data/ (accessed on 13 July 2021).
222. Ebert, A.W.; (Formerly World Vegetable Center, Shanhua, Tainan City, Taiwan). Personal communication, 2015.
223. Crop Genebank Knowledge Base. Characterization. 2021. Available online: https://cropgenebank.sgrp.cgiar.org/index.php/procedures-mainmenu-243/characterization-mainmenu-205 (accessed on 13 July 2021).
224. Engels, J.; Rudebjer, P. Strengthening Institutions and Organizations, and Building Capacity for the Conservation and Use of Agricultural Biodiversity. In *Routledge Handbook of Agricultural Biodiversity*, 1st ed.; Hunter, D., Guarino, L., Spillane, C., McKeown, P.C., Eds.; Routledge: London, UK, 2017; pp. 627–645.
225. ECPGR. AQUAS: Quality System for AEGIS. 2021. Available online: https://www.ecpgr.cgiar.org/aegis/aquas/overview (accessed on 13 July 2021).
226. Crop Genebank Knowledge Base. Quality Management. 2021. Available online: https://cropgenebank.sgrp.cgiar.org/index.php/management-mainmenu-433/quality-management-mainmenu-357 (accessed on 13 July 2021).
227. FAO. WIEWS—World Information and Early Warning System on Plant Genetic Resources for Food and Agriculture. Ex Situ (SDG 2.5.1)—Overview. 2021. Available online: http://www.fao.org/wiews/data/ex-situ-sdg-251/overview/en/ (accessed on 13 July 2021).
228. FAO. Global Information System for PGRFA. 2021. Available online: https://ssl.fao.org/glis/ (accessed on 13 July 2021).
229. ECPGR. EURISCO. 2021. Available online: https://eurisco.ipk-gatersleben.de/apex/f?p=103:1:::::: (accessed on 13 July 2021).

230. Kew. Our Collections. 2021. Available online: https://www.kew.org/science/collections-and-resources/collections (accessed on 13 July 2021).
231. ECPGR. AEGIS—A European Genebank Integrated System. Available online: https://www.ecpgr.cgiar.org/aegis (accessed on 3 August 2021).
232. Pearce, T.R.; Antonelli, A.; Brearley, F.Q.; Couch, C.; Campostrini Forzza, R.; Gonçalves, S.C.; Magassouba, S.; Morim, M.P.; Mueller, G.M.; Nic Lughadha, E.; et al. International collaboration between collections-based institutes for halting biodiversity loss and unlocking the useful properties of plants and fungi. *Plants People Planet* **2020**, *2*, 515–534. [CrossRef]
233. Bramel, P.; Krishnan, S. Systematic assessment for conservation and utilisation of crop genetic resources. In *Sustaining Global Food Security: The Nexus of Science and Policy*; Zeigler, R., Ed.; CSIRO Publishing: Clayton, Australia, 2019; pp. 3–14.
234. Smale, M.; Jamora, N.; Guarino, L. Valuing plant genetic resources in genebanks: Past, present and future. In *Plant Genetic Resources*; Dulloo, M.E., Ed.; Burleigh Dodds Science Publishing: Cambridge, UK, 2021; pp. 35–53. [CrossRef]

plants MDPI

Article

Genebanks at Risk: Hazard Assessment and Risk Management of National and International Genebanks

Theresa Herbold [1,*] and Johannes M. M. Engels [2]

[1] Faculty of Agricultural Sciences, University of Hohenheim, 70599 Stuttgart, Germany
[2] Alliance of Bioversity International and CIAT, 00153 Rome, Italy; j.engels@cgiar.org
* Correspondence: theresa.herbold@gmx.de; Tel.: +49-(0)17639208009

Abstract: Genebanks are crucial for safeguarding global crop diversity but are themselves exposed to several risks. However, a scientific basis for identifying, assessing, and managing risks is still lacking. Addressing these research gaps, this study provides risk analysis for three key risk groups: natural hazards, political risks, and financial risks, carried out on a sample of 80 important national and international genebanks, comprising at least 4.78 million accessions or roughly 65% of the reported total of ex situ conserved accessions worldwide. The assessment tool of Munich Re "Natural Hazards Edition" allowed a location-specific comparison of the natural hazard exposure. Results showed that genebanks in the Asia-Pacific region are most exposed to natural hazards, while institutions in African and some Asian countries are rather vulnerable to political risks. Financing is a major problem for national genebanks in developing countries, whereas the Global Crop Diversity Trust achieved considerable financial security for international genebanks. Large differences in the risk exposure of genebanks exist, making a location- and institution-specific risk assessment indispensable. Moreover, there is significant room for improvement with respect to quality and risk management at genebanks. Transferring risks of genebanks to third parties is underdeveloped and should be used more widely.

Keywords: genebanks; plant genetic resources; hazard assessment; natural hazards; political risks; risk management; risk prevention; risk mitigation; risk transfer; insurance

Citation: Herbold, T.; Engels, J.M.M. Genebanks at Risk: Hazard Assessment and Risk Management of National and International Genebanks. *Plants* **2023**, *12*, 2874. https://doi.org/10.3390/plants12152874

Academic Editor: Yong-Bi Fu

Received: 29 June 2023
Revised: 30 July 2023
Accepted: 1 August 2023
Published: 4 August 2023

1. Introduction

In the context of protecting agrobiodiversity, the conservation of plant genetic resources (PGR) has become an important pillar whereby ex situ conservation realized off-site in genebanks is the principal approach [1,2]. In about 1750 genebanks and collections, nearly 7.4 million plant accessions are maintained worldwide [3].

Genebanks are essential for conserving and making available PGR not only for current progress in plant breeding but also as a treasure for use by future generations. Despite the growing worldwide awareness of this potential, it is often overlooked that genebanks themselves are subject to several challenges and risks that might jeopardize their physical integrity. Past incidents and events resulting in the partial loss of important crop collections or even entire genebanks have shown this drastically. A recent example is the ICARDA genebank, which was originally located in Aleppo, Syria. As a consequence of the Syrian civil war and severe combat operations in Aleppo starting in 2012, the genebank had to be relocated in 2016 to Lebanon and Morocco. Part of the germplasm collection could be restored, with safety duplicates preserved at international genebanks and at the Svalbard Global Seed Vault (SGSV) in Spitsbergen, Norway [4,5]. In 2011, the national genebank of Thailand was flooded, which caused the loss of some of the 20,000 unique rice accessions maintained there [5]. The national genebank of the Philippines at Los Baños was damaged by flooding due to a typhoon in 2006 and hit again by a fire in 2012 [5,6]. Ukraine's seed bank in Kharkiv was at high risk as the city has been a staging ground of military operations since the start of the Russian invasion of Ukraine in February 2022 [7]. In a consolidated effort, the collection was relocated in 2023 [8].

Genebanks are exposed to manifold endogenous and exogenous risks. However, a systematic and comparative assessment of their exposure, risks, and vulnerabilities is still missing. Also, risk management at genebanks—an emerging topic—has been barely studied scientifically. This study aims to fill these research gaps based on a selection comprising the world's most important genebanks.

2. Objectives and Methodology

The objectives of this study were twofold: (1) to compare the most important genebanks worldwide with respect to their exposure to natural hazards and political and financial risks, and (2) based on this risk analysis, to develop risk management strategies for genebanks.

Based on these objectives, two methodological steps have been undertaken:

- Developing a risk analysis and risk management framework suitable for genebanks;
- Applying this framework to a selected, representative sample of the most important germplasm holdings worldwide.

2.1. Risk Analysis and Risk Management Framework

A methodology was developed following definitions, concepts, and frameworks for risk analysis and management used in the scientific literature of environmental hazard appraisals [9] and entrepreneurial risk management approaches [10], which were deemed the most suited for evaluating genebanks. The term *risk management* was used as defined by Wolke [10], involving four steps. These were refined and interpreted for this study as follows:

Step 1: Risk identification

Based on a literature review, expert interviews, and own assessments, the most important risks were classified as follows:

- Natural hazards (exogenous risks);
- Political risks (exogenous risks);
- Financial risks (exo- and endogenous risks).

Step 2: Risk measurement and analysis

The definition of risk as proposed by Dalezios (2017) [9] was used in this study:

$$risk = hazard \times vulnerability \times amount\ of\ elements\ at\ risk$$

The term *hazard* in the above formula is covered by the exposure assessment under Section 3 and is a main part of the study. For exogenous risks, a site-specific quantitative and comparative analysis of the sample of genebanks is provided for natural hazards and a country-specific analysis for political risks. Financing risks are discussed in qualitative and rather general terms due to a lack of data.

The term *vulnerability* of a genebank, i.e., "the extent to which an element at risk can withstand the impact of the hazard" [11] (p. 7), is strongly influenced by its conservation mandate (crops conserved and conservation methods used), the infrastructure (e.g., adherence to building codes), institutional organization, and on-site management. Assessing the vulnerability for each of the sampled genebanks was beyond the scope of this study; therefore, it is covered only in general terms (please note that the OECD [11] states, "Vulnerability . . . is even more difficult to quantify. . . . The scarcity and inconsistencies of vulnerability information often makes it the weakest link in a risk assessment" (pp. 7–8).).

The term *amount of elements at risk* comprises the buildings and facilities, personnel, and the germplasm collections of a given genebank.

Step 3: Risk steering

In this step, the strategies and instruments to control risks are discussed, especially risk prevention measures and—in case risks cannot be avoided or prevented—risk transfer solutions. Risk transfer means to transfer the financial consequences of a risk from the risk owner to a third party through mechanisms like insurance or funds [12].

Step 4: Risk controlling

This step deals with the question of how risks can be controlled to guarantee that they are properly addressed and managed. However, it is not a focus of this study, but it will be included in order to give a comprehensive overview on the management of risks inside genebanks.

Steps 3 and 4 will be combined under the term *risk management* and dealt with in Section 4.

2.2. Methodology for the Selection of Genebanks

This study aimed to cover the world's most important ex situ holdings of plant genetic resources for food and agriculture (PGRFA) in terms of conserved accessions that are maintained by genebanks, including seed and field genebanks, as well as in vitro and cryopreservation facilities at international, regional, and national levels. Currently, only crop-specific rankings of germplasm collections are available, e.g., by the FAO [3] or through the global crop conservation strategies by the Crop Trust [13]), but s no global ranking of the 1750 genebanks worldwide with respect to their overall collection size in terms of number of accessions [14]. Although the number of accessions does not always correspond with the importance of the PGRA stored at a genebank, it is presently one of the least complex and, hence, most operational parameters for a ranking of genebanks and, therefore, used here. Unfortunately, there are still no good measurements for the total diversity that is included in a collection.

To overcome above mentioned challenges, a sampling approach was developed that was orientated towards the organizational structure of genebanks, in particular differentiating between international and regional versus national genebanks. The classification is according to the WIEWS [15]. International genebanks comprise the CGIAR genebanks, the World Vegetable Center and the global safety duplication site SGSV. Regional genebanks have a mandate for conserving PGR in specific geographical world regions (e.g., SADC Plant Genetic Resources Centre (SRGB) in Zambia for Southern Africa). National genebanks refer to holdings at the country level.

Key information for the sampling was compiled from two sources: the WIEWS database [15] and the FAO's second report on *The State of the World's Plant Genetic Resources for Food and Agriculture* [3] as well as the associated country reports. The data analysed in this study stems from the year 2020 and comprises accessions of plant genetic resources (including agricultural crops and its crop wild relatives) conserved under medium- and long-term storage [15]. FAO (2010) and the respective country reports allow identification and confirmation of the relevance of the sampled institutions. In a final step—after the genebanks have been sampled—the precise location of the selected institutions was validated via satellite images of DigitalGlobe/GeoEye.

The identification and selection of genebanks at international and regional levels proved to be without major challenges due to the limited number of institutions worldwide. In contrast to that, the relevant national genebanks were difficult to identify and assess due to the lack of concise and comparable information across the large number of countries, national institutions, and subsidiaries. In the following, the sampling approach of these genebanks is outlined, differentiating between international, regional, and national genebanks.

Selection of international and regional genebanks

International and regional genebanks typically aim for long-term conservation and facilitated access to accessions, predominantly of major food crop genepools that are in the public domain. Consequently, it was deemed most appropriate to comprise in the sample all 13 international genebanks and 8 regional genebanks. In addition, the safety duplication backup facility SGSV at Svalbard in Norway was included and categorized as an international genebank (resulting in 14 international genebanks). Being the world's largest backup facility for seeds designed with the highest security standards: located at 130 m above sea level in a mountain with permafrost conditions and an additional cooling system bringing the seed storage temperature to minus 18 °C [16]. Electricity is provided by a public power plant or in case of power outage by generators [16]. Despite this, climate change is imposing new threats, e.g., in 2022, ice melted, and water entered the entrance section of the facility [17]. Nevertheless it can be regarded as a benchmark with respect to safety characteristics.

Selection of national genebanks

The selection was performed in different steps following specific criteria:

- Only state-managed and publicly funded national genebanks were included in the sample.
- The selection of national genebanks was performed via the identification of the most important agricultural countries using their gross production value for crops (GPV) and the cropping area. The latter was based on two FAO statistics: (1) the total area harvested and (2) the area of land used for agriculture. This approach followed the underlying rationale that a country that is an important agricultural producer is likely to possess a well-established agricultural (research) infrastructure, including facilities for the conservation of plant genetic resources. The first criterium, the GPV, reflects the economic value of the national agricultural sector, while the cropping area refers to the spatial importance of agriculture. The data were accessed via FAOSTAT [18]. Thirty-six countries were selected.
- After selection of the countries, the most important ex situ holdings for PGRFA at the national level had to be identified. For this step, the WIEWS database, the second report on *The State of the World's Plant Genetic Resources for Food and Agriculture*, and relevant country reports were consulted, and an additional web research was conducted, among others, to confirm that the sample comprised only state-managed genebanks. In countries with a decentralized conservation system, more than one location was included in the sample.

The result of the above sampling approach for national genebanks was 58 national genebanks in the selected 36 countries, comprising a total of at least 3,857,013 accessions (Please note: For the Chinese duplication genebank, information about the collection size was not available. The Nottingham Arabidopsis Stock Center (NASC) was included in the sample due to its importance for research and development in plant genetics and breeding, despite the fact that the majority of the accessions is not PGRFA.), and is shown in Figure 1, which indicates that a relatively good geographical distribution was achieved by this sampling approach. Together with the selected international and regional genebanks, the sample consists of 80 genebanks in total (see Appendix A).

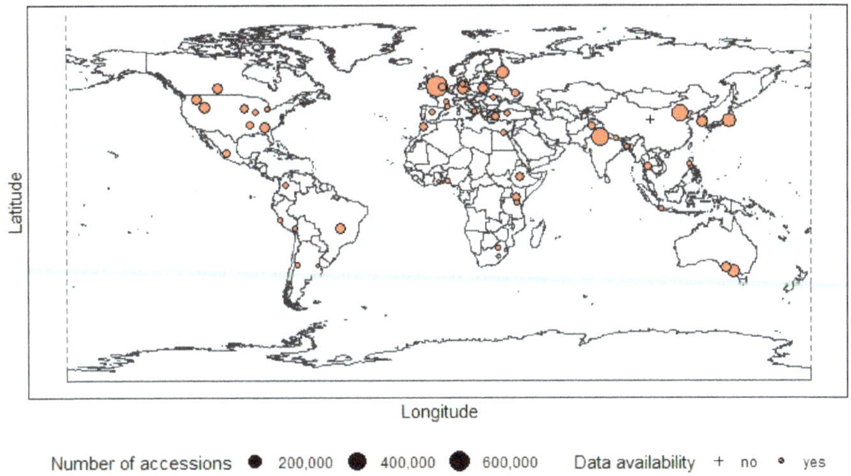

Figure 1. World map of sampled national genebanks and their collection size (circle size corresponds to number of accessions per genebank). If collection size was not available, it is marked with +. Source: adapted from WIEWS (2020) [15] and FAO country reports.

3. Hazard Assessment of Genebanks

3.1. Natural Hazards

The natural hazard assessment was done in cooperation with Munich Reinsurance Company (Munich Re), which provided the results of the location-specific natural hazard assessments of the 80 genebanks using their internal tool "Natural Hazards EditionThe tool allows single-risk assessments for 12 natural hazards and geospatial data analysis. It is based on the database NatCatService [19] and used by Munich Re worldwide for their single risk assessments. (Please note that natural hazard assessment tools are not publicly available. Therefore, presently there is no benchmark available to assess the quality of the "Natural Hazards Edition" tool. In general terms, OECD [11] states in this respect: "The need for better risk assessment data and tools therefore remains high" (p. 3).)

Twelve natural hazards—those considered the most relevant—were analysed and grouped into four categories:

- Geological hazards: earthquake, volcano;
- Hydrological hazards: tsunami, storm surge, river, and flash flood;
- Meteorological hazards: tropical cyclone, extratropical storm, tornado, hailstorm, and lightning;
- Climatological hazards: wildfire.

The natural hazard assessment was done in cooperation with Munich Reinsurance Company (Munich Re), which provided the results of the location-specific natural hazard assessments of the 80 genebanks using their internal tool "Natural Hazards Edition". The tool allows single-risk assessments for 12 natural hazards and geospatial data analysis. It is based on the database NatCatService [19] and used by Munich Re worldwide for their single risk assessments. (Please note that natural hazard assessment tools are not publicly available. Therefore, presently there is no benchmark available to assess the quality of the "Natural Hazards Edition" tool. In general terms, OECD [11] states in this respect: "The need for better risk assessment data and tools therefore remains high" (p. 3).)

The hazards are represented as ordinal data in discrete categories following a multinomial distribution. Categories are defined individually for each hazard. The hazard assessment was conducted in three steps: (1) a single hazard assessment, done individually per hazard and aggregated across the sample of genebanks, (2) a hazard assessment per genebank location, and (3) the calculation of a global risk score per genebank location. Here, only the results of steps 2 and 3 are presented and discussed in detail.

Table 1 presents the results of step 2 exemplarily for international genebanks (for all genebanks, see Supplementary Materials). From a risk management perspective, attention must be especially placed on locations with high to extreme exposure to individual hazards. These have considerable exposure to either one major natural hazard (e.g., earthquake for ICARDA in Lebanon and CIP in Peru) or to multiple hazards (e.g., volcano, flood, and tropical cyclones at IRRI, Philippines, or earthquake, volcano, hail, and lightning at ILRI in Ethiopia). However, because the hazard classes are not directly comparable across hazards, step 3 of the analysis is required for a comparative analysis of the hazard exposure across genebank locations.

Table 1. Natural hazard exposure assessment for international genebanks.

WIEWS Institutional Code	Organization Acronym	Country	Earthquake [1]	Volcano	Tsunami	Storm Surge	Riverflood	Flash Flood	Tropical Cyclone	Extratropical Storm	Tornado	Hailstorm	Lightning	Wildfire
BEL084	ITC	Belgium	1	−1	−1	−1	0	3	−1	2	3 [2]	2	2	1
CIV033	Africa Rice	Côte d'Ivoire	0	−1	−1	−1	0	3	−1	−1	1	2	4	2
COL003	CIAT	Colombia	3	2	−1	−1	0	2	−1	−1	1	3	4	1
ETH013	ILRI	Ethiopia	3	2	−1	−1	0	4	−1	−1	1	5	5	−1
IND002	ICRISAT	India	0	−1	−1	−1	0	4	0	−1	2	2	3	2
KEN056	ICRAF	Kenya	2	2	−1	−1	0	5	−1	−1	1	3	2	−1
LBN002	ICARDA	Lebanon	3	−1	−1	−1	0	3	−1	0	2	4	2	3
MARNA	ICARDA	Morocco	0	−1	−1	−1	0	3	−1	1	1	1	2	−1
MEX002	CIMMYT	Mexico	2	2	−1	−1	0	4	0	−1	1	5	4	1
NGA039	IITA	Nigeria	0	−1	−1	−1	0	2	−1	−1	1	2	4	3
PER001	CIP	Peru	4	−1	−1	−1	0	2	−1	−1	1	1	1	−1
PHL001	IRRI	Philippines	2	3	−1	−1	0	5	4	−1	1	2	4	1
TWN001	World Veg	Taiwan, Province of China	3	−1	−1	−1	0	3	5	0	1	2	2	2
NOR051	SGSV	Norway	0	−1	−1	−1	0	1	−1	3	1	1	/	−1

[1] Note: exposure class varies between hazards. In general, the higher the number, the higher the exposure, with 5 presenting the highest and 1 the lowest exposure class. Minus 1 means no exposure to the hazard. For details, see Appendix B. [2] Highest exposure class per hazard in dark red, second-highest exposure class per hazard in light red. Source: adapted from Munich Re (2022) [19].

For a comprehensive overview of a location's exposure to natural hazards, Munich Re developed a weighted global risk index and a risk score for "ordinary commercial and industrial business" [20]. Because genebanks have similar risk characteristics as "ordinary commercial and industrial business", the index and score can be used here. The global risk index and risk score build upon the hazard zones of the exposure assessment, loss expectations, and expert knowledge of Munich Re to weigh the hazards adequately [20]). As this is confidential information, Munich Re could not disclose it in detail [20]. Despite this limitation, the risk index and score are considered useful for this analysis, as it allows a quantitative comparison of the risk to natural hazards across locations.

The global risk index is a quantitative value calculated as the sum of three individual risk indices that are considered globally the most important ones: the earthquake risk index (comprising earthquake, volcano, and tsunami risks), the storm risk index (comprising tropical cyclone, extratropical storm, hail, tornado, and lightning risks), and the flood risk index (comprising river flood, flash flood, and storm surge risks) [21]. The global risk index ranges from 0 (no risk) to 300 (extreme risk), whereby a risk index of 300 is of theoretical nature, reached only if all three hazard groups would have the maximum value. Therefore, values above 200 are considered highly unlikely in reality [21].

The results of the exposure analysis for the sample of genebanks are shown in Figure 2 (the natural hazard exposure assessment with the respective risk indices and scores are documented for all genebanks in the Supplementary Materials). On average, the analysed genebanks have a risk index of 31.75 (arithmetic mean) and a median of 17. The lowest risk index has been calculated for the subsidiary of the international genebank ICARDA in Morocco (score = 5), and the highest index for the national genebank of the Philippines (PHL129) with a score of 152. Regionally, genebanks in Asia and Oceania are the most exposed, having a global risk index with an arithmetic mean of 57 and 54, respectively. The global risk index for the assessed genebanks on the African continent is the lowest, with an arithmetic mean of 14, followed by Europe with 21 and the American continent with 27. Yet, considerable differences across locations within a given continent exist.

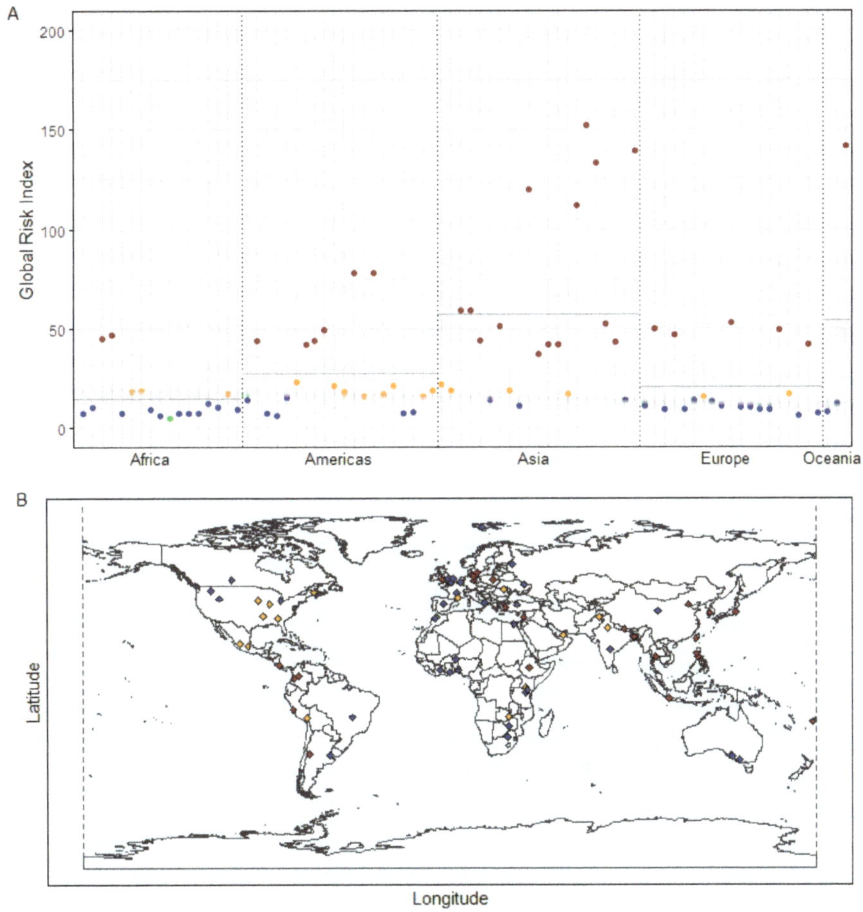

Figure 2. Natural hazard exposure analysis per individual genebank. (**A**) Global risk index (ranging from 0 (no risk) to 300 (extreme risk)), grouped by continent and coloured by global risk score (green = 1 (low), blue = 2 (medium), yellow = 3 (high), red = 4 (extreme)). Horizontal line means continental arithmetic mean of sample. (**B**) World map with global risk score: green = risk score 1 (low), blue = risk score 2 (medium), yellow = risk score 3 (high), red = risk score 4 (extreme). Source: adapted from Munich Re (2022) [19].

Based on the global risk score, 35 genebank locations (44% of the 80 sampled genebanks) were classified with a low (one location) to medium risk (34 locations); 17 genebank locations (21%) have a high and 28 (35%) an extreme risk score. Most of the genebanks with an extreme risk score are located in Asia (14), followed by the Americas (6) and Europe (5). In percentage terms, two-thirds of the assessed Asian genebank locations are classified with an extreme global risk score. In Oceania, 33% of the genebank locations are rated with an extreme risk; in the Americas, 30%, and in Europe, 26%. The lowest share of institutions exposed to extreme risks can be found in Africa, where only 12% of the African genebank locations are classified with the highest risk score 4.

To conclude, from Figure 2, it is evident that the risk exposure varies spatially to a great extent.

Figure 3 shows the risk indices for the three hazard groups. It illustrates that the exposure to earthquakes is most relevant for genebanks in the Americas and, to a lower extent, for Asian and African locations. In Europa, flood risks are the most important, while Asian and Oceanian genebanks are exposed to all three hazard groups, amongst which storm is classified the highest, with extremely high risk indices.

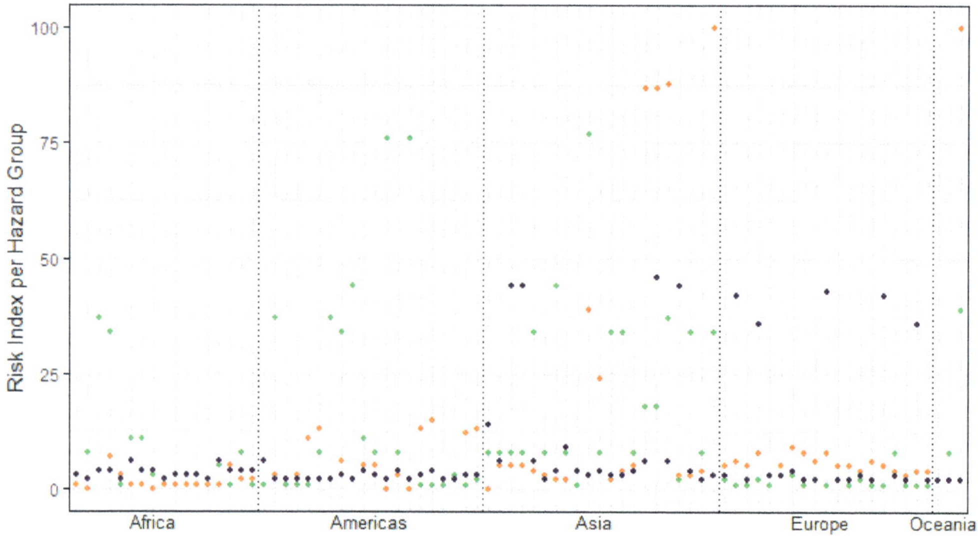

Figure 3. Disaggregated risk index (per hazard group, ranging from 0 (no risk) to 100 (extreme risk)) for sampled genebanks, grouped per continent, and coloured by risk group (green = risk index earthquake, orange = risk index storm, purple = risk index flood). Source: adapted from Munich Re (2022) [19].

Figure 4 gives a detailed analysis for the respective global risk indexes for international genebanks. Six genebanks are classified with the highest risk. Among these, the WorldVeg in Taiwan is the most exposed institution with a global risk index of 139, followed by IRRI in the Philippines (global risk index = 112) and CIP in Peru (global risk index = 78). In addition to Figure 4, three regional genebanks show an extreme high risk score. For the regional genebank CePaCT in Fiji, an overall risk index of 142 has been calculated, one of the highest exposed institutions of the entire group of sampled genebanks. Also, the regional genebanks CATIE in Costa Rica and NORDGEN in Sweden show a high risk, with an index of 49 and 42, respectively. The safety backup facility SVSG in Svalbard shows a comparable low exposure to natural hazards (with a risk index of 9). As its exposure is mainly driven by storm events, this is of minor relevance because the building's infrastructure is mostly located underground and, therefore, has a very low vulnerability to storm.

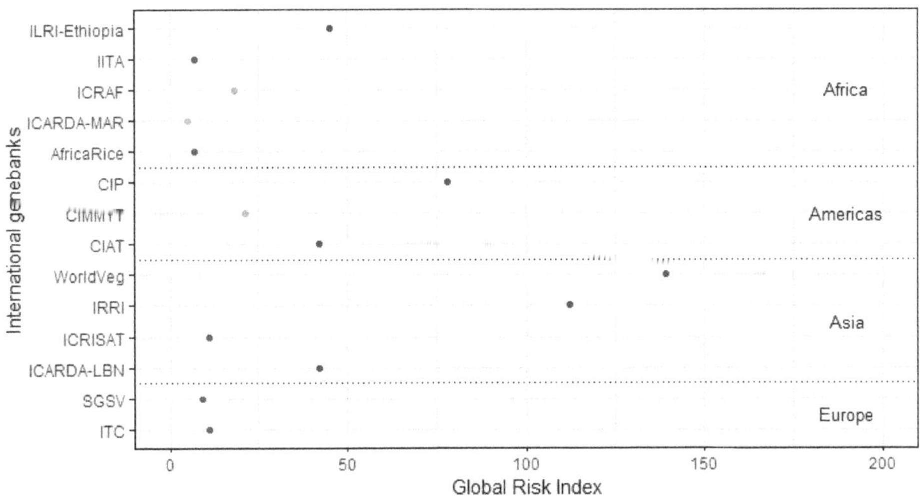

Figure 4. Global risk index for international genebanks, grouped per continent and coloured by global risk score (green = 1 (low), blue = 2 (medium), yellow = 3 (high), red = 4 (extreme)). Source: adapted from Munich Re (2022) [19].

3.2. Political Risks

Due to the complex nature of political risks, they have been analysed at the country level using two indicators: the Worldwide Governance Indicator (WGI), published by the World Bank [22], and the Fragile States Index (FSI), published by The Fund for Peace [23]. To account for possible between-year differences, the six-year average from 2015 to 2020 was calculated for each of the indicators.

The WGI covers six dimensions of governance: (1) voice and accountability, (2) political stability and absence of violence, (3) governance effectiveness, (4) regulatory quality, (5) rule of law, and (6) control of corruption [24]. The indicators are expressed on a scale from −2.5 (weak governance) to +2.5 (strong governance), centred around 0. For this study, all WGI dimensions were considered to be of relevance when assessing political risks for genebanks. Consequently, a total WGI score was calculated as the unweighted average across the six dimensions, also ranging from −2.5 to +2.5.

The FSI by the Fund for Peace assesses the risk of a state failure based on a total score ranging from 1 (low risk) to 120 (high risk). It is calculated as the unweighted sum of twelve indicators covering five aspects of political stability: cohesion, economic, political, social, and cross-cutting indicators [23]. For this study, the total score was used.

The results are illustrated in Figure 5 for the WGI indicator and in Figure 6 for the FSI indicator, using score colours for all 47 countries where the sampled genebanks are located. This gives a visual impression pointing to a good concurrency of the two indicators, despite the methodological differences of the indicators and general methodological challenges associated with quantifying political risks. The concurrency is further confirmed by Figure 7, providing a ranking of the sampled countries for these two indicators. The countries most at risk are ranked highest (left lower corner in Figure 7), rising to the most stable nations (right upper corner). From this, it becomes apparent that the two indicators coincide for the majority of the countries relatively well, i.e., graphically, the countries are close to the plotted line through the origin.

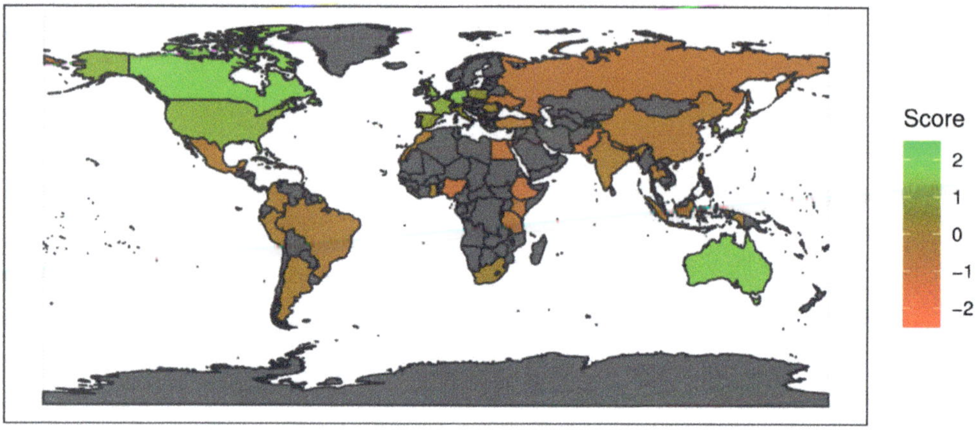

Figure 5. Mean WGI score: Map of all countries comprising the study's genebank sample. Coloured according to the country's average WGI score (six-year average 2015–2020, averaged across the six WGI dimensions) with a scale of −2.5 (low governance) to +2.5 (high governance); in grey = countries not in the sample. Source: adapted from World Bank (n.d.) [22].

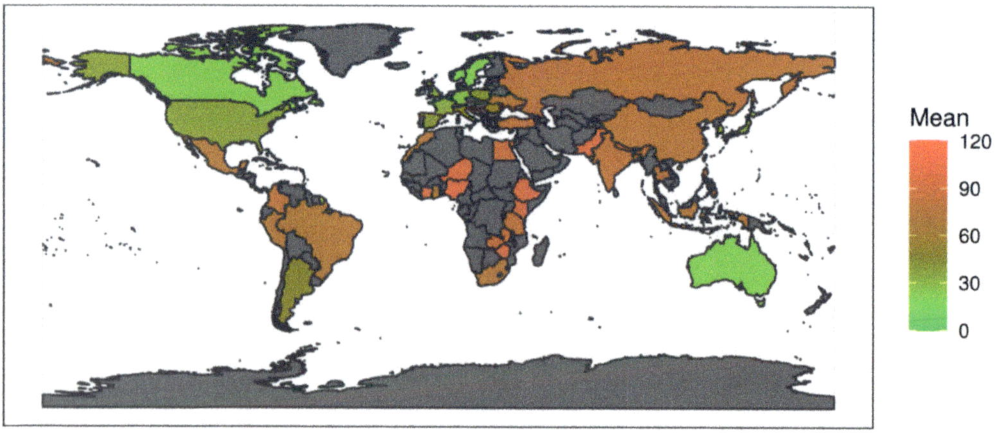

Figure 6. Average FSI total score: Map of all countries comprising the study's genebank sample. Coloured according to the country's average FSI total score (six-year average 2015–2020, unweighted sum of the 10 FSI dimensions) with a scale of 0 (low political risk) to 120 (high political risk); in grey = countries not in the sample. Source: adapted from The Fund for Peace (n.d.) [23].

Overall, African and some Asian countries were assessed with a high political instability, whereas general European (except for Ukraine and the Russian Federation) and North American countries, particularly Canada and USA, show a high political stability. Notably, some important international genebanks are located in countries assessed with a high political risk: IITA in Nigeria, ILRI in Ethiopia, ICARDA in Lebanon, ICRAF in Kenya, and AfricaRice in Côte d'Ivoire.

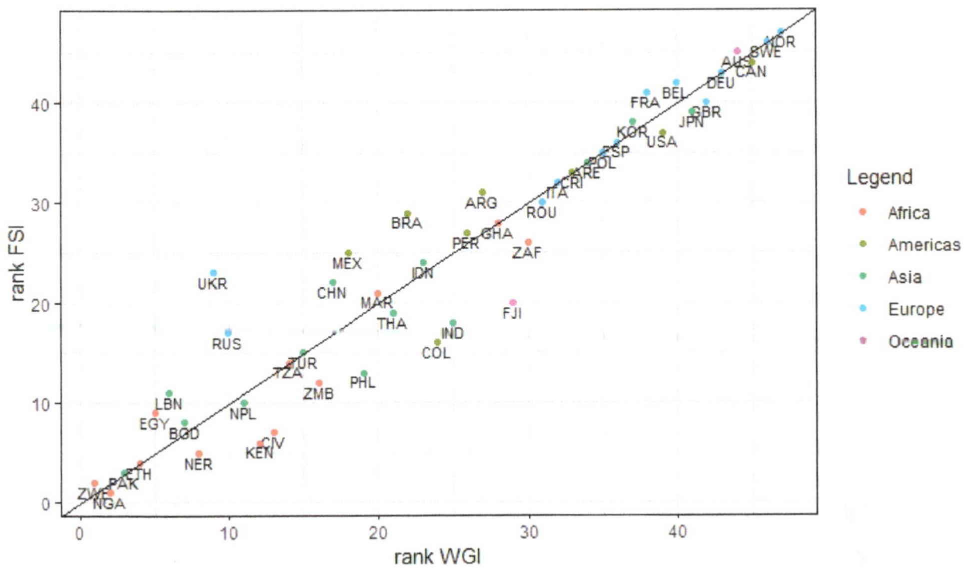

Figure 7. Comparison of the country ranking for political risks by average WGI score and FSI total score for all countries in the study hosting international, regional, and national genebanks, coloured by continent (with 1st rank = high political risk to 49th rank = lowest political risk of the sample). x-axis = country ranking according to the mean WGI score, calculated as the six-year average 2015–2020, averaged across the six WGI dimensions. y-axis = country ranking according to the FSI total score, calculated as the six-year average 2015–2020, unweighted sum of the 12 FSI dimensions. Source: own representation based on data provided by World Bank (n.d.) [22] and The Fund for Peace (n.d.) [23]. Country abbreviations: ARE = United Arab Emirates, ARG = Argentina, AUS = Australia, BEL = Belgium, BGD = Bangladesh, BRA = Brazil, CAN = Canada, CHN = China, CIV = Côte d'Ivoire, COL = Colombia, CRI = Costa Rica, DEU = Germany, EGY = Egypt, ESP = Spain, ETH = Ethiopia, FJI = Fiji, FRA = France, GBR = United Kingdom and Northern Ireland, GHA = Ghana, IDN = Indonesia, IND = India, JPN = Japan, KEN = Kenya, KOR = Republic of Korea, LBN = Lebanon, MAR = Morocco, MEX = Mexico, NER = Niger, NGA = Nigeria, NOR = Norway, NPL = Nepal, PAK = Pakistan, PER = Peru, PHL = Philippines, POL = Poland, POU = Romania, RUS = Russian Federation, SWE = Sweden, THA = Thailand, TUR = Türkiye, UKR = Ukraine, USA = United States of America, ZAF = South Africa, ZMB = Zambia, ZWE = Zimbabwe. Please note: Taiwan (hosting the international genebank WorldVeg) needed to be excluded for the analysis because it is only in the indicator WGI and not in FSI.

3.3. Financial Risks

Financing constraints, implying an insufficient level of funding and non-reliability of funds, are a major threat to sustainable PGR conservation, as they adversely affect genebank operations, functioning, and risk management. Based on the literature review and expert interviews, a tentative qualitative exposure ranking (highest to lowest financial risk) can be given as follows:

1. National genebanks in developing countries;
2. National genebanks in emerging economies and some developed countries with decentralized structures (decentralized structures seem to be more vulnerable, as their funding often comes from different sources, e.g., besides central, also regional governments) and weak national coordination;
3. International CGIAR genebanks;

4. National genebanks in developed countries with centralized management or decentralized structures with strong national coordination.

In general, information on the actual and required budget per genebank, as well as on the nature and provenance of funds, is scarce. Therefore, a thorough analysis of the financial situation of the sampled genebanks could not be carried out. However, it can be noted that the international genebanks have—in particular, through the professional work of the Crop Trust operating the Crop Diversity Endowment Fund—a stable financial backbone, whereas the situation of national genebanks is very diverse. It depends on the organizational and management structure (e.g., centralized vs. decentralized), the overall state budget, and the priority-setting of national governments.

4. Risk Management at Genebanks

After the above-described assessment of risks and hazards, in a further step, strategies and instruments to manage risks should be addressed. There is a global tendency toward an increased exposure to hazards—especially to natural hazards, largely as a consequence of climate change and with respect to political risks due to increased international instability. Therefore, risk steering and controlling are becoming more and more critically important and should become a core activity for secured PGR conservation at genebanks. Hence, the development of criteria and standards for an effective and efficient risk management, as well as respective staff training, will be essential. Two areas are of particular importance: (1) risk prevention and mitigation and (2) risk transfer.

4.1. Specific Risk Prevention and Mitigation Strategies for Genebanks

Risk prevention aims at avoiding emerging and existing risks from materializing, while the objective of risk mitigation is to reduce the impact of hazardous events [25]. Hence, both have a prospective character. For managing specifically exogenous risks at genebanks, two important strategies can be identified: (1) increasing resilience of infrastructure and (2) safety duplication of accessions [6,26–28].

Increasing resilience of infrastructure

Increasing the resilience of infrastructure, e.g., buildings, technical facilities, and IT infrastructure, mitigates the impact of natural hazards, electricity outages, and malfunctioning of technical devices. Based on the literature review, expert interviews, and own assessments, Table 2 has been compiled, which summarizes the most important risk control measures.

Table 2. Infrastructural risk control measures for each of the risk sources.

Risk Source	Risk Control Measure
Natural hazards	Building codes, standards, and practices (e.g., resistance to earthquake, strong wind, and snow load). These are being regulated normally at the national level.
Outage or malfunctioning of technical facilities	Alarm systems (e.g., for open doors, sudden changes in light, temperature, and humidity); early fire, gas, smoke, or water detection; backup equipment or additional rooms available and ready; essential spare parts in storage; qualified staff for repairs or external standby repair services
Fire (ignition point inside the facility)	Detection and mitigation devices: Smoke and fire detection; sprinkler systems; fire extinguishers. Construction measures: fire walls; fire isolation doors; separated compartments; sufficient separation between buildings. Organizational aspects: coordination with external firefighting services like local fire brigades
Fire (ignition point outside the facility, e.g., wildfires)	Fire breaks; fuel load control in the vicinity of the genebank
Power supply cut-offs	Second power line, emergency power generator (for storage rooms, monitoring devices, essential lighting, etc.), lightning rods and deflectors
Theft, vandalism, and terrorism	Alarm systems, locks, surveillance cameras, and sensors to impede the entry of unauthorized people (in addition to security surveillance)
Cyber-attacks to IT [1]	High cybersecurity standards

[1] IT risks are currently underestimated but are expected to become increasingly important in future due to increased digitalization at the genebank level. Source: adapted from CGIAR Genebank Platform (2020) [26], Crop Genebank Knowledge Base (n.d.) [27], Fu (2017) [6], and expert interviews.

For the twelve assessed natural hazards, the infrastructural standards are suggested according to Table 3. Here, only the genebanks that are part of the sample and have been assessed with the highest exposure to the respective natural hazard are mentioned. Especially for these genebanks, the standards are of high priority. However, the suggested measures are obviously not limited to these genebanks only.

Table 3. Infrastructural risk control measures suggested for genebanks with highest exposure, per natural hazard.

Natural Hazard	Risk Control Measures	Examples of Exposed Genebanks [1]
Earthquakes	Earthquake-proof infrastructure [26]	- international genebank CIP, Peru (PER001) - national genebank of Peru (PER066) in Lima and of Japan (JPN183)
Volcanoes	Strengthened roofs and walls, use of shutters on openings and non-flammable materials, fix buildings to foundation, etc. [29]	- international genebank IRRI, Philippines (PHL001) - regional genebank CATIE, Costa Rica (CRI134/CRI142/CRI085) - national genebank of Indonesia (IDN179) and of the Philippines (PHL129)
Tsunami	Tsunami-resistant structures [30]	- international genebank CePaCT, Fiji (FJI049)
Storm surge	Storm surge gates, flood barriers, floor plans for a quick water outflow, shelving above ground level [26,31]	- regional genebank NORDGEN, Sweden (SWE054) - location Poel of national genebank of Germany (DEU271)
River flood	Flood barriers, dikes, spurs, etc., floor plans for a quick water outflow, shelving above ground level, water-proof ink and bags [26,32–35]	- national genebanks of Bangladesh (BGD002, BGD003), Germany (location Gatersleben, DEU146), the Philippines (PHL129), Poland (POL003), Thailand (THA300), and Great Britain (GBR016)
Flash flood	Flood barriers, dikes, spurs, etc., floor plans for a quick water outflow, shelving above ground level [26,32,34]	- national genebank of India (IND001)
Tropical cyclone	Same as against storm surges and floods (e.g., embankment) [36,37], wind-resistant buildings [38], clearing of surroundings (e.g., cutting of trees in proximity to genebank) [39]	- international genebanks WorldVeg, Taiwan (TWN001) and CePaCT, Fiji (FIJ049)
Extratropical storm	Wind-resistant buildings, reinforcing/securing of roofs [40]	- international genebank SGSV (NOR051) - national genebanks of France (FRA139), Germany (DEU271) and Great Britain (GBR004, GBR016)
Tornado	Wind-resistant buildings, safe rooms (e.g., for seed storage rooms), reinforcing/securing of roofs [41,42]	- national genebanks of Canada (CAN025) and the United States of America (USA020, USA970, USA033)
Hailstorm	Hail-resistant roofs and windows [43]	- national genebanks of Colombia (COL017), Ethiopia (ETH085), and the United States of America (USA020)
Lightning	Lightning rod [44]	- national genebank of Pakistan (PAK001)
Wildfire	Fire breaks, non-combustible materials and fire-resistant structures, adequate vegetation [45,46]	- international genebank IITA, Nigeria (NGA039) - regional genebank SRGB, Zambia (ZMB030) - national genebanks of Kenya (KEN212) and Thailand (THA300)

[1] The genebanks listed here are the most exposed locations per hazard amongst the sample, i.e., falling in the highest or second-highest class. This, however, does not mean that for other institutions, no infrastructural control measures are recommended. Source: adapted from Munich Re (2022) [19], risk control measures compiled from the literature cited in each risk measure cell.

It should be highlighted that most of the above-mentioned risk control measures apply to seed genebanks as well as in vitro and cryopreservation facilities. In addition, for in vitro conservation, a high emphasis needs to be put on the control of technical installations and equipment, as specific temperature and light requirements have to be met. A. W. Ebert (personal communication, 25 July 2022) [39] points to the risk of high temperatures above 40 °C; if the air conditioning is defective, this could put the entire in vitro collection in danger. By contrast, for field genebanks, specific risk control measures are needed, as they are exposed to additional hazards (e.g., pests, diseases, theft and animal damages, drought, and flooding). Most of these are difficult to control in the field, so the location of the genebank and protective infrastructure like fences, hail nets, and irrigation are crucial [33,39,47].

If a genebank will be constructed new or renovated it is recommended that a thorough exposure and vulnerability assessment is conducted beforehand and that the respective building codes are applied. As part of such an assessment, the risk of pathogen pressure at a specific location should also be considered, as this can be mitigated substantially through choosing a genebank's location; for instance, in no cropping areas or dry areas to lower the disease pressure on seeds.

Safety duplication of orthodox seeds

Maintaining safety duplicates of accessions at two or even three different locations in another country and possibly on another continent [47] is an important risk management strategy in ex situ conservation. It has proven to be effective already in the past to restore lost accessions or even entire collections (e.g., of ICARDA in Syria). Therefore, it is promoted by stakeholders and researchers and widely applied at national and international genebanks—yet at a varying level [3,6,26,28].

As per the FAO Genebank Standards, most genebanks should have safety duplication arrangements with one or more institutions, including international, regional, and national genebanks, as well as the SGSV [3]. Preferably so-called black box agreements are applied, meaning that the recipient institution conserves the duplicate but has neither rights over it nor further obligations (i.e., is not responsible for viability testing and is not allowed to regenerate, use, or distribute the material if not authorized by the depositor) [13,28].

Genebanks have adopted different strategies: either a system of duplicates (e.g., the IPK in Germany where accessions are safety duplicated only at SGSV) [48] or of triplicates (e.g., the Dutch CGN where accessions are safety duplicated at another national genebank and at SGSV) [3,33,35,39]. Moreover, duplication at another active genebank is a valid approach. This means that the collection is not only stored for conservation but also actively used.

These risk mitigation strategies of safety duplicates come at a certain cost, requiring substantial financial resources, sufficient storage capacities, legal and institutional agreements, and a good documentation and information system [28,33,35,49]. This points to an important implication for the global conservation system: the decision of what material should be safety duplicated requires a prioritization (although it is desirable that all or at least the majority of accessions of a collection are safety duplicated, this is often not possible because of financial constraints). This implies a complex value judgment based on a thorough assessment of the importance and value of individual accessions. At the German genebank IPK and at Plant Gene Resources of Canada, this prioritization is purely governed by logistics, i.e., only recently multiplied material is sent to SGSV [33], but over time, the whole collection will be duplicated there.

Another important limitation is the high level of unintended duplicates within collections. The FAO (2010) estimates that only between 25 to 30% of the accessions in ex situ collections are unique. Therefore, despite recognizing that the ultimate goal should be safety duplicating the whole collection, it is recommended to establish guidelines on how to prioritize accessions for safety duplications based on common principles but flexible according to the respective national context and financial resources. An interesting example is Canada, where results of molecular marker analysis such as accession distinctness are taken into account when prioritizing [49].

With the opening of SGSV in 2008, an important milestone with respect to safety duplication and backup of accessions was achieved. This unique storage facility with a capacity of 4.5 million accessions has the highest safety standards and, therefore, is the world's most important safety backup facility. In February 2023, more than 1.2 million seed samples of more than 5000 plant species coming from 98 institutions in 76 countries have been stored there [16]. The largest numbers of accessions stored are varieties of rice and wheat (each >150,000), followed by barley (close to 80,000), sorghum (>50,000), *Phaseolus* bean species (>40,000), maize (>35,000), cowpea (>30,000), and soybean (>25,000) [16]. About two-thirds of the presently deposited accessions are from the international genebanks. Among national genebanks, the USA, Germany, Canada, and the Netherlands are the main depositors, while for regional genebanks, NORDGEN is the main depositor [16].

Despite the above-mentioned positive development and the considerable progress made during the last two decades with respect to the percentage of safety-duplicated material, important parts of ex situ collections still "remain inadequately safety duplicated" ([3] p. 87). This applies especially to crops that cannot be maintained as seeds, i.e., vegetatively propagated crops or recalcitrant seeds (see below) and to national seed genebanks in some developing countries due to scarce financial resources [3].

The rate of accessions being safety duplicated is in general higher in CGIAR genebanks than at most national institutions [3,16]. This achievement is especially triggered by the Crop Trust, which links its financial support for genebanks to performance targets, including the rate of safety duplications [50]. However, even there, the external reviews conducted between 2017 to 2021 criticised a lack of sufficient safety duplications at some of the CGIAR genebanks, e.g., the ICRAF in Kenya and ICARDA in Morocco and Lebanon [51].

Safety duplications of vegetatively propagated crops and non-orthodox seeds

Vegetatively propagated crops and non-orthodox seeds, comprising intermediate and recalcitrant seeds, are predominantly maintained in field genebanks. The accessions have a high vulnerability, as risk mitigation through infrastructural means is limited to irrigation facilities, hail nets, and fences for their protection. This, however, is much less effective in comparison to seed genebanks [47]. Notwithstanding this constraint, the level of safety duplication is considerably lower compared to orthodox seeds. Therefore, from a risk management perspective, more efforts to secure field genebank accessions are necessary.

For risk mitigation, there are two main ways of safety duplicating these crops, as detailed by FAO (2014) [47]:

- Duplication of field collections at another location (not exposed to the similar risks as the original field genebank);
- Duplication of field genebank accessions under alternative conservation methods, such as in vitro conservation and cryopreservation.

These approaches are especially important, as presently, there is no global backup conservation facility available for field collections, i.e., conserved in vitro or cryopreserved, similar to the SGSV for seed collections.

In vitro conservation and cryopreservation are less susceptible to natural hazards compared to field genebanks, which are directly exposed to environmental risks. In vitro and cryostorage reduce the vulnerability to natural hazards by moving the collection from an exposed outside location to controlled inside conditions. The level of vulnerability, then, depends predominantly on the building infrastructure, the technical facilities securing a controlled environment, and the operating staff. Moreover, a building can be better secured against human-related damages, such as theft, vandalism, and political risks. Furthermore, the accessions maintained in field genebanks are also directly and continuously exposed to pests and diseases. Especially infections with viruses cause severe problems to the genebank, as these might impede their distribution and exchange with other genebanks because of quarantine regulations. During the process of preparing materials for in vitro and cryopreservation storage, viruses can be eliminated through specific treatments; thus, cleaned in vitro/cryopreserved samples can be safely exchanged.

However, there are two major drawbacks to these methods:

- The costs for setting up and introducing material to in vitro and cryopreservation are considerably higher than for field genebanks. However, in the case of cryopreservation, once the system is established, its running costs are relatively low [52]. (Note that in vitro conservation is not suitable for mid-term and long-term conservation. But in vitro is important in connection with cryopreservation, as the plant tissue material first has to be prepared in vitro before it can be stored in liquid nitrogen [39].)
- Both alternative methods require a high level of training to manage conservation appropriately [1,52].

According to Panis et al. (2020) [52], duplicating field genebank accessions in in vitro or cryopreservation at another location is recommended. Cryopreservation is especially suitable for secure long-term conservation, as it requires regeneration only after several hundred years [52]. However, if financial resources are scarce, duplicating the field genebank collection at another distant field location might be an appropriate alternative [28].

4.2. Risk Transfer Strategies

Risk transfer is a key strategy to manage risks and is particularly relevant if risks cannot be prevented. It transfers the financial consequences of a risk from the risk owner to a third party through different mechanisms like insurance schemes or funds [12]. Risk transfer solutions are common in different economic sectors, among others in agricultural production. However, in agrobiodiversity conservation, they are currently hardly applied.

Insurance solutions

The most common and widespread risk transfer solutions are insurance coverages [12]. Insurances are financial agreements to transfer defined risks to a third party, the insurer, against the payment of agreed monetary terms (premium)—and this should be done before the risk materializes [12]. The advantages of insurance covers are that there is a legal entitlement for indemnification and that the insurance company normally carries out a risk assessment, including identifying and requesting risk prevention measures.

There is only limited information available if and to what extent genebank assets are covered by insurance schemes. Based on the literature review [53–55] and expert interviews (F. Begemann (personal communication, 11 August 2022); A. W. Ebert (personal communication, 25 July 2022); U. Lohwasser (personal communication, 15 July 2022); T. van Hintum (personal communication, 28 July 2022) [33,35,39,56]), it can be concluded that in most countries, genebanks are not insured. This is due to the fact that most genebanks are in public ownership, and in case of an emergency, the state is supposed to bail out and rebuild facilities and infrastructure. This premise, however, seems a questionable strategy for the future, considering three trends:

- Natural hazards will increase in frequency and intensity due to climate change, augmenting the exposure of genebanks and other infrastructure [57].
- An increased concentration of ex situ conservation structures at the country and institutional levels, as well as increased numbers of accessions stored, will increase the values at risk in future [3,6].
- Financial constraints of states and decreasing political support for PGR conservation [6]; with the recent increase in interest rates in important economies like the USA and Europe, governmental budget limitations are likely to become more important while financing debts (e.g., as necessary in the aftermath of disasters) will probably become more difficult in future.

Based on these considerations, it is recommendable to include insurance schemes for genebanks as a complementary risk management strategy in future. The relevant issues and challenges in this process are briefly discussed below.

In most cases, insurance coverages are offered by nationally approved insurance companies. Genebanks can access this insurance capacity. Additionally, for genebanks in developing countries, risk pooling insurance instruments for natural hazards might also be relevant. Interesting examples at the regional level are the Caribbean Catastrophe Risk Insurance Facility (CCRIF) and the African Risk Capacity (ARC) (for further information, see [58–60]). Internationally organized facilities like the CGIAR genebanks could potentially also look for an umbrella cover for all its genebanks. This would primarily be offered by globally organized insurance and reinsurance companies.

The insured perils in insurance contracts are usually fire and explosion as well as all major natural hazards (e.g., earthquake, volcano, storm, hail, flood). However, locations with a very high flood exposure, e.g., close to rivers and creeks, might not be eligible for flood coverage. It is important to note that standard exclusions in insurance policies are war, terrorism, and radioactive contamination. As genebanks have been damaged or destroyed by war acts in the past, this is certainly an important limitation.

The most critical insured assets of genebanks are:

- Buildings and storage rooms;
- Technical facilities and equipment (e.g., refrigerated storage facilities, control units, alarm systems, laboratory);
- Germplasm collections.

From a risk management point of view, it is recommendable to insure all asset classes, but it is also possible to select specific ones, e.g., only buildings and storage facilities or germplasm collections. Insured assets are covered for physical loss, damage, or destruction caused by an insured peril.

An essential step in structuring an insurance contract is the valuation of the assets. Based on the valuation, the sums insured per insured asset are defined, and these are the basis for any indemnity paid after a loss event. For the genebank infrastructure (buildings and technical facilities), this is relatively easy to estimate using market or replacement values. However, valuing PGR collections is challenging because PGR accessions are non-tradable items and, hence, do not possess a market or replacement value. For the valuation of PGR collections, two approaches are most suitable:

1. **Replacement value:** In this approach, the cost to replace and rehabilitate any collection lost is determined. In the case of accessions, a replacement is only possible if the accessions are stored as safety duplicates elsewhere and are accessible and viable. Such a replacement exercise was undertaken in the case of the CGIAR genebank ICARDA in Aleppo, whose collections have been restored in Morocco and Lebanon since 2016 using backed-up accessions at other genebanks and SGSV [4,5]. The costs of this operation are, however, not publicly available at present. Another reference is the Dutch genebank, where replacement costs have been recently estimated at €25 to €30 million overall [35]. This would result in a value of €1040 to €1250 per accession. As replacement operations are complex and costly, the respective figures are on the high side.
2. **Costs of conserving accessions:** Using the costs of conserving accessions as an approximation for estimating the value of germplasm collections is an indirect approach. The advantage is that costs are relatively easy to establish [61]. Thereby, the costs of conservation in perpetuity should be used, as they focus on the long-term preservation of plant genetic material. As conservation costs are reasonable, this approach results in a relatively low level of valuation. Koo et al. (2003) [62] also used this approach and collected crop-specific in perpetuity costs at five international CGIAR genebanks. These data—even though dating back to the late 1990s and early 2000s—are the best available. Table 4 compiles the data of Koo et al. (2003) [62] and derives from these present values. These authors worked with different interest rates (2%, 4%, and 6%), which have a considerable impact on the value estimation (cf. Table 4).

Table 4. Costs for in perpetuity conservation at selected CGIAR genebanks, per crop type, as a proxy for collection value.

Crop Type	CGIAR Genebank	Size of Collection		Present Values at Different Interest Rates				Collection Value	
		2001	2020	Reference Year [1]		2021 [2]		2021	
		[No. of Accessions]		2%	6%	2%	6%	2%	6%
				[US $ Per Accession]				[in Million US $]	
Common Bean	CIAT	31,400	32,347	47.1	12.9	76.8	21.0	2.484	0.680
Forages	CIAT	24,184	22,694	83.7	22.9	136.5	37.3	3.097	0.847
Wheat *	CIMMYT	154,912	146,505	22.7	6.3	42.4	11.8	6.208	1.704
Wheat **	CIMMYT	see above	see above	25.9	9.6	48.5	17.9	7.098	2.625
Maize *	CIMMYT	25,086	32,243	151.5	32.3	283.2	60.4	9.132	1.946
Maize **	CIMMYT	see above	see above	260.2	141.0	486.5	263.6	15.686	8.500
Sorghum	ICRISAT	36,721	42,352	47.4	14.3	81.1	24.5	3.434	1.038
Pearl Millet	ICRISAT	21,392	24,373	56.1	15.2	95.9	25.9	2.336	0.632
Chickpea	ICRISAT	17,250	20,764	47.8	14.4	81.8	24.6	1.699	0.510
Pigeonpea	ICRISAT	13,544	13,783	58.7	15.4	100.3	26.4	1.383	0. 363
Groundnut	ICRISAT	15,327	15,622	49.7	14.6	84.9	24.9	1.327	0.389
Rice, cultivated	IRRI	94,564	125,899	25.1	6.3	42.9	10.8	5.397	1.358
Rice, wild	IRRI	4568	5813	37.1	7.5	63.4	12.8	0.368	0.074

* without initial regeneration, ** with initial regeneration. [1] Reference year for CIAT 2000, CIMMYT 1996, ICRISAT and IRRI 1999. [2] Conversion to 2021 figures using OECD producer price index for OECD countries [63]. Source: adapted from Koo et al. (2003) [62]. Collection sizes for 2020 are from WIEWS (2020) [15].

Additionally, Rabenau (2018) [64] collected in perpetuity cost data at the German national genebank IKP in Gatersleben based on the methodology of Koo et al. (2003) [62] (cf. Table 5).

Table 5. Costs of conserving accessions in perpetuity at the IPK Gatersleben, per crop type.

Crop Type	Costs in Perpetuity (2018) [€ per Accession]
Wheat	13.00
Rye	13.00
Soybean—open air	11.98
Soybean—greenhouse	40.77
Chickpea—open air	12.26
Chickpea—greenhouse	29.57
Cabbage	41.88
Cauliflower	35.76
Lettuce	19.50

Source: own representation, adapted from Rabenau (2018) [64].

From both tables, it is obvious that the estimated value per accession varies considerably between crops. Also, the way regeneration is conducted influences the results, e.g., in Table 4, with and without initial regeneration of wheat and maize, and in Table 5, regeneration in open fields vs. greenhouse regeneration for soybean and chickpea.

In summary, to determine the actual value of a specific germplasm collection, it would be best to establish own cost data using the methodology of Koo et al. (2003) [62] and, based on these, to estimate the total value of the collection, e.g., for insurance purposes to establish the insured value of the collection. To account for particular features of the collection, e.g., share of unique accessions, respective loading factors might be used.

Funds

Funds—defined as a pool of money that is allocated for a specific purpose [54]—are another important risk transfer instrument. They can be financed through fees, donations, or financial resources granted by the state and are administered either by state institutions, self-governed bodies, or financial institutions [65,66]. Most funds are designed nationally, but also a supranational or even global scope is possible. Funds are often set up to respond to natural hazards, but any other peril, e.g., nuclear risks or terrorism, can also be covered, depending on the objective of the funders.

Similar to insurance coverages, funds allow the rapid mobilization of financial resources, circumventing time-consuming approval procedures and negotiations for accessing other financing sources after a disaster has occurred [67]. However, as funds are rather difficult and complex to set up, they are usually designed for circumstances where insurance coverages are neither available nor cost-effective.

For designing and implementing a fund, the most critical issues to be addressed include:

- The geographical scope of the fund;
- The assets to be covered;
- Value of the respective assets (as described for insurance solutions);
- The perils covered;
- Size and capacity of the fund;
- Financing of the fund: e.g., either through fees/premiums paid by the participating genebanks, deposits of donors (state or private), or a mixture of both.

In this context, the Global Crop Diversity Trust has taken an active role in establishing in 2021, jointly with the Secretariat of the ITPGRFA, the fund named Emergency Reserve for Genebanks [68]. Although targeting national genebanks in developing countries, the fund is open for national and international seed and field genebank collections, provided a substantial financial need can be demonstrated. Different interventions can be financed, e.g., repairing technical facilities, relocations of collections, or safety duplications of threatened unique accessions [50,69]. This fund, in contrast with the Endowment Fund, uses the financial resources of donors allocated to the fund directly. Therefore, if financial resources are spent, additional money has to be acquired. At present, a target for the financial volume of the fund has not been specified, as the number and scale of future requests are difficult to predict. Notwithstanding the presently limited volume, such a fund is considerable progress towards an improved risk transfer for genebanks.

In addition, other national and regional funds covering the aftermath of natural disasters exist. Two interesting examples that might be of use for genebanks are the Mexican fund FONDEN at the national level and the European Union Solidarity Fund (EUSF) at the regional level (for further information, see European Commission (2019) [70] and World Bank (2012) [71]). FONDEN, established in the late 1990s, is designed to support the quick rehabilitation of public infrastructure after adverse natural events [71]. Exploring the integration of genebanks into these national and regional funds is recommended.

It can be concluded that funds solely or in combination with insurance schemes could be a feasible risk transfer mechanism for genebanks that is worthwhile exploring and developing further.

5. Findings and Conclusions

This study provides a comprehensive risk analysis and risk management framework for genebanks worldwide. Its main findings and recommendations are summarized as follows:

1. The natural hazard exposure analysis for a sample of 80 international, regional, and national genebanks, covering 65% of the world's accessions, shows that risk exposure is highly location-specific and varies considerably between the analysed genebanks. Overall, 35 genebanks (44% of the sampled institutions) show a low to medium risk, while 17 genebanks (21%) present a high and 28 (35%) an extreme risk. Most of the extremely exposed genebanks are located in the Asia-Pacific region (Philippines, Fiji, Taiwan, Japan, and Bangladesh) as well as in South America (Peru) and Europe (UK, Germany, and Poland) (see Appendix B and Table A3). On the other hand, genebanks in Africa tend to have relatively low exposure to natural hazards. Among the international and regional genebanks, the most exposed are CePaCT in Fiji, WorldVeg in Taiwan, IRRI in the Philippines, and CIP in Peru. In contrast, SGSV,

the global backup storage facility of safety duplicates, shows a relatively low risk profile, being only exposed to extratropical storms and rising temperatures affecting the permafrost. As the storage rooms are located underground, the vulnerability of the facility towards storm can be assessed as very low. These findings entail the following consequences for risk management:

- A location- and institution-specific risk assessment is indispensable to define and carry out appropriate risk prevention methods using two main strategies: (1) adequate infrastructural measures like natural hazard-resistant building codes, storage facilities at higher levels (flood prevention), emergency backup generators, and alarm systems; and (2) safety duplication of accessions at another location. Both strategies can be implemented without major obstacles in the conservation of orthodox species but are more complicated when conserving clonal and recalcitrant species in field genebanks.
- Risk transfer solutions like insurance coverages and funds, at present hardly implemented at genebanks, should be considered when developing holistic risk management strategies of genebanks. An important step in this direction was taken by the Global Crop Diversity Trust with the set-up of the Emergency Reserve Fund in 2021. Prices of these solutions vary significantly in line with the site-specific risk exposure.

2. Vulnerability is very site-specific, depending mainly on the quality of infrastructure and risk prevention measures in place. Furthermore, it differs according to the specific conservation methods. Conservation in seed genebanks is the most resilient method compared to in vitro and cryopreservation, as the latter ones imply high technical and technological requirements. Field genebanks have a distinct risk profile and have, among the common conservation methods, the highest vulnerability with respect to natural hazards as well as pest and disease incidences.
3. Assessing the exposure to political risks is challenging due to the complex nature of political risks. Using the two international indicators, the WGI by the World Bank and FSI by The Fund for Peace, this study identified considerable differences in the political stability of countries. Among the most exposed countries of the sample, predominantly located in Africa and Asia, are countries hosting important international genebanks. From a risk management perspective, it would be essential to establish a centralized monitoring system for political risks (e.g., at FAO or the Crop Trust) to be able to take safety measures proactively and on time.
4. The insufficient level of financing has widely been acknowledged as a key limiting factor for genebanks. Yet, information on the actual and required budget per genebank, as well as on the nature and provenance of funds, is scarce and difficult to obtain. More research is necessary; it is recommended to include in the FAO country reports a section about necessary financial resources. In addition, it can be noted that the international genebanks have—in particular through the professional work of the Crop Trust establishing the Crop Diversity Endowment Fund—a stable financial backbone. In contrast, the situation at national genebanks is very diverse. It depends on the organizational structure (centralized vs. decentralized), the overall state budget, and the priority-setting of national governments.

In summary, progress has been made in the last few years to mainstream and strengthen quality and risk management, in particular at international genebanks, with the Crop Trust being a driving force. The CGIAR genebanks, as well as some national genebanks (e.g., in Germany, the Netherlands, the USA, and Canada), are more advanced regarding risk management. However, at many national genebanks, considerable scope for improvement remains. Therefore, a location-specific hazard and vulnerability assessment is recommended in order to define appropriate risk prevention measures and risk management strategies at the genebank level. Any progress in this respect can only be achieved with adequate human and financial resources as well as political support.

Supplementary Materials: The following supporting information can be downloaded at: https://www.mdpi.com/article/10.3390/plants12152874/s1, Table S1: Natural hazard and risk assessment for sampled genebanks.

Author Contributions: The original study was carried out in 2022 by T.H. as a master's thesis with the title "Hazard assessment and risk management of national and international genebanks for a sustainable conservation of plant genetic resources" at the Institute of Farm Management (supervised by Christian Lippert) and the Institute of Plant Breeding, Seed Sciences and Population Genetics (supervised by Karl Schmid) of the University of Hohenheim. The results summarized in this paper were thoroughly reviewed and amended by J.M.M.E. based on his long-term experience in the conservation of plant genetic resources. All authors have read and agreed to the published version of the manuscript.

Funding: This research received no external funding.

Data Availability Statement: The results of the natural hazard exposure analysis provided by Munich Re are made available in the annex and the supporting materials. Public datasets analysed stem from the WIEWS database (https://www.fao.org/wiews/data/ex-situ-sdg-251/overview/en/), the Worldbank's database on the Worldwide Governance Indicator (https://databank.worldbank.org/source/worldwide-governance-indicators), and the database on the Fragile States Index by the Fund for Peace (https://fragilestatesindex.org/).

Acknowledgments: Our special thanks go to Munich Reinsurance Company, namely to Rupert Flatscher and Thomas Zerweck, for providing the output of the internal tool *Natural Hazards Edition*, enabling a location-specific natural hazard assessment. Furthermore, we thank Christian Lippert and Karl Schmid from the University of Hohenheim for their supervision and support in the course of preparing and writing the master's thesis, as well as Andreas Ebert and the reviewers for critically revising the draft version of this paper.

Conflicts of Interest: The authors declare no conflict of interest.

Appendix A

The following presents tables of the selected supranational genebanks (international and regional) (Table A1) as well as of national genebanks and their collection sizes (Table A2).

Table A1. International and regional genebanks and their collection size in number of accessions.

Status	Country	WIEWS Code	Institution Name	Collection Size
International genebanks	Belgium	BEL084	Bioversity International Musa Germplasm Transit Centre (ITA)	1625
	Côte d'Ivoire	CIV033	Africa Rice Center (AfricaRice)	21,815
	Colombia	COL003	Centro Internacional de Agricultura Tropical (CIAT)	66,599
	Ethiopia	ETH013	International Livestock Research Institute (ILRI)	18,641
	India	IND002	International Crop Research Institute for the Semi-Arid Tropics (ICRISAT)	128,691
	Kenya	KEN056	Genetic Resources Unit (ICRAF)	15,157
	Lebanon	LBN002	International Centre for Agricultural Research in Dry Areas (ICARDA, location Lebanon))	151,858
	Morocco	MARNA [1]	International Centre for Agricultural Research in Dry Areas (ICARDA, location Morocco)	n.a. [2]
	Mexico	MEX002	Centro Internacional de Mejoramiento de Maíz y Trigo (CIMMYT)	210,851
	Nigeria	NGA039	International Institute of Tropical Agriculture (IITA)	36,531
	Norway	NOR051	Svalbard Global Seed Vault (SGSV)	n.a. [3]
	Peru	PER001	Centro Internacional de la Papa (CIP)	18,066
	Philippines	PHL001	International Rice Research Institute (IRRI)	132,141
	Taiwan, Province of China	TWN001	World Vegetable Center (WorldVeg)	59,954

Table A1. *Cont.*

Status	Country	WIEWS Code	Institution Name	Collection Size
Regional genebanks	United Arab Emirates	ARE003	International Center for Biosaline Agriculture (ICBA)	14,524
	Costa Rica	CRI085/CRI134/CRI142	Centro Agronómico Tropical de Investigación y Enseñanza (CATIE)	10,972
	Fiji	FJI049	Centre for Pacific Crops and Trees (CePaCT)	2158
	Kenya	KENNA [1]	ICRISAT regional genebank	n.a. [2]
	Niger	NER047	ICRISAT regional genebank	n.a. [2]
	Sweden	SWE054	Nordic Genetic Resource Center (NORDGEN)	33,272
	Zambia	ZMB030	SADC Plant Genetic Resources Centre (SRGB)	11,326
	Zimbabwe	ZWENA [1]	ICRISAT regional genebank	n.a. [2]

[1] Genebank does not possess a WIEWS institutional code. For consistency, the country code + NA was used as a unique identifier. [2] For the ICARDA genebank at Morocco as well as the three regional genebanks of ICRISAT, no collection size was reported at WIEWS [15] or by the FAO [3]. [3] SGSV was included in the study as a benchmark for comparing exposure and risk; hence, information about its collection was due to its exclusive focus on safety backups not being relevant. Source: own representation, data on collection size provided by WIEWS (2020) [15].

Table A2. Selected countries and their assessed national genebanks, including number of sampled genebanks, WIEWS institutional code (if available), and their collection size in number of accessions (of the sampled genebanks).

Continent	Country Code	Country	WIEWS Code	No. of Selected Genebanks	Collection Size
Africa	EGY	Egypt	EGY087	1	14,610
	ETH	Ethiopia	ETH085	1	73,164
	GHA	Ghana	GHA091	1	418
	KEN	Kenya	KEN212	1	51,405
	MAR	Morocco	MAR088	1	69,628
	NGA	Nigeria	NGA010	1	7692
	ZAF	South Africa	TZA016	1	6275
	TZA	United Republic of Tanzania	ZAF062/ZAF064	1	7279
Total Africa				**8**	**230,471**
Americas	ARG	Argentina	ARG1342, ARG1350	2	5025
	BRA	Brazil	BRA003	1	107,537
	CAN	Canada	CAN004, CAN025, CAN064	3	111,157
	COL	Colombia	COL017	1	15,776
	MEX	Mexico	MEX208	1	27,100
	PER	Peru	PER014, PER066	2	6542
	USA	United States of America	USA016, USA020, USA022, USA029, USA033, USA970	6	461,758
Total Americas				**16**	**734,895**
Asia	BGD	Bangladesh	BGD001, BGD002, BGD003	3	22,961
	CHN	China	CHN001, CHNNA [1]	2	351,332 *
	IND	India	IDN179	1	410,565
	IDN	Indonesia	IND001	1	4594
	JPN	Japan	JPN183	1	224,353
	KOR	Republic of Korea	KOR046	1	152,272 *
	NPL	Nepal	NPL069	1	6470
	PAK	Pakistan	PAK001	1	33,003
	PHL	Philippines	PHL129, PHL158	2	6875
	THA	Thailand	THA300	1	31,887
	TUR	Türkiye	TUR001, TUR034	2	38,961

Table A2. *Cont.*

Continent	Country Code	Country	WIEWS Code	No. of Selected Genebanks	Collection Size
		Total Asia		**16**	**1,283,273**
Europe	DEU	Germany	DEU146, DEU159, DEU271	3	150,736
	ESP	Spain	ESP004	1	20,826
	FRA	France	FRA040, FRA139	2	16,143
	GBR	United Kingdom of Great Britain and Northern Ireland	GBR004, GBR006, GBR016, GBR140, GBR247	5	836,237 [2]
	ITA	Italy	ITA436	1	6962
	POL	Poland	POL003	1	76,751
	ROU	Romania	ROM007	1	16,428
	RUS	Russian Federation	RUS001	1	200,717
	UKR	Ukraine	UKR001	1	34,518
		Total Europe		**16**	**1,359,318**
Oceania	AUS	Australia	AUS165, AUS167	2	249,056
Total				**58**	**3,857,013**

[1] Genebank does not possess a WIEWS institutional code. For consistency, the country code + NA was used as a unique identifier. [2] The UK genebanks sampled comprise the Nottingham Arabidopsis Stock Center (NASC), which holds—with a collection size of 684,495 accessions—the largest share of the reported collection size of 836,237 accessions. Source: own representation, data on collection size provided by WIEWS (2020) and—if marked with *—by FAO country reports [72,73]. In case of China, the 351,332 accessions refer only to one genebank; as for the other, no information about the collection size was available.

Appendix B

Table A3 provides an exhaustive risk assessment for the sample of 80 genebanks with regard to the 12 selected natural hazards as well as the risk indexes and scores.

Table A3. Natural hazard exposure for sampled genebanks, including exposure for 12 hazards as well as weighted global risk score and index and the risk scores/indexes for the three hazard groups—earthquake, storm, and flood.

WIEWS Code (*If Available*)	Status	Organization Acronym	Country	Earthquake	Volcano	Tsunami	Storm Surge	Riverflood	Flash Flood	Tropical Cyclone	Extratropical Storm	Tornado	Hailstorm	Lightning	Wildfire	Global risk Score	Global Risk Index	Risk Score Earthquake	Risk Index Earthquake	Risk Score Storm	Risk Index Storm	Risk Score Flood	Risk Index Flood
BEL084	int	ITC	Belgium	1	−1	−1	−1	0	3	−1	2	3	2	2	1	**2**	**11**	1	2	1	5	1	3
CIV033	int	AfricaRice	Côte d'Ivoire	0	−1	−1	−1	0	3	−1	−1	1	2	4	2	**2**	**7**	1	1	1	1	1	3
COL003	int	CIAT	Colombia	3	2	−1	−1	0	2	−1	−1	1	3	4	1	**4**	**42**	4	37	1	2	1	2
ETH013	int	ILRI	Ethiopia	3	2	−1	−1	0	4	−1	−1	1	5	5	−1	**4**	**45**	4	37	1	4	1	4
IND002	int	ICRISAT	India	0	−1	−1	−1	0	4	0	−1	2	2	3	2	**2**	**11**	1	1	1	4	1	4
KEN056	int	ICRAF	Kenya	2	2	−1	−1	0	5	−1	−1	1	3	2	−1	**3**	**18**	2	11	1	1	2	6
LBN002	int	ICARDA-LBN	Lebanon	3	−1	−1	−1	0	3	−1	0	2	4	2	3	**4**	**42**	3	34	1	2	1	3
MARNA [1]	int	ICARDA-MAR	Morocco	0	−1	−1	−1	0	3	−1	1	1	1	2	−1	**1**	**5**	1	1	1	1	1	3
MEX002	int	CIMMYT	Mexico	2	2	−1	−1	0	4	0	−1	1	5	4	1	**3**	**21**	2	11	1	5	1	4
NGA039	int	IITA	Nigeria	0	−1	−1	−1	0	2	−1	−1	1	2	4	3	**2**	**7**	1	1	1	1	1	2
NOR051	int	SGSV	Norway	0	−1	−1	−1	0	1	−1	3	1	1	−999	−1	**2**	**9**	1	1	2	6	1	2
PER001	int	CIP	Peru	4	−1	−1	−1	0	2	−1	−1	1	1	1	−1	**4**	**78**	4	76	1	0	1	2
PHL001	int	IRRI	Philippines	2	3	−1	−1	0	5	4	−1	1	2	4	1	**4**	**112**	3	18	4	87	2	6
TWN001	int	WorldVeg	Taiwan, Province of China	3	−1	−1	−1	0	3	5	0	1	2	2	2	**4**	**139**	3	34	4	100	1	3
ARE003	reg	ICBA	United Arab Emirates	2	−1	−1	−1	500	2	−1	−1	1	1	2	−1	**3**	**22**	2	8	1	0	2	14
CRI134/CRI142/CRI085	reg	CATIE	Costa Rica	3	3	−1	−1	0	2	−1	−1	1	3	4	1	**4**	**49**	4	44	1	2	1	2
FJI049	reg	CePaCT	Fiji	3	−1	500	−1	−999	2	5	−1	1	1	2	1	**4**	**142**	4	39	4	100	1	2
KENNA [1]	reg	ICRISAT-Kenya	Kenya	1	1	−1	−1	0	4	−1	−1	1	1	2	2	**2**	**9**	1	3	1	0	1	4
NER047	reg	ICRISAT-Niger	Niger	0	−1	−1	−1	0	3	−1	−1	1	1	4	2	**2**	**7**	1	1	1	1	1	3
SWE054	reg	NORDGEN	Sweden	0	−1	−1	100	0	1	−1	2	2	2	2	1	**4**	**42**	1	1	1	4	4	36
ZMB030	reg	SRGB	Zambia	2	−1	−1	−1	0	4	−1	−1	1	3	4	3	**3**	**17**	2	8	1	2	1	4

Table A3. *Cont.*

WIEWS Code (*If Available*)	Status	Organization Acronym	Country	Earthquake	Volcano	Tsunami	Storm Surge	Riverflood	Flash Flood	Tropical Cyclone	Extratropical Storm	Tornado	Hailstorm	Lightning	Wildfire	Global risk Score	Global Risk Index	Risk Score Earthquake	Risk Index Earthquake	Risk Score Storm	Risk Index Storm	Risk Score Flood	Risk Index Flood
ZWENA [1]	reg	ICRISAT-Zimbabwe	Zimbabwe	0	−1	−1	−1	0	4	−1	−1	1	3	3	2	**2**	**9**	1	1	1	2	1	4
ARG1342	nat	BBC-INTA	Argentina	0	−1	−1	−1	0	5	−1	1	3	5	4	1	**2**	**14**	1	1	2	6	2	6
ARG1350	nat	BGLACONSULTA	Argentina	3	2	−1	−1	0	2	−1	0	2	3	3	2	**4**	**44**	4	37	1	3	1	2
AUS165	nat	AGG	Australia	1	−1	−1	−1	0	2	−1	1	2	2	2	2	**2**	**8**	1	2	1	2	1	2
AUS167	nat	APG	Australia	2	−1	−1	−1	0	2	−1	1	2	1	1	−1	**2**	**12**	2	8	1	2	1	2
BGD001	nat	BJRI	Bangladesh	2	−1	−1	−1	0	5	0	−1	2	2	5	−1	**3**	**19**	2	8	1	5	2	6
BGD002	nat	BRRI	Bangladesh	2	−1	−1	−1	100	4	0	−1	2	2	5	2	**4**	**59**	2	8	1	5	4	44
BGD003	nat	BARI	Bangladesh	2	−1	−1	−1	100	4	0	−1	2	2	5	2	**4**	**59**	2	8	1	5	4	44
BRA003	nat	CENARGEN	Brazil	0	−1	−1	−1	0	2	−1	−1	1	4	4	2	**2**	**7**	1	1	1	2	1	2
CAN004	nat	PGRC	Canada	0	−1	−1	−1	0	2	−1	1	3	2	2	−1	**2**	**6**	1	1	1	3	1	2
CAN025	nat	CCGB	Canada	0	−1	−1	−1	0	2	−1	1	4	4	4	1	**2**	**15**	1	1	2	11	1	2
CAN064	nat	CPGR	Canada	2	−1	−1	−1	0	2	1	2	3	3	2	−1	**3**	**23**	2	8	2	13	1	2
CHN001	nat	NCGC	China	3	−1	−1	−1	0	5	−1	1	2	4	3	−1	**4**	**44**	3	34	1	4	2	6
CHNNA [1]	nat	/	China	2	−1	−1	−1	0	2	−1	0	2	4	3	1	**2**	**14**	2	8	1	3	1	2
COL017	nat	AGROSAVIA	Colombia	3	−1	−1	−1	0	3	−1	−1	1	6	4	1	**4**	**44**	3	34	2	6	1	3
DEU146	nat	IPK	Germany	0	−1	−1	−1	100	2	−1	2	3	3	2	1	**4**	**50**	[illegible]	1	2	6	4	42
DEU159	nat	IPK	Germany	0	−1	−1	−1	0	2	−1	2	3	2	2	1	**2**	**9**	1	1	1	5	1	2
DEU271	nat	IPK	Germany	0	−1	0	100	0	2	−1	3	3	2	2	1	**4**	**47**	1	2	2	8	4	36
EGY087	nat	NGB	Egypt	2	−1	−1	−1	0	1	−1	−1	1	1	1	−1	**2**	**10**	2	8	1	0	1	2
ESP004	nat	INIA-CRF	Spain	0	−1	−1	−1	0	3	−1	1	2	3	2	2	**2**	**9**	1	1	1	3	1	3
ETH085	nat	EBI	Ethiopia	3	0	−1	−1	0	4	−1	−1	1	6	5	2	**4**	**47**	3	34	2	7	1	4
FRA040	nat	INRAe-CLERMONT	France	1	2	−1	−1	0	3	−1	2	2	4	2	1	**2**	**14**	1	5	1	5	1	3
FRA139	nat	INRAe-VASSAL	France	1	−1	0	−1	0	4	−1	3	2	4	3	−1	**3**	**16**	1	3	2	9	1	4

Table A3. *Cont.*

WIEWS Code (*If Available*)	Status	Organization Acronym	Country	Earthquake	Volcano	Tsunami	Storm Surge	Riverflood	Flash Flood	Tropical Cyclone	Extratropical Storm	Tornado	Hailstorm	Lightning	Wildfire	Global risk Score	Global Risk Index	Risk Score Earthquake	Risk Index Earthquake	Risk Score Storm	Risk Index Storm	Risk Score Flood	Risk Index Flood
GBR004	nat	RBG	UK	0	−1	−1	−1	0	2	−1	3	3	2	2	2	**2**	**13**	1	1	2	8	1	2
GBR006	nat	HRIGRU	UK	1	−1	−1	−1	0	2	−1	2	3	3	2	1	**2**	**11**	1	2	2	6	1	2
GBR016	nat	IBERS-GRU	UK	0	−1	−1	−1	100	3	−1	3	3	1	1	1	**4**	**53**	[illegible]	1	2	8	4	43
GBR140	nat	NASC	UK	1	−1	−1	−1	0	2	−1	2	3	2	1	1	**2**	**10**	1	2	1	5	1	2
GBR247	nat	/	UK	0	−1	−1	−1	0	2	−1	2	3	2	2	2	**2**	**10**	1	1	1	5	1	2
GHA091	nat	PGRRI	Ghana	0	−1	−1	−1	0	2	−1	−1	1	3	5	1	**2**	**7**	1	1	1	3	1	2
IDN179	nat	ICABIOGRAD	Indonesia	3	3	−1	−1	0	4	−1	−1	1	2	5	1	**4**	**51**	4	44	1	2	1	4
IND001	nat	NBPGR	India	2	−1	−1	−1	0	6	−1	−1	2	2	4	−1	**3**	**19**	2	8	1	2	2	9
ITA436	nat	IBBR	Italy	1	−1	−1	−1	0	3	−1	1	2	3	3	−1	**2**	**9**	1	2	1	4	1	3
JPN183	nat	NARO	Japan	4	1	−1	−1	0	3	3	2	2	4	2	1	**4**	**120**	4	77	4	39	1	3
KEN212	nat	GeRRI	Kenya	2	2	−1	−1	0	4	−1	−1	1	3	2	3	**3**	**19**	2	11	1	1	1	4
KOR046	nat	NAC	Republic of Korea	2	−1	−1	−1	0	4	2	1	3	4	2	1	**4**	**37**	2	8	3	24	1	4
MAR 088	nat	INRA CRRAS	Morocco	0	−1	−1	−1	0	2	−1	1	1	1	2	2	**2**	**6**	1	1	1	1	1	2
MEX208	nat	CNRG	Mexico	2	−1	−1	−1	0	3	0	−1	1	5	4	2	**3**	**18**	2	8	1	5	1	3
NGA010	nat	NACGRAB	Nigeria	0	−1	−1	−1	0	3	−1	−1	1	2	4	2	**2**	**7**	1	1	1	1	1	3
NPL069	nat	NAGRC	Nepal	3	−1	−1	−1	0	3	−1	−1	2	5	4	1	**4**	**42**	3	34	1	4	1	3
PAK001	nat	PGRP	Pakistan	2	−1	−1	−1	0	4	−1	−1	2	5	6	−1	**3**	**17**	2	8	1	5	1	4
PER014	nat	E.E.A. Illpa-Puno	Peru	2	−1	−1	−1	0	4	−1	−1	1	5	3	1	**3**	**16**	2	3	1	3	1	4
PER066	nat	UNA	Peru	4	−1	−1	−1	0	2	−1	−1	1	1	1	−1	**4**	**78**	4	75	1	0	1	2
PHL129	nat	IPB-NPGRL	Philippines	2	3	−1	−1	100	5	4	−1	1	2	4	1	**4**	**152**	3	13	4	87	4	46
PHL158	nat	PhilRice	Philippines	3	2	−1	−1	0	5	4	−1	1	2	5	2	**4**	**133**	4	37	4	88	2	6
POL003	nat	IHAR	Poland	0	−1	−1	−1	100	2	−1	2	2	4	2	1	**4**	**49**	1	1	1	5	4	42
ROM007	nat	BRGV Suceava	Romania	2	−1	−1	−1	0	3	−1	1	2	4	3	2	**3**	**17**	2	8	1	4	1	3
RUS001	nat	VIR	Russian Federation	0	−1	−1	−1	0	2	−1	1	2	3	2	−1	**2**	**6**	1	1	1	3	1	2

Table A3. *Cont.*

WIEWS Code (*If Available*)	Status	Organization Acronym	Country	Earthquake	Volcano	Tsunami	Storm Surge	Riverflood	Flash Flood	Tropical Cyclone	Extratropical Storm	Tornado	Hailstorm	Lightning	Wildfire	Global risk Score	Global Risk Index	Risk Score Earthquake	Risk Index Earthquake	Risk Score Storm	Risk Index Storm	Risk Score Flood	Risk Index Flood
THA300	nat	/	Thailand	1	−1	−1	−1	100	4	−1	−1	2	2	5	3	**4**	**52**	1	2	1	3	4	44
TUR001	nat	AARI	Türkiye	3	−1	−1	−1	0	4	−1	1	2	3	2	2	**4**	**43**	3	34	1	3	1	4
TUR034	nat	FCCRI	Türkiye	2	−1	−1	−1	0	2	−1	1	2	3	3	−1	**2**	**14**	2	8	1	4	1	2
TZA016	nat	NPGRC	United Republic of Tanzania	1	2	−1	−1	0	5	−1	−1	1	3	2	−1	**2**	**12**	1	5	1	1	2	6
UKR001	nat	IR	Ukraine	0	−1	−1	−1	0	2	−1	1	2	4	3	−1	**2**	**7**	1	1	1	4	1	2
USA016	nat	S9	USA	0	−1	−1	−1	0	3	1	1	3	5	4	−1	**3**	**17**	1	1	2	13	1	3
USA020	nat	NC7	USA	0	−1	−1	−1	0	4	−1	1	4	6	4	1	**3**	**21**	1	1	2	15	1	4
USA022	nat	W6	USA	0	−1	−1	−1	0	2	−1	1	2	2	2	2	**2**	**7**	1	1	1	2	1	2
USA029	nat	NSGC	USA	1	1	−1	−1	0	1	−1	1	2	2	2	1	**2**	**8**	1	3	1	2	1	2
USA033	nat	SOY	USA	0	−1	−1	−1	0	3	−1	1	4	5	4	−1	**3**	**16**	1	1	2	12	1	3
USA970	nat	DB NRRC	USA	1	−1	−1	−1	0	3	−1	1	4	5	5	1	**3**	**19**	1	2	2	13	1	3
ZAF062/ZAF064	nat	DALRRD/NPGRC	South Africa	0	−1	−1	−1	0	4	−1	0	3	5	4	−1	**2**	**10**	1	1	1	5	1	4

Legend: Earthquakes measured according to the Modified Mercalli scale with −999 = no information available; 0: MM V and below, imperceptible to rather strong; 1 = MM VI, strong; 2 = MM VII, very strong; 3 = MM VIII, destructive; 4 = MM IX and above devastation, major disaster; Volcanoes with the hazard classes −1 = no hazard, 0 = unclassified volcanoes, 1 = minor hazard (>15,000 years return period), 2 = moderate hazard (200 to 15,000 years return period), 3 = high hazard (≤200 years return period); Tsunami with the hazard classes −1 = no hazard, Zone 0 = very low tsunami exposure, Zone 100 = coasts are exposed to a 100-year return period of tsunamis, Zone 500 = coasts are exposed to a 500-year return period, Zone 1000 = coasts are exposed to a 1000-year return period; Storm surge with the hazard classes −1 = no hazard, Zone 100 = coasts are exposed to a 100-year return period (1% annual flood chance), Zone 500 = coasts are exposed to a 500-year return period (0.2% annual flood chance), Zone 1000 = coasts are exposed to a 1000-year return period (0.1% annual flood chance); River flood with the hazard classes −999 = no information available, Zone 0 = minimal flood risk (areas outside the 1% or 0.2% annual flood chance), Zone 100 = areas exposed to a 100-year return period flood event (1% annual flood chance), Zone 500 = areas exposed to a 500-year return period flood event (0.2% annual flood chance); Flash flood with the hazard classes −999 = no information available, Zone 1 = low frequency and intensity of flash floods up to Zone 6 = high frequency and intensity of flash floods; Tropical cyclones with the hazard classes −1 = no hazard with <76 km/h, 0 = 76–141 km/h, 1 = 142–184 km/h, 2 = 185–212 km/h, 3 = 213–251 km/h, 4 = 252–299 km/h, 5 = ≥300 km/h; Extratropical storm with the hazard classes −1 = no hazard, 0 = ≤80 km/h, 1 = 81–120 km/h, 3 = 121–160 km/h, 4 = >200 km/h; Tornado with the hazard classes –999 = no information available, Zone 1 = low frequency and intensity of tornados up to Zone 4 = high frequency and intensity of tornados; Hailstorm with the hazard classes –999 = no information available, Zone 1 = low frequency and intensity of hailstorms up to Zone 6 = high frequency and intensity of hailstorms; Lightning with the hazard classes –999 = no information available, 1 = 0.2–1 lightning strokes per km^2 and year, 2 = 1–4 lightning strokes per km^2 and year, 3 = 4–10 lightning strokes per km^2 and year, 4 = 10–20 lightning strokes per km^2 and year, 5 = 20–40 lightning strokes per km^2 and year, 6 = 40–80 lightning strokes per km^2 and year; Wildfire with the hazard classes −1 = no hazard, zone 1 = low exposure to zone 4 = high exposure; Risk Score: weighted risk score with 1 = low risk, 2 = medium risk, 3 = high risk, 4 = extreme risk; Risk Index: ranging from 0 = no risk to 100 = extreme risk, disaggregated for earthquake, storm, and flood hazards; Global Risk Index: The unweighted sum of the three risk indexes per hazard group (earthquake, storm, and flood), ranging from 0 = no risk to 300 = extreme risk. [1] Genebank does not possess a WIEWS institutional code. For consistency, the country code + NA was used as a unique identifier. Source: own representation, adapted from Munich Re (2022) [19].

References

1. Engels, J.M.M.; Ebert, A.W. A Critical Review of the Current Global Ex Situ Conservation System for Plant Agrobiodiversity. I. History of the Development of the Global System in the Context of the Political/Legal Framework and Its Major Conservation Components. *Plants* **2021**, *10*, 1557. [CrossRef] [PubMed]
2. Louafi, S.; Thomas, M.; Berthet, E.T.; Pélissier, F.; Vaing, K.; Jankowski, F.; Bazile, D.; Pham, J.-L.; Leclercq, M. Crop Diversity Management System Commons: Revisiting the Role of Genebanks in the Network of Crop Diversity Actors. *Agronomy* **2021**, *11*, 1893. [CrossRef]
3. FAO (Ed.) *The Second Report on the State of the World's Plant Genetic Resources for Food and Agriculture*; Commission on Genetic Resources for Food and Agriculture, Food and Agriculture Organization of the United Nations: Rome, Italy, 2010.
4. Bhattacharya, S. Syrian seed bank gets new home away from war. *Nature* **2016**, *538*, 16–17. [CrossRef] [PubMed]
5. Gewin, V. Protecting the Food Protectors: Are Crop Gene Banks in Danger? *Greenbiz*. Available online: https://www.greenbiz.com/article/protecting-food-protectors-are-crop-gene-banks-danger (accessed on 3 June 2015).
6. Fu, Y.-B. The Vulnerability of Plant Genetic Resources Conserved Ex Situ. *Crop Sci.* **2017**, *57*, 2314–2328. [CrossRef]
7. Angel, M. Ukraine's Giant Seed Bank at Risk of Being Lost as War Rages. Available online: https://www.croptrust.org/news-events/in-the-media/ukraines-giant-seed-bank-at-risk-of-being-lost-as-war-rages/ (accessed on 1 June 2022).
8. FAO Ukraine's National Seed Collection Relocates to a Secure Site. Food and Agriculture Organization of the United Nations. Available online: https://www.fao.org/europe/news/detail/ukraine-s-national-seed-collection-relocates-to-a-secure-site/en (accessed on 8 June 2023).
9. Dalezios, N.R. *Environmental Hazards Methodologies for Risk Assessment and Management*; IWA Publishing: London, UK, 2017.
10. Wolke, T. *Risk Management*; Walter de Gruyter GmbH: Berlin, Germany; München, Germany; Boston, MA, USA, 2017.
11. OECD. *Global Modelling of Natural Hazard Risks. Enhancing Existing Capabilities to Address New Challenges*; Organisation for Economic Cooperation and Development (OECD) Global Science Forum: Tokyo, Japan, 2012.
12. Hohl, R.M. *Agricultural Risk Transfer: From Insurance to Reinsurance to Capital Markets*; John Wiley & Sons, Incorporated: Newark, UK, 2019.
13. Crop Trust Global Crop Conservation Strategies. Available online: https://www.croptrust.org/work/projects/global-crop-conservation-strategies/ (accessed on 13 August 2022).
14. Khoury, C.; Laliberté, B.; Guarino, L. Trends in ex situ conservation of plant genetic resources: A review of global crop and regional conservation strategies. *Genet. Resour. Crop Evol.* **2010**, *57*, 625–639. [CrossRef]
15. WIEWS Overview of World Information and Early Warning System on Plant Genetic Resources for Food and Agriculture. Food and Agriculture Organization of the United Nations. Available online: https://www.fao.org/wiews/data/ex-situ-sdg-251/overview/en/ (accessed on 30 April 2022).
16. SGSV Safeguarding Seeds for the Future. Available online: https://www.seedvault.no/ (accessed on 3 March 2023).
17. Crop Trust. (n.d.). Crop Trust Svalbard Global Seed Vault. *Crop Trust*. Available online: https://www.croptrust.org/our-work/svalbard-global-seed-vault/ (accessed on 26 April 2022).
18. FAOSTAT. (n.d.). *FAOSTAT Data*. Available online: https://www.fao.org/faostat/en/ (accessed on 5 July 2022).
19. Munich Re. *Output RiskMapper, Tool Natural Hazards Edition*, 2022; Unpublished Database.
20. Zerweck, T.; Munich Re, Munich, Germany. Personal communication, 2022.
21. Schmieder, J.; Munich Re, Munich, Germany. Personal communication, 2022.
22. World Bank Worldwide Governance Indicators | DataBank. Available online: https://databank.worldbank.org/source/worldwide-governance-indicators (accessed on 31 October 2021).
23. The Fund for Peace Fragile States Index | The Fund for Peace. Available online: https://fragilestatesindex.org/ (accessed on 31 October 2021).
24. Thomas, M.A. What Do the Worldwide Governance Indicators Measure? *Eur. J. Dev. Res.* **2010**, *22*, 31–54. [CrossRef]
25. United Nations Office for Disaster Risk Reduction. UNDRR Terminology. Available online: https://www.undrr.org/terminology (accessed on 12 May 2022).
26. CGIAR Genebank Platform Red Book of Risk. *Risk Identification, Mitigation and Monitoring in CGIAR Genebanks*; CGIAR Genebank Platform Red Book of Risk: Bonn, Germany, 2020; Volume 1.0-002.
27. Crop Genebank Knowledge Base Risk Management. Available online: https://cropgenebank.sgrp.cgiar.org/index.php/management-mainmenu-433/risk-management-mainmenu-236 (accessed on 6 May 2022).
28. Engels, J.M.M.; Ebert, A.W. A Critical Review of the Current Global Ex Situ Conservation System for Plant Agrobiodiversity. II. Strengths and Weaknesses of the Current System and Recommendations for Its Improvement. *Plants* **2021**, *10*, 1904. [CrossRef] [PubMed]
29. Wilson, G.; Wilson, T.M.; Deligne, N.I.; Cole, J.W. Volcanic hazard impacts to critical infrastructure: A review. *J. Volcanol. Geotherm. Res.* **2014**, *286*, 148–182. [CrossRef]
30. Oetjen, J.; Sundar, V.; Venkatachalam, S.; Reicherter, K.; Engel, M.; Schüttrumpf, H.; Sannasiraj, S.A. A comprehensive review on structural tsunami countermeasures. *Nat. Hazards* **2022**, *113*, 1419–1449. [CrossRef]
31. Horn, D.P. Chapter 6—Storm Surge Warning, Mitigation, and Adaptation. In *Coastal and Marine Hazards, Risks, and Disasters*; Shroder, J.F., Ellis, J.T., Sherman, D.J., Eds.; Hazards and Disasters Series; Elsevier: Amsterdam, The Netherlands, 2015; pp. 153–180.

32. BBK Naturgefahr: Hochwasser. Bundesamt für Bevölkerungsschutz und Katastrophenhilfe. Available online: https://www.bbk.bund.de/DE/Themen/Risikomanagement/Baulicher-Bevoelkerungsschutz/Schutz-vor-Naturgefahren/Hochwasser/hochwasser_node.html (accessed on 12 July 2022).
33. Lohwasser, U.; Leibniz-Institut für Pflanzengenetik und Kulturpflanzenforschung IPK, Gatersleben, Germany. Personal communication, 2022.
34. USAID. Flood Control Structures. 2012. Available online: https://www.climatelinks.org/sites/default/files/asset/document/Infrastructure_FloodControlStructures.pdf (accessed on 12 July 2022).
35. van Hintum, T.; Center for Genetic Resources CGN, Wageningen University & Research, Wageningen, The Netherlands. Personal communication, 2022.
36. Dasgupta, S.; Blankespoor, B. Protection from Cyclones: Benefits of Integrating Green and Gray Infrastructure. Available online: https://blogs.worldbank.org/developmenttalk/protection-cyclones-benefits-integrating-green-and-gray-infrastructure (accessed on 8 April 2019).
37. English, E.C.; Friedland, C.J.; Orooji, F. Combined Flood and Wind Mitigation for Hurricane Damage Prevention: The Case for Amphibious Construction. *J. Struct. Eng.* **2017**, *143*, 13. [CrossRef]
38. Jain, N. Hong Kong's Preparedness and Resilient Infrastructure Cut Economic losses from Typhoons. Available online: https://www.undrr.org/news/hong-kongs-preparedness-and-resilient-infrastructure-cut-economic-losses-typhoons (accessed on 4 December 2021).
39. Ebert, A.W.; World Vegetable Center WorldVeg, Tainan, Taiwan. Personal communication, 2022.
40. BBK Naturgefahr: Sturm. Available online: https://www.bbk.bund.de/DE/Themen/Risikomanagement/Baulicher-Bevoelkerungsschutz/Schutz-vor-Naturgefahren/Sturm/sturm_node.html (accessed on 12 July 2022).
41. Federal Emergency Management Agency. Safe Rooms for Tornadoes and Hurricanes. In *Guidance for Community and Residential Safe Rooms, FEMA P-361*, 4th ed.; FEMA: Washington, DC, USA, 2021.
42. Kapur, O.; Mahadevia, A.; Park, J.; Passman, S.; Perotin, M.; Reeder, A.; Seitz, L.; Sheldon, A.; Tezak, S. Wind Retrofit Guide for Residential Buildings. FEMA P-804 2010. Available online: https://www.fema.gov/emergency-managers/risk-management/building-science/publications?name=804&field_keywords_target_id=All&field_document_type_target_id=All&field_audience_target_id=All (accessed on 12 July 2022).
43. BBK Naturgefahr: Hagel. Available online: https://www.bbk.bund.de/DE/Themen/Risikomanagement/Baulicher-Bevoelkerungsschutz/Schutz-vor-Naturgefahren/Hagel/hagel_node.html (accessed on 12 July 2022).
44. BBK Naturgefahr: Gewitter. Available online: https://www.bbk.bund.de/DE/Themen/Risikomanagement/Baulicher-Bevoelkerungsschutz/Schutz-vor-Naturgefahren/Gewitter/gewitter_node.html (accessed on 12 July 2022).
45. Federal Emergency Management Agency. Avoiding Wildfire Damage: A Checklist for Homeowners. Available online: https://www.fema.gov/pdf/hazard/wildfire/wdfrdam.pdf (accessed on 12 July 2022).
46. McCaffrey, S.; McGee, T.K.; Coughlan, M.; Tedim, F. Understanding wildfire mitigation and preparedness in the context of extreme wildfires and disasters. In *Extreme Wildfire Events and Disasters*; Elsevier: Amsterdam, The Netherlands, 2020; pp. 155–174.
47. FAO. *Genebank Standards for Plant Genetic Resources for Food and Agriculture*, Rev. ed.; Food and Agriculture Organization of the United Nations: Rome, Italy, 2014; Available online: https://www.fao.org/3/i3704e/i3704e.pdf (accessed on 28 April 2022).
48. Lohwasser, U.; Leibniz-Institut für Pflanzengenetik und Kulturpflanzenforschung IPK, Gatersleben, Germany. Personal communication, 2023.
49. Fu, Y.-B.; Saskatoon Research and Development Centre, Saskatoon, SK, Canada. Personal communication, 2022.
50. Dempewolf, H.; Global Crop Diversity Trust, Bonn, Germany. Personal communication, 2022.
51. CGIAR. Genebank Platform Genebank Platform (2017–2021) Reviews. Available online: https://www.genebanks.org/resources/genebank-reviews/ (accessed on 22 July 2022).
52. Panis, B.; Nagel, M.; Van Den Houwe, I. Challenges and Prospects for the Conservation of Crop Genetic Resources in Field Genebanks, in In Vitro Collections and/or in Liquid Nitrogen. *Plants* **2020**, *9*, 1634. [CrossRef] [PubMed]
53. Dixon, L.; Barnosky, J.T.; Clancy, N. *Insuring Public Buildings, Contents, Vehicles, and Equipment Against Disasters: Current Practices of State and Local Government and Options for Closing the Insurance Gap*; RAND Corporation: Santa Monica, CA, USA, 2020; Available online: https://www.rand.org/pubs/research_reports/RRA332-1.html (accessed on 13 July 2022).
54. Freeman, P.K.; Martin, L.A.; Linnerooth-Bayer, J.; Mechler, R.; Pflug, G.; Warner, K. *Disaster Risk Management: National Systems for the Comprehensive Management of Disaster Risk and Financial Strategies for Natural Disaster Reconstruction*; Inter-American Development Bank: Washington, DC, USA, 2003.
55. Klafack, M. Öffentliche Gebäude oft Nicht Versichert. Available online: https://www.pfefferminzia.de/elementarschaeden-oeffentliche-gebaeude-oft-nicht-versichert/ (accessed on 13 July 2022).
56. Begemann, F.; Bundesanstalt für Landwirtschaft und Ernährung BLE, Bonn, Germany. Personal communication, 2022.
57. UNDRR; International Science Council. *Hazard Definition and Classification Review—Technical Report*; United Nations Office for Disaster Risk Reduction, 2020. Available online: https://www.undrr.org/publication/hazard-definition-and-classification-review (accessed on 12 May 2022).
58. African Risk Capacity Group. Who We Are. African Risk Capacity Group. Available online: https://www.arc.int/about (accessed on 14 July 2022).

59. African Risk Capacity Group. How ARC Works. African Risk Capacity Group. Available online: https://www.arc.int/how-arc-works (accessed on 14 July 2022).
60. CCRIF SPC. (n.d.). CCRIF SPC [The Caribbean Catastrophe Risk Insurance Facility]. Available online: https://www.ccrif.org/ (accessed on 14 July 2022).
61. Smale, M.; Jamora, N. Valuing genebanks. *Food Secur.* **2020**, *12*, 905–918. [CrossRef]
62. Koo, B.; Pardey, P.G.; Wright, B.D. Costing the Conservation of Genetic Resources: A Comparison of the CIMMYT and ICARDA Genebanks. In *Efficient Conservation Of Crop Genetic Diversity: Theoretical Approaches and Empirical Studies*; Virchow, D., Ed.; Springer: Berlin/Heidelberg, Germany, 2003; pp. 77–87.
63. OECD Producer Price Indices. OECD DATA. Organization for Economic Co-Operation and Development. Available online: https://data.oecd.org/price/producer-price-indices-ppi.htm#indicator-chart (accessed on 22 August 2022)
64. Rabenau, P. *Eine Kostenanalyse zur Feldvermehrung Mit Anschließendem Vergleich Zwischen der Saatgutkühllagerung under der Lagerung von Samen unter Tiefsttemperaturen (−196 °C)*; Justus-Liebig-Universität Gießen: Gießen, Germany, 2018; p. 121.
65. Chen, J. Reserve Fund. Available online: https://www.investopedia.com/terms/r/reservefund.asp (accessed on 14 July 2022).
66. Kurt, D. Emergency Funds. Available online: https://www.investopedia.com/terms/e/emergency_fund.asp (accessed on 9 February 2022).
67. Macaskill, K.; Guthrie, P. Funding mechanisms for disaster recovery: Can we afford to build back better? *Procedia Eng.* **2018**, *212*, 451–458. [CrossRef]
68. Crop Trust; ITPGRFA. Saving Our Seeds: An Emergency Reserve for Genebanks in Peril. Available online: https://www.croptrust.org/fileadmin/uploads/croptrust/wp/wp-content/uploads/2021/11/Crop-Trust-The-Emergency-Reserve-for-Genebanks.pdf (accessed on 22 July 2022).
69. Nnadozie, K. Notification: Establishment of the Emergency Reserve for Genebanks 2021. Available online: https://www.fao.org/3/cb7798en/cb7798en.pdf (accessed on 23 November 2021).
70. European Commission. *Ex Post Evaluation of the European Union Solidarity Fund 2002–2016: Final Report*; European Commission, Directorate General for Regional and Urban Policy. European Institute, London School of Economics & Political Science: London, UK, 2019. Available online: https://data.europa.eu/doi/10.2776/643843 (accessed on 28 June 2023).
71. World Bank. *FONDEN: Mexico's Natural Disaster Fund—A Review*; World Bank: Washington, DC, USA, 2012.
72. Rural Development Administration. *Country Report on the State of Plant Genetic Resources for Food and Agriculture in Republic of Korea*; Food and Agriculture Organization of the United Nations: Suwon, Republic of Korea, 2009.
73. The Ministry of Agriculture, People's Republic of China. *Country Report on the State of Plant GEnetic Resources for Food and Agriculture in China (1996–2007)*; Food and Agriculture Organization of the United Nations: Rome, Italy, 2008.

plants

MDPI

Review

A Performance Management System for Long-Term Germplasm Conservation in CGIAR Genebanks: Aiming for Quality, Efficiency and Improvement

Charlotte Lusty [1,*], Janny van Beem [1] and Fiona R. Hay [2]

[1] Global Crop Diversity Trust, Platz der Vereinten Nationen 7, 53113 Bonn, Germany; janny.vanbeem@croptrust.org
[2] Department of Agroecology, Aarhus University, Forsøgsvej 1, 4200 Slagelse, Denmark; fiona.hay@agro.au.dk
* Correspondence: charlotte.lusty@croptrust.org

Abstract: UN Sustainable Development Goal 2 Target 2.5 focuses on the conservation of genetic diversity in soundly managed genebanks. In examining the term "soundly managed", it becomes quickly evident that there is much more to long-term conservation than placing samples of seeds or other germplasm in long-term conservation conditions. There are several important factors that determine whether germplasm samples will remain viable in storage for long periods of time. To manage these factors efficiently and effectively, genebanks require sound data and quality management systems. The CGIAR Genebank Platform, coordinated by the Crop Trust, put in place a number of mechanisms that enabled effective online reporting, performance management, quality management, audit and external review and validation. These mechanisms do not conform to the usual monitoring systems put in place for research programs and have only been possible thanks to the flexibility of CGIAR in recognising that the genebanks were exceptional. As a result, in the past 10 years, CGIAR genebanks have significantly improved their performance and the conservation status of collections.

Keywords: seed quality management; long-term conservation; quality management system (QMS); performance management; genebanks; standards

Citation: Lusty, C.; van Beem, J.; Hay, F.R. A Performance Management System for Long-Term Germplasm Conservation in CGIAR Genebanks: Aiming for Quality, Efficiency and Improvement. *Plants* **2021**, *10*, 2627. https://doi.org/10.3390/plants10122627

Academic Editors: Johannes M. M. Engels and Andreas W. Ebert

Received: 8 November 2021
Accepted: 25 November 2021
Published: 29 November 2021

1. CGIAR Genebanks, Germplasm Conservation and Genebank Standards

The CGIAR genebanks safeguard some of the largest and most widely used collections of crop diversity in the world, critical to attaining the UN's Sustainable Development Goals (SDG) to end hunger and improve food and nutrition security [1,2]. Most CGIAR genebanks are strategically located in centres of crop diversity, resulting in the collections being founded on a broad and rich representation of diversity from the primary crop genepools and, in some cases, a continued exchange with traditional communities cultivating landraces or even domesticating semi-wild materials. The collections that CGIAR manages have grown over five decades or more in a relatively organic fashion with the gradual introduction and occasional peaks of expansion resulting from collecting missions, donations from partners and materials generated by breeders and researchers. As a result, the collections represent a worldwide diversity of landraces, heritage varieties, crop wild relatives, improved varieties and, to a lesser extent, breeding or research materials for specific mandate crops (Table 1).

Table 1. Crops and accession numbers of CGIAR genebanks in December 2020.

Centre	Crop(s)	Total Accessions in 2020 [3]
AfricaRice	Rice	21,815
Alliance-Bioversity International	Banana	1624
Alliance-International Center for Tropical Agriculture (CIAT)	Beans, cassava, tropical forages	64,635
International Maize and Wheat Improvement Center (CIMMYT)	Maize, wheat	147,842
International Potato Center (CIP)	Potato, sweetpotato, Andean roots and tubers	18,156
International Center for Agricultural Research in Dry Areas (ICARDA)	Dryland cereals, grain legumes, temperate forages	152,609
World Agroforestry (ICRAF)	Trees	14,919
International Crops Research Institute for the Semi-Arid Tropics (ICRISAT)	Sorghum, millets, grain legumes	129,034
International Institute of Tropical Agriculture (IITA)	Cowpea, maize, legumes, banana, cassava, yam	34,774
International Livestock Research Institute (ILRI)	Tropical forages	18,662
International Rice Research Institute (IRRI)	Rice	132,140
Grand Total		**736,210**

The collections and their contents may be reviewed online through the global portal, Genesys (www.genesys-pgr.org last accessed 24 November 2021). The importance of the collections is recognized in international policy through agreements (Article 15) signed in 2006 between each CGIAR Centre and the Governing Body of the International Treaty on Plant Genetic Resources for Food and Agriculture (Plant Treaty) [4]. Centres are obliged to make collections and associated data under their management available under the Multilateral System (MLS) of Access and Benefit-sharing of the Plant Treaty. CGIAR genebanks act as a major source of international germplasm exchange and, together with CGIAR breeding programs, were responsible for close to 90% of the reported distributions under the Plant Treaty [5]. Between 2012 and 2019, the CGIAR genebanks distributed more than 850,000 samples of germplasm to 163 countries in response to requests [3]. Only the US Department of Agriculture National Plant Germplasm System distributes more germplasm, about 250,000 samples yearly, of which about a quarter is distributed internationally.

The UN's 17 SDGs were agreed by world leaders in 2015. SDG 2 has eight targets with the aim to end hunger in the world. Target 2.5 specifically calls on countries and institutions to maintain genetic diversity in food production, including in "soundly managed" seed and plant banks [6].

Over many decades, hundreds of institutes, universities, research groups and communities have carried out research, breeding and collecting and, as a result, have invested in storing seeds in freezers or cold rooms. Which of these multitudes of efforts constitute a formal seed or plant bank ('genebank') is not easily determined. Most countries have formally designated a national genebank under the management of national agricultural research organizations, but there are many other formal and informal collections and gardens, large and small, managed within the public or private sector. Seeds in storage ultimately perish if not planted out and regenerated before viability is lost. The genetic composition and the rare alleles present in a seed sample will be increasingly lost over time as more and more seeds in the sample lose viability. Germplasm conservation in genebanks aims to maintain the viability and genetic integrity of collected samples in storage for the longest time that is biologically possible. Multiple guidelines have been published over the years to share best genebank practices [7–9]. Germplasm processing for conservation involves several critical steps that can have a prolonged and cumulative influence on the viability and quality of the seeds in storage and their use in the future, as well as on the efficiency of future operations [10]. The critical point, therefore, is not defining what is a seed or plant genebank but in determining what is "soundly managed".

Genebanks serve different purposes and clients. There are many different options for localised communities to save seeds for sharing and planting in the near term. For prolonged conservation and for medium to large collections (e.g., more than 10,000 accessions), the requirements in terms of capacity and processes are considerably more extensive. Any institution wishing to manage a genebank today is ably guided by published Genebank Standards, as endorsed by the FAO Commission on Genetic Resources for Food and Agriculture (CGRFA) [4,11], which provide quantitative thresholds, principles, and critical points for genebank management. There is no formal system of monitoring or measuring compliance with the FAO Genebank Standards, but institutes managing collections have an option to undergo certification or accreditation with the International Standards Organization (www.iso.org; last accessed 24 November 2021) or the International Seed Testing Association (for specific tests; www.seedtest.org; last accessed 24 November 2021). A small number of genebanks have pursued a quality management system (QMS) certification through ISO 9001:2015 which applies to businesses or organizations that seek to ensure that their products or services meet customer needs. While there are benefits to the certification, the Genebank QMS described below includes additional aspects that are outside the scope of certification, e.g., adherence to the International Treaty on Plant Genetic Resources for Food and Agriculture (ITPGRFA) (https://www.fao.org/plant-treaty/en/; last accessed 24 November 2021), the International Plant Protection Convention (www.ippc.int; last accessed 24 November 2021) and the Convention on Biological Diversity and its Nagoya Protocol (www.cbd.int/; last accessed 24 November 2021).

So, what evidence is there of sound management or even that reported accessions really exist? Requesting and receiving seeds from genebanks is one way of telling whether they exist and are viable—but relying on feedback from users is leaving it too late and is hardly feasible for very large collections. If taxpayers are to fund and depend on key institutions, such as CGIAR and other international or national genebanks, to safeguard agricultural diversity for future generations, then the large body of stakeholders (including donors, depositors and users) need to have some confidence that the methods in use meet appropriate standards and will achieve long-term conservation objectives. Furthermore, we need to know that institutions are actively improving processes and incorporating appropriate technologies and approaches to conserve germplasm efficiently. Engels and Ebert [12] provide an exhaustive critique of current conservation methods currently deployed in genebanks with many points that should be addressed.

By describing our experience with CGIAR genebanks, we endorse the view that a comprehensive system of quality and performance management is necessary to provide assurance that key genebanks are complying with FAO genebank and other relevant standards and to provide a continuous mechanism for exchanges and improvement. We describe the challenges that lie behind long-term conservation objectives and the need for serious investments to meet such a commitment, perhaps more than many actors anticipated at the time of collecting seeds and putting them in storage. Such long-term conservation objectives can only be managed through long-term and consistent quality management [10]. In our experience, a quality and performance management approach has helped to manage backlogs and to avoid mistakes, losses and duplication of effort, as well as facilitating continuous improvement. The requirements of long-term conservation objectives should be fully understood by all institutions that aim to accommodate such long-term commitments within a predominantly research-oriented program of work.

2. Why Long-Term Conservation Is Not as Easy as One Might Think

The effective conservation of seed germplasm samples with the aim of maintaining genetic integrity involves the complex interaction of various factors, including biological characteristics intrinsic to the species, the quality of the seeds and the conditions of the seeds' storage [10,11,13]. If seeds are stored in cold storage without attention to drying and testing processes, packaging, monitoring and maintenance of conditions, then only with luck will they maintain their viability to be germinated some years later. If the genetic diversity of a collection is to be conserved over many decades and some of the unique and rare traits occurring in the gene pool are to be retained for future research and breeding, then an intensively evidence-based approach to long-term conservation is required. Four main factors explain why long-term conservation efforts specifically for seed collections require considerable investment:

(1) **Biological factors:** Although seed storage has been a fundamental aspect of agricultural practice for millennia, the long-term storage of dry seeds at low temperatures has only been applied scientifically for relatively few decades. Seeds of some species are predicted to remain viable for perhaps hundreds of years under long-term conditions in a genebank (typically at 3–7% moisture content and −20 °C temperature) [13]. Further, many genebanks have now reported the long-term maintenance of high levels of viability, for many (but not all) seed lots (samples) of at least some crop and forage species, over the course of the genebank's existence [10,14,15]. However, there are examples where seed lots have lost viability relatively quickly or abruptly during genebank storage [10,16,17]. As more viability monitoring data becomes available, seeds of more species, particularly those of crop wild relatives, forage and tree species, may be found to be relatively short-lived in conventional genebank storage. In conclusion, the majority of crop species are thought to exhibit orthodox storage behaviour, tolerating drying to low moisture contents and remaining viable in storage for extended periods, but until more evidence is gathered through viability monitoring of actual seeds in storage, we do not know the extent to which certain groups of genotypes or species are not orthodox. Cryopreservation may be more suitable for 'minimally orthodox' seeds with very short lifespans in conventional genebank storage, though it may also be possible to improve their storage potential to some extent. Cryopreservation itself, however, is an expensive commitment.

(2) **Seed quality management:** Seed longevity is a highly plastic trait, affected by factors such as the environment during seed production, the timing of seed harvest and how the seeds are processed after harvest [18–20]; maximizing seed storage potential requires experience, experimental approaches to optimizing protocols and attention to detail. Drying seeds appropriately before storage, maintaining conditions throughout storage and careful practice when repeatedly accessing stored seed lots for distribution or viability monitoring have a large and direct impact on the number of years that seeds remain viable in storage. Different types of collections pose very different challenges. Working with a small, diverse collection that is frequently being accessed poses different challenges to working with a very large single crop collection where efficiencies of scale mean seed lots can be processed at higher throughput, making the whole genebank operation more efficient.

(3) **Sustained data management needs:** Maintaining a seed collection for both active use and conservation demands a major investment in documentation if these two different roles are to be fulfilled efficiently and effectively. All samples should be labelled to ensure that the right seeds are in the right place at the right time. Over long periods of time and many interventions, mislabelling events are inevitable. A study of lettuce accessions conserved at the Centre for Genetic Resources the Netherlands (CGN) found the highest rates of non-authenticity for samples that date back to the earliest collections but even those more recently conserved materials (post-1960) showed around 10% mislabelling [21]. Barcoding is believed to eliminate some mislabelling errors thanks to the automated production of identification labels that reduces the

human error caused by writing labels by hand. The value of barcoding, however, is extremely limited without being fully integrated into a comprehensive seed inventory data management system that follows the complete genebank workflow from the introduction of a new acquisition to the point of its storage, including activities that may be undertaken by associated teams or laboratories, such as health testing and fieldwork. Data on post-harvest treatments, initial viability and the detailed procedures followed at the time of germplasm processing and storage are basic pieces of information required to manage a seed lot in the long-term. The data management system should not only store these data points, or references to them, but should follow the operational workflow providing prompts if data are not entered correctly and assisting quality controls that underpin the smooth running of the genebank [22]. In the long term, staff will change and an effective data management system is fundamental to enabling collection management to pass safely from one pair of hands to another. In the future, such management systems could evolve with smarter tools and algorithms to take some of the burden of data entry and decision-making away from staff, but for now, even the basic functionality is an ambition for most genebanks.

(4) **Sustained infrastructure needs:** Finally, conserving with a long-term perspective means that a one-off investment in a cold room is not enough. Cold rooms may have a service life of up to 50–60 years but other critical equipment such as door seals, cooling systems, incubators, driers, and temperature controls have a shorter service life and require maintenance, replacement and backup on a regular basis.

These four factors apply to all genebanks, but the long-term conservation of 'non-seed' collections is yet more challenging. A relatively recent assessment of vegetatively-propagated crops in ex situ conservation estimated that there are at least 400,000 accessions in genebanks worldwide, of which 75% are held in the field and 15% in tissue culture collections [23].

Field collections are notoriously difficult to maintain over many decades because actively growing accessions need repeated monitoring and intervention throughout the year and, in many cases, yearly planting and harvesting to ensure the continued health of the plants and to manage the effects of weather events and pests and diseases [9]. There is little published information on how long accessions have been successfully maintained in field collections for conservation purposes. There are, however, numerous cases of field collections that have been destroyed due to typhoons, drought, disease, change in land management, loss of interest in "low-income" crops that have been reported to the Global Crop Diversity Trust (Crop Trust). There are, also, a few examples of field collections that have stood the test of time, including stands of cacao and coffee in Turrialba, Costa Rica, maintained by the Tropical Agricultural Research and Higher Education Center (CATIE), which date back to 1947. Many botanic gardens will also include collections and old specimens of crop species, but normally in very low numbers. All such collections remain vulnerable and prone to losses.

Tissue conservation in vitro requires equally, if not more, intensive management and infrastructure. This approach is frequently used more for the propagation of planting materials than for conservation objectives. Although there are significant numbers of accessions reported to be in tissue culture, it is not known what proportion of these is conserved using slow-growth conditions for conservation purposes.

Neither field nor in vitro conservation approaches are sufficiently reliable or optimised to be considered a "long-term" conservation approach [24]. Considerable investment, however, has been made in cryopreservation [25], a technology that is coming of age for certain crops (e.g., potato, garlic, banana, apple) but not yet as a standard technique for all vegetatively-propagated crops and non-orthodox seed crops due to a general lack of capacity both to develop and optimize the specific protocols required for more challenging species, and sometimes genotypes within species, and to implement cryopreservation on a large-scale for whole collections. In 2017, just 17 genebanks were identified as actually using cryopreservation as a conservation procedure [23]. The costs of implementation and staff training remain insurmountable hurdles for most genebanks wishing to take up cryopreservation. A globally coordinated initiative to build capacity would help to secure such collections on a long-term basis [23].

3. Managing Genebanks for Long-Term Conservation Objectives

Even where sound processes and data management are established and conditions for long-term conservation are optimized, it is very common for genebanks to be inadequately resourced to carry out the required processes on all the germplasm in the collection on a continuous basis [10]. Taking account of general guidance on viability monitoring, seeds in long-term conservation should be re-tested for viability every ten years or more frequently if short-lived—that is unless there is evidence that the viability of accessions will remain above the accepted threshold for more than 30 years in which case longer periods between re-testing may be appropriate [11]. Seeds in medium-term storage will potentially need more frequent re-testing. In any case, given such standards, an average genebank managing 30–50,000 (the average number of accessions in 71 surveyed national genebanks worldwide was 41,342 in 2014 [26]) accessions (and perhaps twice as many seed lots) will need to conduct viability tests on thousands of seed lots per year (i.e., at least 3–5000 seed lots or 10% of the collection in long-term storage). Some seed lots will fall below the accepted threshold of viability, triggering a demand for regeneration. An FAO survey of 488 national and international genebanks in 2014 revealed that 5.7% of the collections were regenerated that year and an additional 137,000 accessions needed to be regenerated but were unable to be planted out because of insufficient funds to carry out the work [26]. Likewise, a review of 26 crop conservation strategies identified regeneration backlogs as a critical issue for all types of crop germplasm in all regions. For one of the easiest crops to conserve, wheat, lack of regeneration was described as probably the single greatest threat to the safety of key collections [27]. Lack of funds and trained staff is identified as a perennial constraint. The situation worsens with every passing year, as the number of accessions that require viability re-testing, regenerating and other basic operations increases. The words of Lewis Carroll's Red Queen, "*it takes all the running you can do, to keep in the same place*" could have been written for genebanks.

In 2012, all the CGIAR genebanks, with one exception, had backlogs of materials that required regeneration or processing and were essentially unavailable for distribution unless seeds were tested or regenerated. CGIAR conserved 708,761 accessions in 2012, but only 66% were physically available and 55% safety duplicated in two locations, including the Svalbard Global Seed Vault [28].

The Crop Trust, as coordinator of the CGIAR Genebanks Research Program (2012–2016), put in place a monitoring system for operations across all CGIAR genebanks. The system comprises five elements: (1) performance targets, (2) online reporting, (3) a genebank quality management system (QMS), (4) system-level SOP documentation audit, and (5) external review and validation.

3.1. Performance Targets

Targets related to general operations and performance provide a measurable goal and help display the distance required to get there. There is much discussion on the application of performance targets in private and public sectors, and the gaming or perverse incentives that they provoke, especially if linked to financial rewards. The genebank performance targets (Table 2) follow S.M.A.R.T. principles (specific, measurable, achievable, relevant and time-bound) and are designed to measure the level of backlogs and, therefore, the level of activity and time required for individual genebanks to reach a steady-state of operation where backlogs are reduced to less than 10% of the size of the collection. The Passport Data Completeness Index (PDCI) is used as a key performance indicator to measure the level of completeness of passport data associated with genebank accessions [29]. The PDCI range is between zero and ten and is periodically monitored to assess improvements in data quality over time.

Table 2. Performance targets adopted by CGIAR genebanks in 2014.

Key Performance Indicators and Targets			
No.	**Area**	**Indicator**	**Target**
1	Germplasm availability	% of collection clean of pathogens of quarantine risk, viable, and in sufficient quantity to be immediately available for international distribution from medium-term storage (or local distribution for some tree species).	90% of accessions available
2	Safety duplication	For seed crops: % of collection in long-term storage at two locations and also in Svalbard Global Seed Vault (except for tree species). For clonal crops: % of collection in long-term storage or in vitro in slow growth conditions or in cryopreservation at two locations.	90% of seed accessions safety duplicated 50% of clonal accessions in cryopreservation (intermediate target), 90% of accessions duplicated in vitro
3	Data completeness and availability	Passport Data Completeness Index (PDCI) [29]: quantification of the completeness of the passport data based on the absence or presence of data points (range 0–10)	PDCI > 6
4	Quality management	Implementation of a Quality Management System (QMS)	Eight minimum elements of QMS in place (see QMS section below)

Additional data are also collected on genebank activities and germplasm distribution, but none of these indicators are used as targets for various reasons including consideration of the kind of incentive that would be created. Activities to optimize collections were planned and funded, and annual workplans and reports were designed to monitor the progress in improving the status of the collections.

As a result, in 2020, CGIAR reported the availability of an additional 134,994 accessions or 82% of the total collection [30]. This progress occurred in a context whereby, on average, 100,000 samples were being distributed annually from the genebanks and one of the largest collections, at ICARDA (150,000 accessions), had to be regenerated from safety duplicated samples deposited in the Svalbard Global Seed Vault after the genebank was forced to move from Syria in 2012. There is no doubt that the targets focused efforts and resources towards addressing regeneration and processing backlogs.

3.2. Online Reporting Tool (ORT)

Underpinning the performance targets is a comprehensive monitoring system with some 200 data points related to the status of the collections, germplasm distribution, other genebank services, activities of the genebank with relation to quality and risk management, cryobanking, capacity building events and activities to respond to review recommendations. The data are received from each genebank, reviewed by the Crop Trust as coordinator of the responsible CGIAR program, and subsequently reported to CGIAR and Crop Trust donors and management, as well as the Governing Body of the Plant Treaty and the CGRFA.

Online reporting is hardly a novelty. Many organisations have moved towards digital systems to manage workplace processes, including project management and annual reporting. There are, however, several points to make about the process by which annual reports were submitted by CGIAR genebanks online that contributed to the improved quality of data and a stronger incentive to address performance targets. Firstly, the data fields requested in the online reporting tool (ORT) were not easily answered through existing genebank database systems, which led to considerable time being demanded by genebank teams to manually calculate figures for each submission. For instance, requesting "Total number of accessions with acceptable viability" or "Total number of accessions with acceptable seed number" may require downloading results for seed lots from multiple sources and years and a reconciliation at accession level. This highlighted the lack of a coherent management tool to assess inventory data at the collection level and to respond to such questions with the click of a button. While no such capacity existed, the bespoke ORT software has two specific features that play a significant role in improving the quality of data submitted: (1) online correspondence between the submitter and the reviewer is linked to each individual question, which allows real-time clarifications and revisions to take place and be stored; and (2) a data quality control process is imposed on individual questions so that the report submission is blocked until specific answers are revised. The reporting process is intensive but contributes substantially to data quality regarding the collection status and to a greater understanding of the performance targets and indicators across genebank teams. In addition, trend analysis of ORT responses was used to monitor indicators and other metrics to evaluate the genebank's performance over time. As such, the tool relied on historical analysis to plot trajectories and make funding decisions or take corrective and preventive actions when necessary. This type of dynamic reporting tool underpins the understanding of and adherence to the performance targets and monitoring system.

3.3. *Genebank Quality Management System (QMS)*

In 2014, the CGIAR Genebanks program adopted a QMS approach as a means to formally implement and communicate the standards by which CGIAR genebanks operate. Known as the "Genebank QMS", the approach develops a unique resource of documents, policies and scientific practices that comply with regulatory policy, genebank standards and other relevant standards. The Genebank QMS provides the framework for the management of the genebanks and the way that they are monitored, audited and externally reviewed. It offers distinct advantages over other forms of quality management by being:

- Based on specific genebank standards and performance targets rather than generic standards.
- Holistic and implemented along the entire genebank operation, from acquisition to distribution, rather than for selected procedures or processes.
- Internally driven with collective and individual goals that lead to better technical performance within the individual genebank's operation.
- Efficient in terms of the amount of paperwork required following implementation and establishment of the QMS.
- Dynamic and allowing the integration of topical issues affecting genebank management, including protocol optimization, policy changes and emerging risks.
- Homegrown and easily tailored and rendered practicable to the unique situation of each genebank.
- Adaptable to a network of genebanks with templates and elements that can be shared and harmonized where appropriate across countries, crops and conservation systems.

At the core of the Genebank QMS are eight coordinated elements that serve as building blocks for quality management (Figure 1). Emphasis is placed on the science of conservation to provide an understanding of the underlying principles for adopted processes, including factors affecting seed longevity, cost-efficient ways to conserve genetic resources and use of the collections. The science and the conservation processes are mutually reinforcing, and researchers and curators work together to optimize processes and drive each other forward in the improvement cycle.

Standard operating procedures (SOPs) enable genebanks to comprehensively document, validate, share and improve their operating procedures. The genebanks, therefore, have the possibility to develop and adopt shared policies and approaches for certain processes (e.g., acquisition, distribution, processes for the same crops, etc.) across the system. CGIAR may also put in place certain technical and operational standards and principles that go beyond the published FAO standards, (e.g., CGIAR safety duplicates seed accessions in two locations, uses barcoding for labelling, follows specific standards for cryobanking to name a few examples).

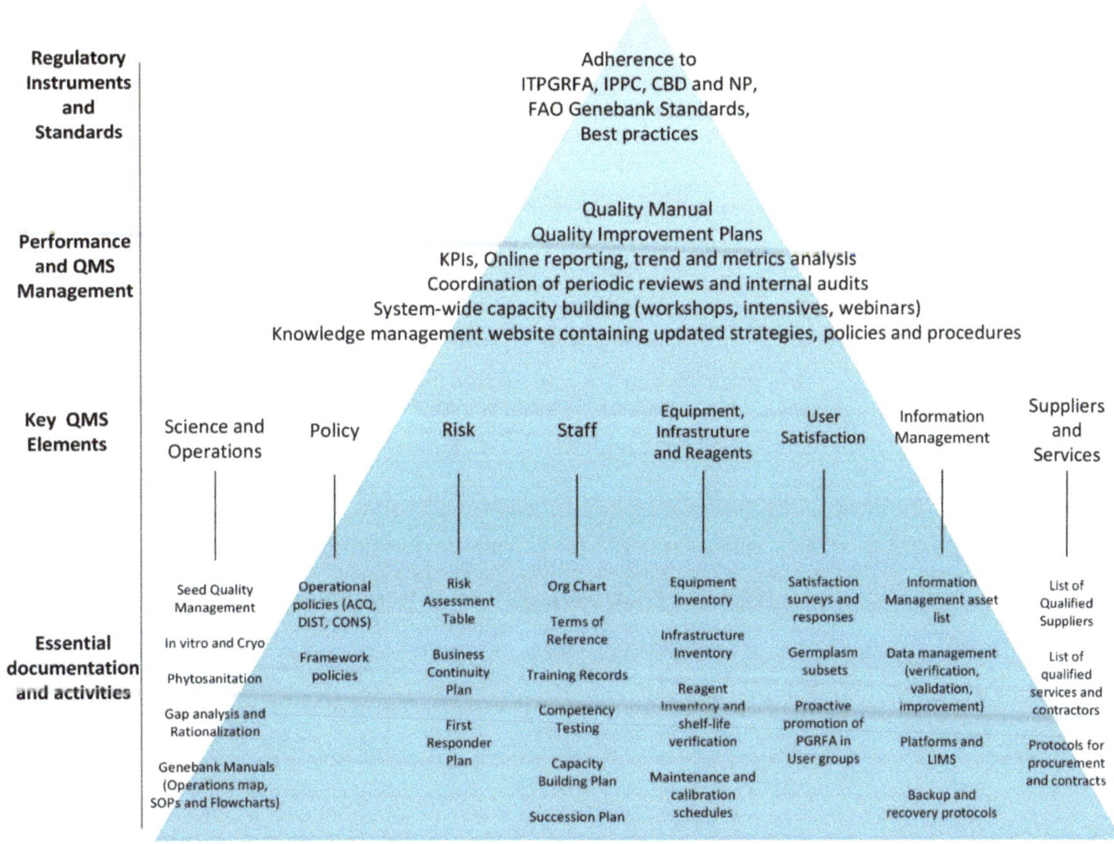

Figure 1. Key elements and regulatory framework of the Genebank QMS.

Each individual Centre has the primary responsibility for developing and implementing its own SOPs and QMS, while at the system level the templates, targets, training and auditing are common across genebanks. Agreement to support and commit time to QMS must be attained from the highest level of Center management and from departments throughout the organization, from human resources to procurement. This is essential to ensure that the relevant staff are able to implement core QMS practices (e.g., calibrating equipment, implementing access controls, ensuring health and safety measures).

Staff have the responsibility of reviewing and updating genebank procedures and are encouraged to use their expertise and knowledge to suggest methods for improvement in their work areas. Centers may appoint a quality manager to coordinate QMS activities and report to upper management about QM progress. The tasks carried out by the quality manager may include overseeing the implementation of QM standards and processes by all staff, managing the audit program, reviewing user satisfaction surveys, identifying critical quality control points and preventive measures, monitoring procedure verification, monitoring equipment inventories, and prioritizing risk management and staff training.

As the QMS prompts staff to document the steps actually implemented, i.e., going beyond the function of generic guidelines and giving details of what happens if expected outcomes do not occur, the SOPs can be highly individual and can vary substantially in detail from one SOP to another and from one genebank to another. As the QMS matures and staff see its worth, so the SOPs become more sophisticated with links to additional policies, instructions, and other appendices. It can be challenging to balance the desire to document every detail against the need to have a document that acts as an easy-to-understand reference source or guide.

In addition, fostering collaborations with genebanks outside the CGIAR is widely considered a key strategic goal. The implementation of a QMS and the development of SOPs and other key documents has provided an opportunity for long-serving staff to share their SOPs and expertise with genebanks with little or no documented procedures. In addition, SOPs can be effective teaching tools to disseminate information on best practices and international policies to partners and those genebanks requiring capacity development. Where genebanks manage the same or similar crops, the respective SOPs may be compared and aligned as a means to improve efficiency or level up processes across locations. This can also occur with genebanks outside CGIAR, who have decided to adopt the genebank QMS (e.g., the World Vegetable Center).

3.4. System-Level SOPs Documentation Audit

Documentation audits are executed on documented information only and are undertaken by independent consultants with quality management and relevant technical expertise. They are the first step in assuring that individual genebank SOPs are properly documented and adhere to relevant treaties, standards and agreed principles. In the Genebank QMS, SOPs, including decision trees, monitoring schedules and related policies, are checked for compliance with the FAO Genebank Standards, the terms and conditions of the standard material transfer agreement, international standards for phytosanitary measures, international rules for seed testing and other relevant best practices that are agreed within the CGIAR genebank community. The system is relatively agile, with recommendations expected to result in improvement within two months of the initial audit and clearance conferred within four months. The resulting SOPs are also subject to language editing and translation for improved clarity. In addition, the SOP template is designed so that the core of the procedures may be extracted to be part of a genebank manual for publication and general circulation.

3.5. External Review and Validation

The first phase of CGIAR genebanks external review began in 2012 and followed a relatively generic approach, whereby two or three experts were asked to make physical visits to individual genebanks and undertake a review based on a small number of relatively broad objectives. The second phase of external review, which started in 2018, however, was able to profit from a vast amount of new information provided in the six years of detailed annual reporting by the genebanks against performance targets and the documentation and auditing of the SOPs. The reviews followed a standard format and involved an extensive document review before the site visit, including review and discussion of the key SOPs, annual report submissions, previous review documents, as well as a self-assessment by the genebank staff.

The review performs an important audit function by validating the actual implementation of specific SOPs by genebank staff on the ground through demonstrations or other evidence. One step in the review process involves checking the inventory data in the live database and the physical samples in the cold rooms of a number of randomly-picked accessions. The format of the reviews allows their implementation even remotely during the pandemic.

Thanks to the QMS and annual reports, the second phase of the review was significantly more evidence-based and in-depth than the first. Weaknesses in data and processes that were not at all evident in the first phase of the review, were easily identified in the second. By contrast, the reviews provided fewer strategic or subjective recommendations on the general direction in which the genebanks were going. The recommendations were incorporated into recommendation action plans (RAP) that feed into the individual genebank's improvement and QMS development.

4. Concluding Remarks

In summary, long-term conservation is a more challenging objective than may be anticipated by many institutions that have stored germplasm in cold rooms. There are several critical requirements of any genebank conserving genetic resources specifically for the long term. Genebank processes involve sustained activities that are basic requirements for long-term conservation, such as viability monitoring and regeneration for seed collections and cryopreservation for vegetatively propagated crops and detailed data management for both. A large number of under-resourced genebanks require more investment to train staff, improve equipment and facilities, support operations and improve data management systems. Key to such investments is the commitment to performance and quality management.

The Genebank QMS does not provide a formal quality certification system, but through the adoption of standard templates and approaches across the group of genebanks and its integration into system-level auditing and review, it assists institutes in developing a recognizable QMS that can be tailored to suit a variety of crops, locations, conservation methods and budgets. Effectively implemented and monitored, the QMS leads to improved administrative, technical and operational performance and the assurance that international standards are being met [31]. Genebank users, regulatory bodies and donors may depend more confidently on the Genebank QMS to recognize and confirm the competence of the genebanks.

After the experience with CGIAR genebanks, we are persuaded that institutes that manage globally important collections for long-term conservation objectives require:

- Either active conservation research projects or association with a university to address research questions concerning recalcitrant or difficult-to-conserve species, improvement of procedures and introduction of new technologies.
- A Genebank QMS with detailed SOPs and system by which they are regularly updated, reviewed, audited and validated.
- A data management system that not only facilitates inventory management but also supports the genebank workflow and quality control and the overall monitoring of the collection.
- An overall monitoring, reporting and review system that incorporates the QMS approach and provides appropriate incentives for genebanks to strive for improvement and efficiency, and to reach and maintain performance targets.
- Adequate facilities and equipment that are accessed only by selected personnel, regularly calibrated and maintained, backed up and replaced prior to reaching the recommended service life.

It is worth noting that such requirements are not easily accommodated within typical research projects or programs. While this may seem an obvious point to make, the fact remains that most genebanks are part of research institutions and are inevitably subject to monitoring and funding systems that are designed to manage research projects. Genebanks are, therefore, obliged to develop budgets and proposals based on research questions and outputs, theories of change and short-term impact for beneficiaries. Within a strongly research-oriented environment, there will be little incentive for genebanks to invest in quality management processes, to share procedures with "competitors" or to seek efficiencies, let alone work towards long-term conservation goals. Since 2012, CGIAR has recognised that the genebank program is unique and requires a bespoke reporting system, which has been further endorsed by two reviews. [32,33]. This support has allowed a unique genebank performance and quality management system to develop and flourish, which we believe has strongly reinforced the long-term conservation objectives of the genebanks and their resilience for the future.

Indeed, CGIAR genebanks have been able to reliably distribute germplasm over the past five decades thanks to their long-term commitment to conserving these collections. They benefit from a high level of operation, modern facilities, field sites, equipment, backup equipment, trained staff and they have had relatively secure funding, at least over the past 10 years. All of which feeds into CGIAR genebanks being secure places for long-term conservation and in some cases the last refuge for landraces and species that have now disappeared from farmers' fields or in situ locations. Indeed, the importance of having multiple layers of good practice and policy, including large-scale safety duplication, production of good quality seeds, sound operating procedures, strong partnership with national agricultural research systems and trained and dedicated staff was illustrated by the reconstitution of the ICARDA collections after the civil war in Syria. If such processes and backups had not been in place, there is no doubt that a considerable amount of unique diversity of wheat, barley, grain legumes and forages originating from the fertile crescent and beyond would have been lost to the world [34].

Author Contributions: Conceptualization, C.L.; writing, C.L., J.v.B. and F.R.H. All authors have read and agreed to the published version of the manuscript.

Funding: The activities described in this paper were undertaken as part of the CGIAR Genebank Platform, funded by the Global Crop Diversity Trust and CGIAR donors.

Data Availability Statement: All data and related information may be found at www.genebanks.org (last accessed 24 November 2021).

Acknowledgments: We thank the guest editors of this special issue, Jan Engels and Andreas Ebert for inviting us to write this article and providing valuable suggestions on the content and writing.

Conflicts of Interest: The authors declare no conflict of interest.

References

1. Byerlee, D.; Dubin, H.J. Crop improvement in the CGIAR as a global success story of open access and international collaboration. *Int. J. Commons* **2010**, *4*, 452–480. [CrossRef]
2. Gollin, D.; Hansen, C.W.; Wingender, A.M. Two blades of grass: The impact of the green revolution. *J. Polit. Econ.* **2021**, *129*, 2344–2384. [CrossRef]
3. Crop Trust. CGIAR Genebank Platform in numbers. Available online: https://www.genebanks.org/resources/genebanks-in-numbers/key-performance/ (accessed on 1 November 2021).
4. FAO. *A Global Treaty for Food Security and Sustainable Agriculture. International Treaty on Plant Genetic Resources for Food and Agriculture*; FAO: Rome, Italy, 2009.
5. Halewood, M.; Jamora, N.; Lopez Noriega, I.; Anglin, N.L.; Wenzl, P.; Payne, T.; Ndjiondjop, M.-N.; Guarino, L.; Kumar, P.L.; Yazbek, M.; et al. Germplasm acquisition and distribution by CGIAR genebanks. *Plants* **2020**, *9*, 1296. [CrossRef] [PubMed]
6. United Nations. Sustainable Development Goals. Available online: https://sdgs.un.org/goals/goal2 (accessed on 1 November 2021).
7. Engels, J.M.M.; Visser, L. (Eds.) *A Guide to Effective Management of Germplasm Collections*; IPGRI Handbooks for Genebanks No. 6; International Plant Genetic Resources Institute: Rome, Italy, 2003; 174p.
8. Reed, B.M.; Engelmann, F.; Dulloo, M.E.; Engels, J.M.M. *Technical Guidelines for the Management of Field and In Vitro Germplasm Collections*; IPGRI Handbooks for Genebanks No. 7; International Plant Genetic Resources Institute: Rome, Italy, 2004; ISBN 92-9043-640-9.
9. Rao, N.K.; Hanson, N.J.; Dulloo, M.E.; Ghosh, K.; Nowella, A.; Larinde, M. *Manual of Seed Handling in Genebanks*; Handbooks for Genebanks No. 8; International Plant Genetic Resources Institute: Rome, Italy, 2006; 147p.
10. Hay, F.R.; Whitehouse, K.J.; Ellis, R.H.; Sackville Hamilton, N.R.; Lusty, C.; Ndjiondjop, M.-N.; Tia, D.; Wenzl, P.; Santos, L.G.; Yazbek, M.; et al. CGIAR genebank viability data reveal inconsistencies in seed collection management. *Glob. Food Sec.* **2021**, *30*, 100557. [CrossRef]
11. FAO. *Genebank Standards for Plant Genetic Resources for Food and Agriculture*; FAO: Rome, Italy, 2014.
12. Engels, J.M.M.; Ebert, A.W. A critical review of the current global ex situ conservation system for plant agrobiodiversity. II Strengths and weaknesses of the current system and recommendations for its improvement. *Plants* **2021**, *10*, 1904. [CrossRef] [PubMed]
13. Whitehouse, K.J.; Hay, F.R.; Lusty, C. Why seed physiology is important for genebanking. *Plants* **2020**, *9*, 584. [CrossRef] [PubMed]
14. Ellis, R.H.; Nasehzadeh, M.; Hanson, J.; Woldemariam, Y. Medium-term seed storage of 50 genera of forage legumes and evidence-based genebank monitoring intervals. *Genet Resour. Crop Evol.* **2018**, *65*, 607–623. [CrossRef]
15. Liu, U.; Cossu, T.A.; Davies, R.M.; Forest, F.; Dickie, J.B.; Breman, E. Conserving orthodox seeds of globally threatened plants ex situ in the Millennium Seed Bank, Royal Botanic Gardens, Kew, UK: The status of seed collections. *Biodivers. Conserv.* **2020**, *29*, 2901–2949. [CrossRef]
16. Walters, C.; Wheeler, L.M.; Grotenhuis, J.M. Longevity of seeds stored in a genebank: Species characteristics. *Seed Sci. Res.* **2005**, *15*, 1–20. [CrossRef]

17. Van Treuren, R.; Bas, N.; Kodde, J.; Groot, S.P.C.; Kik, C. Rapid loss of seed viability in ex situ conserved wheat and barley at 4 degrees C as compared to-20 degrees C storage. *Conserv. Physiol.* **2018**, *6*, coy033. [CrossRef] [PubMed]
18. Hay, F.R.; Probert, R.J. Advances in seed conservation of wild plant species: A review of recent research. *Conserv. Physiol.* **2013**, *1*, cot030. [CrossRef] [PubMed]
19. Kameswara Rao, N.; Dulloo, M.E.; Engels, J.M.M. A review of factors that influence the production of quality seed for long-term conservation in genebanks. *Genet. Resour. Crop Evol.* **2017**, *64*, 1061–1074. [CrossRef]
20. Leprince, O.; Pellizzaro, A.; Berriri, S.; Buitink, J. Late seed maturation: Drying without dying. *J. Exp. Bot.* **2017**, *68*, 827–841. [CrossRef]
21. Van de Wouw, M.; Van Treuren, R.; Van Hintum, T. Authenticity of old cultivars in genebank collections: A case study on lettuce. *Crop Sci.* **2011**, *51*, 736–746. [CrossRef]
22. Hay, F.R.; Sershen, N. New technologies to improve the ex situ conservation of plant genetic resources. In *Plant Genetic Resources: A Review of Current Research and Future Needs*; Dulloo, M.E., Ed.; Burleigh Dodds Science Publishing: Cambridge, UK, 2021. [CrossRef]
23. Acker, J.P.; Adkins, S.; Alves, A.; Horna, D.; Toll, J. *Feasibility Study for a Safety Back-Up Cryopreservation Facility. Independent Expert Report*; Bioversity International: Rome, Italy, 2017; p. 100.
24. Engels, J.M.M.; Ebert, A.W. A critical review of the current global ex situ conservation system for plant agrobiodiversity. I. History of the development of the global system in the context of the political/legal framework and its major conservation components. *Plants* **2021**, *10*, 1557. [CrossRef] [PubMed]
25. Ballesteros, D.; Fanega-Sleziak, N.; Davies, R. Cryopreservation of seeds and seed embryos in orthodox-, intermediate, and recalcitrant-seeded species. *Methods Mol. Biol.* **2021**, *2180*, 663–682. [CrossRef] [PubMed]
26. FAO. Assessment of the Implementation of the Second Global Plan of Action for Plant Genetic Resources for Food and Agriculture 2012–2014. In Proceedings of the Sixteenth Regular Session of the Commission on Genetic Resources for Food and Agriculture, Rome, Italy, 30 January–3 February 2017.
27. Khoury, C.; Laliberté, B.; Guarino, L. Trends in ex situ conservation of plant genetic resources: A review of global crop and regional conservation strategies. *Genet. Resour. Crop Evol.* **2010**, *57*, 625–639. [CrossRef]
28. Crop Trust. 2012 Annual report of the CGIAR Genebank Platform. Available online: https://www.genebanks.org/wp-content/uploads/2017/01/2012-Annual-report-GENEBANKS-CRP-final.pdf (accessed on 1 November 2021).
29. Van Hintum, T.; Menting, F.; Van Strien, E. Quality indicators for passport data in *ex situ* genebanks. *Plant Genet. Resour.* **2011**, *9*, 478–485. [CrossRef]
30. Crop Trust. 2020 Annual Report of the CGIAR Genebank Platform. Available online: https://www.genebanks.org/wp-content/uploads/2021/06/2020-Genebank-Platform-Annual-Report.pdf (accessed on 1 November 2021).
31. To, W.M.; Yu, B.T.W.; Lee, P.K.C. How quality management system components lead to improvement in service organizations: A system practitioner perspective. *Adm. Sci.* **2018**, *8*, 73. [CrossRef]
32. Jackson, M.T.; Borja Tome, M.J.; Ford-Lloyd, B. *Evaluation of CGIAR Research Support Program for Managing and Sustaining Crop Collections: Genebanks CRP*; Summary Report; CGIAR: Montpellier, France, 2017.
33. CGIAR-Crop Trust System Level Review of Genebank Costs and Operations. Report October 2020. Available online: https://www.genebanks.org/wp-content/uploads/2020/11/GCO-Report_261020.pdf (accessed on 1 November 2021).
34. Westengen, O.T.; Lusty, C.; Yazbek, M.; Amri, A.; Asdal, A. Safeguarding a global seed heritage from Syria to Svalbard. *Nat. Plants* **2020**, *6*, 1311–1317. [CrossRef] [PubMed]

plants

MDPI

Review

Integrating Genomic and Phenomic Approaches to Support Plant Genetic Resources Conservation and Use

Gayle M. Volk [1,*], Patrick F. Byrne [2], Clarice J. Coyne [3], Sherry Flint-Garcia [4], Patrick A. Reeves [1] and Chris Richards [1]

1 United States Department of Agriculture, Agricultural Research Service, National Laboratory for Genetic Resources Preservation, Fort Collins, CO 80521, USA; Pat.Reeves@usda.gov (P.A.R.); Chris.Richards@usda.gov (C.R.)
2 Department of Soil and Crop Sciences, Colorado State University, Fort Collins, CO 80523, USA; patrick.byrne@colostate.edu
3 United States Department of Agriculture, Agricultural Research Service, Western Regional Plant Introduction Station, Pullman, WA 99164, USA; Clarice.Coyne@usda.gov
4 Plant Genetics Research Unit, United States Department of Agriculture, Agricultural Research Service, Columbia, MO 65211, USA; Sherry.Flint-Garcia@usda.gov
* Correspondence: gayle.volk@usda.gov

Abstract: Plant genebanks provide genetic resources for breeding and research programs worldwide. These programs benefit from having access to high-quality, standardized phenotypic and genotypic data. Technological advances have made it possible to collect phenomic and genomic data for genebank collections, which, with the appropriate analytical tools, can directly inform breeding programs. We discuss the importance of considering genebank accession homogeneity and heterogeneity in data collection and documentation. Citing specific examples, we describe how well-documented genomic and phenomic data have met or could meet the needs of plant genetic resource managers and users. We explore future opportunities that may emerge from improved documentation and data integration among plant genetic resource information systems.

Keywords: breeding; database; genebank; genotype; phenotype

Citation: Volk, G.M.; Byrne, P.F.; Coyne, C.J.; Flint-Garcia, S.; Reeves, P.A.; Richards, C. Integrating Genomic and Phenomic Approaches to Support Plant Genetic Resources Conservation and Use. *Plants* **2021**, *10*, 2260. https://doi.org/10.3390/plants10112260

Academic Editors: Johannes M. M. Engels and Andreas W. Ebert

Received: 1 October 2021
Accepted: 20 October 2021
Published: 22 October 2021

1. Introduction

Genebanks offer a broad range of plant genetic diversity for use in research and breeding programs. For decades, crop researchers have collected phenotypic trait data on genebank accessions. New high-throughput technologies facilitate the collection of phenomic data (large-scale, often multi-dimensional, phenotypic datasets). Similarly, smaller-scale DNA marker data have been eclipsed by more comprehensive genomic data for characterizing collection genetic diversity. With these large datasets, the challenges of digital information management are becoming as important as managing the physical collection of germplasm. Databases that integrate data types, promote standardized data collection and documentation methods, incorporate appropriate analytical tools and provide user-friendly access will help curators and users of plant genetic resources (PGR) to manage, locate and identify diversity of agronomic and horticultural importance (Figure 1). Tanksley and McCouch [1] and, more recently, the Divseek Initiative advocate for more effective use of genebank diversity; however, the genetic resources stored in genebanks are still underutilized [2,3]. This largely stems from the challenge of identifying useful accessions in large and diverse ex situ collections.

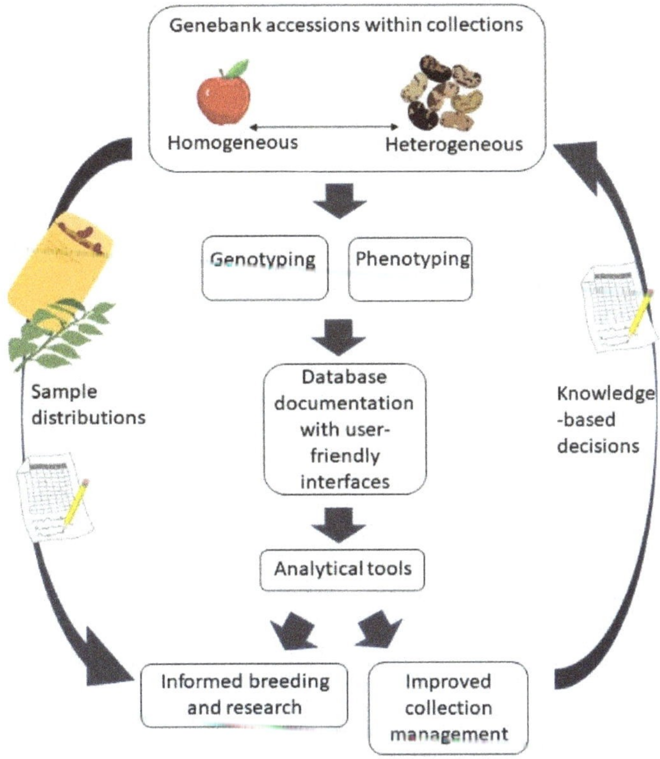

Figure 1. Diagram of selected genebank processes. Genebank accessions range from homogeneous varieties to heterogeneous seedlots. Genomic and phenomic data are collected, documented and analyzed. With this information, improved collection management is possible and specific samples can be requested for breeding and research.

Genebank accessions are genetically complex, with a range of genetic profiles. Homozygous/homogeneous seed samples include inbred lines and accessions derived by single seed descent. These include elite cultivars (e.g., wheat [*Triticum aestivum* L.] variety 'Jagger' and maize [*Zea mays* L.] inbred line 'B73'). An example of homozygous/heterogeneous accessions are landraces of self-pollinating crops that are comprised of an assortment of different genotypes. Heterozygous/homogeneous accessions include clonally maintained but originally outcrossing crops (e.g., 'Granny Smith' apple [*Malus domestica* Borkh.]). Heterozygous/heterogeneous accessions are represented by wild species accessions and outcrossing landraces. As wild species accessions are regenerated, the extent of heterozygosity may decrease [4]. An understanding of the relative level of within-accession genetic variation is important to successfully regenerate genebank accessions, collect and document phenotypic and genotypic data, and use genebank materials in breeding and research programs.

2. Customer and Stakeholder Needs for Plant Genetic Resources

Plant breeders, geneticists, biologists, and educators have different levels of genetic knowledge and access to different kinds of field and laboratory facilities and software tools. Nevertheless, there is an overall need for a wide range of PGR, especially novel germplasm (i.e., materials with genetic variation that are not readily available in existing breeding populations or cultivars) that is viable, disease-free, and has acceptable legal conditions for use [5,6]. To access these resources, an easily navigated search and request system is necessary. Characterization and evaluation data based on standardized methods and rating systems help narrow down the search to materials that meet objectives.

Improved web interfaces that provide genebank inventory identities, along with genotype and phenotype data, can guide accession selection. Many users of genebank materials would find it helpful to know the allelic states of major genes (or associated molecular markers) that control important crop-specific traits (e.g., *Rht*, *Ppd*, and *Vrn* genes for plant stature and flowering time in wheat [7]). Plant breeders would be aided by the availability of traits from wild or unadapted germplasm introgressed into an adapted genetic background or introduced "exotic" germplasm that has been adapted to particular locations via recurrent selection and crosses with adapted germplasm. Examples include products of pre-breeding/genetic enhancement programs such as the Germplasm Enhancement of Maize project (see description below) or panels of synthetic hexaploids in wheat [8].

DNA sequence data for accessions [9], as well as a reference genome for the crop of interest (e.g., [10,11]), are increasingly available. Whole-genome estimates of genetic diversity and population structure of germplasm panels (e.g., [12]), and genome-wide association study (GWAS) results for relevant traits and germplasm (e.g., [13]), can generate data for identifying the accessions best suited for particular purposes. In addition, detailed phenotypic evaluation data for both grower-oriented and consumer-targeted traits (e.g., unpersoned aerial vehicle (UAV) data of accessions under heat or drought stress; in-depth data on health-promoting properties) will have a more limited number of users, but at least a subset of customers would be able to leverage these types of datasets. Plant breeders will benefit from genomics-assisted breeding software to facilitate the introgression of desired genomic regions into breeding material and for genomic selection [5,14].

3. Genomic and Phenomic Approaches Help Meet Customer and Stakeholder Needs for Plant Genetic Resources

While germplasm collections are extraordinarily valuable for geneticists and breeders, there are several challenges for efficient use of these collections. For example, it is often unclear how accessions are related genetically or how adapted they are to a given environment. Many accessions were received decades ago, sometimes with scant or contradictory passport data (e.g., pedigree, provenance, and place of origin). Modern genomic and phenomic tools can be applied to collections to guide germplasm choice for target traits and environments.

3.1. Genomic Tools for Elucidating Germplasm Relationships

Genomic tools such as high-density single-nucleotide polymorphism (SNP) genotyping and whole-genome resequencing can describe the diversity in collections and reveal relationships among accessions. Nearly 4400 samples representing approximately 2500 inbred maize lines known as the "Ames Panel" in the USDA collection were genotyped with 680,000 SNPs (using genotyping by sequencing; GBS) and evaluated for a core set of traits in a field trial at three locations in 2010 [15,16]. Analyses revealed the population structure within the collection: popcorn and sweet corn accessions formed distinct subpopulations separate from the remaining temperate germplasm, as did the tropical germplasm [16]. There was also a clear pattern based on geographic origin. Within U.S. germplasm, genetic differentiation occurred from north to south consistent with adaptation to day length. Tropical lines contained the highest level of diversity in the collection, consistent with previous reports. These data also described the diversity of various subsets, relative to the whole collection. For example, the 282-inbred line Goodman association panel [17] captured 75% of the diversity of the whole collection, and the founders of the nested association mapping population [18] captured 57% of the diversity in the whole collection, attesting to their value in exploring maize inbred diversity. Inbred lines from US and Canadian public breeding programs represented 83% of the diversity in the collection, while private germplasm with expired Plant Variety Protection (PVP) certificates contained only 45% of the diversity in the collection, reflecting a focus on maintaining the three main heterotic patterns in temperate maize. The success of these endeavors was due in part to the homogeneous nature of the inbred lines that were assessed. Future progress for other crops is dependent upon having tools available for heterogeneous accession types.

3.1.1. Germplasm Enhancement of Maize (GEM) Program for Pre-Breeding

It is difficult to introgress unadapted accessions into desirable backgrounds for target environments. Evaluating all accessions for a crop across multiple environments can require vast resources and time. The Germplasm Enhancement of Maize (GEM) program is a collaboration between private industry, USDA and university partners that aims to increase the diversity of US maize germplasm [19]. Fifty-one maize landrace accessions with agronomic merit and high yield potential [20] were the starting materials for the GEM Program beginning in 1995. In the GEM traditional protocol, an exotic accession is crossed with an elite corn belt dent inbred line from a private cooperator, marked only as belonging to the stiff stalk or non-stiff stalk heterotic group. The F1 is then crossed to a second inbred line in the same heterotic group by either the same or different cooperator. The resulting progeny are self-pollinated for several generations and undergo evaluation/selection each season to eliminate disease or insect susceptibility. A moderate number of selected S2 families are testcrossed to the opposite heterotic group to make hybrids for yield trials in a small number of locations. The best 10 families (selection intensity ~3%) are self-pollinated and testcrossed with numerous testers for yield trials in a larger number of locations. The GEM Program typically releases approximately 10 GEM lines per year, which are used primarily for corn breeding by private companies. This program continues to be strongly supported by small and large multi-national corn breeding companies, demonstrating its value to the industry. This approach for pre-breeding could be extended to other crops where there is a need for a broader genetic base and where public and private sectors agree to expend resources for the common good of the crop community.

In addition to the traditional GEM breeding protocol, the Allelic Diversity project was started in 2005 to create a resource for gene discovery, allele mining, and genomic research beginning with ~600 maize landraces, regardless of *a priori* agronomic merit. Each landrace accession was crossed and backcrossed into both a stiff stalk (PHB47) and non-stiff stalk (PHZ51) background. Inbred lines were then produced by either doubled haploid methods [21] or self-pollination. To date, approximately 500 inbreds have been released to the public as a new genomic resource. The Allelic Diversity inbreds have been used to conduct GWAS for flowering time and plant height [22], root system architecture [23], and kernel composition traits where novel loci have been identified [24] reflecting the wide diversity of this unique resource.

3.1.2. Subsets for Allele Mining

The origin of the core subset concept traces its roots to the 1970s and 1980s when genebank holdings were continuing to increase from institutional exchanges and active collecting missions for landrace and wild material [25,26]. The influx of new and diverse accessions was vital for filling gaps in global collections, but the rise in holdings had the unintended effect of making management decisions relating to monitoring and regeneration more difficult. It also made the selection process more complicated for users, who had to navigate thousands or tens of thousands of accessions with little more than a species designation and basic passport information.

Breeding programs make use of collection subsets. For example, the Ames Panel of maize was used for GWAS of numerous traits [27–30]. Subsets of the Ames Panel are often created based on adaptation to a specific environment; the Wisconsin Diversity Panel consists of lines that flower and mature within the shorter growing season of the northern United States [31]. The Wisconsin Diversity Panel has been used for several GWAS experiments [32–37]. Prior to the availability of genomic data, subsetting the Ames Panel would have required either phenotypic datasets from common garden experiments or relied on curator knowledge of the entire collection.

It has become easier to collect genomic data for genebank collections than to perform phenotypic evaluations. For genebank accessions that are genetically homogeneous, it is acceptable to sample only a single or several individuals for whole-genome genotyping. In contrast to the maize example, genotyping one or several individuals for a heterogeneous accession is of limited utility because (1) their multi-locus genotypes cannot be retrieved because all individuals are different, (2) small samples provide poor estimates of allele frequencies, and (3) rare alleles will not be discovered for all loci.

For cataloging genetic variation in heterogeneous accessions, whole-genome pooled sequencing offers a cost-effective alternative to independent sequencing of multiple individuals (e.g., [38]; Figure 2). Pooled sequencing captures whole-genome sequence diversity segregating among the individuals of an accession. The resulting pan-genome-like data structure can be interrogated using genomic position or sequence homology (e.g., BLAST). With adequate sampling (~100 individuals per pool) and adequate sequencing depth (~1x per individual), pooled sequencing allows estimation of allele frequencies genome-wide with an accuracy equal to or better than sequencing individuals, at a lower cost [39]. As collection-wide genotyping projects commence [3], pooled sequencing data can be accumulated one accession at a time for all accessions of a species, with each completed dataset improving insight into the genetic structure of a collection. Allele frequencies, or probabilities of recovering haplotypic variants, are delivered to the genebank customer in the form of heterogeneous germplasm. Providing the means for users to query allele frequencies at any locus of interest in advance of requesting germplasm would be of enormous utility.

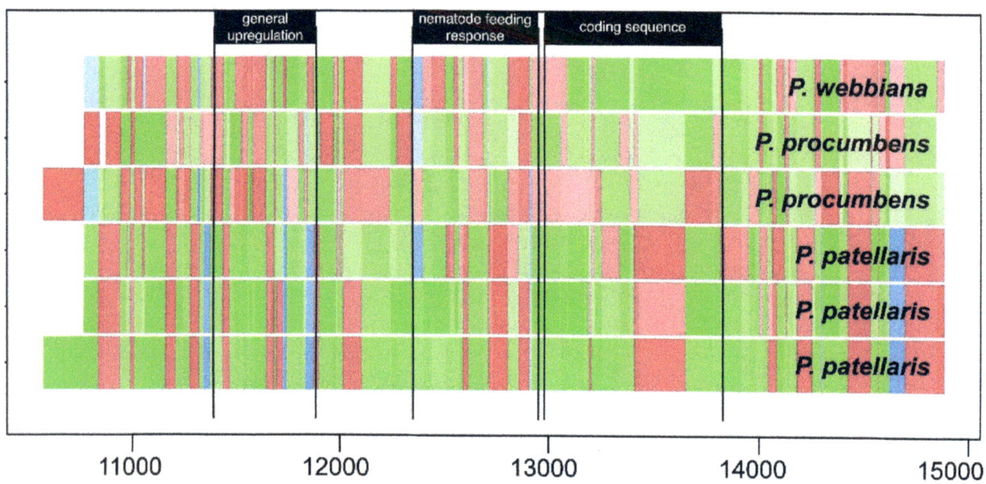

Figure 2. Allele mining in the tertiary gene pool of sugar beet [*Beta vulgaris* L.]. The cyst nematode resistance gene $Hs1^{pro-1}$ from *Patellifolia procumbens* (C. Sm.) A.J. Scott, Ford-Lloyd & J.T. Williams. is conveyed to sugar beet in a large translocation. A promoter sequence activated by nematode feeding makes this locus a target for engineering the disease response in sugar beet [40]. Major allele frequency differences across the genomic region containing $Hs1^{pro-1}$, recovered from the pooled sequencing pan-genomes of six wild *Patellifolia* spp. populations, are shown. Colors indicate different allelic variants, shading within a color indicates variant frequency within the pool (lower = lighter). Substantial variation in the nematode responsive region can be mined from these populations. Whole-genome pooled sequencing data from 202 cultivars, breeding lines, wild relatives, and genebank accessions of *Beta vulgaris* is available under NCBI BioProject PRJNA563463.

Modern crop improvement programs augment traditional crosses and field evaluations with transgenic techniques and gene editing. To enable these approaches, comprehensive sequence data for genebank accessions is important. Whole-genome pooled sequencing datasets support this use by providing information on the full complement of sequences available for individuals in an accession, ensuring gene editing targets are present and avoiding off-target effects [41]. Pooled sequencing data that have been processed into genome-wide SNP or short haplotype frequencies describe accession diversity more practically for breeding programs and correlate logically with phenotypic characterization and evaluation data held in genebank databases, which are usually measured at the population, not individual, level [42].

3.1.3. Gene Discovery—GWAS Using Genebank Collections

Genome-wide association analysis has enabled PGR to contribute more extensively to marker/gene discovery in most species. Diversity subsets from genebank collections are a readily available resource to perform GWAS using a range of markers from simple sequence repeats (SSRs) to SNPs from GBS or SNP arrays [15]. Whole-genome sequencing (WGS) of diversity panels is underway (e.g., rice [*Oryza sativa* L.] [43]; soybean [*Glycine max* (L.) Merr.] [44]; chickpea [*Cicer arietinum* L.] [45]) and perhaps in the near future WGS will be available for an entire crop collection. Statistical approaches and their software implementations are improving [46]. GWAS is becoming possible for any variable trait that can be precisely phenotyped in genebank accessions, thus increasing the value of genebank collections.

3.2. Phenomics in Applied Breeding

Plant breeders and researchers require an understanding of the phenotypic/phenomic data that are available. This includes documentation about how the data were collected, such as the numbers of individuals sampled and experimental field designs (particularly for heterogeneous accessions), as well as the use of standardized descriptors and ontologies. Most of the traits important to breeders exhibit significant genotype × environment interactions so the full environmental context under which phenotypic measurements were made is necessary [47,48]. High-throughput phenotyping (HTPP) in plant breeding programs (e.g., [49]) provides a model for HTPP of large numbers of PGR accessions across multiple environments by genebanks and cooperators in the future [50]. Efforts are underway to improve data presentation and availability, an example is the AgBioData project, which will harmonize data storage and interoperability across databases [51].

Beginning in the 1990s, an evolution occurred from simple, formal phenotypic descriptors to the machine-readable ontologies available today, wherein the controlled vocabulary includes not only trait definitions but also relationship terms, so that complex phenotypic information can be processed by computer. Phenotypic descriptors for PGR have been published for over 100 crops under the auspices of the CGIAR centers IPGRI and Bioversity International (now The Alliance of Bioversity International and CIAT). Bioversity used these to develop crop ontologies and the Crop Ontology Curation Tool [52]. These ontologies are not available for all crops, such as pea, *Pisum sativum* L., and many fruits. However, pea and other new crop ontologies are under development. Other harmonious crop ontologies can be found on Planteome [53] and AgroPortal [54]. The goals of all these efforts are data interoperability to assist in meeting FAIR (findability, accessibility, interoperability, and reusability) data standards [55]. As genebanks collect data using internationally recognized standard formats, data become more accessible and specific accessions can be selected for use in breeding programs. This also facilitates accession comparisons among different genebanks.

4. Genebank Curation Needs

High-quality genebank collections contain well-curated passport, phenotypic, and genotypic data. Acquiring these data is challenging due to resource limitations and the vast size of most genebank collections. Success depends upon partnerships between genebanks and user communities. As they become available, genomic and phenomic data can help guide collection management and improve the value of the collection. These data permit the identification of collection gaps, which can be filled with new acquisitions [56]. They also can help curators maintain the genetic integrity of accessions. Knowledge of within-accession variation helps define optimal regeneration and storage strategies to minimize the effects of genetic drift [4].

Genomic and phenomic data can provide knowledge applicable to curation, such as whether accessions are correctly assigned to taxon, whether they are redundant, or whether they differ when they should not [56]. They can help quantify changes accumulated during ex situ conservation or changes occurring in situ, in the wild, since the accession was collected. Genomic and phenomic data can be used to ensure that cultivar identities are consistent with names used in other genebanks and user communities. At times, curators and their advisory groups must make difficult management decisions because resources are insufficient to accommodate ever-expanding collections. Genomic and phenomic data can help guide decisions to ensure that the purpose and goals of genebank collections are met, even if some accessions must be eliminated. Finally, genomic and phenomic data can help guide users in selecting the best genebank materials for their purposes.

5. Genomic and Phenomic Approaches Help Meet Curation Needs

5.1. Genomic Data Improve Collection Management

Availability of accession-level genomic data for genebank collections has facilitated taxonomic identification of unusual or hybrid species. Lentils, which include *Lens* species and *Lens culinaris* Medik. sub-species, are difficult for non-taxonomists to identify. Wong et al. [57] used sequence data from GBS to correct species and sub-species identifications and identify interspecific hybrids of wild accessions of lentil.

Genotyping the maize Ames Panel revealed several interesting findings related to curation [16]. First, to assess intra-accession variability, approximately 2200 duplicate inbreds (as determined by accession name) were genotyped. Of these, 98% were determined to be at least 0.99 identical by state (IBS), the threshold used to determine that two accessions were "identical". Redundant accessions could be removed in order to simplify curation efforts. Numerous accessions were determined to be isogenic or nearly isogenic (>0.97 IBS). For example, 50 inbreds had IBD > 0.97 with B73, a historically important inbred line. Genebank managers and stakeholders can weigh costs and benefits when determining which collections of near isogenic lines should be maintained.

Genomic data have provided useful information about genetic gaps in collections. Previously, geographic coverage was the primary criterion for targeting plant acquisitions and assembling core collections. Correlation of geographic gap-filling and genomic gap-filling has shown the advantage of using genomic data. The IPK genebank analyzed GBS data for 21,405 barley accessions to identify gaps in the collection. They found "a pronounced under-representation of some regions of the world in IPK's wild barley (*Hordeum vulgare* subsp. *spontaneum* C. Koch.) collection" [58]. Duplicates were also identified, most likely from merger of former East and West German collections. The International Potato Center genebank genotyped 250 potato (*Solanum tuberosum* L.) accessions with a 12K SNP array and identified putative misclassified accessions [59]. GBS data for 441 lettuce accessions showed that crisp head lettuce (*Lactuca sativa* L.) PGR in the USDA collection lack genetic diversity, hindering genetic advances in this important crop [60].

5.2. Phenomic Data Improve Collection Use

Access to phenotypic/phenomic data allows users to select specific materials based on traits of interest, such as resistance to specific diseases, resistance to abiotic stress, or quality components. Replicated, multi-year trials using standardized methods are preferred [61], although often a single data point for a single year may be all that is available. Uploaded images of fruits, seeds or other parts of interest, with scale bars and color standards, provide valuable information to determine accession heterogeneity, trueness-to-type (at the accession or species level; Figure 3), and value to breeding programs [62]. Comparison of images can determine if accession phenotypes are consistent with original submission descriptions or other historical records.

Figure 3. Montage of fruit of *Malus angustifolia* (Aiton) Michx. accessions in the USDA-NPGS apple collection in Geneva, NY. The red fruits are not characteristic of the species. Taxon assignment will be confirmed using genotyping by sequencing [63].

The traits addressed by HTPP technologies are broad, from well-established disease assessment [64,65], to tree and plant architecture [66,67], to protein content in wheat [49]. Satellite images of wheat [68], UAVs [69], tractor mounted (e.g., [70]) and handheld detectors (e.g., [71,72]), and between-row ultracompact robots [73] comprise a set of new technologies potentially valuable for HTPP of genebank accessions.

Although genebank HTPP is in early stages, examples of completed studies are emerging that efficiently provide needed evaluation data to the user community. A diversity panel of lentil was phenotyped in seven countries for three years, providing insight into photothermal interactions to guide future production expansions [74]. HTPP deployed in the Canadian lentil trials (six environments) contributed to a data depth that was not possible in the non-HTPP sites [71]. This type of data analysis could be adapted for other genebank tasks such as seed germination scans. The International Potato Center deployed a UAV remote sensing and multi-spectral camera to accurately predict maturity and productivity of potatoes [75]. In the future, UAV and robots could collect field data for large collections at multiple timepoints in the growing season [76].

Several examples are available for HTPP of seeds. A true HTPP seed imaging platform with a conveyor belt and barcode reader was developed to record the color, size and shape dimensions of the small-seeded lentil [77]. A similar approach for soybean lacks the third dimension and automation aspects [78]. A mobile application was developed for grain width and length HTPP in the field [79]. Maize ear/kernel HTPP has two published platforms [80,81]. Commercial software was used to scan pulse crop PGR seed images, but throughput is limited as seed must be manually separated [82]. SmartGrain software developed for rice eliminates this impediment by excluding overlapping grains and removes awns and pedicels [83].

6. Data Integration for Plant Genetic Resource Curation and Use

Genebanks have implemented inventory management databases that associate accessions with passport, image, phenotypic and genotypic data. Resource limitations have resulted in phenotypic and genotypic data that are not comprehensive within or across crops. There are also a number of challenges in associating and retrieving heterogeneous accessions with their corresponding phenotypic/phenomic and genotypic/genomic data within databases. These challenges must be overcome to ensure well-documented, standardized data are available for improving genebank management and utility.

To efficiently exploit the variation contained in genebank collections, there have been efforts to develop sophisticated information management systems. These systems primarily provide database, analytical and decision support capabilities to genomics driven breeding projects. Capabilities include managing and tracking physical and digital descriptor data and, increasingly, integrating genomic sequence data, field observations of phenotypic data, including digital image and spectral data. Platforms include the Genomic Open-source Breeding Informatics Initiative [84], Breeding Insight [85], Excellence in Breeding [86], and Breedbase [87–89]. A common API (application programming interface) defines protocols for interoperability among breeding applications and databases (BrAPI, [90]). Informatics projects focused on defined sets of crops such as Seeds of Discovery [91,92] and Agent [93] provide important data integration among genomic and phenotypic projects. These projects also promote collaborative networks to connect genebank collection information with populations from breeding and pre-breeding projects. Software platforms such as Germinate [94] and the IPK's Bridge web portal [95] focus on data visualization and integrating gene bank genomic and phenotypic data.

Crop-specific or clade-oriented genomic databases (e.g., MaizeGDB) offer numerous visualizations and capabilities connected to genebank accessions. Maize SNP data were used to establish an IBS relationship matrix among 2800 inbred lines, and this matrix was used to populate a tool called "TYPSimSelector" [96,97], where the user can query for inbred lines that are most closely related or most diverged from the inbred line of interest. Users can search for a replacement for an accession because of a defect in a trait, such as disease or insect susceptibility.

Looking forward, data curation will likely involve the use of persistent identifiers (PIDs) such as Plant Introduction (PI) numbers or Digital Object Identifiers (DOI) for sample identification. Through the adoption of BrAPI, the GRIN-Global information management system can become interoperable with Breeding Insight and crop-specific databases. Several databases, including Bridge IPK [98] and Germinate [99], have already successfully merged some components of genebank and breeding information management.

7. Future Prospects

Current and future genebank users, as well as curation teams, will increasingly require access to high-quality genomic and phenomic data. Access to genetic diversity information could help ensure phenotypic data are collected from an adequate number of individuals. Integration of genomic and phenomic data for heterogeneous accessions requires special attention to DNA polymorphism and elevated or complex patterns of phenotypic variance. Technological advances in information management systems have focused primarily on specific crops. Future efforts should consider the relative heterogeneity of genebank accessions and how it can be effectively managed when data are collected. Systems must be user-friendly and widely applicable to diverse customers. In addition, they must be adaptable and scalable, to support new genomic and phenomic technologies. Future research should focus on ensuring that tools are available to effectively use genomic and phenomic data for informed curation decisions.

Increasingly sophisticated artificial intelligence methods and sequence data lead to the question of whether every accession must be phenotyped to choose accessions for gene(s) or trait(s) of interest. Advances in statistical prediction may change how characterization data can select germplasm for further evaluation. FIGS (Focused Identification of Germplasm

Strategy) applies machine learning (ML) algorithms and environmental data to identify candidate accessions associated with a trait of interest [100]. In Bari et al. [100], non-phenotyped wheat accessions carrying rust resistance were predicted based on the trait-environment relationship in other accessions. Similarly, ML approaches can be used to assemble germplasm subsets that maximize variation of haplotype blocks associated with phenotypic variation [101–103]. Genomic prediction is an increasingly useful tool in specialty crops such as pea [104] and well-resourced collections such as barley [105]. A training population with both genotypic and phenotypic values is the basis for developing a statistical model. The model then predicts the phenotypic values of lines using only genotypic data. A genomic selection application in sorghum (*Sorghum bicolor* (L.) Moench) demonstrated the power of this approach for selecting accessions for high biomass from a large PGR collection [64]. Improvement of genomic prediction algorithms and increasing genomic data coverage across PGR accessions will advance their use by genebank managers and genebank users alike [106,107].

Funding: This research was funded in part by USDA-NIFA Higher Education Challenge Program grant 2020-70003-30930, USDA-ARS, and a Griffin Foundation grant to Colorado State University.

Acknowledgments: The authors appreciate Katheryn Chen for providing Figure 1 illustrations. We thank Moira Sheehan, Peter Bretting, Jan Engels, and Andreas Ebert for pre-submission reviews. USDA is an equal opportunity provider and employer. Mention of trade names or commercial products in this article is solely for the purpose of providing specific information and does not imply recommendation or endorsement by the US Department of Agriculture.

Conflicts of Interest: The authors declare no conflict of interest.

References

1. Tanksley, S.D.; McCouch, S.R. Seed banks and molecular maps: Unlocking genetic potential from the wild. *Science* **1997**, *277*, 1063–1066. [CrossRef]
2. McCouch, S.; Baute, G.J.; Bradeen, J.; Bramel, P.; Bretting, P.K.; Buckler, E.; Burke, J.M.; Charest, D.; Cloutier, S.; Cole, G.; et al. Feeding the future. *Nature* **2013**, *499*, 23–24. [CrossRef]
3. McCouch, S.; Navabi, K.; Abberton, M.; Anglin, N.L.; Barbieri, R.L.; Baum, M.; Bett, K.; Booker, H.; Brown, G.L.; Bryan, G.J.; et al. Mobilizing crop biodiversity. *Mol. Plant.* **2020**, *13*, 1341–1344. [CrossRef]
4. Richards, C.M.; Lockwood, D.R.; Volk, G.M.; Walters, C. Modeling demographics and genetic diversity in ex situ collections during seed storage and regeneration. *Crop Sci.* **2010**, *50*, 2440–2447. [CrossRef]
5. Ebert, A.W. The role of vegetable genetic resources in nutrition security and vegetable breeding. *Plants* **2020**, *9*, 736. [CrossRef] [PubMed]
6. Smith, S.; Nickson, T.E.; Challender, M. Germplasm exchange is critical to conservation of biodiversity and global food security. *Agron. J.* **2021**, *113*, 2969–2979. [CrossRef]
7. Grogan, S.M.; Brown-Guedira, G.; Haley, S.D.; McMaster, G.S.; Reid, S.D.; Smith, J.; Byrne, P.F. Allelic variation in developmental genes and effects on winter wheat heading date in the U.S. Great Plains. *PLoS ONE* **2016**, *11*, e0152852. [CrossRef] [PubMed]
8. Rosyara, U.; Kishii, M.; Payne, T.; Sansaloni, C.P.; Singh, R.P.; Braun, H.J.; Dreisigacker, S. Genetic contribution of synthetic hexaploid wheat to CIMMYT's spring bread wheat breeding germplasm. *Sci. Rep.* **2019**, *9*, 12355. [CrossRef] [PubMed]
9. Gaffney, J.; Tibebu, R.; Bart, R.; Beyene, G.; Girma, D.; Ardo Kane, N.; Mace, E.S.; Mockler, T.; Nickson, T.E.; Taylor, N.; et al. Open access to genetic sequence data maximizes value to scientists, farmers, and society. *Global Food Secur.* **2020**, *26*, 100411. [CrossRef]
10. Tao, Y.; Luo, H.; Xu, J.; Cruickshank, A.; Zhao, X.; Teng, F.; Hathorn, A.; Wu, X.; Liu, Y.; Shatte, T.; et al. Extensive variation within the pan-genome of cultivated and wild sorghum. *Nat. Plants* **2021**, *7*, 766–773. [CrossRef] [PubMed]
11. Torkamaneh, D.; Lemay, M.-A.; Belzile, F. The pan-genome of the cultivated soybean (PanSoy) reveals an extraordinarily conserved gene content. *Plant Biotechnol. J.* **2021**, *19*, 1852–1862. [CrossRef]
12. Faye, J.M.; Maina, F.; Akata, E.A.; Sine, B.; Diatta, C.; Mamadou, A.; Marla, S.; Bouchet, S.; Teme, N.; Rami, J.F.; et al. A genomics resource for genetics, physiology, and breeding of West African sorghum. *Plant Genome* **2021**, *14*, e20075. [CrossRef] [PubMed]
13. Bassett, A.; Kamfwa, K.; Ambachew, D.; Cichy, K. Genetic variability and genome-wide association analysis of flavor and texture in cooked beans (*Phaseolus vulgaris* L.). *Theor. Appl. Genet.* **2021**, *134*, 959–978. [CrossRef] [PubMed]
14. Marsh, J.I.; Hu, H.; Gill, M.; Batley, J.; Edwards, D. Crop breeding for a changing climate: Integrating phenomics and genomics with bioinformatics. *Theor. Appl. Genet.* **2021**, *134*, 1677–1690. [CrossRef] [PubMed]
15. Elshire, R.J.; Glaubitz, J.C.; Sun, Q.; Poland, J.A.; Kawamoto, K.; Buckler, E.S.; Mitchell, S.E. A robust, simple Genotyping-by-Sequencing (GBS) approach for high diversity species. *PLoS ONE* **2011**, *6*, e19379. [CrossRef] [PubMed]

16. Romay, M.C.; Millard, M.J.; Glaubitz, J.C.; Peiffer, J.A.; Swarts, K.L.; Casstevens, T.M.; Elshire, R.J.; Acharya, C.B.; Mitchell, S.E.; Flint-Garcia, S.A.; et al. Comprehensive genotyping of the USA national maize inbred seed bank. *Genome Biol.* **2013**, *14*, R55. [CrossRef]
17. Flint-Garcia, S.; Thuillet, A.-C.; Yu, J.; Pressoir, G.; Romero, S.M.; Mitchell, S.E.; Doebley, J.; Kresovich, S.; Goodman, M.M.; Buckler, E.S. Maize association population: A high-resolution platform for quantitative trait locus dissection. *Plant J.* **2005**, *44*, 1054–1064. [CrossRef]
18. McMullen, M.D.; Kresovich, S.; Villeda, H.S.; Bradbury, P.; Li, H.; Sun, Q.; Flint- Garcia, S.; Thornsberry, J.; Acharya, C.; Bottoms, C.; et al. Genetic properties of the maize nested association mapping population. *Science* **2009**, *325*, 737–740. [CrossRef]
19. Pollak, L.M. The history and success of the public-private project on Germplasm Enhancement of Maize (GEM). *Adv. Agron.* **2003**, *78*, 45–87.
20. Salhuana, W.; Pollak, L. Latin American Maize Project (LAMP) and Germplasm Enhancement of Maize (GEM) project: Generating useful breeding germplasm. *Maydica* **2006**, *51*, 339–355.
21. Brenner, E.A.; Blanco, M.; Gardner, C.; Lübberstedt, T. Genotypic and phenotypic characterization of isogenic doubled haploid exotic introgression lines in maize. *Mol Breed.* **2012**, *30*, 1001–1016. [CrossRef]
22. Vanous, A.; Gardner, C.; Blanco, M.; Martin-Schwarze, A.; Lipka, A.E.; Flint-Garcia, S.; Bohn, M.; Edwards, J.; Lübberstedt, T. Association mapping of flowering and height traits in Germplasm Enhancement of Maize Doubled Haploid (GEM-DH) lines. *Plant Genome* **2018**, *11*, 1–14. [CrossRef]
23. Ma, L.; Qing, C.; Frei, U.; Shen, Y.; Lübberstedt, T. Association mapping for root system architecture traits under two nitrogen conditions in germplasm enhancement of maize doubled haploid lines. *Crop J.* **2020**, *8*, 213–226. [CrossRef]
24. Vanous, A.; Gardner, C.; Blanco, M.; Martin-Schwarze, A.; Wang, J.; Li, X.; Lipka, A.E.; Flint-Garcia, S.; Bohn, M.; Edwards, J.; et al. Stability analysis of kernel quality traits in exotic-derived doubled haploid maize lines. *Plant Genome* **2019**, *12*, 1–14. [CrossRef]
25. Frankel, O.H. Genetic perspectives of germplasm conservation. In *Genetic Manipulation: Impact on Man and Society*; Arber, W., Illemensee, K., Peacock, W.J., Starlinger, P., Eds.; Cambridge University Press: Cambridge, UK, 1984; pp. 161–170.
26. Brown, A. The case for core collections. In *The Use of Plant Genetic Resources*; Brown, A.H.D., Frankel, O.H., Marshall, D.R., Williams, J.T., Eds.; Cambridge University Press: Cambridge, UK, 1989; pp. 136–156.
27. Zila, C.T.; Samayoa, L.F.; Santiago, R.; Butrón, A.; Holland, J.B. A genome-wide association study reveals genes associated with Fusarium ear rot resistance in a Maize Core Diversity Panel. *G3 Genes | Genomes | Genetics* **2013**, *3*, 2095–2104. [CrossRef]
28. Peiffer, J.A.; Romay, M.C.; Gore, M.A.; Flint-Garcia, S.A.; Zhang, Z.; Millard, M.J.; Gardner, C.A.C.; McMullen, M.D.; Holland, J.B.; Bradbury, P.J.; et al. The genetic architecture of maize height. *Genetics* **2014**, *196*, 1337–1356. [CrossRef]
29. Lu, F.; Romay, M.C.; Glaubitz, J.C.; Bradbury, P.J.; Elshire, R.J.; Wang, T.; Li, Y.; Li, Y.; Semagn, K.; Zhang, X.; et al. High-resolution genetic mapping of maize pan-genome sequence anchors. *Nat. Commun.* **2015**, *6*, 6914. [CrossRef] [PubMed]
30. Xue, S.; Bradbury, P.J.; Casstevens, T.; Holland, J.B. Genetic architecture of domestication-related traits in maize. *Genetics* **2016**, *204*, 99–113. [CrossRef] [PubMed]
31. Hansey, C.N.; Johnson, J.M.; Sekhon, R.S.; Kaeppler, S.M.; de Leon, N. Genetic diversity of a maize association population with restricted phenology. *Crop Sci.* **2011**, *51*, 704–715. [CrossRef]
32. Gage, J.L.; White, M.R.; Edwards, J.W.; Kaeppler, S.; de Leon, N. Selection signatures underlying dramatic male inflorescence transformation during modern hybrid maize breeding. *Genetics* **2018**, *210*, 1125–1138. [CrossRef]
33. Gustafson, T.; Leon, N.; Kaeppler, S.; Tracy, W. Genetic analysis of resistance in the Wisconsin Diversity Panel of maize. *Crop Sci.* **2018**, *58*, 1853–1865. [CrossRef]
34. Mazaheri, M.; Heckwolf, M.; Vaillancourt, B.; Gage, J.L.; Burdo, B.; Heckwolf, S.; Barry, K.; Lipzen, A.; Ribeiro, C.B.; Kono, T.J.Y.; et al. Genome-wide association analysis of stalk biomass and anatomical traits in maize. *BMC Plant Biol.* **2019**, *19*, 45. [CrossRef] [PubMed]
35. O'Connor, C.H.; Qiu, Y.; Coletta, R.D.; Renk, J.S.; Monnahan, P.J.; Noshay, J.M.; Gilbert, A.; Anderson, A.N.; McGaugh, S.E.; Springer, N.M.; et al. Population level variation of transposable elements in a Maize Diversity Panel. *bioRxiv* **2020**. [CrossRef]
36. Renk, J.S.; Gilbert, A.M.; Hattery, T.J.; O'Connor, C.H.; Monnahan, P.J.; Anderson, N.; Waters, A.J.; Eickholt, D.; Flint-Garcia, S.A.; Yandeau-Nelson, M.D.; et al. Genetic architecture of kernel compositional variation in a maize diversity panel. *Plant Genome* **2021**, in press. Available online: https://www.biorxiv.org/content/10.1101/2021.03.29.436703v1.full (accessed on 20 October 2021). [CrossRef]
37. Burns, M.J.; Renk, J.S.; Eickholt, D.P.; Gilbert, A.M.; Hattery, T.J.; Holmes, M.; Anderson, N.; Waters, A.J.; Kalambur, S.; Flint-Garcia, S.A.; et al. Predicting moisture content during maize nixtamalization using machine learning with NIR spectroscopy. *Theor. Appl. Genet.* **2021**, *134*, 3743–3757. [CrossRef] [PubMed]
38. Rubinstein, M.; Eshed, R.; Rozen, A.; Zviran, T.; Kuhn, D.N.; Irihimivitch, V.; Sherman, A.; Ophir, R. Genetic diversity of avocado (*Persea americana* Mill.) germplasm using pooled sequencing. *BMC Genom.* **2019**, *20*, 379. [CrossRef]
39. Schlötterer, C.; Tobler, R.; Kofler, R.; Nolte, V. Sequencing pools of individuals—Mining genome wide polymorphism data without big funding. *Nat. Genet.* **2014**, *15*, 749–763. [CrossRef] [PubMed]
40. Kumar, A.; Harloff, H.J.; Melzer, S.; Leineweber, J.; Defant, B.; Jung, C. A rhomboid like protease gene from an interspecies translocation confers resistance to cyst nematodes. *New Phytol.* **2021**, *231*, 801–813. [CrossRef]
41. Danilevicz, M.F.; Fernandez, C.G.T.; Marsh, J.I.; Bayer, P.E.; Edwards, D. Plant pangenomics: Approaches, applications and advancements. *Curr. Opin. Plant Biol.* **2020**, *54*, 18–25. [CrossRef]

42. Galewski, P.; McGrath, J.M. Genetic diversity among cultivated beets (*Beta vulgaris*) assessed via population based whole genome sequences. *BMC Genom.* **2020**, *21*, 189. [CrossRef]
43. Lv, Q.; Li, W.; Sun, Z.; Ouyang, N.; Jing, X.; He, Q.; Wu, J.; Zheng, J.; Zheng, J.; Tang, S.; et al. Resequencing of 1,143 indica rice accessions reveals important genetic variations and different heterosis patterns. *Nat. Commun.* **2020**, *11*, 4778. [CrossRef] [PubMed]
44. Torkamaneh, D.; Laroche, J.; Valliyodan, B.; O'Donoughue, L.; Cober, E.; Rajcan, I.; Vilela Abdelnoor, R.; Sreedasyam, A.; Schmutz, J.; Nguyen, H.T.; et al. Soybean (*Glycine max*) Haplotype Map (GmHapMap): A universal resource for soybean translational and functional genomics. *Plant Biotechnol. J.* **2021**, *19*, 324–334. [CrossRef]
45. Varshney, R.K.; Thudi, M.; Roorkiwal, M.; He, W.; Upadhyaya, H.D.; Yang, W.; Bajaj, P.; Cubry, P.; Rathore, A.; Jian, J.; et al. Resequencing of 429 chickpea accessions from 45 countries provides insights into genome diversity, domestication and agronomic traits. *Nat. Genet.* **2019**, *51*, 857–864. [CrossRef]
46. Tibbs Cortes, L.; Zhang, Z.; Yu, J. Status and prospects of genome-wide association studies in plants. *Plant Genome* **2021**, *14*, e20077. [CrossRef] [PubMed]
47. Crossa, J. From genotype × environment interaction to gene × environment interaction. *Curr. Genom.* **2012**, *13*, 225–244. [CrossRef]
48. Kang, M.S. Using genotype-by-environment interaction for crop cultivar development. *Adv. Agron.* **1997**, *62*, 199–252. [CrossRef]
49. Sandhu, K.S.; Mihalyov, P.D.; Lewien, M.J.; Pumphrey, M.O.; Carter, A.H. Combining genomic and phenomic information for predicting grain protein content and grain yield in spring wheat. *Front. Plant Sci.* **2021**, *12*, 170. [CrossRef]
50. Rebetzke, G.J.; Jimenez-Berni, J.; Fischer, R.A.; Deery, D.M.; Smith, D.J. High-throughput phenotyping to enhance the use of crop genetic resources. *Plant Sci.* **2019**, *282*, 40–48. [CrossRef] [PubMed]
51. Harper, L.; Campbell, J.; Cannon, E.K.; Jung, S.; Poelchau, M.; Walls, R.; Andorf, C.; Arnaud, E.; Berardini, T.Z.; Birkett, C.; et al. AgBioData consortium recommendations for sustainable genomics and genetics databases for agriculture. *Database* **2018**, *2018*, bay088. [CrossRef]
52. Shrestha, R.; Matteis, L.; Skofic, M.; Portugal, A.; McLaren, G.; Hyman, G.; Arnaud, E. Bridging the phenotypic and genetic data useful for integrated breeding through a data annotation using the Crop Ontology developed by the crop communities of practice. *Front. Physiol.* **2012**, *3*, 326. [CrossRef]
53. Cooper, L.; Meier, A.; Laporte, M.A.; Elser, J.L.; Mungall, C.; Sinn, B.T.; Cavaliere, D.; Carbon, S.; Dunn, N.A.; Smith, B.; et al. The Planteome database: An integrated resource for reference ontologies, plant genomics and phenomics. *Nucleic Acids Res.* **2018**, *46*, D1168–D1180. [CrossRef]
54. Jonquet, C.; Toulet, A.; Arnaud, E.; Aubin, S.; Yeumo, E.D.; Emonet, V.; Graybeal, J.; Laporte, M.A.; Musen, M.A.; Pesce, V.; et al. AgroPortal: A vocabulary and ontology repository for agronomy. *Comput. Electron. Agric.* **2018**, *144*, 126–143. [CrossRef]
55. Wilkinson, M.D.; Dumontier, M.; Aalbersberg, I.J.; Appleton, G.; Axton, M.; Baak, A.; Blomberg, N.; Boiten, J.W.; da Silva Santos, L.B.; Bourne, P.E.; et al. The FAIR Guiding Principles for scientific data management and stewardship. *Sci. Data* **2016**, *3*, 10018. [CrossRef]
56. Bretting, P.K.; Widrlechner, M. Genetic markers and plant genetic resource management. *Plant Breed. Rev.* **1995**, *13*, 11–86. [CrossRef]
57. Wong, M.M.; Gujaria-Verma, N.; Ramsay, L.; Yuan, H.Y.; Caron, C.; Diapari, M.; Vandenberg, A.; Bett, K.E. Classification and characterization of species within the genus *Lens* using genotyping-by-sequencing (GBS). *PLoS ONE* **2015**, *10*, e0122025. [CrossRef] [PubMed]
58. Milner, S.G.; Jost, M.; Taketa, S.; Mazón, E.R.; Himmelbach, A.; Oppermann, M.; Weise, S.; Knüpffer, H.; Basterrechea, M.; König, P.; et al. Genebank genomics highlights the diversity of a global barley collection. *Nat. Genet.* **2019**, *51*, 319–326. [CrossRef] [PubMed]
59. Ellis, D.; Chavez, O.; Coombs, J.; Soto, J.; Gomez, R.; Douches, D.; Panta, A.; Silvestre, R.; Anglin, N.L. Genetic identity in genebanks: Application of the SolCAP 12K SNP array in fingerprinting and diversity analysis in the global in trust potato collection. *Genome* **2018**, *61*, 523–537. [CrossRef] [PubMed]
60. Park, S.; Kumar, P.; Shi, A.; Mou, B. Population genetics and genome-wide association studies provide insights into the influence of selective breeding on genetic variation in lettuce. *Plant Genome* **2021**, e20086. [CrossRef]
61. Watts, S.; Migicovsky, Z.; McClure, K.A.; Yu, C.H.J.; Amyotte, B.; Baker, T.; Bowlby, D.; Burgher-MacLellan, K.; Butler, L.; Donald, R.; et al. Quantifying apple diversity: A phenomic characterization of Canada's Apple Biodiversity Collection. *Plants People Planet* **2021**. [CrossRef]
62. Gross, B.L.; Volk, G.M.; Richards, C.M.; Forsline, P.L.; Fazio, G.; Thomas, C.T. Identification of "duplicate" accessions within the USDA-ARS National Plant Germplasm System *Malus* collection. *J. Am. Soc. Hort. Sci. 2012. 137*, 333–342. [CrossRef]
63. Gottschalk, C.; (USDA-ARS, Kearneysville, WV, USA). Personal communication, 2021.
64. Yu, X.; Li, X.; Guo, T.; Zhu, C.; Wu, Y.; Mitchell, S.E.; Roozeboom, K.L.; Wang, D.; Wang, M.L.; Pederson, G.A.; et al. Genomic prediction contributing to a promising global strategy to turbocharge gene banks. *Nat. Plants 2016. 2*, 16150. [CrossRef]
65. Sankaran, S.; Mishra, A.; Ehsani, R.; Davis, C. A review of advanced techniques for detecting plant diseases. *Comput. Electron. Agric.* **2010**, *72*, 1–13. [CrossRef]
66. Sangjan, W.; Sankaran, S. Phenotyping architecture traits of tree species using remote sensing techniques. *Trans. ASABE* **2021**, in press. [CrossRef]

67. Zhang, C.; Craine, W.A.; McGee, R.J.; Vandemark, G.J.; Davis, J.B.; Brown, J.; Hulbert, S.H.; Sankaran, S. High-throughput phenotyping of canopy height in cool-season crops using sensing techniques. *Agron. J.* **2021**, *113*, 3269–3280. [CrossRef]
68. Zhang, Z.; (Department of Crop and Soil Science, Washington State University, Pullman, WA, USA). *Personal communication*, 2021.
69. Sankaran, S.; Khot, L.R.; Espinoza, C.Z.; Jarolmasjed, S.; Sathuvalli, V.R.; Vandemark, G.J.; Miklas, P.N.; Carter, A.H.; Pumphrey, M.O.; Knowles, N.R.; et al. Low-altitude, high-resolution aerial imaging systems for row and field crop phenotyping: A review. *Eur. J. Agron.* **2015**, *70*, 112–123. [CrossRef]
70. Andrade-Sanchez, P.; Gore, M.A.; Heun, J.T.; Thorp, K.R.; Carmo-Silva, A.E.; French, A.N.; Salvucci, M.E.; White, J.W. Development and evaluation of a field-based high-throughput phenotyping platform. *Funct. Plant Biol.* **2013**, *41*, 68–79. [CrossRef]
71. Nielsen, K.M. Using UAV-Based Imagery to Determine Volume, Groundcover, and Growth Rate Characteristics of Lentil (*Lens culinaris* Medik.). Ph.D. Thesis, University of Saskatchewan, Saskatoon, Saskatchewan, 2020.
72. Crain, J.L.; Wei, Y.; Barker III, J.; Thompson, S.M.; Alderman, P.D.; Reynolds, M.; Zhang, N.; Poland, J. Development and deployment of a portable field phenotyping platform. *Crop Sci.* **2016**, *56*, 965–975. [CrossRef]
73. Zhang, Z.; Kayacan, E.; Thompson, B.; Chowdhary, G. High precision control and deep learning-based corn stand counting algorithms for agricultural robot. *Auton. Robot.* **2020**, *44*, 1289–1302. [CrossRef]
74. Wright, D.M.; Neupane, S.; Heidecker, T.; Haile, T.A.; Chan, C.; Coyne, C.J.; McGee, R.J.; Udupa, S.; Henkrar, F.; Barilli, E.; et al. Understanding photothermal interactions will help expand production range and increase genetic diversity of lentil (*Lens culinaris* Medik.). *Plants People Planet* **2021**, *3*, 171–181. [CrossRef]
75. Silva-Díaz, C.; Ramírez, D.A.; Rinza, J.; Ninanya, J.; Loayza, H.; Gómez, R.; Anglin, N.L.; Eyzaguirre, R.; Quiroz, R. Radiation interception, conversion and partitioning efficiency in potato landraces: How far are we from the optimum? *Plants* **2020**, *9*, 787. [CrossRef]
76. Nguyen, G.N.; Norton, S.L. Genebank phenomics: A strategic approach to enhance value and utilization of crop germplasm. *Plants* **2020**, *9*, 817. [CrossRef] [PubMed]
77. Halcro, K.; McNabb, K.; Lockinger, A.; Socquet-Juglard, D.; Bett, K.E.; Noble, S.D. The BELT and phenoSEED platforms: Shape and colour phenotyping of seed samples. *Plant Methods* **2020**, *16*, 49. [CrossRef] [PubMed]
78. Baek, J.; Lee, E.; Kim, N.; Kim, S.L.; Choi, I.; Ji, H.; Chung, Y.S.; Choi, M.S.; Moon, J.K.; Kim, K.H. High throughput phenotyping for various traits on soybean seeds using image analysis. *Sensors* **2020**, *20*, 248. [CrossRef] [PubMed]
79. Komyshev, E.; Genaev, M.; Afonnikov, D. Evaluation of the SeedCounter, a mobile application for grain phenotyping. *Front. Plant Sci.* **2017**, *7*, 1990. [CrossRef]
80. Makanza, R.; Zaman-Allah, M.; Cairns, J.E.; Eyre, J.; Burgueño, J.; Pacheco, Á.; Diepenbrock, C.; Magorokosho, C.; Tarekegne, A.; Olsen, M.; et al. High-throughput method for ear phenotyping and kernel weight estimation in maize using ear digital imaging. *Plant Methods* **2018**, *14*, 1–13. [CrossRef] [PubMed]
81. Miller, N.D.; Haase, N.J.; Lee, J.; Kaeppler, S.M.; de Leon, N.; Spalding, E.P. A robust, high-throughput method for computing maize ear, cob, and kernel attributes automatically from images. *Plant J.* **2017**, *89*, 169–178. [CrossRef]
82. Eizenga, G.C.; McClung, A.M.; Huggins, T.D. Registration of two *Oryza sativa* tropical japonica germplasm lines selected for panicle architecture and grain size traits. *J. Plant Regist.* **2021**, *15*, 573–582. [CrossRef]
83. Tanabata, T.; Shibaya, T.; Hori, K.; Ebana, K.; Yano, M. SmartGrain: High-throughput phenotyping software for measuring seed shape through image analysis. *Plant Physiol.* **2012**, *160*, 1871–1880. [CrossRef]
84. Genomics Open-Source Breeding Informatics Initiative. Available online: http://cbsugobii05.biohpc.cornell.edu/wordpress/ (accessed on 28 September 2021).
85. Breeding Insight. Available online: https://breedinginsight.org/ (accessed on 29 September 2021).
86. Excellence in Breeding Platform. Available online: https://excellenceinbreeding.org/ (accessed on 29 September 2021).
87. Breedbase. Available online: https://breedbase.org (accessed on 29 September 2021).
88. Hershberger, J.; Morales, N.; Simoes, C.C.; Ellerbrock, B.; Bauchet, G.; Mueller, L.A.; Gore, M.A. Making waves in Breedbase: An integrated spectral data storage and analysis pipeline for plant breeding programs. *Plant Phenome J.* **2021**, *4*, e20012. [CrossRef]
89. Morales, N.; Bauchet, G.J.; Tantikanjana, T.; Powell, A.F.; Ellerbrock, B.J.; Tecle, I.Y.; Mueller, L.A. High density genotype storage for plant breeding in the Chado schema of Breedbase. *PLoS ONE* **2020**, *15*, e0240059. [CrossRef]
90. BrAPI. The Breeding API. Available online: https://brapi.org/ (accessed on 29 September 2021).
91. Pixley, K.V.; Salinas-Garcia, G.E.; Hall, A.; Kropff, M.; Ortiz, C.; Bouvet, L.; Suhalia, A.; Vikram, P.; Singh, S. CIMMYT's Seeds of Discovery Initiative: Harnessing Biodiversity for Food Security and Sustainable Development. *Indian J. Plant Genet. Resour.* **2018**, *31*, 1–10. [CrossRef]
92. Seeds of Discovery. Available online: https://seedsofdiscovery.org/ (accessed on 29 September 2021).
93. Agent. Global access to plant genetic resources. Available online: https://agent-project.eu/ (accessed on 29 September 2021).
94. Germinate. Available online: https://germinateplatform.github.io/get-germinate/ (accessed on 29 September 2021).
95. BRIDGE web portal. Available online: https://bridge.ipk-gatersleben.de/ (accessed on 29 September 2021).
96. Andorf, C.M.; Cannon, E.K.; Portwood, J.L.; Gardiner, J.M.; Harper, L.C.; Schaeffer, M.L.; Braun, B.L.; Campbell, D.A.; Vinnakota, A.G.; Sribalusu, V.V.; et al. MaizeGDB update: New tools, data and interface for the maize model organism database. *Nucleic Acids Res.* **2016**, *44*, D1195–D1201. [CrossRef] [PubMed]
97. MaizeGDB TypSimSelector. Available online: https://www.maizegdb.org/TYPSimSelector (accessed on 29 September 2021).

98. König, P.; Beier, S.; Basterrechea, M.; Schüler, D.; Arend, D.; Mascher, M.; Stein, N.; Scholz, U.; Lange, M. BRIDGE–a visual analytics web tool for barley genebank genomics. *Front. Plant Sci.* **2020**, *11*, 701. [CrossRef] [PubMed]
99. Raubach, S.; Kilian, B.; Dreher, K.; Amri, A.; Bassi, F.M.; Boukar, O.; Cook, D.; Cruickshank, A.; Fatokun, C.; El Haddad, N.; et al. From bits to bites: Advancement of the Germinate platform to support prebreeding informatics for crop wild relatives. *Crop Sci.* **2021**, *61*, 1538–1566. [CrossRef]
100. Bari, A.; Street, K.; Mackay, M.; Endresen, D.T.F.; De Pauw, E.; Amri, A. Focused identification of germplasm strategy (FIGS) detects wheat stem rust resistance linked to environmental variables. *Genetic Resources Crop Evol.* **2012**, *59*, 1465–1481. [CrossRef]
101. Thachuk, C.; Crossa, L.; Franco, J.; Dreisigacker, S.; Warburton, M.; Davenport, G.F. Core Hunter: An algorithm for sampling genetic resources based on multiple genetic measures. *BMC Bioinform.* **2009**, *10*, 243. [CrossRef] [PubMed]
102. Belzile, F.; Abed, A.; Torkamaneh, D. Time for a paradigm shift in the use of plant genetic resources. *Genome* **2020**, *63*, 189–194. [CrossRef]
103. Reeves, P.A.; Tetreault, H.M.; Richards, C.M. Bioinformatic extraction of functional genetic diversity from heterogeneous germplasm collections for crop improvement. *Agronomy* **2020**, *10*, 593. [CrossRef]
104. Al Bari, M.A.; Zheng, P.; Viera, I.; Worral, H.; Szwiec, S.; Ma, Y.; Main, D.; Coyne, C.J.; McGee, R.; Bandillo, N. Harnessing genetic diversity in the USDA pea (*Pisum sativum* L.) germplasm collection through genomic prediction. *bioRxiv* **2021**.
105. Jiang, Y.; Weise, S.; Graner, A.; Reif, J.C. Using genome-wide predictions to assess the phenotypic variation of a barley (*Hordeum* sp.) gene bank collection for important agronomic traits and passport information. *Front. Plant Sci.* **2021**, *11*, 2180. [CrossRef]
106. Allier, A.; Teyssèdre, S.; Lehermeier, C.; Moreau, L.; Charcosset, A. Optimized breeding strategies to harness genetic resources with different performance levels. *BMC Genom.* **2020**, *21*, 349. [CrossRef] [PubMed]
107. Bayer, P.E.; Petereit, J.; Danilevicz, M.F.; Anderson, R.; Batley, J.; Edwards, D. The application of pangenomics and machine learning in genomic selection in plants. *Plant Genome* **2021**, *e20112*. [CrossRef]

plants

MDPI

Review

Advancing the Conservation and Utilization of Barley Genetic Resources: Insights into Germplasm Management and Breeding for Sustainable Agriculture

Andrea Visioni [1,*], Boris Basile [2], Ahmed Amri [1], Miguel Sanchez-Garcia [1] and Giandomenico Corrado [2,*]

1 International Center for Agricultural Research in the Dry Areas (ICARDA), Rabat 10100, Morocco; a.amri@cgiar.org (A.A.); m.sanchez-garcia@cgiar.org (M.S.-G.)
2 Department of Agricultural Sciences, University of Naples Federico II, 80055 Portici, Italy; boris.basile@unina.it
* Correspondence: a.visioni@cgiar.org (A.V.); giandomenico.corrado@unina.it (G.C.)

Abstract: Barley is a very important crop particularly in marginal dry areas, where it often serves as the most viable option for farmers. Additionally, barley carries great significance in the Western world, serving not only as a fundamental crop for animal feed and malting but also as a nutritious food source. The broad adaptability of barley and its ability to withstand various biotic and abiotic stresses often make this species the sole cereal that can be cultivated in arid regions. The collection and utilization of barley genetic resources are crucial for identifying valuable traits to enhance productivity and mitigate the adverse effects of climate change. This review aims to provide an overview of the management and exploitation of barley genetic resources. Furthermore, the review explores the relationship between gene banks and participatory breeding, offering insights into the diversity and utilization of barley genetic resources through some examples such as the initiatives undertaken by ICARDA. Finally, this contribution highlights the importance of these resources for boosting barley productivity, addressing climate change impacts, and meeting the growing food demands in a rapidly changing agriculture. The understanding and utilizing the rich genetic diversity of barley can contribute to sustainable agriculture and ensure the success of this vital crop for future generations globally.

Keywords: *Hordeum vulgare*; genetic diversity; crop improvement; biodiversity; climate resilience; landraces; participatory breeding approaches; genetic resource centers

Citation: Visioni, A.; Basile, B.; Amri, A.; Sanchez-Garcia, M.; Corrado, G. Advancing the Conservation and Utilization of Barley Genetic Resources: Insights into Germplasm Management and Breeding for Sustainable Agriculture. *Plants* **2023**, *12*, 3186. https://doi.org/10.3390/plants12183186

Academic Editors: Johannes M. M. Engels and Andreas W. Ebert

Received: 6 July 2023
Revised: 30 August 2023
Accepted: 4 September 2023
Published: 6 September 2023

1. Introduction

1.1. The Diversity and Main Uses of Barley: An Outline of the Plant Domestication, Adaptation, Agronomic Classifications, and Health Claims

Barley (*Hordeum vulgare* L.) is one of the most adaptable crops belonging to the Poaceae family, serving as a vital source of food, feed, and malt for the brewing industry on a global scale [1]. This self-fertile species is characterized by a diploid genome (2n = 2x = 14), although tetraploid (2n = 4x = 28) and hexaploid (2n = 6x = 42) species are also described in *Hordeum*. A high-quality genome sequence was achieved following the establishment of the International Barley Genome Sequencing Consortium (IBSC) in 2006. This endeavor was based on two main approaches: the BAC-by-BAC strategy, which exploited a large set of genetically mapped BAC clones, and whole-genome shotgun sequencing, synergized with deep RNA-Seq analysis [2,3]. A physical map encompassing 4.98 Gbp of the haploid barley genome size (5.10 Gbp) has been successfully constructed, with a substantial portion (exceeding 3.90 Gbp) aligned and anchored to a high-resolution genetic map [2]. The genome was found to be rich in repeat sequences (more than 80% of the total genome) [2]. Later, chromosome conformation capture sequencing (Hi-C) was employed to order and orient super-scaffolds. The final chromosome-scale assembly represents about 95% of the barley genome [3]. At extra-nuclear level, it was found that chloroplast sequences from cultivated and wild barley were closely related [4,5]. The mitochondrial genomes of wild and cultivated barley were also compared, showing only minimal differences in terms of sequence and structure [6]. As expected, fragmented organelle sequences are reported in the barley nuclear genome [3,6].

Barley has been cultivated and selectively bred for thousands of years, resulting in abundant genetic and morphological diversity [7,8]. Although there are significant taxonomic disagreements, particularly for the delimitations within the various species of the *Hordeum* genus, all the cultivated forms are classified as *H. vulgare* subsp. *vulgare*, with the wild progenitor considered conspecific (*H. vulgare* subsp. *spontaneum* (K. Koch) Asch. et Graebn.); hereinafter *H. spontaneum*). This classification is supported by phylogenetic analysis and the ability of these plants to interbreed [9]. A key morphological distinction between wild and cultivated barley is the characteristic of the rachis (i.e., the main axis of the spike). Wild barley exhibits a brittle rachis and when it reaches maturity, the spike releases its triplets as dispersal units. On the other hand, cultivated varieties possess a tough rachis that retain the intact spike, providing a fundamental trait for domestication.

Among cultivated forms, various types are commonly distinguished. These distinctions are often based on characteristics such as growth habit, row type (two-rowed or six-rowed), and the presence/absence of hulls [10,11]. All the species of the genus have three one-flowered spikelets, with one central spikelet and two lateral spikelets arranged alternately at each rachis node. In two-rowed and wild barley, the lateral spikelets are smaller in size and are sterile, while in six-rowed barley cultivars, all three spikelets are fertile and capable of developing into grains. The two-rowed spikes of the wild barley provide an evolutionary advantage in nature, favoring seed dispersal and germination. The presence of a six-rowed spike was determined by a single allele called *vrs1* (previously known as *v* for *vulgare*), which is recessive to the dominant allele responsible for the two-rowed spike (*Vrs1*). Throughout domestication, the *Vrs1* gene has undergone independent mutations multiple times, resulting in the emergence of the six-rowed phenotype [12]. This is accompanied by the presence of *vrs5*, an allele that enhances lateral grain fill characteristics to modern six-row cultivars [13]. Recent studies have unveiled the molecular basis of five major row-type genes (*Vrs1* to *Vrs5*), all involved in the regulation of the arrangement and development of spikelets [14].

Coming to the different types of barley, hulled barley is encased in a tough outer layer, while hulless (naked) barley, scientifically known as *Hordeum vulgare* L. var. *nudum* Hook. f., lacks this outer layer. In relation to adaptability, winter barley is more suitable to colder climates and typically sown in the fall, while spring barley thrives in warmer climates and is usually sown in the spring in higher latitudes and in fall in the Mediterranean climate. Furthermore, variants can be classified based on agronomic characteristics. Finally, certain types of barley are particularly suitable for malting due to their content of amylase and proteins. Protein content in barley seeds ranges from 8% to 30% of the total seed mass [15–18]. Generally, for malting total protein, it should be from 9 to 12%, while alpha amylase activity should be higher than 150 U/g [19]. These varieties are widely used in the production of malt for the brewing industry, syrups, and other non-alcoholic beverages. Other types of barley are better suited for food production (e.g., baked goods, soups, stews, breakfast cereals, but also couscous, bulgur, and other types of mixed grain-based products commonly used in various cuisines), animal feed (primarily for livestock such as cattle, pigs, and poultry), and various biotechnological applications (e.g., enzyme purification, cosmetics, pharmaceuticals) [1,10,11].

Recently, barley has gained renewed attention as a food source particularly with the advancement of biofortified varieties that are enriched with beta glucans [20], a type of soluble fiber found also in oat (*Avena sativa*) and some other cereals. Beta glucans are known for their ability to lower cholesterol levels, regulate blood sugar levels, and promote digestive health. They have been linked to a reduced risk of heart disease, improved immune function, and enhanced weight management [21,22]. By incorporating biofortified barley into common food products (e.g., bread, biscuits, cereal breakfast), it would be possible to provide consumers with an affordable and enjoyable way to integrate the health benefits of beta glucans into their daily diet. A scientific opinion on beta-glucans has been provided by the European Food Safety Authority (EFSA) [23]. The claim pertaining to maintaining normal blood LDL-cholesterol concentrations was assessed with a favorable outcome, while the claim regarding an increase in satiety, which should lead to a reduction in energy intake, is not supported by a confirmed cause-and-effect relationship between the consumption of beta-glucans and sustained satiety. The claim for the reduction of post-prandial glycemic responses was confirmed by a cause-and-effect relationship, with a recommended wording "Consumption of beta-glucans from oats or barley contributes to the reduction of glucose rise after a meal". It is recommended to consume 4 g of beta-glucans from oats or barley for every 30 g of available carbohydrates per meal to achieve this effect. Finally, the claim regarding an improvement in digestive function was deemed insufficiently defined because it lacks clarity (i.e., the claim does not specify the targeted nutrients) [23].

1.2. Aim of the Review

Barley possesses extensive phenotypic and genetic diversity, due to its long history of cultivation and adaptation to diverse environments worldwide, which is preserved in governmental or international gene banks and germplasm collections in research institutions, universities, breeding companies, and agricultural organizations across the globe (Figure 1).

Figure 1. Morphological variations in barley head types: (**a**) examples of two-rowed and (**b**) examples of six-rowed ears.

The progress made in genomics and biotechnology has facilitated the identification and characterization of genes and molecular markers associated with essential agronomic traits in barley [24]. These advancements have not only led to the development of new barley varieties but have also deepened our understanding of the diversity and utility of barley genetic resources [25,26]. For example, exome sequencing and a combination of genome-wide analyses of a set of barley geographically diverse landraces revealed significant associations between days to heading and plant height with seasonal temperature and dryness variables, suggesting that these traits were major drivers of environmental adaptation in the sampled germplasm. A further detailed analysis of flowering time genes showed that patterns of single and multiple haplotypes exhibit strong geographical structure that have contributed to the wide ecogeographical adaptation of barley [27].

This review aims to provide an assessment of the genetic resources of barley, including its wild relatives and cultivated varieties. It will explore the management and exploitation of these genetic resources, highlighting the importance of strategies such as gene bank establishment and core collections' development. The review will also delve into the management and exploitation of barley genetic resources, considering participatory breeding approaches and the importance and limitations of including local knowledge and perspectives. Finally, the historical contributions of the International Center for Agricultural Research in the Dry Areas (ICARDA) in the exploration and utilization of barley genetic resources will be highlighted. We conclude by discussing how advanced genomics and phenomics tools can be harnessed and exploited by institutions in an equitable manner, prioritizing principles of fairness and equal access among different institutions or organizations globally involved in the conservation of barley genetic resources.

2. Genetic Resources of Barley

There is little doubt that environmental changes, agricultural intensification, urbanization, land degradation, and economic factors have significantly contributed to loss of traditional farming practices and, hence, crop diversity in barley, as well as in many other crop species [28,29]. To limit this loss, several political, economic, social, and scientific strategies should be implemented. Among them, the documentation, conservation, and characterization of barley's genetic resources are central to promote the continued use of barley's genetic diversity. The achievement of these activities includes a variety of approaches and procedures, which are summarized below.

2.1. In Situ Conservation of Barley Genetic Resources: Main Natural Habitats and Traditional Cultivation Areas Worldwide

In situ conservation of genetic resources involves their preservation in their habitats [30]. Specifically, this approach aims at maintaining the genetic diversity of plant populations in their natural environment, but it also requires the protection of traditionally cultivated areas from land degradation and over-exploitation [30]. Barley is conserved on-farm in various regions where it has adapted to specific environmental conditions. These regions are often areas with a long history of cultivation (mainly within traditional farming systems) or are represented by the natural habitats of its wild relatives.

Briefly, it is possible to identify the following main agro-ecological regions in which barley has thrived and adapted: the Near Eastern area (i.e., Fertile Crescent including Asia Minor and the Caucasus), the European–Siberian area, the Ethiopian area, the East Asiatic area, and New World centers (Americas and Oceania). In each area, the combination of geographic isolation and significant climatic variations has led to the emergence of an ample set of agro-ecological groups of varieties that also include further subdivisions [8].

The Fertile Crescent, which is considered the area where barley was domesticated, is home to a diverse range of wild and cultivated varieties of barley [24]. This region exhibited a vertical zonality and a diverse range of environmental and climatic factors, leading to the development of various barley types, including two-rowed hulled and naked forms, as well as six-rowed barley with different spike densities. Barley is probably the primary cereal crop in regions with less than 300 mm of annual rainfall, where there are significant fluctuations in rainfall in terms of amounts and distribution. In the Middle East and North Africa, barley is mainly grown as animal feed. The two-row barley is possibly the most common form of barley currently found in Syria and Turkey, while six-row forms are predominant in Jordan, Tunisia, and Lebanon. Landraces have been described in virtually every area of the Fertile Crescent, such as Syria, Tunisia, Lebanon, Jordan, Iraq, and Turkey [7,31–34]. In North African countries, six-row landraces remain prevalent, primarily due to limited adoption of improved varieties and the perception of barley as a risk-averse crop among the majority of farmers. These farmers often rely on traditional practices and show minimal or no uptake of certified seeds and other agricultural inputs.

It is also worth adding the widespread occurrence of wild barley, which can be found across a vast geographic range spanning from Morocco to China [35,36]. Among the 45 taxa of the *Hordeum* genus, wild barley is the sole wild representative in the primary gene pool, and it is the only wild species utilized for genetic improvement in cultivated barley [9,37]. Wild barley (*H. spontaneum*) is normally considered a weed in barley and wheat fields, but it can also be used as forage for livestock. This plant presents a significant amount of genetic variation related to various traits such as biomass, yield, nitrogen content, drought and salinity tolerance, and resistance to diseases [38].

The Ethiopian highlands are another important region for in situ conservation, with several barley landraces that are considered important for food security and livelihoods in the region [39]. In Ethiopia, barley is predominantly cultivated as landraces across all regions by subsistence farmers with limited (if any) use of fertilizers, pesticides, or herbicides. Barley is considered a major source of protein in highlands also because of limited alternative crops. The recent introduction of malting varieties has led to a shift among farmers towards cultivating malting barley to boost their incomes, resulting in an overall diversification in the types of barley being grown, with an increasing focus on meeting the demands for both animal feed and human consumption. Ethiopian landraces have been instrumental in the development of powdery mildew-resistant barley varieties in Northwestern Europe. This resistance is conferred by a naturally occurring recessive resistance allele at the *Mlo* gene (mlo-11), most likely present in landraces from the highlands in the southwest region of Ethiopia collected during German expeditions in the 1930s [40].

In the East Asiatic area, the Himalaya is another important center of diversity for barley, and the region is home to several wild and cultivated varieties of the crop. Among the former, it is necessary to cite *Hordeum agriocrithon* Åberg, a wild species discovered by Åberg in 1938 known because of its resistance to abiotic and abiotic stress [41]. Among cultivated varieties in Central Asia, noteworthy mentions include the Tibetan, Nepalese, Ladakhi, Bhutanese, Kinnauri, Kumaoni, and Sikkim barley [42–47]. These groups of varieties are named after the regions or ethnic groups where they are traditionally grown or have originated from. Although thorough comparative evaluations are not available, each of these barley varieties may have distinct characteristics, including variations in growth habits, yield potential, grain quality, and adaptation to specific environmental and cultural factors such as altitude and culinary traditions (including the production of specific beverages) [42–47]. It had been reported that East Asiatic barley is generally characterized by short straw, early maturity, and a high temperature requirement during ripening [26].

The Americas and Oceania have a relatively recent history of barley development compared to other regions and are dominated by six-row types. In South America, barley is a valuable staple crop and even today represents an important source of food and income for local communities. Although introduced, this crop is embedded in some traditional cultures, playing a role in traditional festivals and rituals. For example, the Puno region (Peru) is home to the Feast of the Virgen de la Candelaria, a festival that includes drinking of chicha, a traditional fermented beverage made from the two-rowed Andean barley [48]. The Andean region is home to unique varieties of barley because of their adaptation to specific environmental conditions. The two-rowed Andean barley, often classified as *Hordeum vulgare* var. *coeleste*, is well suited to the high altitude and cool temperatures of the region [49]. The six-rowed Andean barley is more common in the lower altitude areas and is often used for animal feed or to produce malt for the brewing industry. The Andean regions are also the origin of *H. chilense* Roem. et Schult., a wild species of barley, known for its crossability with the genera *Triticum* and *Secale* [50,51].

2.2. Ex Situ Conservation of Barley Genetic Resources: The Role of the Gene Banks

Ex situ conservation of barley includes establishment of gene banks, botanical gardens, and other facilities that store plant genetic resources. Figure 2 visually reports the key steps and core activities to collect and manage the plant germplasm. Even though different approaches can be used, barley is mainly conserved as seed material, whose longevity is usually highly increased by reducing seed moisture (3–7%) and temperature (around −18 °C) during storage [52,53].

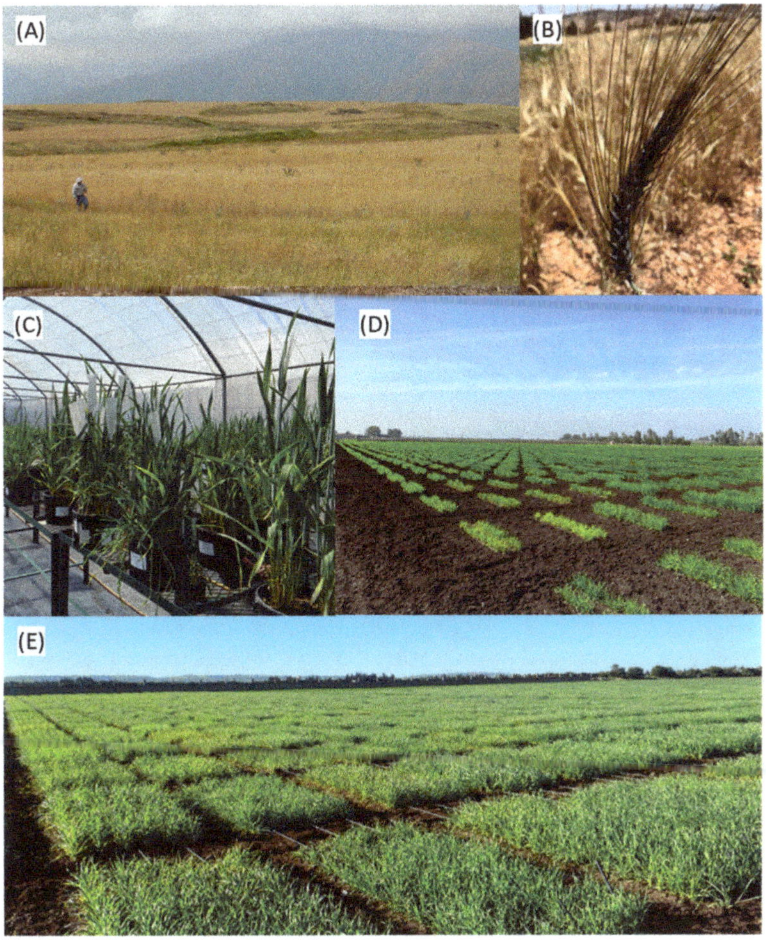

Figure 2. Illustrated workflow for barley germplasm collection and utilization. This figure provides a visual representation of the step-by-step process involved in collecting and utilizing barley germplasm for breeding and research purposes. (**A**) Expedition: Expeditions are organized to collect barley samples from diverse geographical regions, with the highlands of Georgia provided as an example. These expeditions may encompass remote and challenging environments, ensuring a broad range of genetic diversity is captured. (**B**) Collection: Diverse barley accessions, exemplified by the barley with dark-colored grains in the picture, are identified, collected, and documented. This stage also involves carefully cataloging the barley samples for further utilization at experimental stations. (**C**,**D**) Characterization, multiplication, and regeneration: Collected barley accessions are cultivated often under optimum field conditions or less frequently, under greenhouses, and are characterized for major descriptors and agronomic traits. This step also ensures an abundant supply of seeds for distribution to users for evaluation for breeders sought traits. (**E**) Field trial: This panel depicts a field trial specifically focused on drought tolerance. Barley accessions with potential drought tolerance traits undergo rigorous field evaluations under drought-stress conditions. The primary objective of this stage is to identify promising genotypes for further breeding programs and the development of climate-resilient barley varieties.

Barley is well-represented in gene banks, evidenced by the considerable number of accessions available. The species is considered among the 14 most important crops for food security by the International Union for Conservation of Nature (IUCN) Red List of Threatened Species, a comprehensive inventory of the conservation status of plant and animal species worldwide. However, more diversity can be collected targeting more wild relative species and populations with adaptive traits to climate change. According to the Second State of the World's Plant Genetic Resources for Food and Agriculture report published by the Food and Agriculture Organization of the United Nations (FAO) (http://www.fao.org/3/i1500e/i1500e.pdf; accessed on 1 May 2023), there are a total of 466,531 barley accessions (including all taxa) in ex situ collections, and this genus ranks third following *Triticum* and *Oryza*. Nonetheless, as for many other plant species, the level of duplication is not well defined, with estimates indicating that around 120,000 accessions could be distinct (http://www.fao.org/3/i1500e/i1500e.pdf; accessed on 1 August 2023).

A global inventory of barley genetic resources in ex situ collections was performed through the development of the Global Strategy for the Ex Situ Conservation and Use of Barley Germplasm. This collaborative approach was led by the ICARDA, the Global Crop Diversity Trust, and the Leibniz Institute of Plant Genetics and Crop Plant Research (IPK) (Gatersleben, Germany), with the support of CGIAR and FAO, and the contribution from international barley experts (https://www.genebanks.org/resources/publications/barley-strategy-2008/; accessed on 1 August 2023). The inventory revealed that there are 47 major barley gene banks (holding more than 500 accessions), with a total of 402,000 accessions [11]. The major ones include the following: the Plant Gene Resources of Canada (PGRC), Saskatoon, with around 40,000 accessions, the USDA National Small Grains Collection in Aberdeen (Idaho, USA) and the Recursos Genéticos e Biotecnologia of EMBAPRA (Brazil), the ICARDA-CGIAR, with each having around 30,000 accessions, the John Innes Centre in UK and IPK in Germany, each with holding exceeding 20,000 accessions, and the NordGen's, with more than 12,000 barley accessions [26,54]. PGRC is a gene bank that maintains a global collection consisting of wild, cultivated varieties and genetic resources for specific traits of interest. Many of the barley accessions are six rowed, although there are also several naked types. The collection includes Canadian and North American varieties, as well as a diverse range of accessions from around the world. The USDA maintains a large gene bank of barley (wild, cultivated, breeding line, etc.) from around the world, with an estimated 18% of national origin. The Recursos Genéticos e Biotecnologia of the Brazilian Agricultural Research Corporation focuses on various aspects of genetic resources and biotechnology, including crop improvement. ICARDA, an international research organization and member of the CGIAR group, has the global mandate for the improvement in barley productivity in dry areas, and maintains a unique and diversity rich collection with more than 32,400 accessions, with approximately two-thirds of landraces [55]. While major descriptors have been used to characterize the ICARDA's accessions, only a subset of 2400 accessions have been genotyped thus far. NordGen's collection by a large part (over 11,000) is made of cultivated germplasm, with a focus on locally adapted landraces (mainly to the Nordic and Arctic regions). The collection also includes breeding lines, mutants, and hundreds of wild relatives. The IPK besides a large collection of barley, has developed a deeply phenotyped core collection of 1000 genotypes [56].

3. Leveraging Barley Resources: Utilization, Participatory Activities, and Capacity Enhancement in Breeding

3.1. Participatory Breeding Approaches to Empower Farmers for Genetic Diversity Conservation and Variety Selection

Participatory approaches involve local communities in the conservation and use of plant genetic resources. Farmers play a central role in identifying, selecting, and preserving local varieties, landraces, and wild relatives, thereby contributing to the overall genetic diversity and adaptive potential of the barley germplasm. Participatory Plant Breeding (PPB) refers to a breeding approach that involves farmers and other stakeholders by actively engaging them in variety cultivation, evaluation, and selection. In barley, PPB has long been an important complementary process within the broader context of preserving and exploiting biodiversity [57]. A key feature is that this approach is expected to ensure that breeding outcomes are context specific and able to address local challenges and needs, including the resistance or resilience to stresses prevalent in specific environments. On the other hand, PPB faces challenges in not being supported by large, centralized seed production and marketing systems. These systems are currently dominated by a small number of corporations not only for barley, but also for a number of arable crops. A main obstacle in integrating the PPB more widely in barley is the requirement for a different seed-managing system that can accommodate the agrobiodiversity fostered by PPB over time and across different locations. Small seed companies should be sustained by a model that goes beyond traditional profit-driven approaches, recognizing the value of farmers' knowledge and prioritizing long-term benefits. Since 1991, ICARDA has been gradually decentralizing barley selection work to national programs, paving the way for experiments in several countries, including Syria, Egypt, Jordan, Tunisia, Morocco, Yemen, and Eritrea [34]. These programs have generated valuable insights, including the capacity of farmers to handle breeding material and properly select traits of interest, and also confirmed the feasibility and potential of the approach [34]. Briefly, this experience indicated significant differences between the lines selected by breeders on research stations and those chosen by farmers in their fields and a limited phenotypic correlation between research stations and farmers' fields. Moreover, farmers primarily focused on grain yield as a selection criterion, while also considering other traits including grain filling, straw yield, and the color of straw and leaves, due to the crop's importance as animal feed [34,58]. Notably, disease resistance received much less emphasis from farmers compared to breeders [34].

PPB, despite demonstrating its efficiency, has not been widely embraced mainly because of the reluctance in accepting the "implicit paradigm shift regarding seed sovereignty" [59]. To address this, Evolutionary Participatory Breeding (EPB) was proposed as an alternative. This approach involves planting mixtures of diverse genotypes of the same crop in farmers' fields, preferably using early segregating generations that should maximize (allelic) diversity, especially for the trait(s) of interest. The creation of a diverse, mixed population, along with the repeated sowing without the active selection of individual genotypes, represents the main conceptual and technical differences within PPB. Over successive crop cycles in the same environment, the genetic composition of the whole population will differ from the originally planted seeds because it is expected that the population will gradually evolve and adapt to the specific environmental conditions such as soil type, fertility, agronomic practices, rainfall, temperature, and more. As climatic conditions differ from year to year, the genetic makeup of the population should be considered always dynamic, but the fraction of genotypes more adapted to that environment will progressively become more prevalent, while the relatively less amenable genotypes will represent a genetic reservoir to guarantee resilience to extreme conditions and long-term adaptation [59]. In essence, EPB retains the main advantages of PPB but is more effective in reintroducing an ample diversity in farmers' fields without necessarily relying on the long-term support of a scientific institution and formal breeding activities.

The improvement in landraces through participatory selection also encompasses the enhancement of seed quality through seed cleaning and treatment against seed-borne diseases, which have proven to be effective approaches for increasing yields of landraces and contribute to the on-farm conservation of landraces [60].

Irrespective of the approach employed, the active involvement of farmers in the breeding process strongly decreased the risk of developing cultivars that do not meet their preferences or needs. Participatory plant breeding is crucial to preserve germplasm and sustain cultivation for indigenous communities in agroecological niches or marginal lands but also to address needs of specific stakeholders such as smallholder famers, local food processing and brewing industries, consumer groups and associations focused on food quality, safety, and nutrition, and more generally, of agricultural systems that do not rely heavily on off-farm input. To provide an example in cereals, Morocco has experienced a significant increase in wheat productivity through the replacement of local landraces with contemporary cultivars. However, the campaign to replace cultivars halted in the early 1990s, and therefore, most of the currently cultivated varieties were released decades ago. Although breeding programs have continuously released improved varieties, their adoption has been typically low [61], and only recently, we are observing a rising trend towards the adoption of new cultivars, mainly because traditional farming methods are merging with modern agricultural techniques. Coming to barley, it is estimated that landraces still make up over 70% of the cultivated material. Even if different political and socio-economic factors contribute to this situation, this phenomenon has also technical reasons [62]. It has been highlighted that there is clear misalignment between Moroccan farmers' preferences and North African breeders' targets, suggesting that the development of participatory weighted selection (PWS) indices could be implemented to increase the likelihood of acceptance of new releases [61]. Briefly, PWS indices are numerical representation that combine multiple data in a weighted way (i.e., not all traits are given equal importance) taking into account the priorities and preferences of the end-users and are designed to allow local knowledge to be incorporated into the selection process.

Participatory Variety Selection (PVS) is an approach that aims to expedite the adoption of new improved varieties in farmers' fields by involving farmers in the selection process [63]. PVS complements ongoing, centralized varietal development efforts by offering farmers a broader, often preselected range of germplasm options to evaluate and adopt under their own local conditions. The primary contrast between PVS and PPB revolves around the extent of farmers' engagement throughout the breeding program. PVS focuses on directly addressing farmers' requirements, which commonly encompass agronomic and quality traits, with the aim to ensure that the attributes of interest are immediately selected and tailored to their preferences. This approach has proven to be efficient in disseminating new varieties and addressing farmers' specific requirements that may not be recognized through conventional non-participatory methods of varietal development [64]. Successful implementations of PVS have been reported for maize in Africa and for rice in South Asia [65–67]. In barley, PVS has been fruitfully employed for the selection of rainfed barley in Iran [68], malt barley in North–West Ethiopia [69,70], and food barley in Ethiopia [71]. While it should be acknowledged that PVS was not designed as a conservation approach, it is important to recognize that the implementation of PVS can inadvertently impact the conservation of traditional local germplasm. Moreover, in addition to the common restrictions related to resource availability and long-term economic sustainability of participatory and decentralized efforts mentioned before, recognized limits of PVS are issues related to the scale and representativeness (including gender equity considerations) of the stakeholders, and the level of their involvement, which may be a simple visit or visual assessment of experimental stations. Finally, it is accepted that PVS like any participatory selection process may compromise some elements of scientific rigor of plant breeding, such as statistical analysis and replications, posing intrinsic limitations on the evaluation and comparison of the varieties.

3.2. Key Policy and Legal Frameworks to Sustain the Conservation and Use of Barley Genetic Resources

In response to the increasing concerns on the loss of diversity in barley and in other crops, international efforts are underway to promote the sustainable use and conservation of barley genetic resources. These frameworks provide guidelines, regulations, and incentives to promote the effective management of plant germplasm. They also facilitate access to genetic resources for research, breeding, and development purposes while ensuring fair benefits for all stakeholders involved. The Convention on Biological Diversity (CBD) established in 1992 and the International Treaty on Plant Genetic Resources for Food and Agriculture (ITPGRFA, also known as the "Plant Treaty"), adopted by FAO in 2001, are arguably the most important international agreements that aim to promote the conservation and sustainable use of biodiversity, including plant genetic resources. They recognize the sovereignty of countries over their genetic resources and are calling the parties to contribute to their conservation and sustainable use and for a fair and equitable sharing of the benefits arising from their use. These two binding agreements call for the development of national strategies and action plans and of national policies and legislations facilitating access to and exchange of genetic resources. An important component is the establishment of legal instruments to regulate access and benefit-sharing in a transparent and fair way. They involve the implementation of material transfer agreements, such as the Standard Material Transfer Agreement (SMTA). This is a private contract developed by the ITPGRFA that defines the terms and conditions for the exchange and use of genetic materials ensuring that countries and organizations that contribute genetic resources are recognized and rewarded for their efforts. Barley and its related *Hordeum* species are part of the Multilateral System of the Plant Treaty, encompassing a total of 26 crops. Moreover, access to germplasm derived from breeding activities, categorized as "Material under Development," is governed by the SMTA. However, when it comes to collecting additional accessions, two options are available: utilizing the SMTA or adhering to the requirements of the Nagoya Protocol, which involve Prior Informed Consent and Mutually Agreed Terms. The Nagoya Protocol is a supplementary agreement to the CBD for the implementation of measure to access GR and the fair and equitable sharing of benefits arising from their utilization [72]. Just as example, at ICARDA, both genetic resources and material under development are distributed using the SMTA. In the case of the latter, the SMTA also includes provisions for acknowledging the sources of germplasm, sharing information, and sending seed samples to ICARDA.

3.3. A Selection of Initiatives and Partnerships for Barley Genetic Resources Management

Collaborative efforts between national and international organizations, research institutions, and farming communities are critical for the sustainable management and use of barley genetic resources. Key milestones in international initiatives for barley genetic resources management and exploitation include (i) the International Barley Genetics Symposium (IBGS) initiated in 1963, which can be considered among the first international platforms for researchers, breeders, and other stakeholders, to share knowledge and advancements in barley genetics and breeding, (ii) the International Barley Genetic Resources Network (IBGRNet), a global network of barley gene banks established in 1986, and (iii) the Barley Genome Project, launched in 2006 and coordinated by the International Barley Genome Sequencing Consortium (IBGS) [73,74].

A prominent on-going initiative is currently the AGENT project (Access to Genetic Resources and Digitization of Plant Genetic Resources). AGENT aims to unleash the untapped potential of the vast biological material stored in gene banks worldwide by leveraging FAIR (Findable, Accessible, Interoperable, and Reusable) international data standards and an open digital infrastructure dedicated to the management of plant genetic resources. The project aims to promote a more systematic effort to exploit Plant Genetic Resources (PGR) and advocate for generating extensive genotypic and phenotypic data for PGR stored in gene banks (GB), thus facilitating their educated selection and utilization in breeding and agriculture. AGENT consortium focuses on the following: (i) the establishment of an actively cooperating gene bank network, (ii) collecting new data and working on agreed standards and protocols for the use of passively stored GB information, (iii) generating new genotypic information for European barley and wheat collections in order to establish a roadmap and pave the way for a complete global wheat and barley biodiversity atlas, (iv) using this extensive genotypic information to evaluate the quality and redundancy of existing PGR collections as a basis for new quality control and management pathways, (v) establishing coordinated PGR training populations for phenotyping of independent collections as a foundation for a pan-European, genome-wide prediction, and (vi) mining new and historic genotypic and phenotypic information to drive the discovery of genes, traits, and knowledge for future missions. Furthermore, at a larger scale, the project aims to increase data density on collections and disseminate the societal impact of PGR to provide the community with a new database and novel data-mining tools to facilitate a well-informed selection of PGR for different purposes (https://www.agent-project.eu/; accessed on 1 July 2023).

An important on-going initiative focused on scaling the impacts of previous efforts in the field of crop wild relatives (CWR) and participatory research is BOLD-DIIVA-PR II project (Dissemination of ICARDA Varieties through Participatory Research). ICARDA breeding programs for durum wheat, barley, and lentil have successfully integrated the use of CWR that possess unique genetic traits that can enhance the adaptability and resilience of cultivated varieties to drought and other environmental stresses prevalent in the regions where ICARDA operates [75]. Building upon the achievements of the DIIVA-PR project, the current project aims to further advance the utilization of CWR-derived lines of durum wheat and barley developed by DIIVA-PR II. The project operates in countries such as Morocco, Tunisia, Ethiopia, Senegal, Nigeria, and Sudan. The objectives of BOLD-DIIVA-PR II are multifold. By involving local farming communities in participatory actions, this project ensures that the developed varieties align with farmers' preferences and requirements. Additionally, landraces of barley and durum wheat are assessed in the fields for various traits of interest. The best-performing germplasm is then incorporated into breeding programs, considering both performance and farmer preferences. The project strives to generate scientific articles that document the performance of CWR- and landrace-derived elites, facilitate the release of improved varieties in partner countries, and enhance the capacity of local partners in conducting field evaluations. Furthermore, the project promotes open data sharing, ensuring that valuable breeding materials and evaluation data are freely accessible. By engaging stakeholders at various stages, from breeding to evaluation and release, the project strengthens collaboration between research institutions, international organizations, and farming communities, ultimately contributing to the sustainable management and utilization of barley genetic resources.

Finally, although the activity of the following institutions is not limited to barley, it is worth acknowledging the efforts of the European Cooperative Programme for Plant Genetic Resources (ECPGR) Working Group on Barley and the U.S. National Plant Germplasm System (NPGS). The former plays a pivotal role in the conservation and sustainable utilization of barley genetic resources across Europe. This working group brings together experts, scientists, breeders, and stakeholders from various European countries who are dedicated to the conservation, evaluation, and improvement in barley diversity. Within the NPGS, a collaborative effort to preserve the genetic diversity of economically significant plant species, the Plant Genetic Resources Unit (PGRU) and its partners are responsible for collecting, preserving, and distributing barley genetic material worldwide. NPGS collections are managed by the Agricultural Research Service (ARS) of the United States Department of Agriculture (USDA).

4. Exploring and Utilizing Barley Genetic Resources at ICARDA: A Historical Appraisal

ICARDA, a member of the CGIAR, was established in 1977 and holds the global mandate for the improvement in barley. This means that this non-profit agricultural research institute is recognized as the leading organization responsible for conducting research, developing strategies, and implementing programs aimed at enhancing the genetic potential and overall performance of barley crops worldwide, in addition to ensuring the conservation of barley genetic resources. Table 1 presents the composition of the barley collection at ICARDA. Besides cultivated and wild barley, the collection comprises the following: *H. murinum* L., adapted to saline environments and serving as genetic resource for studying and improving salt tolerance [76]; *H. bulbosum* Sieber ex Kunth, a perennial species of applied interest as a source of traits to improve biotic stress resistance [77]; *H. marinum* Huds., primarily found in coastal and maritime habitats, of interest to gain insights into the adaptation to saline environments [78]; *H. brevisubulatum* Link and *H. geniculatum* All., two halophytes studied as a potential source of tolerance to drought and other abiotic stress such as alkalinity [79,80]; and *H. turkestanicum* R.E. Regel and *H. hrasdanicum* (part of *H. murinum* subsp. *leporinum* (Link) Arcang.) [81] thought to be adapted to diverse environments, hence holding potential to provide insights into barley responses to abiotic stress [82,83].

Table 1. Composition of the barley collection at ICARDA.

Taxon	Common Name	Accessions (n)
Hordeum vulgare subsp. *vulgare*	Cultivated barley	30,215
Hordeum vulgare subsp. *spontaneum*	Wild barley	2005
Hordeum murinum	wall barley or mouse barley	284
Hordeum bulbosum	bulbous barley	197
Hordeum marinum	sea barley	54
Hordeum brevisubulatum	short-awned barley	16
Hordeum turkestanicum	syn: *H. brevisubulatum* subsp. *turkestanicum* Tzvelev	6
Hordeum geniculatum	kneed barley	4
Hordeum hrasdanicum	synonym: *H. murinum* subsp. *leporinum*	3
Other		6
Grand total		32,790

The Genetic Resources Unit (GRU) was established in 1983 to safeguard the genetic resources of all the ICARDA mandate crops, through germplasm exploration, gathering, evaluation, and conservation [84]. The GRU also actively works to support ICARDA and NARS breeders through various research activities. They include phenotyping and genotyping resources to identify critical resistance traits that can be used in breeding programs; pre-breeding initiatives to introgress adaptive traits from wild relatives and landraces into elite germplasm of mandated crops with the goal of ensuring yield stability, quality, and nutritional attributes, contributing to the development of new open access tools for electronic data capture, analysis, and decision support. In addition to the core activities of the ICARDA GRU and GB, several collaborative projects have demonstrated the efficiency of using CWR in both pre-breeding and breeding programs for rainfed cereal-based production systems.

For barley, the routine pre-breeding activities at ICARDA focus on the identification and use of landraces and CWR, such as *H. spontaneum* and *H. bulbosum*, as sources of disease resistance, abiotic stress tolerance, and end-use quality. ICARDA's work on landraces began in Syria in 1984 [85], at the time host nation of the ICARDA headquarters. The work continued by testing pure lines extracted from landrace populations with the aim of deploying useful traits into adapted genetic backgrounds [32,86]. The objectives of utilizing landrace genetic variability and adaptability were twofold: first, to develop new varieties for Syria, where conventional breeding had encountered challenges, and second, to develop a methodology for landrace utilization adaptable to different crops and regions, particularly for breeders from developing countries with limited resources [32]. The inclusion of CWR became an integral part of ICARDA GRU's strategy in the 1980s when researchers focused their interest on mining beneficial alleles in untapped germplasm, particularly *H. spontaneum*, due to its crossability with *H. vulgare*. In 2004, ICARDA released six barley cultivars with drought tolerance derived from *H. spontaneum* for use in Syria [87,88]. With the advent of molecular marker technology, crosses between elite cultivars and *H. spontaneum* were also used to dissect complex traits such as drought stress [89]. For example, a study using a subset of the same germplasm tested in multi-environment trials showed that there was variation in grain yield under drought stress among barley genotypes [90], and some of the lines derived from *H. spontaneum* consistently exhibited superior specific adaptation to the range of severe stress conditions employed [90]. Other studies focused on the identification of QTL for agronomic traits in Mediterranean environments [89], dryland characters [91], straw quality [92], powdery mildew [93], and leaf scald [94]. In addition to abiotic stress tolerance, disease resistance plays a crucial role in the development of new breeding lines, particularly for smallholder farmers who often face challenges in accessing and affording fungicides. For instance, a screening of 307 accessions of *H. spontaneum* from different origins, introgression lines derived from crosses between different barley cultivars (Morex, Golden Promise, Vada, and Emir), and lines derived from interspecific crosses with both *H. spontaneum* and *H. bulbosum*, was carried out in two cropping seasons at four experimental stations in Morocco [95]. The germplasm was evaluated for field reactions to powdery mildew, leaf scald, leaf rust, and the net form of net blotch. Specifically, powdery mildew resistance was screened at key locations under natural occurrence, while for the other diseases, plants were tested with mixtures of different isolates collected at different locations in Morocco using spreader rows. Among the materials tested, only three accessions of *H. spontaneum* showed high resistance levels to all four diseases, while 23 other accessions and 16 *H. bulbosum*-derived lines showed resistance to a combination of two to four diseases. Despite the good resistance levels found for individual diseases, the agronomic performance of all lines never surpassed that of the checks used, highlighting the importance of crossing selections from landraces and CWR derivatives with elite adapted materials [95].

The GRU is also working on the nutritional content of barley germplasm. For instance, the beta-glucan and microelement content (iron, zinc, selenium) were recently assessed in a panel of cultivated barley and *H. spontaneum* accessions [96]. The results showed that *H. spontaneum* accessions have higher beta-glucan content, while no substantial differences were found in microelements, except for a few *H. spontaneum* lines that showed higher combined contents of iron, zinc, and selenium. These accessions are currently being used in interspecific crosses to develop biofortified barley germplasm and eventually new varieties.

The choice of accessions from ICARDA, as well as from other gene banks, has been typically carried out using random sampling or core collections. The former involves selecting accessions without any specific criteria, while core collections are designed to capture a significant proportion of the genetic diversity within 10% of the total accessions [97]. To further enhance the efficiency and representativeness of core collections, the Generation Challenge Program (GCP), a global research initiative launched in 2004, introduced the concept of a reference set based on molecular markers and genomics technologies. The GCP proposed developing a reference set that would encompass 70% of the genetic diversity found within the core collection but in a more manageable subset, consisting of only 10% of the core collection. Without undermining the vastness and usefulness of the whole genetic resources available in gene banks, statistical and computational techniques to analyze large datasets and identify patterns or relationships between traits and genetic markers are today deemed necessary to construct 'best bet' subsets. These are carefully selected subsets of non-overlapping genetic resources that have a high probability of possessing the desired traits with a variability adequate to achieve the best interpretability and exploitability of the feature of interest.

The Focused Identification of Germplasm Strategy (FIGS) is an approach developed by ICARDA in collaboration with the Vavilov Institute in Russia and GRDC-Australia to efficiently identify and prioritize crop germplasm collections based on specific target environments and traits [98]. It was developed in the late 1990s and has since been widely adopted in plant genetic resources management. FIGS involves a systematic process of collecting and characterizing germplasm resources, particularly landraces and wild relatives, to identify those with desirable traits that are well adapted to specific agroecological conditions. The strategy considers factors such as climate, soil, and farming systems to define target environments, and traits of interest may include tolerance to abiotic stresses (e.g., drought, heat) or biotic stresses (e.g., pests, diseases), as well as specific quality attributes. Technically, FIGS is based on an algorithm that matches plant traits with geographic and agro-climatic information of the places where samples were collected. It creates 'best-bet' trait-specific subsets of material by passing accession-level information, especially agro-climatic site information, through a series of filters that increase the chances of finding the adaptive trait of interest. For instance, by utilizing eco-geographic parameters linked to the original collection sites, climate modelling techniques were employed for trait mining among barley accessions for resistance to net blotch (*Pyrenophora teres* Drechs. f. *teres*), illustrating the potential of using freely available databases to enhance the efficiency of field screening trials [99]. Recently the FIGS has been improved by combining it with machine learning methods for a predictive characterization of GB materials for agronomic traits and for grain morphological parameters. The high predictability of models for all traits was later used for predicting the non-evaluated accessions and assigning probabilities for the characterization traits. These results were used as predictive characterization [98]. A further step towards the fine tuning of FIGS approach using machine learning has been recently carried out [100]. The study compared two different subsets of potentially scald-resistant germplasm, one selected with GCP while the second was selected using the FIGS approach. Seedling and adult plant evaluation in multi-environmental trials for scald resistance was performed for both sets, and results showed that the FIGS approach was able to capture higher percentages of resistant accessions compared to the GCP subset. Furthermore, machine learning models tuned on training sets were used to predict scald response on test sets. All models efficiently identified resistant accessions with specificities higher than 0.88 but showed different performances between isolates at the seedling and to field populations at the adult plant stage.

5. Conclusions and Perspectives

The procedures and approaches for the conservation and use of barley genetic resources described here highlight some crucial aspects that need to be addressed, such as the need to align conservation strategies with continuous progress in cultivation practices, particularly in developing countries [101,102]. Moreover, the demand to unlock genetic resources for breeding remains strong due to the ongoing development of improved varieties also suitable for smallholder farmers [101]. These aspects center around identifying emerging trends in the local diffusion of barley varieties, understanding the adoption of new technologies, and recognizing the changing preferences of farmers [32,103].

In barley, the creation and distribution of improved varieties in various areas of the world is still one of the key goals of the breeding efforts of international institutions [104]. While developing countries continue to face well-known specific conditions in agriculture (such as limited resources, vulnerable agro-ecological environments, and unstable socio-economic contexts), new improved varieties of certain crops, including barley, are becoming more widely adopted than in the past [105,106]. Varieties with enhanced traits such as yield, disease resistance, and quality attributes, as well as adaptation to specific agro-ecological conditions, are offering advantages to farmers, leading to higher crop yields and potentially higher income. Correspondingly, traditional farming methods are increasingly integrating modern agricultural techniques systems [107]. This convergence blends knowledge and agro-techniques, resulting in more efficient yet not strictly traditional farming systems [107]. In this globally changing scenario, it is important to consider the potential negative implications of a widespread shift towards improved varieties on the conservation of landraces [103]. To address the dilemma between the loss of barley biodiversity and the promotion of large-scale adoption of new varieties, it is therefore necessary to strengthen the conservation of barley genetic resources using complementary ex situ and in situ approaches. Conservation efforts will continue to be central, but strategies and the support of local farming systems need to be updated to effectively address changing needs [108]. Considering also the threats faced by natural habitats and traditional farming systems due to climate change, urbanization, and land-use change, a shift towards an enhanced ex situ conservation will have to represent a substantial transformation in the conservation strategies of barley [109–111]. This could involve the adoption of cutting-edge technologies for seed storage, including automated or robotic high-density storage facilities, the expansion of conserved barley material diversity, and the enhancement of documentation and data management through advanced systems and digital databases, while also minimizing redundancy. Additionally, this endeavor entails the development of more efficient, standardized, and transferable approaches for characterizing and sharing genetic resources among research institutions and breeding programs.

While in situ conservation will always remain crucial for maintaining the dynamic interactions between barley and its natural environment, in the future, it will be necessary to strengthen policies to define and manage controlled environments where barley genetic material can be stored under specific conditions. It is also necessary to further recognize the importance of biodiversity hotspots worldwide. These should be given special attention, included in protected areas, and better managed for the conservation of biodiversity.

Relevant advancements are expected to increase our understanding of the genetic variation of barley germplasm [112]. Briefly, this is being expanded by the continuous emergence of new genomics tools, including advancements in long-read DNA sequencing technologies, bioinformatics analysis, and the digitalization of bioresources [113]. Omics-based approaches on barley genetic resources can help unravel the molecular mechanisms underlying important adaptive traits and provide a deeper understanding of how genetic variation influences the phenotype [112,114]. However, these technologies often require substantial investments in infrastructure, equipment, and expertise, making them less accessible to researchers and breeding programs with limited resources. This may result in disparities in the ability to fully exploit the genetic potential of barley germplasm across different regions or institutions. Staying updated and integrating new techniques into existing workflows requires continuous training and investment in research and development, which may be a logistical and financial burden for some institutions or breeding programs. The challenges posed by highly technological approaches in the characterization of barley germplasm make it difficult to predict how they will be tackled. There are two potential approaches: a more centralized style or an increase in networks. The centralized approach would involve establishing a handful of laboratories with innovative, highly automated, advanced phenotypic and genotypic technologies, which should set and promote coordination and standardization across the globe. On the other hand, expanding networks and ensuring widespread access to cutting-edge technologies should be implemented if the aim is to foster collaboration, inclusivity, and capacity building. It is likely that a combination of both approaches would be the most effective strategy, leveraging the strengths of centralized expertise and distributed knowledge networks. Ultimately, the success will depend not only on willingness to collaborate, but on investment in capacity building to ensure the effective and shared utilization of barley genetic resources.

To enhance the effectiveness of conservation efforts, it is also crucial to establish a strong connection between the conservation and utilization of barley genetic resources. As an example, an approach to achieve this is by improving the Focused Identification of Germplasm Strategy (FIGS) [101,115]. Furthermore, it is essential to strengthen pre-breeding initiatives by incorporating climate-resilient and nutritionally rich genes from both landraces and wild relatives. By tightly linking conservation with the utilization of genetic resources, we can effectively harness the potential of landraces and wild relatives to develop climate-resilient and nutritionally superior barley lines.

In conclusion, now more than ever, it is an integrated approach that can ensure that genetic resources are not only preserved but also actively utilized to address the challenges of sustainable agriculture and food security [116]. Therefore, we believe that a complementary genotypic and phenotypic evaluation of gene bank collections will be the next crucial step towards tapping into the vast untapped biodiversity of barley genetic resources. From an applied perspective, this approach will allow for a targeted selection of accessions for pre-breeding purposes from the massive pool of currently available genetic resources. Considering the needs of traditional farming, we believe that it may be necessary to go beyond grain yield improvement as indicator of profitability, lifting the selection focus from allele mining to a more comprehensive genome-wide selection strategy for adaptability. Implementing these proposed activities requires a long-term commitment, and therefore, only sustained funding and collaborative efforts (involving multiple institutions and stakeholders) may embrace this strategy effectively. The commitment to ongoing research, collaboration, and the utilization of advanced breeding techniques will be key in ensuring the sustainable and efficient use of genetic resources for the development of improved barley varieties with enhanced grain yield and other desirable traits.

Author Contributions: Writing—original draft preparation, A.V., B.B., A.A., M.S.-G. and G.C.; writing—review and editing, A.V., B.B. and G.C. All authors have read and agreed to the published version of the manuscript.

Funding: This work was partially supported by the European Union's Horizon 2020 Research and Innovation Program (AGENT project: grant agreement no. 862613) and DIIVA-PR II: Dissemination of ICARDA Varieties through Participatory Research funded by Crop Diversity Trust (CONT-0817).

Data Availability Statement: Not applicable.

Conflicts of Interest: The authors A.V., A.A., and M.S.G. have affiliations with institutions that are cited within the content, including but not limited to collaborative research projects and joint publications. Every effort has been made to maintain objectivity and ensure the accuracy of the information presented.

References

1. Newman, R.K.; Newman, C.W. *Barley for Food and Health: Science, Technology, and Products*; John Wiley & Sons: New York, NY, USA, 2008.
2. Mayer, K.; Waugh, R.; Langridge, P.; Close, T.; Wise, R.; Graner, A.; Matsumoto, T.; Sato, K.; Schulman, A.; Muehlbauer, G. A physical, genetic and functional sequence assembly of the barley genome. *Nature* **2012**, *491*, 711–716. [PubMed]
3. Mascher, M.; Gundlach, H.; Himmelbach, A.; Beier, S.; Twardziok, S.O.; Wicker, T.; Radchuk, V.; Dockter, C.; Hedley, P.E.; Russell, J. A chromosome conformation capture ordered sequence of the barley genome. *Nature* **2017**, *544*, 427–433. [CrossRef] [PubMed]
4. Poulsen, C. The barley chloroplast genome: Physical structure and transcriptional activity in vivo. *Carlsberg Res. Commun.* **1983**, *48*, 57–80. [CrossRef]
5. Middleton, C.P.; Senerchia, N.; Stein, N.; Akhunov, E.D.; Keller, B.; Wicker, T.; Kilian, B. Sequencing of chloroplast genomes from wheat, barley, rye and their relatives provides a detailed insight into the evolution of the Triticeae tribe. *PLoS ONE* **2014**, *9*, e85761. [CrossRef] [PubMed]
6. Hisano, H.; Tsujimura, M.; Yoshida, H.; Terachi, T.; Sato, K. Mitochondrial genome sequences from wild and cultivated barley (*Hordeum vulgare*). *BMC Genom.* **2016**, *17*, 824. [CrossRef] [PubMed]
7. Badr, A.; Rabey, H.E.; Effgen, S.; Ibrahim, H.; Pozzi, C.; Rohde, W.; Salamini, F. On the origin and domestication history of barley (*Hordeum vulgare*). *Mol. Biol. Evol.* **2000**, *17*, 499–510. [CrossRef] [PubMed]
8. Knüpffer, H.; Terentyeva, I.; Hammer, K.; Kovaleva, O.; Sato, K.; Yasuda, S. Ecogeographical diversity—A Vavilovian approach. In *Diversity in Barley (Hordeum vulgare)*; von Bothmer, R., van Hintum, T.J.L., Knüpffer, H., Sato, K., Eds.; Eslevier: Amsterdam, The Netherlands, 2003; pp. 53–76.
9. Blattner, F.R. Taxonomy of the genus *Hordeum* and barley (*Hordeum vulgare*). In *The Barley Genome*; Stein, N., Muehlbauer, G.J., Eds.; Springer: Berlin/Heidelberg, Germany, 2018; pp. 11–23.
10. Slafer, G.A.; Molina-Cano, J.L.; Savin, R.; Araus, J.L.; Romagosa, I. *Barley Science: Recent Advances from Molecular Biology to Agronomy of Yield and Quality*; Food Products Press: New York, NY, USA; London, UK; Oxford, UK, 2002.
11. Ullrich, S.E. (Ed.) *Barley: Production, Improvement, and Uses*; World Agriculture Series; John Wiley & Sons: Hoboken, NJ, USA, 2011; Volume 12.
12. Komatsuda, T.; Pourkheirandish, M.; He, C.; Azhaguvel, P.; Kanamori, H.; Perovic, D.; Stein, N.; Graner, A.; Wicker, T.; Tagiri, A. Six-rowed barley originated from a mutation in a homeodomain-leucine zipper I-class homeobox gene. *Proc. Natl. Acad. Sci. USA* **2007**, *104*, 1424–1429. [CrossRef] [PubMed]
13. Zwirek, M.; Waugh, R.; McKim, S.M. Interaction between row-type genes in barley controls meristem determinacy and reveals novel routes to improved grain. *New Phytol.* **2019**, *221*, 1950–1965. [CrossRef]
14. Koppolu, R.; Schnurbusch, T. Developmental pathways for shaping spike inflorescence architecture in barley and wheat. *J. Integr. Plant Biol.* **2019**, *61*, 278–295. [CrossRef]
15. Qi, J.-C.; Zhang, G.-P.; Zhou, M.-X. Protein and hordein content in barley seeds as affected by nitrogen level and their relationship to beta-amylase activity. *J. Cereal Sci.* **2006**, *43*, 102–107. [CrossRef]
16. Fox, G. Chemical composition in barley grains and malt quality. In *Genetics and Improvement of Barley Malt Quality*; Zhang, G., Li, C., Eds.; Springer Science & Business Media: Berlin/Heidelberg, Germany, 2010.
17. Vieira, E.; Rocha, M.A.M.; Coelho, E.; Pinho, O.; Saraiva, J.A.; Ferreira, I.M.; Coimbra, M.A. Valuation of brewer's spent grain using a fully recyclable integrated process for extraction of proteins and arabinoxylans. *Ind. Crops Prod.* **2014**, *52*, 136–143. [CrossRef]
18. Gupta, M.; Abu-Ghannam, N.; Gallaghar, E. Barley for brewing: Characteristic changes during malting, brewing and applications of its by-products. *Compr. Rev. Food Sci. Food Saf.* **2010**, *9*, 318–328. [CrossRef] [PubMed]

19. Bouhlal, O.; Affricot, J.R.; Puglisi, D.; El-Baouchi, A.; El Otmani, F.; Kandil, M.; Hafidi, A.; Keser, M.; Sanchez-Garcia, M.; Visioni, A. Malting quality of ICARDA elite winter barley (*Hordeum vulgare* L.) germplasm grown in Moroccan middle atlas. *J. Am. Soc. Brew. Chem.* **2022**, *80*, 401–412. [CrossRef]
20. Sakellariou, M.; Mylona, P.V. New uses for traditional crops: The case of barley biofortification. *Agronomy* **2020**, *10*, 1964. [CrossRef]
21. Brown, G.D.; Gordon, S. Fungal β-glucans and mammalian immunity. *Immunity* **2003**, *19*, 311–315. [CrossRef] [PubMed]
22. Hong, F.; Yan, J.; Baran, J.T.; Allendorf, D.J.; Hansen, R.D.; Ostroff, G.R.; Xing, P.X.; Cheung, N.-K.V.; Ross, G.D. Mechanism by which orally administered β-1, 3-glucans enhance the tumoricidal activity of antitumor monoclonal antibodies in murine tumor models. *J. Immunol.* **2004**, *173*, 797–806. [CrossRef] [PubMed]
23. EFSA Panel on Dietetic Products, Nutrition and Allergies (NDA). Scientific Opinion on the substantiation of health claims related to beta-glucans from oats and barley and maintenance of normal blood LDL-cholesterol concentrations (ID 1236, 1299), increase in satiety leading to a reduction in energy intake (ID 851, 852), reduction of post-prandial glycaemic responses (ID 821, 824), and "digestive function" (ID 850) pursuant to Article 13 (1) of Regulation (EC) No 1924/2006. *EFSA J.* **2011**, *9*, 2207.
24. Stein, N.; Muehlbauer, G.J. (Eds.) *The Barley Genome*; Springer: Berlin/Heidelberg, Germany, 2018.
25. Pourkheirandish, M.; Komatsuda, T. The importance of barley genetics and domestication in a global perspective. *Ann. Bot.* **2007**, *100*, 999–1008. [CrossRef]
26. Von Bothmer, R.; van Hintum, T.; Knüpffer, H.; Sato, K. (Eds.) *Diversity in Barley (Hordeum vulgare)*; Elsevier: Amsterdam, The Netherlands, 2003.
27. Russell, J.; Mascher, M.; Dawson, I.K.; Kyriakidis, S.; Calixto, C.; Freund, F.; Bayer, M.; Milne, I.; Marshall-Griffiths, T.; Heinen, S. Exome sequencing of geographically diverse barley landraces and wild relatives gives insights into environmental adaptation. *Nat. Genet.* **2016**, *48*, 1024–1030. [CrossRef]
28. McNeill, J.R. *Something New under the Sun: An Environmental History of the Twentieth-Century World*; W. W. Norton: New York, NY, USA, 2001.
29. Russell, J.R.; Ellis, R.P.; Thomas, W.T.; Waugh, R.; Provan, J.; Booth, A.; Fuller, J.; Lawrence, P.; Young, G.; Powell, W. A retrospective analysis of spring barley germplasm development from 'foundation genotypes' to currently successful cultivars. *Mol. Breed.* **2000**, *6*, 553–568. [CrossRef]
30. Brush, S.B. (Ed.) The issues of in situ conservation of crop genetic resources. In *Genes in the Field: On-Farm Conservation of Crop Diversity*; IPGRI: Rome, Italy; IDRC: Ottawa, ON, Canada; Lewis Publishers: Boca Raton, FL, USA, 2000; pp. 3–26.
31. Russell, J.; Booth, A.; Fuller, J.; Baum, M.; Ceccarelli, S.; Grando, S.; Powell, W. Patterns of polymorphism detected in the chloroplast and nuclear genomes of barley landraces sampled from Syria and Jordan. *Theor. Appl. Genet.* **2003**, *107*, 413–421. [CrossRef] [PubMed]
32. Ceccarelli, S.; Grando, S. Barley landraces from the Fertile Crescent. In *Genes in the Field: On-Farm Conservation of Crop Diversity*; Brush, S.B., Ed.; IPGRI: Rome, Italy; IDRC: Ottawa, ON, Canada; Lewis Publishers: Boca Raton, FL, USA, 2000; pp. 51–76.
33. Weltzien, E. Differentiation among barley landrace populations from the Near East. *Euphytica* **1989**, *43*, 29–39. [CrossRef]
34. Ceccarelli, S.; Grando, S.; Bailey, E.; Amri, A.; El-Felah, M.; Nassif, F.; Rezgui, S.; Yahyaoui, A. Farmer participation in barley breeding in Syria, Morocco and Tunisia. *Euphytica* **2001**, *122*, 521–536. [CrossRef]
35. Harlan, J.R.; Zohary, D. Distribution of wild wheats and barley: The present distribution of wild forms may provide clues to the regions of early cereal domestication. *Science* **1966**, *153*, 1074–1080. [CrossRef] [PubMed]
36. Zohary, D.; Harlan, J.R.; Vardi, A. The wild diploid progenitors of wheat and their breeding value. *Euphytica* **1969**, *18*, 58–65. [CrossRef]
37. Von Bothmer, R.; Komatsuda, T. Barley origin and related species. In *Barley: Production, Improvement and Uses*; Ullrich, S.E., Ed.; John Wiley & Sons: Hoboken, NJ, USA, 2011; pp. 14–62.
38. Nevo, E. Origin, evolution, population genetics and resources for breeding of wild barley, Hordeum spontaneum, in the Fertile Crescent. In *Barley: Genetics, Biochemistry, Molecular Biology and Biotechnology*; Shewry, P.R., Ed.; CABI: Wallingford, UK, 1992; pp. 19–43.
39. Negassa, M. Patterns of phenotypic diversity in an Ethiopian barley collection, and the Arussi-Bale Highland as a center of origin of barley. *Hereditas* **1985**, *102*, 139–150. [CrossRef]
40. Jørgensen, I.H. Discovery, characterization and exploitation of Mlo powdery mildew resistance in barley. *Euphytica* **1992**, *63*, 141–152. [CrossRef]
41. Konishi, T. Genetic diversity in *Hordeum agriocrithon* E. Åberg, six-rowed barley with brittle rachis, from Tibet. *Genet. Resour. Crop Evol.* **2001**, *48*, 27–34. [CrossRef]
42. Yadav, R.K.; Gautam, S.; Palikhey, E.; Joshi, B.K.; Ghimire, K.H.; Gurung, R.; Adhikari, A.R.; Pudasaini, N.; Dhakal, R. Agro-morphological diversity of Nepalese naked barley landraces. *Agric. Food Secur.* **2018**, *7*, 86. [CrossRef]
43. Witcombe, J. Two-rowed and six-rowed wild barley from the Western Himalaya. *Euphytica* **1978**, *27*, 601–604. [CrossRef]
44. Konishi, T.; Yano, Y.; Fukushima, Y. Genetic variation in barley landraces from Bhutan. *Genet. Resour. Crop Evol.* **1993**, *40*, 33–38. [CrossRef]
45. Lata, S.; Negi, P.S.; Samant, S.S.; Seth, M.K.; Shrama, S. Documentation of traditional alcoholic beverages and their indigenous utilization pattern by Kinnaura tribes of Himachal Pradesh, North Western Himalaya. *Indian J. Tradit. Knowl.* **2021**, *20*, 1002–1013.
46. Joshi, M. An ethnobotanical study of the Kumaon region of India. *Econ. Bot.* **1971**, *25*, 414–422.

47. Tamang, J.P.; Okumiya, K.; Kosaka, Y. Cultural adaptation of the Himalayan ethnic foods with special reference to Sikkim, Arunachal Pradesh and Ladakh. *Himal. Study Monogr. Kyoto Univ.* **2010**, *11*, 177–185.
48. Hardwick, W.A.; van Oevelen, D.E.; Novellie, L.; Yoshizawa, K. Kinds of beer and beerlike beverages. In *Handbook of Brewing*; Hardwick, W.A., Ed.; Marcel Dekker: New York, NY, USA; Basel, Switzerland, 1995; pp. 53–86.
49. Danial, D.; Parlevliet, J.; Almekinders, C.; Thiele, G. Farmers' participation and breeding for durable disease resistance in the Andean region. *Euphytica* **2007**, *153*, 385–396. [CrossRef]
50. von Bothmer, R.; Jacobsen, N.; Baden, C.; Jorgensen, R.B.; Linde Laursen, I. *An Ecogeographical Study of the Genus Hordeum*; International Board for Plant Genetic Resources: Rome, Italy, 1991.
51. Martín, A.; Martínez-Araque, C.; Rubiales, D.; Ballesteros, J. *Triticale: Today and Tomorrow*; Guedes-Pinto, H., Darvey, N., Carnide, V.P., Eds.; Springer: Dordrecht, The Netherlands, 1996; pp. 57–72.
52. Dickie, J.B.; Ellis, R.H.; Kraak, H.; Ryder, K.; Tompsett, P. Temperature and seed storage longevity. *Ann. Bot.* **1990**, *65*, 197–204. [CrossRef]
53. Probert, R.; Adams, J.; Coneybeer, J.; Crawford, A.; Hay, F. Seed quality for conservation is critically affected by pre-storage factors. *Aust. J. Bot.* **2007**, *55*, 326–335. [CrossRef]
54. Knüpffer, H. Triticeae genetic resources in ex situ genebank collections. In *Genetics and Genomics of the Triticeae*; Muehlbauer, G., Feuillet, C., Eds.; Springer: New York, NY, USA, 2009; pp. 31–79.
55. Shaw, P.D.; Weise, S.; Obreza, M.; Raubach, S.; McCouch, S.; Kilian, B.; Werner, P. Database solutions for genebanks and germplasm collections. In *Plant Genetic Resources for the 21st Century: The OMICS Era*; Ghamkhar, K., Williams, W.M., Brown, A.H.D., Eds.; CRC Press: Boca Raton, FL, USA, 2023; pp. 285–309.
56. Oppermann, M.; Weise, S.; Dittmann, C.; Knüpffer, H. GBIS: The information system of the German Genebank. *Database* **2015**, *2015*, bav021. [CrossRef]
57. Ceccarelli, S.; Grando, S. Participatory plant breeding. *Cereals* **2009**, *3*, 395–414.
58. Ceccarelli, S.; Grando, S.; Singh, M.; Michael, M.; Shikho, A.; Al Issa, M.; Al Saleh, A.; Kaleonjy, G.; Al Ghanem, S.; Al Hasan, A. A methodological study on participatory barley breeding II. Response to selection. *Euphytica* **2003**, *133*, 185–200. [CrossRef]
59. Ceccarelli, S.; Grando, S. From participatory to evolutionary plant breeding. In *Farmers and Plant Breeding: Current Approaches and Perspectives*; Westengen, O.T., Winge, T., Eds.; Routledge: London, UK, 2019; pp. 231–244.
60. Amri, A.; Yazbek, M.; Shehadeh, A.; Nawar, M.F.; Tsivelikas, A.; Chandrashekhar, B. ICARDA efforts to promote in situ/on-farm conservation of dryland agrobiodiversity. *Indian J. Plant Genet. Resour.* **2016**, *29*, 265–267. [CrossRef]
61. Alary, V.; Yigezu, Y.A.; Bassi, F.M. Participatory farmers-weighted selection (PWS) indices to raise adoption of *durum* cultivars. *Crop Breed. Genet. Genom.* **2020**, *2*, e200014. [CrossRef]
62. Reeves, J. *Quantitative Analysis of Data from Participatory Methods in Plant Breeding*; Bellon, M.R., Reeves, J., Eds.; CIMMYT: Mexico City, Mexico, 2002.
63. Westengen, O.T.; Winge, T. (Eds.) New perspectives on farmer–breeder collaboration in plant breeding. In *Farmers and Plant Breeding: Current Approaches and Perspectives*; Routledge: London, UK, 2019; pp. 3–16.
64. Wilkus, E.L.; Francesconi, G.N.; Jäger, M. Rural seed sector development through participatory varietal selection: Synergies and trade-offs in seed provision services and market participation among household bean producers in Western Uganda. *J. Agribus. Dev. Emerg. Econ.* **2017**, *7*, 174–196. [CrossRef]
65. Abebe, G. Participatory selection of drought tolerant maize varieties using mother and baby methodology: A case study in the semi arid zones of the central rift valley of Ethiopia. *Afr. Crop Sci. Conf. Proc.* **2005**, *7*, 1479–1485.
66. Singh, Y.; Nayak, A.; Sharma, D.; Gautam, R.; Singh, R.; Singh, R.; Mishra, V.; Paris, T.; Ismail, A. Farmers' participatory varietal selection: A sustainable crop improvement approach for the 21st century. *Agroecol. Sustain. Food Syst.* **2014**, *38*, 427–444. [CrossRef]
67. Walker, T.S. Participatory Varietal Selection, Participatory Plant Breeding, and Varietal Change. Background Paper for the World Development Report 2008. Available online: https://documents1.worldbank.org/curated/en/829781468778811451/pdf/413790Veriatal0selection01PUBLIC1.pdf (accessed on 23 August 2023).
68. Mohammadi, R.; Mahmoodi, K.N.; Haghparast, R.; Grando, S.; Rahmanian, M.; Ceccarelli, S. Identifying superior rainfed barley genotypes in farmers' fields using participatory varietal selection. *J. Crop Sci. Biotechnol.* **2011**, *14*, 281–288. [CrossRef]
69. Aynewa, Y.; Dessalegn, T.; Bayu, W. Participatory evaluation of malt barley (*Hordeum vulgare* L.) genotypes for yield and other agronomic traits at North-West Ethiopia. *Wudpecker J. Agric. Res.* **2013**, *2*, 218–222.
70. Ferede, M.; Demsie, Z. Participatory evaluation of malt barley (*Hordium disticum* L.) varieties in barley-growing highland areas of Northwestern Ethiopia. *Cogent Food Agric.* **2020**, *6*, 1756142. [CrossRef]
71. Buli, W.M. Participatory varietal selection of food barley (*Hordeum vulgare* L.) genotypes in Ethiopia. *Int. J. Res. Stud. Agric. Sci.* **2021**, *7*, 32–36.
72. Ebert, A.W.; Engels, J.M.; Schafleitner, R.; Hintum, T.v.; Mwila, G. Critical Review of the Increasing Complexity of Access and Benefit-Sharing Policies of Genetic Resources for Genebank Curators and Plant Breeders—A Public and Private Sector Perspective. *Plants* **2023**, *12*, 2992. [CrossRef]
73. Bockelman, H.E.; Valkoun, J. Barley germplasm conservation and resources. In *Barley: Improvement, Production, and Uses*; Ullrich, S., Ed.; Wiley-Blackwell: Oxford, UK, 2011; pp. 144–159.

74. Schulte, D.; Close, T.J.; Graner, A.; Langridge, P.; Matsumoto, T.; Muehlbauer, G.; Sato, K.; Schulman, A.H.; Waugh, R.; Wise, R.P. The International Barley Sequencing Consortium—At the threshold of efficient access to the barley genome. *Plant Physiol.* **2009**, *149*, 142–147. [CrossRef]
75. El Haddad, N.; Sanchez-Garcia, M.; Visioni, A.; Jilal, A.; El Amil, R.; Sall, A.T.; Lagesse, W.; Kumar, S.; Bassi, F.M. Crop Wild relatives crosses: Multi-location assessment in durum wheat, barley, and lentil. *Agronomy* **2021**, *11*, 2283. [CrossRef]
76. Isayenkov, S.; Hilo, A.; Rizzo, P.; Tandron Moya, Y.A.; Rolletschek, H.; Borisjuk, L.; Radchuk, V. Adaptation strategies of halophytic barley *Hordeum marinum* ssp. *marinum* to high salinity and osmotic stress. *Int. J. Mol. Sci.* **2020**, *21*, 9019. [CrossRef] [PubMed]
77. Devaux, P. The *Hordeum bulbosum* (L.) method. In *Doubled Haploid Production in Crop Plants: A Manual*; Maluszynski, M., Kasha, K.J., Forster, B.P., Szarejko, I., Eds.; Kluwer: Dordrecht, The Netherlands; Boston, MA, USA; London, UK, 2003; pp. 15–19.
78. von Bothmer, R.; Jacobsen, N. Origin, taxonomy, and related species. In *Barley*; Rasmusson, D.C., Ed.; American Society of Agronomy, Crop Science Society of America, Soil Science Society of America, Publishers: Madison, WI, USA, 1985; pp. 19–56.
79. Li, R.; Zhang, J., Wu, G.; Wang, H.; Chen, Y.; Wei, J. HbCIPK2, a novel CBL-interacting protein kinase from halophyte *Hordeum brevisubulatum*, confers salt and osmotic stress tolerance. *Plant Cell Environ.* **2012**, *35*, 1582–1600. [CrossRef] [PubMed]
80. Eliáš, P.; Dítě, D.; Šuvada, R.; Píš, V.; Ikrényi, I. *Hordeum geniculatum* in the Pannonian Basin: Ecological requirements and grassland vegetation on salt-affected soils. *Plant Biosyst.* **2013**, *147*, 429–444. [CrossRef]
81. Von Bothmer, R. Distribution and habitat preferences in the genus *Hordeum* in Iran and Turkey. In *Annalen des Naturhistorischen Museums in Wien*; Serie B für Botanik und Zoologie; Naturhistorisches Museum: Bern, Switzerland, 1996; pp. 107–116.
82. Hou, Y.-C.; Liu, Q.; Long, H.; Wei, Y.-M.; Zheng, Y.-L. Characterization of low-molecular-weight glutenin subunit genes from *Hordeum brevisubulatum* ssp. *turkestanicum*. *Biol. Bull.* **2006**, *33*, 35–42. [CrossRef]
83. Preston, C.; Holtum, J.A.; Powles, S.B. On the mechanism of resistance to paraquat in *Hordeum glaucum* and *H. leporinum*: Delayed inhibition of photosynthetic O_2 evolution after paraquat application. *Plant Physiol.* **1992**, *100*, 630–636. [CrossRef]
84. Ceccarelli, S.; Valkoun, J.; Erskine, W.; Weigand, S.; Miller, R.; Van Leur, J.A.G. Plant genetic resources and plant improvement as tools to develop sustainable agriculture. *Exp. Agric.* **1992**, *28*, 89–98. [CrossRef]
85. Ceccarelli, S.; Erskine, W.; Hamblin, J.; Grando, S. Genotype by environment interaction and international breeding programmes. *Exp. Agric.* **1994**, *30*, 177–187. [CrossRef]
86. Weltzien, E. Barley collecting and evaluation in Syria and Jordan. *FAO Plant Genet. Resour. Newsl.* **1982**, *52*, 5–6.
87. Hajjar, R.; Hodgkin, T. The use of wild relatives in crop improvement: A survey of developments over the last 20 years. *Euphytica* **2007**, *156*, 1–13. [CrossRef]
88. Ceccarelli, S. Utilization of landraces and *Hordeum spontaneum* in barley breeding for dry areas at ICARDA. *Barley Wheat Triticale Newsl.* **1984**, *3*, 8–11.
89. Baum, M.; Grando, S.; Backes, G.; Jahoor, A.; Sabbagh, A.; Ceccarelli, S. QTLs for agronomic traits in the Mediterranean environment identified in recombinant inbred lines of the cross 'Arta' × *H. spontaneum* 41-1. *Theor. Appl. Genet.* **2003**, *107*, 1215–1225. [CrossRef] [PubMed]
90. Lakew, B.; Eglinton, J.; Henry, R.J.; Baum, M.; Grando, S.; Ceccarelli, S. The potential contribution of wild barley (*Hordeum vulgare* ssp. *spontaneum*) germplasm to drought tolerance of cultivated barley (*H. vulgare* ssp. *vulgare*). *Field Crops Res.* **2011**, *120*, 161–168. [CrossRef]
91. Baum, M.; Grando, S.; Ceccarelli, S.; Backes, G.; Jahoor, A. Localization of Quantitative Trait Loci for dryland characters in barley by linkage mapping. *Chall. Strateg. Dryland Agric.* **2004**, *32*, 191–202.
92. Grando, S.; Baum, M.; Ceccarelli, S.; Goodchild, A.; El-Haramein, F.J.; Jahoor, A.; Backes, G. QTLs for straw quality characteristics identified in recombinant inbred lines of a *Hordeum vulgare* × *H. spontaneum* cross in a Mediterranean environment. *Theor. Appl. Genet.* **2005**, *110*, 688–695. [CrossRef] [PubMed]
93. Dreiseitl, A. High diversity of powdery mildew resistance in the ICARDA wild barley collection. *Crop Pasture Sci.* **2017**, *68*, 134–139. [CrossRef]
94. Genger, R.; Williams, K.; Raman, H.; Read, B.; Wallwork, H.; Burdon, J.; Brown, A. Leaf scald resistance genes in *Hordeum vulgare* and *Hordeum vulgare* ssp. *spontaneum*: Parallels between cultivated and wild barley. *Aust. J. Agric. Res.* **2003**, *54*, 1335–1342. [CrossRef]
95. Rehman, S.; Amouzoune, M.; Hiddar, H.; Aberkane, H.; Benkirane, R.; Filali-Maltouf, A.; Al-Jaboobi, M.; Acqbouch, L.; Tsivelikas, A.; Verma, R.P.S. Traits discovery in *Hordeum vulgare* sbsp. *spontaneum* accessions and in lines derived from interspecific crosses with wild *Hordeum* species for enhancing barley breeding efforts. *Crop Sci.* **2021**, *61*, 219–233. [CrossRef]
96. Elouadi, F.; Amri, A.; El-baouchi, A.; Kehel, Z.; Salih, G.; Jilal, A.; Kilian, B.; Ibriz, M. Evaluation of a set of *Hordeum vulgare* subsp. *spontaneum* accessions for β-Glucans and microelement contents. *Agriculture* **2021**, *11*, 950.
97. Knüpffer, H.; van Hintum, T. Summarised diversity-The barley core collection. In *Diversity in Barley (Hordeum vulgare)*; von Bothmer, R., van Hintum, T.J.L., Knüpffer, H., Sato, K., Eds.; Elsevier: Amsterdam, The Netherlands, 2003; pp. 259–267.
98. Azough, Z.; Kehel, Z.; Benomar, A.; Bellafkih, M.; Amri, A. Predictive characterization of ICARDA genebank barley accessions using FIGS and machine learning. In *Intelligent Environments 2019*; Workshop Proceedings of the 15th International Conferende on Intelligent Environments; Muñoz, A., Ouhbi, S., Minker, W., Echabbi, L., Navarro-Cía, M., Eds.; IOS Press: Amsterdam, The Netherlands, 2019; pp. 121–129.
99. Endresen, D.T.F.; Street, K.; Mackay, M.; Bari, A.; De Pauw, E. Predictive association between biotic stress traits and eco-geographic data for wheat and barley landraces. *Crop Sci.* **2011**, *51*, 2036–2055. [CrossRef]

100. Hiddar, H.; Rehman, S.; Lakew, B.; Verma, R.P.S.; Al-Jaboobi, M.; Moulakat, A.; Kehel, Z.; Filali-Maltouf, A.; Baum, M.; Amri, A. Assessment and modeling using machine learning of resistance to scald (*Rhynchosporium commune*) in two specific barley genetic resources subsets. *Sci. Rep.* **2021**, *11*, 15967. [CrossRef] [PubMed]
101. Bohra, A.; Kilian, B.; Sivasankar, S.; Caccamo, M.; Mba, C.; McCouch, S.R.; Varshney, R.K. Reap the crop wild relatives for breeding future crops. *Trends Biotechnol.* **2022**, *40*, 412–431. [CrossRef] [PubMed]
102. Galluzzi, G.; Seyoum, A.; Halewood, M.; López Noriega, I.; Welch, E.W. The Role of genetic resources in breeding for climate change: The case of public breeding programmes in eighteen developing countries. *Plants* **2020**, *9*, 1129. [CrossRef] [PubMed]
103. Dhehibi, B.; Rudiger, U.; Moyo, H.P.; Dhraief, M.Z. Agricultural technology transfer preferences of smallholder farmers in Tunisia's arid regions. *Sustainability* **2020**, *12*, 421. [CrossRef]
104. Chávez-Dulanto, P.N.; Thiry, A.A.; Glorio-Paulet, P.; Vögler, O.; Carvalho, F.P. Increasing the impact of science and technology to provide more people with healthier and safer food. *Food Energy Secur.* **2021**, *10*, e259. [CrossRef]
105. Kosmowski, F.; Alemu, S.; Mallia, P.; Stevenson, J.; Macours, K. *Shining a Brighter Light: Comprehensive Evidence on Adoption and Diffusion of CGIAR-Related Innovations in Ethiopia*; SPIA Synthesis Report; Standing Panel on Impact Assessment (SPIA): Rome, Italy, 2020.
106. Krishna, V.V.; Lantican, M.A.; Prasanna, B.; Pixley, K.; Abdoulaye, T.; Menkir, A.; Bänziger, M.; Erenstein, O. Impact of CGIAR maize germplasm in Sub-Saharan Africa. *Field Crops Res.* **2023**, *290*, 108756. [CrossRef] [PubMed]
107. Birhanu, M.Y.; Jensen, N. Dynamics of improved agricultural technologies adoption: The chicken and maize paradox in Ethiopia. *Sustain. Futures* **2023**, *5*, 100112. [CrossRef]
108. Mascher, M.; Schreiber, M.; Scholz, U.; Graner, A.; Reif, J.C.; Stein, N. Genebank genomics bridges the gap between the conservation of crop diversity and plant breeding. *Nat. Genet.* **2019**, *51*, 1076–1081. [CrossRef]
109. Zair, W.; Maxted, N.; Brehm, J.M.; Amri, A. Ex situ and in situ conservation gap analysis of crop wild relative diversity in the Fertile Crescent of the Middle East. *Genet. Resour. Crop Evol.* **2021**, *68*, 693–709. [CrossRef]
110. Braverman, I. Conservation without nature: The trouble with in situ versus ex situ conservation. *Geoforum* **2014**, *51*, 47–57. [CrossRef]
111. Longin, C.F.H.; Reif, J.C. Redesigning the exploitation of wheat genetic resources. *Trends Plant Sci.* **2014**, *19*, 631–636. [CrossRef]
112. Milner, S.G.; Jost, M.; Taketa, S.; Mazón, E.R.; Himmelbach, A.; Oppermann, M.; Weise, S.; Knüpffer, H.; Basterrechea, M.; König, P. Genebank genomics highlights the diversity of a global barley collection. *Nat. Genet.* **2019**, *51*, 319–326. [CrossRef]
113. Ghamkhar, K.; Williams, W.M.; Brown, A.H.D. (Eds.) *Plant Genetic Resources for the 21st Century: The OMICS Era*; CRC Press: Boca Raton, FL, USA, 2023.
114. Mochida, K.; Shinozaki, K. Advances in omics and bioinformatics tools for systems analyses of plant functions. *Plant Cell Physiol.* **2011**, *52*, 2017–2038. [CrossRef]
115. Bari, A.; Street, K.; Mackay, M.; Endresen, D.T.F.; De Pauw, E.; Amri, A. Focused identification of germplasm strategy (FIGS) detects wheat stem rust resistance linked to environmental variables. *Genet. Resour. Crop Evol.* **2012**, *59*, 1465–1481. [CrossRef]
116. Adenle, A.A.; Wedig, K.; Azadi, H. Sustainable agriculture and food security in Africa: The role of innovative technologies and international organizations. *Technol. Soc.* **2019**, *58*, 101143. [CrossRef]

plants MDPI

Review

The Role of Home Gardens in Promoting Biodiversity and Food Security

Helena Korpelainen

Department of Agricultural Sciences, Viikki Plant Science Centre, University of Helsinki, P.O. Box 27, FI-00014 Helsinki, Finland; helena.korpelainen@helsinki.fi; Tel.: +358-29-4158383

Abstract: Plant genetic resources provide the basis for sustainable agricultural production, adaptation to climate change, and economic development. Many present crop plants are endangered due to extreme environmental conditions induced by climate change or due to the use of a limited selection of plant materials. Changing environmental conditions are a challenge for plant production and food security, emphasizing the urgent need for access to a wider range of plant genetic resources than what are utilized today, for breeding novel crop varieties capable of resilience and adaptation to climate change and other environmental challenges. Besides large-scale agricultural production, it is important to recognize that home gardens have been an integral component of family farming and local food systems for centuries. It is remarkable how home gardens have allowed the adaptation and domestication of plants to extreme or specific ecological conditions, thus contributing to the diversification of cultivated plants. Home gardens can help in reducing hunger and malnutrition and improve food security. In addition, they provide opportunities to broaden the base of cultivated plant materials by harboring underutilized crop plants and crop wild relative species. Crop wild relatives contain a wide range of genetic diversity not available in cultivated crops. Although the importance of home gardens in conserving plant genetic resources is well recognized, there is a risk that local genetic diversity will be lost if traditional plant materials are replaced by high-yielding modern cultivars. This paper provides an overview of home gardens and their present role and future potential in conserving and utilizing plant genetic resources and enhancing food and nutritional security under global challenges.

Keywords: biodiversity; home gardens; crop wild relatives; food security; plant genetic resources

Citation: Korpelainen, H. The Role of Home Gardens in Promoting Biodiversity and Food Security. *Plants* **2023**, *12*, 2473. https://doi.org/10.3390/plants12132473

Academic Editors: Andreas W. Ebert, Johannes M. M. Engels and Igor G. Loskutov

Received: 12 May 2023
Revised: 22 June 2023
Accepted: 26 June 2023
Published: 28 June 2023

1. Introduction

Plant genetic resources are the basis of sustainable agricultural production, adaptation to climate change and economic development. Besides being critical for food security, they are sources for many other products as well, such as animal feed, fiber, ornamentals, and energy as well as other ecosystem services. Plant genetic diversity contains raw materials for improving the capacity of crops to adapt to climate change and other environmental challenges. It is estimated that, for instance, certain major agricultural crops, e.g., wheat, maize, rice, and soy, may lose up to 25% of their present yield by 2050 due to climate change if varieties adapted to changing conditions are not available by then [1]. Besides decreasing the average yield, climatic drivers, such as drying and warming trends, extreme temperature and precipitation, and carbon dioxide fertilization, are expected to increase yield variability [1]. Despite being crucial for food production, plant biodiversity remains as a relatively little exploited means of breeding crops adapted to new climates [2]. To ensure the continued availability of valuable genetic materials, plant genetic resources can be conserved in wild habitats or on-farm conditions in agroecosystems (both forms of in situ) or outside those (ex situ) in germplasm collections and gene banks. Home gardens are a type of agroecosystem, which are typically small but globally prevalent.

Food production on small plots adjacent to human settlements is the oldest form of cultivation throughout the world [3,4]. The beginning of modern agriculture can be dated back to such subsistence production systems that began in small garden plots around the household as far back as in Early Mesopotamia (10,000 BC) [5]. Still today, these gardens continue to play an important role in local food systems by providing food and income for households [6]. They are an especially prevalent feature of local food systems and the agricultural landscape in developing countries [4]. They provide a continuous supply of fresh vegetables for family use, and they may have an important role in crisis and post-crisis situations [4]. In addition, home gardens have a remarkable role among indigenous communities for food, medicine, and cultural practices, and their deterioration can lead to the erosion of traditional knowledge and practices [4]. Home gardens are not limited to rural settings; in fact, they create the most common form of urban agriculture [7].

Home gardens often consist of multi–layer systems of trees, vegetables, fruits, field crops, spices, herbs, and ornamental and medicinal plants around homesteads [8,9]. Similar to large-scale plant production practices, home gardens are a cropping system composed of soil, crops, weeds, pathogens, and insects, which convert resource inputs, i.e., solar energy, water, nutrients, labor, etc. into food, feed, fuel, fiber, and pharmaceuticals [10]. Although the structure, functions, and contributions of home gardens vary between geographic regions [4], they have some common features, as follows: (1) they are located near a residence, (2) they contain a high diversity of plants, (3) production is supplemental rather than the main source of family consumption and income, (4) they occupy a small area, and (5) the production system is suitable for anybody at some level [6,11].

The vast accessibility of home gardens also provides opportunities for innovative plant production through testing different kinds of plant materials, such as wild materials and local varieties, and through conducting crossing experiments and even fundamental research. The groundbreaking investigations on peas by the monk Gregor Mendel, presented in 1865, were performed in the home garden of a monastery in Brünn and resulted in the formulation of the genetic laws of inheritance that have greatly facilitated the science of genetics and plant breeding [12]. The present paper explores the benefits of home gardens, their importance for food and nutritional security under global challenges, and especially their role and potential in conserving and utilizing genetic resources, as well as enhancing the diversity and adaptation of crop plants.

2. Home Gardens Supporting Food and Nutritional Security, and Providing Other Benefits

With an increasing world population, rising urbanization, decreasing arable land, and weather extremes due to climate change, global agriculture is under pressure. Home gardens have an important supporting role when responding to the challenges agricultural production is facing. It is widely recognized that home gardens have multiple functions. They contribute to providing food, proper nutrition, medicine, and other useful products, and they fulfill social and cultural needs, being part of traditional knowledge and practices, while providing different ecosystem services, helping in mitigating climate change effects, and contributing to economic needs and sustainable livelihoods [4,13–17]. It is understandable that attempts to further improve the productivity of these widespread, often eco-friendly, culturally important, and sustainable agricultural practices have been initiated [18–21].

The most typical function of home gardens is to provide a regular supply of fresh vegetables, which are a very important part of a good diet as they contain various nutrients [22]. Independent production not only saves money but ensures access to a healthy diet that contains adequate macro- and micronutrients. Thus, homestead production of food provides households with direct access to important nutrients that may not be readily available or within their economic reach [4,7]. Cultivation in home gardens has been shown to be associated, for instance, with reduced hunger and malnutrition, improved health, and dietary diversity and balance, including, e.g., an increased consumption of vitamin-A-rich fruits and vegetables, pulses, and other fruits and vegetables [17,23]. Thus, home gardens offer a viable means to improve household food security. It is notable that many uncultivated, as well as neglected and underutilized species and cultivars could make an important contribution to the dietary diversity of local communities [24,25]. In addition, home gardens can be a source of additional income if the household sells a portion of the garden's produce.

Malnutrition and poor health status are a common problem in developing countries. Specifically, malnutrition among women of reproductive age increases the risk of mortality during pregnancy and puts their newborn children at risk of long-term deficiencies [26]. To overcome this problem, home gardens are considered as a possible solution by contributing to poverty alleviation, improved health, and lower maternal and infant mortality rates [27,28]. In developed countries, home gardening can also decrease the risk of obesity and unhealthy diets [29]. It has been reported that vegetable consumption increased significantly but obesity and metabolic risk decreased as a result of home gardening activities [30]. School gardening projects are an effective way to introduce both plant production and healthier diets [31].

Because of the ongoing global food crisis and rising food prices, there is an increased emphasis on adopting more resilient food systems and strengthening local food production to mitigate the emerging adverse effects [32]. Food production and livelihood enhancement through home gardens can contribute to this effort [7,33]. Such gardens have endured over time as an integral part of local food systems and the agricultural landscape all over the world. Furthermore, home gardens are not limited to rural areas, but they allow the production of fresh, healthy, and inexpensive food in urban settings as well. Whether involving home gardens or other small-scale production systems, urban agriculture can have beneficial environmental and societal impacts, such as a reduction in the urban heat island effect, improved local air quality, improved stormwater management, increased pollinator populations, and climate mitigation services such as carbon sequestration [34].

3. The Importance of Plant Genetic Resources

Plant genetic resources are considered to include cultivars, landraces, crop wild relatives (CWR), ecotypes, and genetic stocks (Table 1). They provide the basis for sustainable agricultural production, adaptation to climate change and economic development [35–37]. The cultivated crop plants contain only a small part of all plant genetic diversity. An important consequence of crop domestication and bottlenecks created by breeding is that the current gene pool is relatively narrow for most crops, with rather little variation remaining for traits related to resilience and nutritional value. Thus, crop diversification and the development of improved varieties are necessary means for maintaining and stabilizing yields and product qualities, and this is increasingly important under changing climate conditions.

Table 1. Types of plant genetic resources.

Type	Description
Cultivars	Varieties produced by plant breeders, usually uniform and adapted to high farm management standards
Landraces	Varieties developed over time in traditional farming systems, usually variable and adapted to local conditions
Crop wild relatives	Wild taxa within the same genus as a crop
Ecotypes	Populations of wild forms of domesticated species or their wild relative species, or other wild material; specific adaptations
Genetic stocks	Material generally used by research or breeding programs resulting in specific information on a gene or character, or other data of value for breeding and research

It is recognized that many present crop plants are endangered due to extreme environmental conditions induced by climate change or due to the use of a limited selection of plant materials [38,39]. The increased environmental variability is a challenge for plant production and food security. Without sufficient genetic resources, it will be difficult or even impossible to develop crops that contain important traits, such as pest and disease resistances and the ability to withstand drought, extreme temperatures, and other environmental challenges [2,39,40]. The base of limited genetic resources available in widely used crop plants can be widened by including CWRs, which are taxa related to crops that can potentially donate genetic material with beneficial traits, as well as other underutilized sources. CWRs and wild-harvested plants contain a wide range of genetic diversity and adaptations not available otherwise. Besides being an important part of biodiversity, they carry socioeconomic value and enhance food security [41,42].

In response to the increasing visibility of CWRs in international political agendas since the early 1990's, a good number of projects have been created and various tools and guidelines have been developed at local, regional, and global levels to enhance their use [13]. It is evident that CWRs are an indispensable asset for breeding programs for expanding and improving the gene pool of cultivated varieties, when novel traits are needed for improvement, e.g., resistance to pathogens and herbivores, nutritional properties, the ability to withstand waterlogging and resilience under changing climatic conditions. Although CWRs have been used for plant breeding for several decades and they have contributed a wide range of beneficial traits, the proportion of CWRs utilized for breeding purposes has, so far, been small. The vast majority of the CWR reservoir remains unexplored [43]. Major obstacles for tapping the genetic diversity of CWRs to improve crops are the hybridization barriers between undomesticated germplasm and the crop, increasing along with divergence, and the extensive and time-consuming pre-breeding work typically required for transferring desirable genetic material to new varieties [44]. On the other hand, recent improvements in the speed of breeding, and high-throughput genotyping and phenotyping make the use of CWR genetic resources more attractive for utilization [41]. Indeed, there are many examples of CWR genes being used to improve crops, such as, wheat, maize, rice, barley, potato, cassava, and legumes [45,46]. Among positive outcomes, the use of CWR genes may lead to a reduced use of pesticides, and sturdier plants which can better manage in competition against weeds, followed by a reduced application of herbicides. Furthermore, improved drought resistance would help saving water by reducing irrigation, and plants with a more efficient use of nutrients need smaller amounts of fertilizers.

An unexpected situation concerning plant genetic resources is their underuse, not overexploitation that threatens their existence [40]. If not being actively used, farmers' crop varieties as well as those bred by professional plant breeders will not be maintained through continued selection. Rather, they will degrade and may eventually disappear. Yet, such currently underutilized crop plant materials may be able to contribute to climate adaptation and thus are certainly worth further attention. Especially in the case of landraces, home gardens may have an important role in the conservation and utilization of plant genetic resources [15,19,21,24]. The diversity of home gardens is determined by sociocultural (e.g., traditional knowledge and practices) and economic factors, and by climatic and other environmental features [22,47,48]. It is important to learn more of the present role and further potential of home gardens in the management and conservation of a wide range of unique genetic resources for food and on-farm agriculture. So far, many landraces and cultivars, and rare and endangered species have been preserved in home gardens [8,20,49].

Paleoethnobotanical research provides an opportunity to learn about plant use in daily life in an area's ancient history [50]. It appears that the division between wild and cultivated (or semi-domesticated) food plants was not clearly distinct, since many wild species are thought to have been exposed to various levels of intervention and human management during growth cycles [51]. For instance, Mesoamerican farmers of the village of Joya de Cerén circa 600 CE clearly managed the landscape around their settlements in a manner that would have allowed for harvest from both agricultural and non-agricultural species simultaneously [50]. All weed species recovered from these fields have known uses in nutrition, medicine, or other purposes.

4. The Role of Home Gardens in Enhancing the Diversity, Adaptation, and Conservation of Crop Plants

It is remarkable that home gardens have allowed the adaptation and domestication of plants to extreme or specific ecological conditions [24], thus contributing to the diversification of cultivated plants. In such cases, plants develop morphological and physiological characteristics, allowing their adaptation to new or unfavorable habitats. Consequently, home gardens may contain unique and rare locally evolved or developed genetic diversity. An interesting process is introgression, which is used as a breeding tool [52] but which sometimes occurs in home gardens [53]. While such introgression may be beneficial to the productivity of home gardens and perhaps even larger-scale production, it has been clearly proven that gene flow from crop taxa may have a substantial impact on the evolution of wild populations [54]. It has been shown that at least 12 important crops hybridize with wild relatives in some part of their agricultural distribution [54]. These crops include wheat, rice, maize, soybean, barley, cotton, sorghum, millet, beans, rapeseed, sunflower and sugar cane. Crop wild relatives used in cultivation or through introgressive hybridization provide an additional source of plant diversity. Table 2 shows a few examples of evolutionary events observed in local home gardens. It is notable that most such events remain unrecorded for the wider plant breeding community. Further investigations on the composition of home gardens would probably reveal an interesting range of plant materials with adaptations enabling their cultivation in different and changing environmental conditions and use as a material in further breeding, thus promoting resilience and food security.

Landraces developed over time in traditional farming systems, including home gardens, are an underutilized source of genetic variation. One of these species is the common bean (*Phaseolus vulgaris*), domesticated in Mesoamerica and the Andes, but its secondary center of genetic diversity probably extended to Brazil, China, and Europe [55]. After domestication, *P. vulgaris* has become an important crop plant, especially in developing countries. The genetic diversity of Mesoamerican landraces of *P. vulgaris* has been studied, and it has been discovered to possess a very high genetic diversity, which is expected to allow adaptation to diverse environmental conditions, e.g., [56,57]. The proper identification of these novel sources of genetic variation and their use in local breeding efforts can justify and further enhance the conservation of locally adapted beans' genetic resources in countries where a robust conservation strategy is still missing [56]. In addition, the utilization of wild relatives with specific adaptation traits, such as disease resistances, may be a useful addition to *P. vulgaris* breeding programs. It has been reported in Cuba that a *P. vulgaris* landrace 'Negrito' has superior resistance to diseases and harsh weather conditions [53]. Other examples of cultivars discovered in home gardens include several drought-resistant *Allium* landraces from Cuba and salinity-resistant tomato (*Lycopersicon esculentum*) from Guatemala [5].

Most domesticated plant taxa mate with wild relatives somewhere in the world, and gene flow from crop taxa may have a substantial impact on the evolution of wild populations [53,54]. For instance, introgression from the wild tomato *L. esculentum* var. *cerasiforme* to tomato *L. esculentum* has been detected in Cuba [58,59], resulting in interesting variation, including intermediate forms valuable for plant breeding because of fruit characters or disease tolerance. In maize (*Zea mays*), introgressive hybridization and further selection of races have been shown to be common events in the contemporary maize evolution in Cuba [60]. Comparably, introgression among lima bean (*Phaseolus lunatus*) landraces showing intermediate characteristics [61], and introgression between modern varieties and landraces of squash (*Cucurbita moschata*) [60] have been found in Cuban farmers' fields.

Plant domestication most likely began around human settlements, and the domestication processes of wild plants has continued in home gardens. For instance, the date palm (*Phoenix dactylifera*), which is a widely cultivated species in small-scale home gardens as well as in large plantations and possesses a great number of cultivars, was one of the first fruit trees to be domesticated around 6800–6300 BCE, followed by a complex history of breeding and use [62]. However, such domestication processes have rarely been demonstrated empirically [63]. For instance, for the majority of polyploid crops, it remains uncertain to what extent hybridization and polyploidization preceded domestication or were precipitated by human activities [63]. Some of the crop wild relatives brought into use as a vegetable include the annual green amaranth *Amaranthus viridis* [24], biannual cabbage *Brassica oleracea* [64], and perennial watercress *Nasturtium officinale* [65]. Some herbs, which are considered only marginally important, may have considerable use potential, such as sorrels (genus *Rumex*). Sorrels have been utilized for thousands of years as food, herbal preparations and as a source of different colors of dyes [66]. They are mostly consumed through wild foraging or growing in home gardens. A few types of sorrels are available commercially, including wild types and a few cultivars [66]. There are also wild plants with unusually high tolerance to harsh conditions, e.g., the palm tree *Medemia argun*, which is highly tolerant to drought and heat [67,68]. Its fruit is not considered palatable, and its presently known utilization possibilities are based on the use of its strong fibrous leaves and woody stems for sheltering purposes. However, it is believable that *M. argun* has a wider use potential than presently recognized and it could be a good plant with highly interesting adaptive traits for small-scale home garden production in challenging conditions [67,68].

Table 2. Reported examples of evolutionary events observed in local home gardens.

(a) Landraces Adapted to Specific Conditions				
Species	**Landrace**	**Region**	**Reference**	**Comments**
Allium sp.	Several landraces	Cuba	Esquivel et al., 1988 [58]	Superior drought resistance
Lycopersicon esculentum (tomato)	Several landraces	Guatemala	Esquivel et al., 1988 [58]	Superior salinity resistance
Phaseolus vulgaris (common bean)	Negrito	Cuba	Esquivel and Hammer 1992 [53]	Superior resistance to diseases and harsh weather
(b) Cultivated Plants Developed via Introgression				
Taxon 1	**Taxon 2**	**Region**	**Reference**	**Comments**
Lycopersicon esculentum(tomato)	*L. esculentum* var. *cerasiforme* (wild tomato)	Cuba	Esquivel and Hammer 1991 [59]	Fruit characters and disease tolerance
Phaseolus lunatus landraces (lima bean)	*P. lunatus* landraces	Cuba	Castiñeiras et al., 1991 [61]	Heterosis in seed characters
Zea mays landraces (maize)	*Z. mays* landraces	Cuba	Hatheway 1957 [60]	Hybridization typical in maize evolution
(c) Crop Wild Relatives Brought into Use				
Species		**Region**	**Reference**	**Comments**
Amaranthus viridis (green amaranth)		India	Barbhuiya et al., 2016 [24]	Annual herb/leafy vegetable
Brassica oleracea (cabbage)		Romania	Papp et al., 2013 [64]	Biannual plant/diverse vegetable
Nasturtium officinale (watercress)		Nepal	Gautam et al., 2006 [65]	Aquatic perennial/leafy vegetable

5. How to Improve the Conservation of Biodiversity in Home Gardens?

A good approach to enhance the role of home gardens in promoting biodiversity and food security is to utilize both landraces and wild materials to answer breeding needs created by climate change with higher temperatures and more frequent drought periods. An example is hop (*Humulus lupulus*), which has a long home garden history due to its use in beer production [69]. Future breeding efforts with different quality and adaptation targets are expected to utilize wild hop populations and landraces present in many regions [70]. In fact, good candidates of hop cultivars for growth in warm climates have been found [71]. In addition, it is possible to improve the management practices of genetic resources in home gardens, which would result in a combination of better productivity and superior maintenance of genetic diversity, specifically through the introduction of new crops, improved varieties, and specific characteristics [18]. At the same time, home gardens have a role as important sites of experimentation, plant introduction, and crop improvement as well as refuges for unique genetic diversity [18]. Furthermore, precise molecular characterization is desirable for plant genetic resources, including, to some extent, home garden materials as well. Molecular marker techniques and DNA sequencing allow direct surveys of variation at the DNA level, thereby excluding all environmental influence. Nowadays, these analyses can be performed effectively, even at very early growth stages. Therefore, they have marginalized other methods in genotypic identification [41]. With the onset of DNA and 'omics' analyses (i.e., analyses of complete genetic or molecular profiles of organisms based on genomics, transcriptomics, proteomics, or metabolomics), the knowledge of genetic diversity has increased dramatically, as also our understanding of issues, such as domestication, adaptation, and genetic erosion.

Besides the important role of home gardens in conserving plant genetic resources, they contribute to the conservation of biodiversity. These two components are integrated. Especially if plant materials with adequate tolerance and resistance to different biotic stresses are available, the need for the use of pesticides will be reduced, followed by beneficial effects on the biodiversity of the ecosystem in question. This also provides interesting agroecological research possibilities, for instance, the assessment of above- and belowground biodiversity, for which validated and precise tools, such as DNA metabarcoding, are available. It has been shown that the fungal and bacterial biodiversity includes microorganisms, which are potentially beneficial for plant production, e.g., [72,73]. However, the structure and function of fungal and bacterial communities and their interaction and impact on plant performance are still understudied [73].

The ongoing process of evolution in home gardens can be enhanced through selection by farmers, to obtain suitable, adapted plant types to be grown under prevailing and upcoming production conditions. Figure 1 outlines the role of home gardens in conserving plant genetic resources, this being especially important for presently underutilized plant material. Nevertheless, it is important to raise broader awareness of the urgency of conserving plant genetic resources. Although the importance of home gardens in conserving plant genetic resources is well recognized, there is a risk that local genetic diversity will be lost if traditional plant materials are replaced by high-yielding modern cultivars. This type of tendency has been observed, for instance, in Bulgaria, where, however, the local crop diversity of home gardens has remained well preserved until now [21]. Yet, overall, home gardens, when properly managed, have a definite role and importance in the long-term conservation of plant genetic resources.

The conservation of plant genetic resources is based on international agreements, yet each country has a national responsibility to conserve its own genetic resources. One of the goals listed in the draft of the post-2020 Global Biodiversity Framework by the United Nations Environment Programme is that "genetic diversity of wild and domesticated species is safeguarded, with at least 90 per cent of genetic diversity within all species maintained" [74]. Achieving this challenging target will require well-coordinated conservation action at national, regional, and international levels. For decades, gene banks have played a key role in conservation measures when attempting (a) to sustainably conserve the broadest range of genetic diversity found in the target species maintained as population samples or accessions, (b) to characterize and evaluate this diversity to aid selection for utilization, and (c) to make the accessions available to the users [75]. However, it is evident that complementary in situ conservation action is needed. In fact, a systematic application of in situ conservation measures is estimated to at least double the diversity available to users [75].

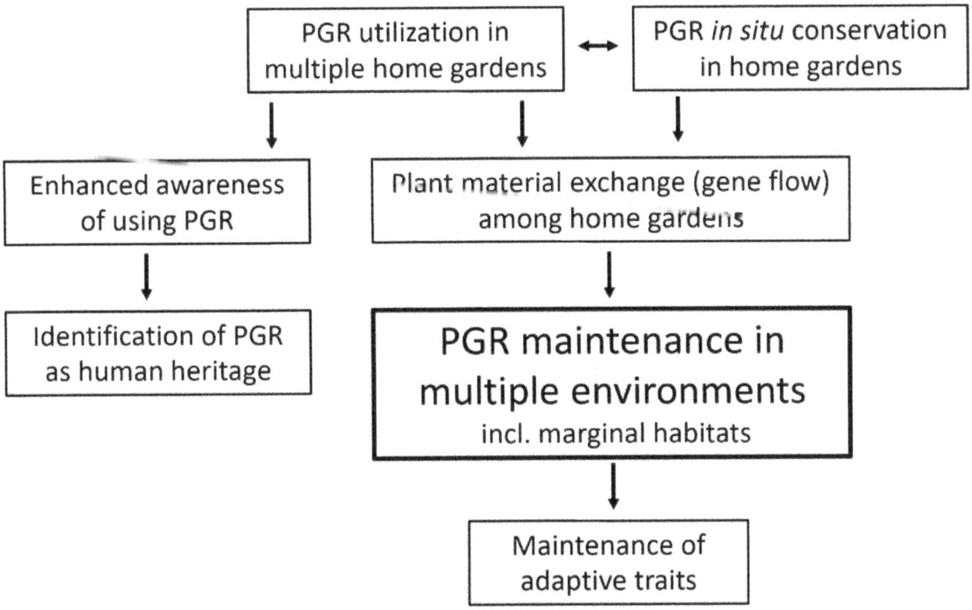

Figure 1. The role of home gardens in conserving plant genetic resources (PGR) in multiple environments through PGR utilization, material exchange and identification as cultural heritage.

Home gardens are typically small and privately operated. Despite that, they can contribute to in situ conservation through on-farm conservation measures. Their impact on the conservation of plant genetic resources can be enhanced through better structured and coordinated management measures, in combination with ex situ conservation. In addition, it is important to recognize that home gardens and their plant genetic resources are dynamic systems. Firstly, they are influenced by cultural, social, and economic factors, as home gardens are typically managed by persons having different goals and preferences, which change over time. Secondly, biological evolution functions in on-farm production systems, resulting in changes in the composition of genetic resources over time, possibly generating new adaptations. Such changes are especially important under ongoing climate change. Therefore, besides increasing conservation through using the existing genetic diversity, home gardens and other on-farm systems have an important complementary role in relation to the static ex situ conservation through dynamic evolutionary processes. This is especially important for underutilized crops, which are often neglected in plant breeding activities.

In summation, measures enhancing the conservation of plant diversity present in home gardens include (1) systematic monitoring and documentation of the diversity, dynamic changes in diversity due to ongoing evolution, threats, and the conservation status of plant materials, (2) improving knowledge on valuable traits and genetic characteristics, (3) expanding gene bank coverage of home garden materials, and (4) increasing the availability of these plant materials to both formal and on-farm crop improvement programs.

Funding: This research received no external funding.

Data Availability Statement: Not applicable.

Acknowledgments: Open access funding provided by University of Helsinki.

Conflicts of Interest: The author declares no conflict of interest.

References

1. IPCC. *Climate Change 2014: Synthesis Report. Contribution of Working Groups I, II and III to the Fifth Assessment Report of the Intergovernmental Panel on Climate Change*; IPCC: Geneva, Switzerland, 2014.
2. Henry, R.J. Innovations in plant genetics adapting agriculture to climate change. *Curr. Opin. Plant Biol.* **2020**, *56*, 168–173. [CrossRef] [PubMed]
3. Niñez, V.K. Household gardens: Theoretical and policy considerations. *Agric. Syst.* **1987**, *23*, 167–186. [CrossRef]
4. Galhena, D.H.; Freed, R.; Maredia, K.M. Home gardens: A promising approach to enhance household food security and wellbeing. *Agric. Food Secur.* **2013**, *2*, 8. [CrossRef]
5. Vasey, D.E. *An Ecological History of Agriculture, 10,000 B.C.–A.D.10,000*; Iowa State University Press: Ames, IA, USA, 1992.
6. Marsh, R. Building on traditional gardening to improve household food security. *Food Nutr. Agric.* **1998**, *22*, 4–14.
7. Eigenbrod, C.; Gruda, N. Urban vegetable for food security in cities: A review. *Agron. Sustain. Dev.* **2015**, *35*, 483–498. [CrossRef]
8. Kumar, B.M.; Nair, P.K.R. The enigma of tropical homegardens. In *New Vistas in Agroforestry. Advances in Agroforestry*; Nair, P.K.R., Rao, M.R., Buck, L.E., Eds.; Springer: Dordrecht, The Netherlands, 2004; Volume 1, pp. 135–152.
9. Salako, V.K.; Fandohan, B.; Kassa, B.; Assogbadjo, A.E.; Idohou, A.F.R.; Gbedomon, R.C.; Chakeredza, S.; Dulloo, M.E.; Glele Kakaï, R. Home gardens: An assessment of their biodiversity and potential contribution to conservation of threatened species and crop wild relatives in Benin. *Genet. Resour. Crop Evol.* **2014**, *61*, 313–330. [CrossRef]
10. Fresco, L.O.; Westphal, E. A hierarchical classification of farm systems. *Exp. Agric.* **1988**, *24*, 399–419. [CrossRef]
11. Mitchell, R.; Hanstad, T. *Small Homegarden Plots and Sustainable Livelihoods for the Poor*; LSP Working Paper 11; FAO: Rome, Italy, 2004.
12. Weiling, F. Historical study: Johann Gregor Mendel 1822–1884. *Am. J. Med. Genet.* **1991**, *40*, 1–25. [CrossRef]
13. Calvet-Mir, L.; Gomez-Baggethun, E.; Reyes-Garcia, V. Beyond food production: Ecosystem services provided by home gardens. A case study in Vall Fosca, Catalan Pyrenees, Northeastern Spain. *Ecol. Econ.* **2012**, *74*, 153–160. [CrossRef]
14. Clarke, L.W.; Li, L.; Jenerette, G.D.; Yu, Z. Drivers of plant biodiversity and ecosystem service production in home gardens across the Beijing Municipality of China. *Urban Ecosyst.* **2014**, *17*, 741–760. [CrossRef]
15. Cruz-Garcia, G.S.; Struik, P.C. Spatial and seasonal diversity of wild food plants in home gardens of northeast Thailand. *Econ. Bot.* **2015**, *69*, 99–113. [CrossRef] [PubMed]
16. Setiani, S.; Setiawan, E.; Huang, W.-C. Taneyan Lanjang shared home gardens and sustainable rural livelihoods of ethnic Madurese in Madura Island, Indonesia. *Sustainability* **2022**, *14*, 5960. [CrossRef]
17. Sileshi, M.; Sieber, S.; Friedrichs, K.; Rybak, C.; Feyisa, B.W.; Lana, M.A. Adoption and impact of kitchen garden on food and nutritional security of farming households in Tanzania. *Ecol. Food Nutr.* **2022**, *61*, 651–668. [CrossRef] [PubMed]
18. Engels, J. Home gardens—A genetic resources perspective. In *Home Gardens and In Situ Conservation of Plant Genetic Resources in Farming Systems, Proceedings of the Second International Home Gardens Workshops, Witzenhausen, Germany, 17–19 July 2001*; Watson, J.W., Eyzaguirre, P.B., Eds.; IPGRI: Rome, Italy, 2002; pp. 3–9.
19. Galluzzi, G.; Eyzaguirre, P.; Negri, V. Home gardens: Neglected hotspots of agro-biodiversity and cultural diversity. *Biodivers. Conserv.* **2010**, *19*, 3635–3654. [CrossRef]
20. Ebert, A.W. The role of vegetable genetic resources in nutrition security and vegetable breeding. *Plants* **2020**, *9*, 736. [CrossRef] [PubMed]
21. Ivanova, T.; Bosseva, Y.; Chervenkov, M.; Dimitrova, D. Enough to feed ourselves!—Food plants in Bulgarian rural home gardens. *Plants* **2021**, *10*, 2520.
22. Rybak, C.; Mbwana, H.A.; Bonatti, M.; Sieber, S.; Müller, K. Status and scope of kitchen gardening of green leafy vegetables in rural Tanzania: Implications for nutrition interventions. *Food Secur.* **2018**, *10*, 1437–1447. [CrossRef]
23. Girard, A.W.; Self, J.L.; McAuliffe, C.; Olude, O. The effects of household food production strategies on the health and nutrition outcomes of women and young children: A systematic review. *Paediatr. Perinat. Epidemiol.* **2012**, *26* (Suppl. S1), 205–222. [CrossRef] [PubMed]
24. Barbhuiya, A.R.; Sahoo, U.K.; Upadhyaya, K. Plant diversity in the indigenous home gardens in the Eastern Himalayan region of Mizoram, Northeast India. *Econ. Bot.* **2016**, *70*, 115–131. [CrossRef]
25. Singh, A.; Dubey, R.K.; Bundela, A.K.; Abhilash, P.C. The trilogy of wild crops, traditional agronomic practices, and UN-Sustainable Development Goals. *Agronomy* **2020**, *10*, 648. [CrossRef]
26. Keats, E.C.; Das, J.K.; Salam, R.A.; Lassi, Z.S.; Imdad, A.; Black, R.E.; Bhutta, Z.A. Effective interventions to address maternal and child malnutrition: An update of the evidence. *Lancet Child Adolesc. Health* **2021**, *5*, 367–384. [CrossRef] [PubMed]
27. Kim, C.; Mansoor, G.F.; Paya, P.M.; Ludin, M.H.; Ahrar, M.J.; Mashal, M.O.; Todd, C.S. Review of policies, data, and interventions to improve maternal nutrition in Afghanistan. *Matern. Child Nutr.* **2020**, *16*, e13003. [CrossRef] [PubMed]
28. Dubey, S.D. Impact of planned and non-planned kitchen gardening for improvement of nutrition and economical benefits of societies. *J Pharm. Phytoch.* **2021**, *10* (Suppl. S1), 502–504.
29. Knapp, M.B.; Hall, M.T.; Mundorf, A.R.; Partridge, K.L.; Johnson, C.C. Perceptions of School-Based Kitchen Garden Programs in Low-Income, African American Communities. *Health Promot. Pract.* **2019**, *20*, 667–674. [CrossRef]

30. Gatto, N.M.; Martinez, L.C.; Spruijt-Metz, D.; Davis, J.N. LA sprouts randomized controlled nutrition, cooking and gardening programme reduces obesity and metabolic risk in Hispanic/Latino youth. *Pediatr. Obes.* **2017**, *12*, 28–37. [CrossRef]
31. Rochira, A.; Tedesco, D.; Ubiali, A.; Fantini, M.P.; Gori, D. School gardening activities aimed at obesity prevention improve body mass index and waist circumference parameters in school-aged children: A systematic review and meta-analysis. *Child. Obes.* **2020**, *16*, 154–173. [CrossRef]
32. Gliessman, S. Why is there a food crisis? *Agroecol. Sust. Food Syst.* **2022**, *46*, 1301–1303. [CrossRef]
33. Lal, R. Home gardening and urban agriculture for advancing food and nutritional security in response to the COVID-19 pandemic. *Food Sec.* **2020**, *12*, 871–876. [CrossRef]
34. Siegner, A.; Sowerwine, J.; Acey, C. Does urban agriculture improve food security? Examining the nexus of food access and distribution of urban produced foods in the United States: A systematic review. *Sustainability* **2018**, *10*, 2988.
35. Engels, J.M.M., Ebert, A.W. A critical review of the current global ex situ conservation system for plant agrobiodiversity. I. History of the development of the global system in the context of the political/legal framework and its major conservation components. *Plants* **2021**, *10*, 1557. [CrossRef]
36. Swarup, S.; Cargill, E.J.; Crosby, K.; Flagel, L.; Kniskern, J.; Glenn, K.C. Genetic diversity is indispensable for plant breeding to improve crops. *Crop Sci.* **2021**, *61*, 839–852. [CrossRef]
37. Korpelainen, H. Plant genetic resources for food and agriculture: Novel materials for adapting to changing environmental conditions. *Ann. Acad. Sci. Fenn.* **2023**, *in press*. [CrossRef]
38. Khoury, C.K.; Brush, S.; Costich, D.E.; Curry, H.A.; de Haan, S.; Engels, J.M.M.; Guarino, L.; Hoban, S.; Mercer, K.L.; Miller, A.J.; et al. Crop genetic erosion: Understanding and responding to loss of crop diversity. *New Phytol.* **2022**, *233*, 84–118. [CrossRef] [PubMed]
39. Miedaner, T.; Juroszek, P. Climate change will influence disease resistance breeding in wheat in Northwestern Europe. *Theor. Appl. Genet.* **2021**, *134*, 1771–1785. [CrossRef] [PubMed]
40. Anderson, R.; Bayer, P.E.; Edwards, D. Climate change and the need for agricultural adaptation. *Curr. Opin. Plant Biol.* **2021**, *56*, 197–202. [CrossRef]
41. Hübner, S.; Kantar, M.B. Tapping diversity from the wild: From sampling to implementation. *Front. Plant Sci.* **2021**, *12*, 626565. [CrossRef]
42. Maxted, N.; Kell, S.P. *Establishment of a Global Network for the In Situ Conservation of Crop Wild Relatives: Status and Needs*; FAO: Rome, Italy, 2009.
43. Engels, J.M.M.; Thormann, I. Main challenges and actions needed to improve conservation and sustainable use of our crop wild relatives. *Plants* **2020**, *9*, 968. [CrossRef]
44. Viruel, J.; Kantar, M.B.; Gargiulo, R.; Hesketh-Prichard, P.; Leong, N.; Cockel, C.; Forest, F.; Gravendeel, B.; Pérez-Barrales, R.; Leitch, I.J.; et al. Crop wild phylorelatives (CWPs): Phylogenetic distance, cytogenetic compatibility and breeding system data enable estimation of crop wild relative gene pool classification. *Bot. J. Linn. Soc.* **2020**, *195*, 1–33. [CrossRef]
45. Dempewolf, H.; Baute, G.; Anderson, J.; Kilian, B.; Smith, C.; Guarino, L. Past and future use of wild relatives in crop breeding. *Crop Sci.* **2017**, *57*, 1070–1082. [CrossRef]
46. Kamenya, S.N.; Mikwa, E.O.; Song, B.; Odeny, D.A. Genetics and breeding for climate change in Orphan crops. *Theor. Appl. Genet.* **2021**, *134*, 1787–1815. [CrossRef]
47. Gautam, R.; Suwal, R.; Sthapit, B.R. Securing family nutrition through promotion of home gardens: Underutilized production systems in Nepal. *Acta Hortic.* **2009**, *806*, 99–106. [CrossRef]
48. Gbedomon, R.C.; Fandohan, A.B.; Salako, V.K.; Idohou, A.F.; Kakaï, R.G.; Assogbadjo, A.E. Factors affecting home gardens ownership, diversity and structure: A case study from Benin. *J. Ethnobiol. Ethnomed.* **2015**, *11*, 56. [CrossRef] [PubMed]
49. Suwardi, A.B.; Navia, Z.I.; Mubarak, A.; Mardudi, M. Diversity of home garden plants and their contribution to promoting sustainable livelihoods for local communities living near Serbajadi protected forest in Aceh Timur region, Indonesia. *Biol. Agric. Hortic.* **2023**, *in press*. [CrossRef]
50. Slotten, V.; Lentz, D.; Sheets, P. Landscape management and polyculture in the ancient gardens and fields at Joya de Cerén, El Salvador. *J. Anthrop. Arch.* **2020**, *59*, 101191. [CrossRef]
51. Bharucha, Z.; Pretty, J. The roles and values of wild foods in agricultural systems Philos. *Trans. Roy. Soc. B Biol. Sci.* **2010**, *365*, 2913–2926. [CrossRef] [PubMed]
52. Burgarella, C.; Barnaud, A.; Kane, N.A.; Jankowski, F.; Scarcelli, N.; Billot, C.; Vigouroux, Y.; Berthouly-Salazar, C. Adaptive introgression: An untapped evolutionary mechanism for crop adaptation. *Front. Plant Sci.* **2019**, *10*, 4. [CrossRef] [PubMed]
53. Esquivel, M.; Karl Hammer, K. The Cuban homegarden 'conuco': A perspective environment for evolution and in situ conservation of plant genetic resources. *Gen. Res. Crop Evol.* **1992**, *39*, 9–22. [CrossRef]
54. Ellstrand, N.C.; Prentice, H.C.; Hancock, J.F. Gene flow and introgression from domesticated plants into their wild relatives. *Annu. Rev. Ecol. Syst.* **1999**, *30*, 539–563. [CrossRef]
55. Santalla, M.; De Ron, A.M.; De La Fuente, M. Integration of genome and phenotypic scanning gives evidence of genetic structure in Mesoamerican common bean (*Phaseolus vulgaris* L.) landraces from the southwest of Europe. *Theor. Appl. Genet.* **2010**, *120*, 1635–1651. [CrossRef]
56. Jiménez, O.R.; Korpelainen, H. Microsatellite markers reveal promising genetic diversity and seed trait associations in common bean landraces (*Phaseolus vulgaris* L.) from Nicaragua. *Plant Gen. Res. Charact. Util.* **2012**, *10*, 108–118. [CrossRef]

57. Jiménez, O.R.; Korpelainen, H. Preliminary evaluation of F_1 generation derived from two common bean landraces (*Phaseolus vulgaris*) from Nicaragua. *Plant Breed.* **2013**, *132*, 205–210. [CrossRef]
58. Esquivel, M.; Shagarodsky, T.; Krieghoff, K.; Rodriguez, B.; Hammer, K. Collecting plant genetic resources in Cuba. Report of the second mission 1986. *Genet. Resour. Crop Evol.* **1988**, *36*, 437–449. [CrossRef]
59. Esquivel, M.; Hammer, K. The cultivated species of the Solanaceae family in Cuba. In *Solanaceae III: Taxonomy, Chemistry, Evolution*; Hawkes, J.G., Lester, R.N., Nee, M., Estrada-Ramos, N., Eds.; Royal Botanic Gardens: Kew, UK, 1991; pp. 357–364.
60. Hatheway, W.H. *Races of Maize in Cuba*; National Academy of Sciences, National Research Council: Washington, DC, USA, 1957; Volume 453.
61. Castiñeiras, L.; Esquivel, M.; Lioi, L.; Hammer, K. Origin, diversity and utilization of the Cuban germplasm of common bean (*Phaseolus vulgaris* L.). *Euphytica* **1991**, *57*, 1–8. [CrossRef]
62. Zohary, D.; Hopf, M. *Domestication of Plants in the Old World*, 3rd ed.; Oxford University Press: Oxford, UK, 2000.
63. Hughes, C.E., Govindarajulu, R.; Robertson, A.; Filer, D.L.; Harris, S.A.; Bailey, C.D. Serendipitous backyard hybridization and the origin of crops. *Proc. Natl. Acad. Sci. USA* **2007**, *104*, 14389–14394. [CrossRef]
64. Papp, N.; Birkás-Frendl, K.; Farkas, A.; Pieroni, A. An ethnobotanical study on home gardens in a Transylvanian Hungarian Csángó village (Romania). *Genet. Resour. Crop Evol.* **2013**, *60*, 1423–1432. [CrossRef]
65. Gautam, R.; Suwal, R.; Shrestha, P. Status of home gardens of Nepal: Findings of baseline survey conducted in four sites of Home Garden Project. In *Home Gardens in Nepal, Proceedings of the Workshop on "Enhancing the Contribution of Home Garden to On-farm Management of Plant Genetic Resources and to Improve the Livelihoods of Nepalese Farmers: Lessons Learned and Policy Implications", Pokhara, Nepal, 6–7 August 2004*; Gautam, R., Sthapit, B.R., Shrestha, P.K., Eds.; LI-BIRD, Bioversity International and SDC: Pokhara, Nepal, 2006; pp. 54–65.
66. Korpelainen, H.; Pietiläinen, M. Sorrel (*Rumex acetosa* L.): Not only a weed but a promising vegetable and medicinal plant. *Bot. Rev.* **2020**, *86*, 234–246. [CrossRef]
67. Elshibli, S.; Korpelainen, H. Genetic profiling of the critically endangered palm species *Medemia argun* using newly developed chloroplast DNA markers. *Plant Ecol. Divers.* **2018**, *11*, 185–192. [CrossRef]
68. Elshibli, S.; Korpelainen, H. Genetic diversity and population structure of *Medemia argun* (Mart.) Wurttenb. ex H.Wendl. based on genome-wide markers. *Front. Ecol. Evol.* **2021**, *9*, 687188. [CrossRef]
69. Edwardson, J.R. Hops—Their botany, history, production and utilization. *Econ. Bot.* **1952**, *6*, 160–175. [CrossRef]
70. Korpelainen, H.; Pietiläinen, M. Hop (*Humulus lupulus* L.): Traditional and present use, and future potential. *Econ. Bot.* **2021**, *75*, 302–322. [CrossRef]
71. Eriksen, R.L.; Rutto, L.K.; Dombrowski, J.E.; Henning, J.A. Photosynthetic activity of six hop (*Humulus lupulus* L.) cultivars under different temperature treatments. *Hortscience* **2020**, *55*, 403–409. [CrossRef]
72. Xia, Z.; He, Y.; Korpelainen, H.; Niinemets, Ü.; Li, C. Sex-specific interactions shape root phenolics and rhizosphere microbial communities in Populus cathayana. *For. Ecol. Manag.* **2022**, *504*, 119857. [CrossRef]
73. Liu, M.; Wang, J.; Zhao, W.; Korpelainen, H.; Li, C. Females face more positive plant-soil feedback and intersexual competition under adequate nitrogen conditions compared to males in Populus cathayana. *Sci. Total Environ.* **2023**, *874*, 162479. [CrossRef] [PubMed]
74. UNEP. *Draft Post-2020 Global Biodiversity Framework*; United Nations Environment Programme: Nairobi, Kenya, 2022; Available online: https://www.unep.org/resources/publication/1st-draft-post-2020-global-biodiversity-framework (accessed on 6 June 2023).
75. Maxted, N.; Magos Brehm, J. Maximizing the crop wild relative resources available to plant breeders for crop improvement. *Front. Sustain. Food Syst.* **2023**, *7*, 1010204. [CrossRef]

plants

MDPI

Review

Plant Diversity Conservation Challenges and Prospects—The Perspective of Botanic Gardens and the Millennium Seed Bank

Elinor Breman [1,*], Daniel Ballesteros [1], Elena Castillo-Lorenzo [1], Christopher Cockel [1], John Dickie [1], Aisyah Faruk [1], Katherine O'Donnell [2], Catherine A. Offord [3], Samuel Pironon [4], Suzanne Sharrock [2] and Tiziana Ulian [1]

1 Royal Botanic Gardens, Kew, Wakehurst, Ardingly, Haywards Heath, West Sussex RH17 6TN, UK; D.Ballesteros@kew.org (D.B.); E.CastilloLorenzo@kew.org (E.C.-L.); C.Cockel@kew.org (C.C.); j.dickie@kew.org (J.D.); A.Faruk@kew.org (A.F.); t.ulian@kew.org (T.U.)

2 Botanic Gardens Conservation International, Descanso House, 199 Kew Road, London TW9 3BW, UK; katherine.o'donnell@bgci.org (K.O.); Suzanne.Sharrock@bgci.org (S.S.)

3 The Australian Plant Bank, Australian Institute of Botanical Science, Australian Botanic Garden, Mount Annan, Sydney, NSW 2567, Australia; Cathy.Offord@botanicgardens.nsw.gov.au

4 Royal Botanic Gardens, Kew, Kew Green, Richmond, Surrey TW9 3AE, UK; S.Pironon@kew.org

* Correspondence: e.breman@kew.org

Abstract: There is a pressing need to conserve plant diversity to prevent extinctions and to enable sustainable use of plant material by current and future generations. Here, we review the contribution that living collections and seed banks based in botanic gardens around the world make to wild plant conservation and to tackling global challenges. We focus in particular on the work of Botanic Gardens Conservation International and the Millennium Seed Bank of the Royal Botanic Gardens, Kew, with its associated global Partnership. The advantages and limitations of conservation of plant diversity as both living material and seed collections are reviewed, and the need for additional research and conservation measures, such as cryopreservation, to enable the long-term conservation of 'exceptional species' is discussed. We highlight the importance of networks and sharing access to data and plant material. The skill sets found within botanic gardens and seed banks complement each other and enable the development of integrated conservation (linking in situ and ex situ efforts). Using a number of case studies we demonstrate how botanic gardens and seed banks support integrated conservation and research for agriculture and food security, restoration and reforestation, as well as supporting local livelihoods.

Keywords: biodiversity; long-term conservation; plant populations; strategic collecting; exceptional species; collection quality; safety duplication; seed longevity; seed viability; viability monitoring; integrated conservation

Citation: Breman, E.; Ballesteros, D.; Castillo-Lorenzo, E.; Cockel, C.; Dickie, J.; Faruk, A.; O'Donnell, K.; Offord, C.A.; Pironon, S.; Sharrock, S.; et al. Plant Diversity Conservation Challenges and Prospects—The Perspective of Botanic Gardens and the Millennium Seed Bank. *Plants* **2021**, *10*, 2371. https://doi.org/10.3390/plants10112371

Academic Editors: Johannes M. M. Engels and Andreas W. Ebert

Received: 29 September 2021
Accepted: 23 October 2021
Published: 3 November 2021

Publisher's Note: MDPI stays neutral with regard to jurisdictional claims in published maps and institutional affiliations.

1. Introduction

We are living in a time of unprecedented change. In the past 50 years, the human population has doubled, while the global economy has quadrupled, and global trade has increased ten-fold [1]. This has resulted in a concomitant increase in demand for resources and energy, the consumption of which is having profound impacts on the natural world. Human economic activities are driving climate change and biodiversity loss, both of which are mutually reinforcing, further exacerbating the problem [2,3]. In 2020, for the first time, environmental issues dominated the top five global risks by likelihood, as identified by the World Economic Forum, with biodiversity loss, climate action failure and extreme weather also all in the top five risks with greatest impact [4].

Human actions have significantly altered 75% of the land surface of our planet and led to an increase in the rate of biodiversity loss unparalleled in human history [1]. In relation to plants, this has resulted in a ~50% reduction in plant biomass relative to pre-human levels [5], and to ~40% of plant species being threatened with extinction [6]. Conservation interventions are urgently required to help reverse these trends and ensure that adequate plant diversity is available for both current and future generations. All life depends on plants. We rely on them, and the fungi that support them, for our foods, materials, medicines, and for regulating, supporting and cultural ecosystem services. The Dasgupta Review [7] demonstrates how our economies, livelihoods and wellbeing are dependent on nature, and embedded within nature, and calls for transformative change to put the sustainable use of nature at the heart of our economies.

In the face of a sixth mass extinction [8], the conservation of plant diversity has never been more important for the future of the planet and people. Protecting plants in their natural environment (i.e., in situ conservation) is the primary approach for species conservation, but as many of the threats to their continued existence (e.g., climate and land use change, invasive species, pollution) do not respect the boundaries of protected areas a safety back up is required. Conserving plants away from these threats (i.e., ex situ conservation), in botanic gardens (as living collections of plants) and seed banks (as propagules), is crucial if we are to stop and even reverse the extinction trend and preserve plant diversity for current use and for future generations. Storing representative germplasm from plants (any part of the plant that can be used to regenerate a new individual) collected from the wild ex situ facilitates the utilisation of this material for a variety of purposes without impacting the wild population. So not only does it protect the plant material from the threats faced in situ, it provides a readily accessible resource for use.

The conservation of living plant material in these human-made repositories has been practiced for centuries, but in recent decades has greatly expanded and become increasingly refined. To date, more than 105,600 wild plant species are conserved as seed and/or living collections in botanic gardens [9]. Their conservation takes the form not only of the living plant, seed or tissue, but also encompasses the knowledge and data associated with the individual samples and collections, and the ability to communicate this to a wide variety of audiences to improve conservation outcomes.

Botanic gardens have a long history of conserving plant diversity through their living collections. These are generally raised from material collected in the wild (Figure 1), and often include threatened species. However, the primary method of ex situ conservation for plant genetic resources today is seed banking (Figure 1). While the function and name of the seed bank (genebank, germplasm bank, biobank) may vary, the general concept remains the same—using controlled environments (drying and cooling) to preserve a broad diversity of plant germplasm for immediate and future use. This methodology underpins global food security, forestry, horticulture and ecological restoration as well as conservation. While the scale of the seed bank, the size and diversity of the seed accessions held, the post-harvest handling and storage procedures employed, and the extent of environmental controls used may vary, the purpose remains the same—to preserve high-quality, viable germplasm until required for use [10].

However, a wide variety of plant diversity, referred to as 'exceptional species' [11,12], cannot be stored for the long-term under conventional seed banking methods (commonly drying to 15% relative humidity at 15 °C, followed by storage at −20 °C in airtight containers). Examples include gametophytes of non-seed-bearing plants (e.g., bryophytes and pteridophytes), germplasm with desiccation and/or freezing intolerance (e.g., *Quercus* seeds), and species with desiccation and freezing tolerant germplasm (both spores and seeds) that is short-lived under conventional seed bank conditions. Such 'exceptional species' can form a large part of the global flora; for example, it is estimated that at least 8% of all flora, and 18% of tropical and sub-tropical flora, are likely to have desiccation sensitive (recalcitrant) seeds [13]. Other options for the conservation of such species include tissue culture, cryopreservation, nursery-based plant collections and seed orchards (Figure 1).

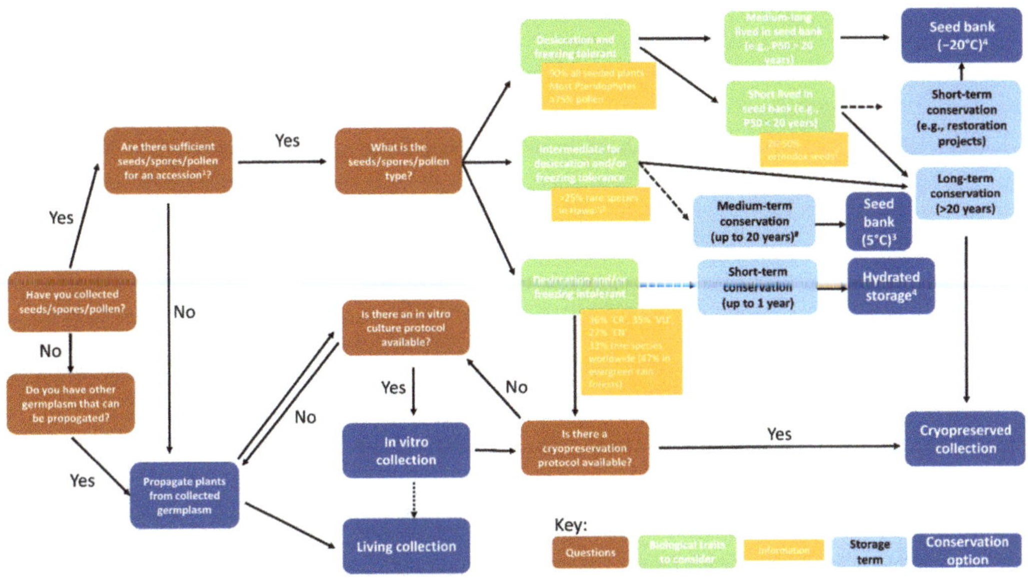

Figure 1. Decision tree showing conservation options depending on the type of germplasm conserved and some of the biological traits associated with different germplasm. Short- and medium-term storage options are indicated by dashed arrows. In vitro collections can be a source of germplasm for living collections (dotted arrow). CR: critically endangered species, EN: endangered species, VU: vulnerable species as per IUCN Red List. # refers only to desiccation tolerant but freezing sensitive seeds. [1] [14–17], [2] [12,18], [3] [19], [4] [20].

Botanic Gardens Conservation International (BGCI) was established in 1987 with a particular focus on identifying threatened species in botanic garden collections and supporting ex situ conservation efforts for these species. There are presently over 3000 botanic gardens located in 185 countries included in BGCI's GardenSearch database [21], many of which are actively engaged in conservation action for threatened plant species. Recently, efforts have been focused on establishing consortia of gardens with common conservation interests to establish decentralised metacollections. The increasing use of metacollections across botanic gardens and arboreta has led to a step change in the importance of these collections for plant conservation [22,23].

The Millennium Seed Bank (MSB) of the Royal Botanic Gardens, Kew (Kew), was established in 2000 building on more than three decades of research at Kew into seed biology and conservation. The MSB's mission is safeguarding wild plant diversity and enabling its sustainable utilisation through global partnership. To this end, the Millennium Seed Bank Partnership (MSBP), consisting of seed banks associated with botanic gardens, agricultural, forestry or research institutes, and government organisations around the world, has been collaboratively conserving native floras. To date, more than 97 countries and territories and over 250 organisations have been involved. When referring to seed material collected of wild origin stored within seed banks for long-term conservation, we use the term 'accession'. Each accession represents material collected from an individual population (unless maternal lines are stored individually) at a given time, and multiple accessions of an individual species may be held.

With the increased focus on plant conservation provided by the Global Strategy for Plant Conservation (GSPC, 2011–2020), notably Target 8 [24], and the continued call for plant conservation to support the UN Sustainable Development Goals [25], the number of botanic gardens working in conservation, and the number of seed banks conserving wild species has greatly increased. However, there is a disparity between the location of biodiverse, threatened habitats and the sites of these conservation centres [26]. The policy framework provided by the Convention of Biodiversity and its Nagoya Protocol [27] is crucial for ensuring that plant germplasm collected from the wild and stored, often in remote locations, is managed and utilised to ensure equitable benefit-sharing.

In this review, we focus on the challenges and prospects facing the long-term conservation of plant diversity in botanic gardens' living collections and seed banks. Having evaluated the unique role that both forms of conservation play, we highlight the synergies afforded when they are co-located in the same botanic garden and the importance of networks for enabling plant conservation. We provide examples of how wild plant collections in both botanic garden living collections and seed banks support plant conservation in situ in the context of restoration, agriculture, forestry and livelihoods, with a focus on the challenges and prospects of these programmes in relation to long-term plant conservation ex situ. We finish by signposting three key areas for future development: funding security for ex situ collections; exploiting technological advances, specifically to enable the conservation of 'exceptional species'; and maintaining and developing networks to ensure the best outcomes for the world's flora.

2. Ex Situ Conservation

2.1. Botanic Gardens

The practice of cultivating plants in specialised gardens has been around for thousands of years. However, the first 'true' botanic gardens with an underlying scientific foundation were the physic gardens of Italy created in the 16th century [28]. These gardens were purely for the study of the healing properties of medicinal plants and by the end of the century had spread to universities and apothecaries throughout Europe [28].

Botanic gardens experienced a change in usage during the 17th and 18th centuries. This was the age of exploration and the beginnings of international plant trade. Gardens such as Kew and the Real Jardín Botánico de Madrid were set up to try and cultivate 'new' species that were being brought back from expeditions to the tropics. Not only did these gardens promote and encourage botanical exploration, but they also helped found new gardens in the tropical regions to help cultivate these newly 'discovered' plant species, as well as providing a garden environment that ex-patriots recognised from home. During the 19th and 20th century, municipal and civic gardens were created around the world. However, many of these gardens were pleasure gardens with very few of them having any scientific programmes. Meanwhile, especially amongst university gardens and building on the work started by Linnaeus in the 18th century, collecting, naming and describing plant diversity became a major focus of activity and their diverse living and herbarium collections supported the teaching and practice of taxonomy.

In the last 50 years, there has been an exceptional growth in the establishment of botanic gardens, with 60% of the gardens in existence today being established since the mid-20th century [29]. This growth has been particularly notable in China, where new gardens are being developed across the country with the aim of using local plant diversity to support economic development. Similarly, Indonesia plans to have a botanic garden in every province to function as a botanic resource centre to support conservation and sustainable development. Around the world, botanic gardens have seen a revival as scientific institutions due to the emergence of the conservation movement and the recognition of the importance, not only of their diverse collections, but also of the taxonomic and horticultural knowledge they possess, vital for the conservation, management and restoration of plant diversity. A botanic garden today can be defined as "an institution holding documented collections of living plants for the purposes of scientific research, conservation, display and education" [30].

BGCI underpins its work with three databases. The first, GardenSearch [21], is a directory of the world's botanic gardens compiling information on their location and facilities, while PlantSearch [9] documents the plant collections held by these gardens. In order to identify threatened species in collections, BGCI has also developed a third database—ThreatSearch [31], which lists global, regional and national red list assessments for plants from a range of sources. It is the most comprehensive database of conservation assessments of plants.

At the present time, GardenSearch includes details on 3715 botanical institutions, of which 3038 are botanic gardens (the remainder being a combination of seed banks (many of which are located within or connected to botanic gardens and maintain seed accessions of wild plant diversity (e.g., Kew's MSB, which has its own listing), zoos and private collections). Over 350 of the botanic gardens listed in GardenSearch have established seed banks for the conservation of wild plant species [32]. PlantSearch presently contains 1,559,119 records, representing 634,235 taxa (species, varieties, sub-species and cultivars) held by 1186 institutions. An analysis of these records (excluding cultivars) carried out by Mounce et al. in 2017 [26], revealed that the global network of botanic gardens houses over 100,000 species of the 350,699 accepted plant species on The Plant List in 2013 [33]. These species are from nearly 10,000 genera, representing 30% of global plant species diversity, and over 41% of known threatened species. The botanic garden network is also rich in expertise, with more than 60,000 botanical specialists employed, covering plant sciences, taxonomy, education and specialist horticulture [34].

Most of the plant diversity held by botanic gardens exists in their living collections and not all of this diversity is being maintained for conservation purposes. Research, display, education and public outreach are also important functions of botanic gardens that are supported by the living collections. It is estimated that, globally, some 750 million people visit botanic gardens annually [34]. Therefore, as well as being important for conservation, the living collections of botanic gardens provide an important resource for educating and informing the public about the importance of conservation issues. Species that have memorable economic, ecological, or cultural stories are particularly useful in this regard, and help botanic gardens combat plant blindness [35].

As well as focusing on wild species, there are numerous examples where conservation by botanic gardens plays an important complementary role to that of the agricultural and forestry sectors [34]. For example, important collections of non-timber trees, such as fruit and nut species, exist in botanic gardens, and these might otherwise 'fall through the cracks' of plant conservation [36]. Examples include the breadfruit collection at the National Tropical Botanic Garden in Hawaii and the tropical fruit collection at Fairchild Tropical Botanic Garden in Florida.

Metacollections—Enhancing the Conservation Value of Living Collections

Some endangered plants are considered 'exceptional species' and do not store well in seed banks. Conservation efforts therefore rely heavily on living plant collections. Such collections need to include a high level of genetic diversity to ensure that the species can adapt and survive in the face of future changing environmental pressures. Curating a genetically diverse seed bank collection is relatively easy and affordable, but much more challenging for living plant collections. Balancing space and cost limitations while maximizing the number of individuals an institution can sustainably curate in its living plant collection is critical. The number of individual plants that are needed to capture a population's genetic diversity varies considerably from species to species. This is illustrated by recent work on palms and cycads at the Montgomery Botanical Center [37]. For instance, for the rare Key Thatch Palm, *Leucothrinax morrisii*, curating 15 individuals can conserve as much as 83% of a population's genetic diversity [38]. However, for the rare Sinkhole Cycad, *Zamia decumbens*, curating 30 individuals only conserves about 35% of the species' genetic diversity. It may take more than 205 individuals to capture 77% of known genetic diversity for this cycad [39].

Given that any number of botanic gardens may hold collections of the same species, the conservation value of such collections can be considerably enhanced by combining these holdings into a network of collections—or a metacollection [40]. Metacollections are envisioned as common resources held by separate institutions but stewarded collaboratively for research and conservation purposes [22,23]. Networking multiple collections into a single metacollection increases potential coverage within a taxonomic group, allows broader access to greater diversity, dilutes risk of loss, and can reduce maintenance costs by reducing duplication and redundancy across collections. Like any collection, a metacollection can be of any scope or taxonomic level; however, the approach is particularly useful in the case of taxa that are only represented by a few individuals per garden—such as many tree taxa. The metacollection strategy was adopted by zoos over 40 years ago and is now embodied in the successful species management programmes for zoo animals. Gardens have been less formalised in networking plant genetic resources, but adapting zoo methods for plant collections is now yielding some important advances [41]. Established examples of botanical metacollections include the American Public Gardens Association's Multisite Collections [42], BGCI's Global Conservation Consortia [43] and the Center for Plant Conservation (CPC) National Collection [44].

As well as promoting the conservation value of living collections through the metacollection approach, in recent years BGCI has been active in supporting and encouraging the establishment of seed banks in botanic gardens as a complement to living collections to ensure the long-term conservation of native plant diversity. Such initiatives include the Global Seed Conservation Challenge (described below), training and capacity building activities and the provision of small grants for seed banking. The on-going development of an accession-level module as part of the PlantSearch database is also aimed at supporting a more cost-effective and coordinated approach to the conservation of threatened species across the botanic garden community.

2.2. Seed Banks

In addition to the >350 wild plant seed banks established by botanic gardens [21], between 710 and 1750 genebanks exist, and maintain between 5.4 and 7.4 million accessions of plant genetic resources for food and agriculture, as seed, in vitro, DNA and cryopreserved collections [45,46]. Out of all these banks, two prominent examples are the Kew's MSB (UK), and the Global Seed Vault (Svalbard, Norway); both exceptional in the extent of their international nature. The former is the world's largest repository of wild plant genetic diversity, storing almost 997,000 accessions representing over 40,000 species, while the latter holds the world's most diverse collection of food crop seeds with over 1.1 million seed accessions of around 4000 species. While many types of seed banks exist, for the purposes of this review we are focusing on those banking seeds collected from wild plant populations for long-term conservation, using the MSB and its global partnership, the MSBP, together with other seed banks in the BGCI network as our examples.

2.2.1. Botanic Garden Seed Banks

Of the botanic gardens listing seed banks as part of their facilities in GardenSearch, the majority are located in the global north, particularly in Europe and North America. Interest in using seed banks to conserve wild plants is relatively recent. Spain was one of the first countries to focus on collection of wild flora, creating its seed bank in 1966, at the Department of Plant Biology of the Polytechnical University in Madrid. Several Spanish botanic gardens then created seed banks focused on conserving wild flora in the regions in which they were based. Together, 10 of these seed banks formed the Spanish Network of genebanks for wild plants (Red Española de Bancos de Germoplasma de Plantas Silvestres) in 2002 [47].

The number of seed banks in botanic gardens has doubled in the last 20 years and now the botanic garden community has some of the largest and most sophisticated seed banks in the world. As well as the MSB, such seed banks include the Germplasm Bank of Wild Species, located at the Kunming Institute of Botany in China which presently conserves over 11,000 species, and the Australian PlantBank that holds more than 12,000 seed accessions from almost 5300 plant species—437 of which are threatened species (G. Errington pers. comm.) (Figure 2). Of the records in PlantSearch, seed bank accessions represent 67,270 taxa in at least 100,218 accessions held in over 80 institutions. An analysis of seed bank data from PlantSearch and ThreatSearch [31] indicates that less than 10% of taxa in seed bank collections are globally threatened.

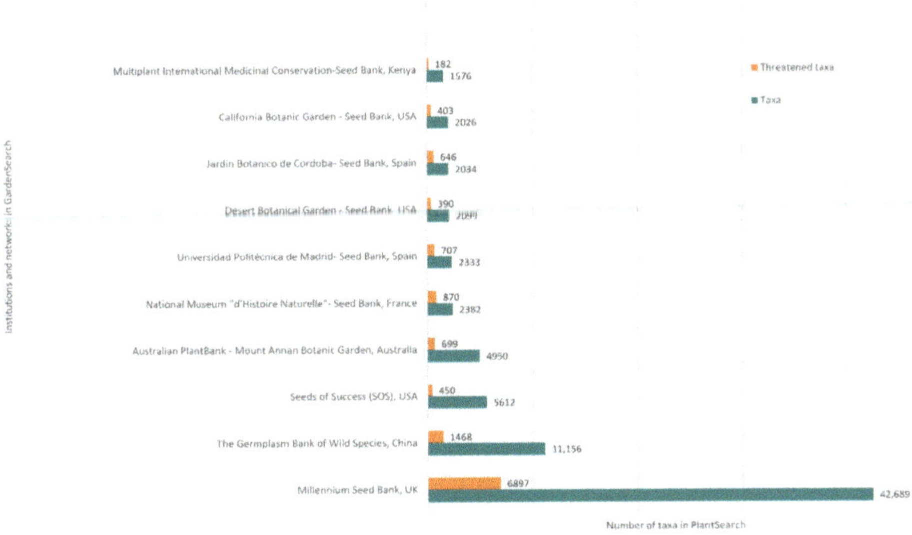

Figure 2. Seed collection holdings of the largest ten wild plant seed banks by the number of taxa stored based on PlantSearch data. Threatened taxa includes species categorised as threatened (including global, regional and national assessments) from data in ThreatSearch.

The skill sets developed in botanic gardens for the management of living collections and seed bank collections complement one another, and, together with the taxonomic skills of herbarium staff, provide the knowledge required for successful long-term ex situ conservation of plants. Excellent taxonomic and identification skills ensure that the correct plant material is sought and stored. Understanding of phenology and seed biology increases the success of fieldwork and storage. Propagation, both laboratory and nursery based, is vital for turning seeds back into plants that can be used. In addition, the education and outreach roles of botanic gardens enable the stories relating to the plants conserved, and the need for conservation itself, to be communicated to the public. It is not surprising, therefore, that the number of wild plant seed banks located in botanic gardens has seen a dramatic increase in recent years.

Various manuals and guidelines have been created by the botanic garden seed bank community to ensure the quality of seed bank collections. An important aspect of the expansion of conservation efforts in Australia, is the progressive development of a suite of guidelines that capture the latest science and practice for seed banking and associated conservation activities—the Australian Germplasm Guidelines [48]. The publication is a collaboration of over 70 seed banks, botanic gardens and other organisations throughout Australia providing expertise in planning, collecting and management. The knowledge and skills shared in these guidelines are largely underpinned by contributions from Australian botanic gardens staff and their international collaborators. Of particular note in the latest edition is the expansion in information and techniques available for conservation of 'exceptional species' [12]. A major step forward is that requirements to conserve these species ex situ are being progressively understood and workflows have been developed to assist the fast-tracking of conservation efforts. This includes the latest techniques for selection of appropriate germplasm to conserve ex situ, an important step for the success of ex situ conservation of 'exceptional species' which require significant research effort [49]. At the regional level, the European Native Seed Conservation Network (ENSCONET), a consortium of organisations interested in native species conservation, created seed collecting and processing manuals for practitioners, available in nine European languages [50,51].

2.2.2. The Millennium Seed Bank

The advantages of seed banking for the long-term preservation of wild plant diversity include the ability to store a wide range of genetic diversity in the form of seed accessions from populations, in a relatively small space (~40,000 species stored in 300 m^2 at the MSB) and at a relatively low cost compared to other ex situ options (e.g., in vitro, field genebanks, etc.). Holding well documented collections also enables the current and future use of this germplasm resource.

Kew has been working in seed banking since the 1970s, developing a seed banking programme for UK native species in the 1990s, and increasing its support of international plant conservation through seed banking since the 1990s. All the partners who have worked with the MSB since 2000, forming the MSBP, enter access and benefit sharing agreements with Kew to ensure equitable sharing of benefits and prior informed consent (PIC) relating to use of materials stored at MSB. They receive on-going support from the MSB in relation to development of seed banking facilities in their countries, training in seed conservation techniques, and in targeting and banking their most important parts of their floras. Many also undertake joint research programmes with Kew. The MSB works with a range of seed banking institutes, from those in botanic gardens (27%), to forestry and agricultural genebanks (20%), universities and other research organisations (32%), governmental (17%) and other organisations (4%). For example, the Royal Botanic Gardens and Domain Trust joined the MSBP in 2003 and, in 2013, opened the Australian Plantbank, a purpose-built conservation centre that incorporates the seed bank, alongside cryostorage, tissue culture and a well-developed nursery, with associated staff including a well-developed conservation focused science programme.

The facilities at the MSB were built to last for 500 years, and seed accessions stored there should be viable for decades to centuries (depending on the species—some are short lived even under ideal storage conditions—and the quality of the accession). The long-term nature of this storage (>10 years) sets conservation seed banking apart from other seed storage initiatives with short (<5 years e.g., restoration and regeneration) and medium (5–10 years, e.g., plant breeding) storage needs, and defines the conditions of storage (Figure 1) [20].

The storage procedures and monitoring of long-term conservation collections of wild plant germplasm applied at the MSB, and botanic gardens seed banks generally, vary in some respects from those employed in other types of genebanks which follow the Food and Agriculture Organization of the United Nations (FAO) and the International Seed Testing Association (ISTA) guidelines [20,52], but are based on practice developed in this sector. Changes result from the diversity of wild seed types that are handled and stored (non-uniform seed material), and the long-term nature of the conservation collections (10–100 s of years). Storage conditions consist of drying to 15% equilibrium relative humidity and storing in glass containers or trilaminate foil bags at −20 °C. Protocols are adapted for short-lived species and for micro-seeds (<0.2 mm in length [53]). The MSBP Seed Conservation Standards [54] were developed to ensure high quality collections are made and stored across the partnership. The quality of an accession is assessed through the number of seeds in the accession together with their viability and longevity. The genetic representativeness of individual seed accessions and storing multiple accessions from different populations of the same species is also important. Seeds are stored in the country of origin with up to half of the collection sent to the MSB for duplicate storage, spreading risk by splitting the collections between two geographically separate facilities. Where no adequate facilities exist in country for the long-term storage of seed accessions, the whole collection can be sent to the MSB, and half returned on establishment of appropriate facilities in country. In some cases, for example, where legislation prevents the movement of national germplasm, seeds remain in the country of origin and are duplicated nationally, while data are shared with the MSB. Updates relating to the viability of accessions and requests for use of material are communicated to partners that store the duplicates.

A conservation seed accession consists of three items: the seeds; an herbarium voucher from the population collected; the data associated with the accession (field, processing, germination, etc). The herbarium voucher enables identification of the seed material to be verified, and taxonomic changes to be tracked, while also providing a valuable ecological and historical record.

Collection Size

The quality of seed collected from wild plants varies significantly. A cut test to check seed quality prior to collecting is recommended. This enables an estimate of the number of potentially viable seeds in the population to be made, which will determine if sufficient high-quality seeds are available to enable a collection of 10,000 seeds to be made without impacting on the regeneration success of the wild population. The amount of seed collected should not exceed 10–20% of the seed available on the day of collection [55–57]. Collecting 10,000 seeds is recommended by the MSB to enable duplication between seed banks, routine seed bank activities and seed supply over the intended lifespan of the collection [14], it also allows for genetic attrition over time due to germination failure, disease and active use [58]. It is, however, recognised that collections of this size will not be possible for many rare and threatened species, and for those with restricted distributions, and a median of 1000 seeds per accession is more common and often still adequate for conservation purposes (Figure 1) [15,58]. For very rare plants, a seed collection of any size provides options for the future.

Accessions undergo X-ray analysis after cleaning and prior to banking to determine the proportion of full, potentially viable seeds within the sample. Many seed banks do not have access to an X-ray machine but can still undertake a cut-test to determine the quality of the seed. The number of seeds in an accession is identified either by direct count (for small accessions <500 seeds) or by weight (five samples of 50 seeds weighed, remainder of accession weighed, calculation performed). The original seed number is then adjusted using the X-ray (or cut-test) results to provide an estimate of the number of potentially viable seeds in the accession. Prior to sampling an accession, it is thoroughly mixed to ensure that seeds representative of the whole accession are utilised.

Genetic Representativeness

Conservation seed collections should contain genetic diversity that is representative of the population from which they were made, and collections should be made from enough populations to represent the genetic diversity of the species across its range. Multi-year collections from the same population may also be needed to capture the genetic diversity of annual or short-lived species. Typically, those making conservation collections aim to collect from at least 50 plants across the population, and to spread the number of seeds collected per plant evenly between individuals [59]. For large shrubs and trees, it is also recommended to collect from across the canopy [60]. The genetic diversity of wild plant species is generally unknown, sampling strategies based on predictive models for the capture of alleles across a population with increasing sampling effort, taking into consideration factors such as the population structure and inherent species' traits (e.g., pollination syndrome) are helping improve previous rules of thumb [58,61]. The needs of the end user (e.g., restoration, plant health research, plant breeding, etc.) must also be considered when developing a sampling strategy, for some uses maternal lines should be banked separately (e.g., UK National Tree Seed Project, [62]). Both seed and living plant collections offer opportunities to add back diversity to wild populations that have had their extant populations reduced [63,64].

Germination

Germination tests are the most effective method for checking the viability of a seed accession, they also provide a protocol (set of test conditions) for turning the seed back into a plant. At the MSB, all accessions of sufficient size (see below) undergo an initial germination test post banking. This represents a significant task, with over 40,000 species banked from 190 countries each requiring individual germination conditions to be assessed. Re-tests occur every 10 years; however, if longevity is known to be short or seen to be declining, the re-test interval is reduced to 5 years.

For accessions with an adjusted (see below) seed quantity of >2500 seeds, 50 full seeds are used for the test and up to five initial tests with varying conditions may be performed. This requires an over-sow calculation should the adjusted seed number be lower than the original seed number. For example, if 45 out of 50 seeds were full in the X-ray test (90% potentially viable), a total of 56 seeds (50/0.9 to the nearest whole number) would be required per test to account for potentially empty seeds. The number of seeds per test and number of tests decreases below this accession size, until for accessions with <259 seeds no test is performed. In no instance should more than 10% of an accession be used, this ensures sufficient seeds remain in the collection for monitoring and use over the predicted lifespan of the collection in storage. The MSB's use of 50 seeds in periodic germination (viability) tests mirrors the FAO recommendation for distribution. It is less than the 200 usually recommended by ISTA, to both avoid excessive depletion of wild species collections, which are generally smaller than those in crop genebanks, and also to make best use of limited staff time.

Anticipating Loss of Viability and Decline of Longevity

The longevity of an accession in storage is dependent on a variety of factors, including species characteristics and genetics, the point of seed development at the time of harvest, post-harvest handling and storage conditions employed. Monitoring of viability in storage is vital, as declines in viability represent a loss of genetic diversity from an accession, and management decisions around recollecting or regeneration will be required.

The MSB uses analysis of the results of periodic viability testing of accessions to fit survival curves, extrapolation of which can be used to predict when the viability of each accession would reach regeneration level (75% of initial viability, variation from FAO standards, allowing for wild species issues) [14]. Approaching that level would trigger a management decision; and in practice regeneration would be a very rare event, with a request to re-collect in the source country, if possible, being the preferred option. For a significant proportion of the MSB's seed accessions made since 2000, there are not yet sufficient viability test data points to permit fitting of survival curves, from which to estimate likely regeneration/recollection intervals. The relatively sparse, usable real-time survival data have been supplemented by an on-going programme of comparative accelerated ageing experiments across diverse species (see [65,66]). While the possibility that the causes of death of individual seeds are different under accelerated ageing conditions from those in long-term storage in a seed bank, the data provide a relative ranking of species' likely storage longevity. These are used to inform decisions on monitoring period (reduced from 10 years to 5 where a species' seeds are suspected to be relatively short-lived). In addition, accessions from taxa known or predicted to be short-lived under traditional seed bank conditions have a subsample backed-up under cryogenic storage (liquid nitrogen, see below, Figure 1, and [67]), though more research is needed to confirm the expected improved survival at ultra-low temperatures.

Genetic Integrity

The MSB at present does not engage in any direct routine assessment of the genetic integrity of accessions or its decline. With the particular issues attached to regeneration of very diverse species from many countries, we do not engage in regeneration, except in certain circumstances, mostly for UK native species. Instead, so far as is practicable, we rely on the correlation between loss of viability and decrease in genetic diversity; and efforts are focused on making collections in the field of the highest viability, transferring them as quickly as possible to optimum storage conditions, followed by regular monitoring of viability in real time. Recollection from the wild is the preferred method of replacing an accession, from the original population if it still exists, or from an alternative population if that is possible. Analysis of the survival of MSB accessions over periods varying from 10 to >40 years are giving preliminary indications that 80–85% of accessions are not yet showing any detectable loss in viability and thus are assumed to remain at or close to original levels of genetic diversity and representative of the populations from which they were sampled. For accessions that do show a loss in viability, or no initial viability, recollection is recommended, or required. Since 1984, the average number of species recollection requests per annum is 26.

Seed Supply

Seeds are available to bona fide individuals representing recognised organisations for non-commercial purposes (e.g., research on seed biology, morphology or germination), as defined by a material transfer agreement through the MSB Seed List [68]. The only collections available for distribution are those with: a verified name; adjusted seed quantity >1050; a germination test within the last 10 years; permission from the donor or project; not covered by CITES; and not held at the MSB under quarantine conditions in compliance with current UK Plant Health Regulations. Up to 60 seeds per accession are supplied, the number depending on the adjusted seed quantity of the accession, following the FAO minimum recommendation (30–50 seeds [20]). However, the standard does not cite specific population genetic research in support of the recommendation.

Staffing

Wild species seed accessions at the MSB currently amount to ~97,000, of ~40,000 diverse species; with an addition of around 3000 accessions per annum. Collection curation consists of accessioning/databasing, drying, cleaning, banking, viability testing (initial and periodic), sample distribution, etc.; and needs a team of a Seed Collections Manager plus 16 staff, with a variety of skills and experience: and, unless growth of the collection is to lead to unmanageable backlogs of processing and periodic viability testing, the team needs the addition of a seed collections assistant every 2–3 years.

3. Challenges and Prospects

While the importance of living collections and seed banks has been clearly demonstrated, together with the wider work carried out by botanic gardens in this sphere (e.g., outreach and education), there remain challenges to the long-term conservation of plant diversity using these ex situ conservation options.

3.1. Geographic and Taxonomic Biases in Collections

While the number of botanic gardens working in conservation, and the number of seed banks conserving wild species, has greatly increased in recent years, there remains a disparity between the location of biodiverse, threatened habitats and the sites of these conservation centres (Figure 3) [26]. Based on the records in BGCI's PlantSearch, the distribution of botanic gardens appears disproportionately temperate, with 93% of plant species conserved in the northern hemisphere—mainly in Europe and North America [26]. BGCI's GardenSearch similarly shows that two thirds of the 551 gardens that provide records for plant conservation programmes are based in high-income economy countries as defined by the World Bank [34]. While this may represent a slightly distorted view, as there are more countries in the northern hemisphere, and gardens located in the northern hemisphere have greatest data sharing capacity, it highlights the need to support and establish botanic gardens in the tropics. This need is emphasised by the finding that an estimated 76% of species not currently held in living collections are tropical in origin [26], and that the majority of the world's plant conservation collections are located outside the most biodiverse regions, with only a third occurring within the 36 global biodiversity hotspots [34].

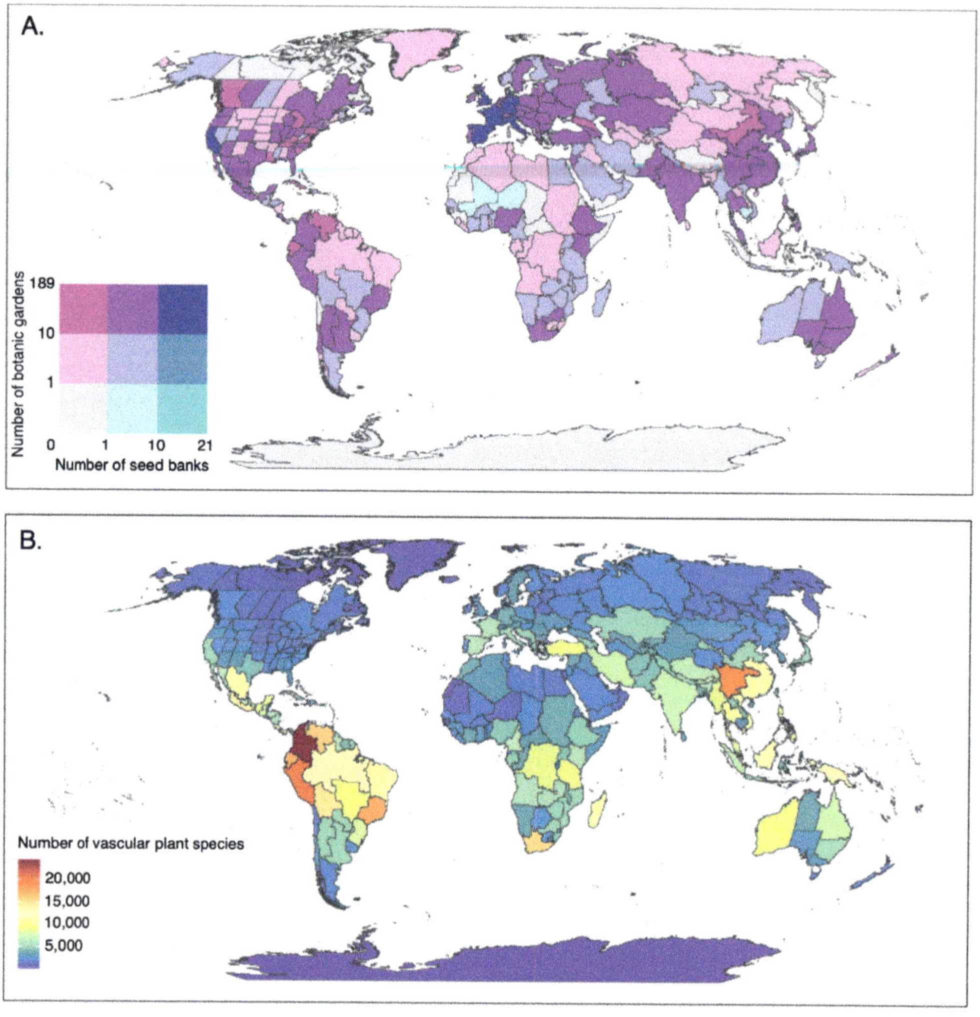

Figure 3. Geographic distribution of (**A**) ex situ conservation capacity and (**B**) centres of plant species richness. Ex situ conservation capacity is represented by the numbers of seed banks and botanic gardens found in each botanical region of the world according to level 3 of the World Geographical Scheme for Recording Plant Distributions (WGSRPD) [69]. Numbers of seed banks per region were extracted from the Millennium Seed Bank Partnership (MSBP) and Botanic Gardens Conservation International (BGCI), whereas numbers of botanic gardens per region were extracted from only the latter. For direct comparison, plant species richness was also extracted for each botanical region (level 3 WGSRPD) from the World Checklist of Vascular Plants [70].

Seed collecting programmes of the MSB have typically targeted areas of high biodiversity and conservation need, originally in temperate and dry regions and more recently in tropical regions. To date, 71% of MSB partner countries occur within the 36 biodiversity hotspots—but only 27% of countries where the partner is a botanic garden lie within a biodiversity hotspot compared to 78% of those where partners are not botanic gardens, and 76% of countries with multiple partners, both within and outside botanic gardens. However, the disparity between collections-based institutes in the northern and southern hemisphere is still evident within the partnership, with only 18% of MSB partner countries occurring in the southern hemisphere (22% for non-botanic gardens partners, 14% for countries with both botanic garden and other institutes as partners, 0% for countries with only botanic garden partners). When looking at the economy of the MSB partner countries, as defined by the World Bank, a similar pattern emerges, with botanic garden only partner countries occurring in only upper-middle and high income brackets, those with partners within and outside the botanic garden sector also include some lower income countries, and only those countries with partners exclusively outside of botanic gardens also include low income countries.

In addition to the geographical gaps in collections, botanic gardens and collections-based institutes, there are phylogenetic biases in the species that are conserved in living collections and seed banks. While 30% of plant diversity has been found to be conserved in botanic garden collections, representing 59% of all plant genera, there are certain groups of plants that are less well represented. For vascular plants 93% of families and 50% of genera are conserved, but for non-vascular plants only 5% of genera are held [26]. Furthermore, certain plant families and genera are favoured, not least because of their ability to be readily propagated and grown under prevailing conditions, but also because of their horticultural appeal. Most species cultivated in botanic gardens, particularly larger longer-lived species, are represented by an average of two to three individuals, and plants are often clonally reproduced and shared, meaning that the genetic diversity of the wild species or even population is not represented. In addition, as individuals and not populations are conserved, and the number of individuals that can be housed is limited, genetic bottlenecks can arise [71]. To be effective as conservation collections, 10–100 s of individuals of known wild origin, collected from across the ecological and geographic range of the species, are required [34], highlighting the importance of metacollections.

Furthermore, duplication between institutions is desirable to mitigate for loss due to attrition, pests and disease outbreaks, natural disasters and theft [58].This is one important reason why many botanic gardens have developed conservation seed banks and are linking accessions to form metacollections of a given species or genus. For the latter, it is preferable if the material held represents germplasm from separate collecting efforts of different populations to maximise global coverage and relies on excellent record keeping tracking the wild origins of shared collections.

Of increasing concern are gaps in ex situ collections of plants with known uses, many of which are also threatened. For example, a study of the neglected and useful plants of Mexico found gaps in the conservation of wild edible plants. Although 2598 wild plant species (more than 10% of the Mexican flora) are conserved ex situ as seeds in Mexico, with duplicates stored at the MSB, only 62 seed accessions of 21 species from the most important groups of neglected and underutilized plant species mentioned in the review have been safeguarded in the seed banks [72]. In addition, the lack of coverage of these species in ex situ collections means that the associated research needed for species propagation (e.g., germination requirements, dormancy issues, etc.) at a scale to support agriculture and restoration activities is also missing.

Although 75% of all embryophyte plant families are recorded as being conserved in botanic gardens [26], it is well known that not all of plant diversity can be stored as living or seed collections (Figure 1, [26,73]). A further issue with seed banks is that the material is in stasis, offering a snapshot of the genetic diversity of a population at the time of collecting. While this is helpful for resurrection studies (e.g., [74]) it means that material is no longer evolving. Living collections have a similar problem; while their plants can adapt and evolve, it is to conditions often outside their native range.

Small accession size (low seed number) can be an issue for seed banks too when working with very threatened or rare species, which tend to enable only small seed accessions to be made from relatively few individuals. One option to overcome this is to employ multiple-year collecting to increase accession size, or to grow material on from germination tests in order to harvest additional seed (regeneration) or produce tissues (e.g., shoot tips, somatic embryos, etc.) that can be preserved in tissue culture or cryopreserved accessions (Figure 1). Ex situ collections of 'exceptional species' often inadequately capture the diversity needed to represent diversity likely to be lost in the wild, even more so that for non-'exceptional species'. The increasing discriminating power and lowering costs of molecular techniques means that they can be routinely added to conservation workflows to increase the diversity of species within seed and genebanks, and for end use in translocation (movement of plants between different sites) and restoration [75]. This is particularly helpful in situations where germplasm is held in living collections, where each plant held represents significant cost to the managing organisation, minimising duplication, while holding high diversity. For example, at the Australian Plantbank, *Rhodomyrtus psidioides* and other critically threatened rainforest species, are being held as living collections, in nursery-held pots, in gardens, field genebanks and in tissue culture [76]. These can be seen as intermediate steps to the lower-cost long-term ex situ conservation goals of securing appropriately diverse germplasm of these species in seed banks, cryostorage and as important elements of metacollections [22,23].

3.2. Cryopreserved Plant Collections

For conventional seed banking, an increasing challenge is the inability to store material from 'exceptional species' for the long-term [11,12]. While living collections can overcome this difficulty (Figure 1), issues around the amount of genetic diversity conserved and the possibility of genetic erosion, hybridisation or problems associated with pathogens and pests remain [77–79]. Cryopreservation is increasingly recommended as a solution (Figure 1), enabling the long-term preservation of a diversity of plant materials (e.g., cells, spores, pollen, shoot tips, seed embryos, whole seeds) and taxa (from algae to bryophytes, ferns, cycads and orchids) and the storage of relatively comprehensive genetic diversity of the population sampled on a relatively small space [12]. Plant cryopreservation is based in the use of ultra-low temperatures (typically those provided by liquid nitrogen, <-130 °C) and often chemical protectants to preserve cells and tissues without the formation of lethal intracellular ice. There are different approaches that can be used which are mainly based on the vitrification (i.e., 'ice-free' solidification) of the cell cytoplasm while protecting its physicochemical properties and the structural integrity of tissues [12,77].

However, unlike conventional seed banking, plant cryopreservation does not have a universal formula that can be used to preserve a wide range of plant taxa and tissues, and cryopreservation protocols must often be developed and adapted at the species or variety level [20] (Figure 1). In addition, the level of success of many protocols is lower than that applied to germination standards set for conventional seed bank collections. For example, a cryopreservation protocol is considered successful when regeneration is accomplished in at least 20–40% of the preserved samples [80,81], for long-term conservation seed banking storage is considered successful if levels of germination are above 75% initially and do not drop below 85% of initial test levels on subsequent testing [20]. These discrepancies, in both the lack of a universal method for plant cryopreservation and in the levels of initial percentage of plant regeneration between cryopreserved plant collections and conventional seed banking, have often been a barrier for the establishment of cryopreserved collections of wild species within seed banks. However, this has not been the case for crop plants [12,82] and we think similar standards of success should be applied to wild and threatened 'exceptional species' [83]. In this regard, cryopreserved plant accessions of wild species can be viewed like the elements of metacollections described in a previous section of this paper. For example, in the case of the rare Sinkhole Cycad, *Zamia decumbens*, indicated above, over 60% of known genetic diversity for this cycad could likely be preserved by combining the curation, in vitro culture and/or cryopreservation of 30 maternal lines (35% of known genetic diversity for this cycad) and the preservation of pollen from 100–200 individuals. This metacollection of cryopreserved, in vitro cultured and whole plant germplasm would conserve a high genetic diversity of this rare cycad in a way that, if seed banking is not easy, living metacollections alone would find challenging.

Cryopreserved plant accessions are relatively common for the long-term conservation of certain crop species that are propagated vegetatively or have desiccation sensitive seeds [77]. However, cryopreserved collections of wild plant species are not common but are increasingly being considered and created within conventional seed banks and botanic gardens. Examples include the CryoBiobank of the Cincinnati Zoo and Botanical Garden, created in the late 1980s and holding the oldest, largest and most diverse collection of cryopreserved plant cells and tissues [84], the MSB, which cryopreserves short-lived seeds and spores, the USDA/ARS National Laboratory for Genetic Resources Preservation that hold seeds and spores of the CPC network, and diverse Australian genebanks [85] (see Box 1). However, we need to expand the number and scale of these types of collections, particularly in tropical areas, where the proportion of species that cannot be banked using traditional seed banking is larger [12,84,86]. Cryopreserved crop collections offer a great source of knowledge not only in the techniques that can be applied but also in solutions for challenges that arise from the management of globally cryopreserved collections [87]. Some historical cryopreserved wild plant collections have provided data to evaluate not only the costs and challenges of preserving wild plant species in vitro and stored in liquid nitrogen, but also on the stability and longevity of the preserved samples [12,84,88].

Nevertheless, there are some aspects to consider if we want to increase the number and scale of cryopreserved plant collections of wild species. Firstly, the 'fear' of using liquid nitrogen technologies needs to be reduced, as this is often the first barrier to the development of basic cryopreserved plant collections across conservation institutions. For example, cryopreserved collections for short-lived desiccation tolerant seeds, fern spores, and desiccation tolerant pollen can be created with minimal investment and training, as dry seed, spore and/or pollen collections can be stored in liquid nitrogen with relatively low technical requirements [67]. Secondly, conventional seed banks and botanic gardens need to invest in infrastructure and specialised training to increase the taxonomic and geographic variation of cryopreserved collections of wild plant species (the scale of the investment will depend on the scale of the cryobank desired). Thirdly, cryobiotechnological research needs to be increased if more species and tissues are to be successfully cryopreserved [12,83–85,89–91]. Fourthly, we must strengthen networking between wild species genebanks and crop genebanks to facilitate the preservation of wild species collections at the regional level in their crop genebank cryobank facilities, particularly in tropical areas where funding, a stable supply of liquid nitrogen and training for the development of wild species cryopreserved collections may be challenging [84].

Box 1. Meeting the challenges of seed banking and conservation of 'exceptional species' in Australia.

Australia has a large, diverse flora and many endemic species. The negative effects of climate change—such as decades of frequent drought and bushfires—coupled with habitat loss, rapid spread of invasive weeds and disease, have resulted in an increased level of threat to native flora [92]. The unprecedented scale and intensity of the 'Black Summer' fires of 2019/20, burning more than 10 M ha of land in south-eastern Australia across 11 Australian bioregions and 17 major native vegetation groups [93], has undoubtedly brought many species closer to extinction. Botanic gardens are being activated to help monitor post-fire recovery and to collect germplasm for ex situ conservation.

There are presently ten conservation seed banks in Australia, mainly in botanic gardens, that hold 68% of threatened flora represented by at least one accession [92]. Efforts have largely focused on dryland species, due to the expectation of desiccation sensitivity of seeds of rainforest. Several studies have explored the seed storage potential of Australian rainforest flora [49,94] and these results were combined with other data sets to develop a key for determining the seed storage potential of untested rainforest species [49]. This key will help us to understand the complex biology of 'exceptional species' and recognise species that need conservation efforts beyond the traditional seed bank.

A number of Australian crop wild relatives fall into the exceptional category, including *Citrus*, *Syzygium* [95] and *Macadamia* species. *Macadamia* is Australia's only indigenous crop grown at large scale, and current efforts are focusing on the twin challenges of securing the remaining germplasm of the four *Macadamia* species in the wild (which are all threatened), while ensuring the availability of material for inclusion in breeding programmes. Alternative conservation methods such as tissue culture and field genebanking are available for such species [96]. The further development of cryostorage techniques for 'exceptional species' conservation is an increasing focus of collaborative research due to the great potential of this technique as a long-term conservation option [85,97,98].

The recent environmental disasters in Australia have greatly increased the imperative for ex situ conservation of all species, but particularly for those of fire-impacted east coast rainforests, including the relictual Gondwanan rainforests. These forests rarely burn, and the species are often poorly adapted to fire. The fires may have left these forests 'susceptible to regeneration failure and landscape-scale decline' [93]. Many species are already under pressure from habitat loss due to clearing, the effects of invasive weeds and diseases. An added, looming existential threat to rainforest Myrtaceae species, is the recent incursion of Myrtle Rust fungus (*Austropuccinia psidii*). This disease has spread rapidly to more than 358 species in Australia since its unfortunate introduction in 2010 and has been found more recently in New Zealand. Myrtle Rust is decimating a number of once common rainforest species, such as the Scrub Turpentine (*Rhodamnia rubescens*), and Native Guava (*Rhodomyrtus psidioides*), with imminent annihilation expected for a number of species [99]. If the plants are not killed outright, the disease often affects the flowers and fruit and therefore collecting of seeds from the wild is usually not an option. The disease can be controlled in cultivation; however, progress of the disease may outpace germplasm capture for ex situ conservation such is the scale of the problem [76,100].

3.3. The Importance of Networks

Developing and maintaining active networks of institutions focused on a shared goal can be a significant tool for nations to meet their commitments to international plant conservation and restoration targets [101]. Well established networks aimed at tackling plant conservation issues exist at different levels: locally, nationally, regionally, and globally. Networks operating at the national level often have a highly targeted approach to plant conservation. An example is the Mexican native species Biodiversity Nurseries Network (REVIVE), implemented jointly with the Seed Reserve (RESEM) in the Veracruz State. Their mission is to increase the diversity of native species growing in nurseries for restoration purposes. Initially most nurseries only worked with native *Pinus* species, but REVIVE now have more than 200 native species from different Mexican ecosystems available as seedlings or seeds for distribution [102]. National networks can also feed into regional ones. ENSCONET is a regional network of institutes with an interest in native species conservation through seed banking that includes the Italian seed banking network (RIBES) [103] and part of the Spanish seed bank network (Red Española de Bancos de Germoplasma de Plantas Silvestres) [47] and Mediterranean network (GENMEDA) [104] as part of their current membership. Collectively, the membership has contributed to ensuring 62.7% of European threatened species are in long-term conservation [105].

Global networks have the greatest diversity of associated organisations and have the potential of having the greatest global impact. Under the umbrella of the MSBP, botanic garden seed banks, agricultural genebanks and forestry seed centres are brought together with a shared purpose, enabling the development of more collaborative and complementary conservation programmes. Similarly, BGCI's Global Seed Conservation Challenge (GSCC), a network of over 200 botanic gardens involved in seed banking, supports seed banking through provision of training, resources and funding while challenging botanic gardens to conserve more threatened species in seed banks. Since 2015 the number of taxa conserved as seed added to PlantSearch has doubled. Through the GSCC fieldwork fund, 120 species have been collected, including 45 threatened with extinction. These global networks and consortia can make a significant contribution towards global conservation and restoration targets but do require significant levels of resourcing to be effective at a global scale.

The existence of a maintained network can increase reactivity and dynamism amidst political and environmental instability. Climate change is leading to an increase in extreme weather events, including drought and related wildfires, hurricanes, and flooding. In 2012, the US endured several environmental disasters including the burning of two million acres of sagebrush in four western States, and widespread damage to native plant communities responsible for stabilizing soils and filtering water on the East Coast by Hurricane Sandy. These events led to the creation of the National Seed Strategy for Rehabilitation and Restoration to provide a more coordinated approach and response to these large-scale events [106]. Ecological restoration is often constrained by a lack of the large quantities of seed required. The Strategy focuses on the establishment of a nationwide network of native seed collectors, farmers and growers, nurseries and seed storage facilities to supply adequate quantities of appropriate seed, together with a network of restoration ecologists. The vision of the strategy is 'The right seed in the right place at the right time'. A progress report in 2021 showed 380 partners are involved in the resulting network and have together invested $167 M in the programme [106].

More recently, the 2019 'Black Summer' bushfire season impacted 67–83% of globally significant forests and woodlands of Australia, and decimated >50% of known populations or ranges for over 800 vascular plant species native to Australia (see also Box 1) [93]. Australian botanic gardens have long had a strong focus on the conservation of native and, particularly, threatened species. This was fostered by the formation of the Australian Network for Plant Conservation in the early 1990s, followed by various partnerships with the MSB from the late 1990s, which in turn enabled the establishment of the Australian Seed Bank Partnership (ASBP) [107]. These networks and partnerships supported the development of seed banking capacity in each State and Territory, and firmly placed botanic gardens as providers of plants and services for conservation of Australian native species. The ASBP enabled a rapid response post-fire in the form of habitat assessments and collecting seeds from remaining individuals found in refugia, as well as long-term monitoring of habitats of affected species over the coming years. ASBP is working not only with plant conservation consortia across Australia but remains part of the global network of the MSBP. This continued partnership provides further security for the Australian seed collections, with over 9000 Australian species duplicated to the MSB over the past 20 years and forming an integral part of their collections, with the potential for repatriation when required. This ability of networks to enable reactivity and dynamism in an ever-changing world will be increasingly important for plant conservation and responses to the biodiversity crises in the coming decades.

In relation to cryopreservation of 'exceptional species' plant cryobanks are well established in many national and international crop centres, such as Bioversity International, the International Center for Tropical Agriculture (CIAT), the International Potato Center (CIP), the International Institute of Tropical Agriculture (IITA), the World Agroforestry Centre (ICRAF), the Global Network on Cacao Genetic Resources Conservation and Use

(CacaoNet), the USDA/ARS Agricultural Genetic Resources Preservation Research, the Leibniz Institute of Plant Genetics and Crop Plant Research (IPK), where vegetatively propagated or desiccation sensitive seed crops are preserved in liquid nitrogen [77,87]. Some of these cryobanks are well interconnected through global networks and initiatives such as the CGIAR [77,87]. Similarly, for the forestry sector, nursery stands of economically important tree species are general practice across many agroforestry centres, allowing for the use, conservation, improvement, and distribution of germplasm [108]. Coordinated networks for the conservation of 'exceptional' wild species are less prominent; however, there are notable examples. At a global scale, BGCI in collaboration with Valerie Pence at the Cincinnati Zoo and Botanical Gardens is working to link knowledge, resources and projects globally to conserve threatened 'exceptional species' in a systematic way [109]. Within this collaboration they are promoting the creation of an Exceptional Plant Conservation Network (EPCN), that aims to share resources, including a list of 'exceptional species', information and links to information on the species and alternative conservation technologies, ways to link to other researchers working with 'exceptional species', as well as other supplemental information [110]. The EPCN is planned as an open resource that will benefit from the input of all researchers who are working on the conservation of threatened 'exceptional' plants. Moreover, BGCI's Global Conservation Consortia coordinates networks of institutions and experts towards the conservation of priority threatened plant groups, such as maples, oaks, magnolias, and dipterocarps. These networks help to develop and implement comprehensive strategies for in situ and ex situ conservation efforts (including cryopreservation) and dissemination of species recovery knowledge. At the national scale, networks such as the CPC in the USA, promote research and the use of tissue culture and cryopreservation techniques as alternative storage methods to conventional seed banking for the conservation of threatened 'exceptional species' [57].

3.4. Data Management and Access

Data sharing is key for enhancing global plant conservation collaborations, enabling exchange of knowledge across networks and the establishment of metacollections. For both living collections and seed bank accessions, the associated data, gathered through the collecting, processing and/or growing activities, represents a wealth of potential for collection management, research, restoration and conservation action [111,112]. Typically, data shared by botanic gardens is related to taxonomy, distribution, conservation status, plant availability in gardens, uses and a brief description of the plant. Data captured by seed banks is generally focused on seed traits such as seed storability, viability and/or germination, morphology, collection location, and seed availability.

A challenge for data sharing, particularly for wild species seed banks, is the incompatibility of different data management systems across botanic gardens and seed banks. Data sharing for crop species is more advanced in relation to unifying accession and trait data under an open access system (e.g., Genesys [113]). The MSBP network goes some way to tackling this issue through the development of the MSBP Data Warehouse [114]. Developed in 2015 as an online resource for partners, this platform aggregates data from partner's in-country collections with the duplicates held at the MSB. Currently, the MSBP Data Warehouse holds data of 230,166 seed accessions, 2074 X-ray images and 220,295 germination data, and partners can access the majority of this seed related data from across the global network. It also provides links to other RBG Kew online resources such as Plants of the World Online (POWO) and the Seed Information Database (SID). Processing data for duplicate accessions is repatriated through the system, highlighting accession quality and enabling reflection on existing processes to aid future collecting.

Alongside the use of networks as a hub for data and knowledge exchange, the development of shared network systems can be a way of accelerating institutional data onto globally accessible platforms. In 2005 ENSCONET developed ENSCOBASE [115], an open access database where members of the network can upload data on native European wild species within their seed bank. Publicly accessible platforms such as ENSCOBASE and

PlantSearch can be used to measure progress towards international targets such as the GSPC by tracking collections of threatened species [105]. They also connect accessions directly to conservationists, educators, horticulturalists, researchers, policy makers and many others who are working to conserve and understand plant diversity, and data can be used to prioritise conservation of threatened species not held in ex situ collections. This prioritisation can be implemented by individuals, organisations or networks at the local, national, regional or global level depending on the species' distribution. The number of seed banks uploading data to PlantSearch has doubled in the last five years; however, only around 80 of the 355 botanic gardens with seed banks upload their seed accession data to PlantSearch in addition to information on their living collections. Institution level resourcing to facilitate upload of data is an issue common to many conservation-oriented data sharing platforms—however, the more data these platforms contain, the better-informed conservation actions will be.

The MSB Data Warehouse, ENSCOBASE and PlantSearch are among the very few examples of initiatives to share data across and beyond the botanic garden network. A wide range of different systems are used by botanic gardens to maintain accession-level data on the plants in their collections, but unlike the plant genetic resource sector, data sharing and the adoption of common data standards has not yet been a priority. While some botanic gardens (e.g., RBG Edinburgh) provide on-line access to collection catalogues, this tends to be the exception rather than the rule. Legitimate concerns about potential theft keep curators of living collections from sharing their full catalogues. However, an initiative from the German Botanic Garden Network (gardens4science) aims to develop a data portal giving accesses to the local databases of the living collections of more than 10 gardens, starting with bromeliads and cacti [116].

Sharing data relating to species that are highly threatened or have the potential for exploitation also presents a challenge for seed banks. Accession data typically have locality and use documented within databases, and whilst this provides opportunity for use in research and conservation, it can lead to misuse and may exacerbate illegal trafficking of material for commercial gains [117]. The MSBP Data Warehouse ensures compliance with agreed use of associated data from donating institutions in two ways. The first is by sharing an online view containing ~85% of the offline database, with any requests for data sensitivity accounted for. The second is the ability to restrict locality information through 'fuzzy mapping', where coordinate data can be restricted, or resolution decreased on the interactive map. ENSCOBASE uses an alternative solution, and only provides location data at the country and biogeographic region level.

Language can be a limiting factor in the sharing of information and knowledge. Although English is generally accepted as the language of science, the greatest plant diversity lies in regions where English is not the native language, and these are also the areas experiencing the greatest threats to plant species survival [118,119]. The use of multi-language tools and ensuring that information generated for a specific location's flora is available in the local language will help relieve this issue. There are good examples of developing such material: the ENSCONET seed collecting manual is available in nine European languages; and the ColPlanta website [120] for Columbian plant and fungi information is accessible in English and Spanish [121]. Similarly, global aggregators such as the Global Biodiversity Information Facility (GBIF) [122] have the capacity to switch to a variety of commonly used languages.

Finally, we must also acknowledge the global disparity to internet access and the impact that this has on successful data exchange, particularly for those within global networks. Areas where there are high biodiversity and associated threats to plants, such as much of Africa and Papua New Guinea sit relatively low in the rankings of internet users per population size [123]. For these key areas, the development and maintenance of national and/or regional networks remain important, as they serve as a way of ensuring continued support and knowledge exchange across multiple players in plant conservation within high biodiverse ecoregions. The inclusion of expansive knowledge and technology transfer programmes that enhances capacity of local collections alongside global repositories can also mitigate the issue of access, for example, the various standard and bespoke training programmes run by BGCI and MSB.

3.5. The Importance and Challenges of Material Sharing

We have already articulated the importance of duplicating seed collections at two geographically distinct seed banks and joining living collections in metacollection strategies. This insurance policy is increasingly important as environmental change becomes less predictable and impacts greater areas. For many critical plant species that are in living collections, a singular locality can dramatically increase the risk of a collection being compromised either through total loss or gradually, through genetic erosion. Furthermore, use of seed material for research can also contribute towards species conservation [112] and to finding solutions to global challenges, such as food security [124].

There are various challenges in relation to sharing of material globally. One relates to colonial histories and biopiracy, where historical imbalances have led to a lack of trust in material sharing, particularly at the international scale, potentially to the detriment of global plant conservation. Limiting access to physical material and associated data can hamper research progress, potentially impacting the long-term conservation of endemic floras. As plant species' distributions straddle national boundaries, these limitations can also impact the conservation of species with regional or global distributions, making a comprehensive assessment of their risk of extinction and overall management difficult.

In 1993 we witnessed the first step change in recording consent through the ratification of the Convention on Biological Diversity (CBD) [125]. The ratified members of the CBD recognise the sovereign right of countries over their genetic material and any sharing of materials across national borders must take place in the context of PIC and under an agreement on the terms of transfer of the material, including any subsequent access and benefit arising from its use. The Nagoya Protocol provides a legal framework for the access and benefit sharing of biological diversity [27]. Benefit sharing negotiated between parties can include, but is not limited to, access to accessions and associated data, augmentation of national collections, transfer of technology, training, joint research activities and in the case of commercialisation, monetary exchange.

The implementation of the Nagoya Protocol (currently ratified by 132 countries) requires botanic gardens in both provider and user countries to understand the modalities involved in collecting and storing plants and seeds outside their national boundaries. In Europe and beyond, many gardens have joined the International Plant Exchange Network (IPEN) and/or endorsed the Kew Principles on Access and Benefit Sharing. These initiatives include Codes of Conduct which guide how material in collections can be used and shared in line with the Nagoya Protocol. The principles of Access and Benefit Sharing also apply within countries and have shone a light on 'ownership' of natural resources, especially those which occur on land owned or managed by indigenous communities. In Mexico for example, the national botanic garden network has developed a Code of Conduct and best practices for collecting seeds with indigenous communities, and a similar code is under development in Australia [126].

The establishment of a policy group at RBG Kew with legal expertise to develop, maintain and record agreements has ensured its compliance with the CBD and CITES. The majority of MSB partnerships are developed through bilateral agreements, where the terms of material use (i.e., Material Transfer Agreement) are clearly outlined. Continued communication between the two acting parties ensures benefits are transferred (e.g., germination protocols) and PIC is sought for third party use. Exceptions exist, for example, the Adapting Agriculture to Climate Change project [124] was governed through the Multilateral System (MLS) of access and benefit-sharing under the International Treaty on Plant Genetic Resources for Food and Agriculture (ITPGRFA). Negotiations for this type of relationship require further trust-building and ought to incorporate the use of the ABS clearing house as part of the project to ensure compliance.

Bilateral (or multilateral) agreements are typically made at the national level, and so tend not to capture the diversity of stakeholders required to deliver highly impactful conservation outputs. The inclusion of indigenous and/or local communities in relation to access (i.e., of land, knowledge and material) and benefit sharing (e.g., monetary/non-monetary and national/international) is an increasingly important aspect to consider [127], particularly in countries where well-established networks exist. Examples include Australia, New Zealand, Canada, and the United States. An added challenge is the practical management of data relating to indigenous knowledge, a subject currently actively debated within the museums and humanities sectors [128].

Living material (e.g., live plants and seeds), have the potential to carry a variety of bacteria, viruses and fungi as part of their microbiome [129,130]. The relationship can be beneficial (i.e., required for germination or growth), commensal or pathogenic. Therefore, sharing of material will inevitably carry some level of risk, notably with regards to the introduction and spread of novel pathogens. Since its emergence from commercial nurseries and plantations in South and Central America, the pathogen responsible for Myrtle rust (*Austropuccinia psidii*) has expanded its international range rapidly, spreading into the Australasian continent within the past decade and affecting native populations of Myrtaceae [131–133]. The World Trade Organisation Agreement on the Application of Sanitary and Phytosanitary Measures [134] provides the international regulatory system for plant health and aims to prevent the introduction and spread of harmful diseases or pests. Within this agreement are a set of reference standards on which nations can build a legal framework that relates to plant health and trade. For example, the EU Plant Health Regime provides a framework governing the movement of plants from non-EU countries. Import of some species that are deemed high-risk are completely banned, whilst others require accompanying phytosanitary certificates [135]. Aside from selected high-risk material (e.g., *Malus, Aegilops*, etc.) requiring 'plant passports', plants or plant products can move freely within EU member states. Therefore, any changes within the political landscape, for example, the UK's exit from the EU, will inevitably have an impact on the ease of material access and sharing. Additionally, processes of disease and pest screening prior to issuing phytosanitary certificates can be flawed or limited by the capacity and effectiveness of a country's screening measures [136]. The type of material being shared will also need to be considered as the level of regulation can vary greatly (e.g., pollen versus seeds).

While seed is generally considered to be relatively 'clean' with respect to pests and diseases, and its movement is generally less restricted by phytosanitary restrictions than other types of plant material, seed collected from wild populations may present a 'trojan horse' for transmission of pathogens. Recent research on seeds of African eggplant wild relatives found the seed contained potato spindle tuber viroid, both a first finding for this host and a first record in several countries. Like seeds, pollen international exchange is considered safe, as harmful pests and diseases are rarely transferred through pollen [77]. However, pollen may be a vector for viruses and other pathogens [137] and specific phytosanitary restrictions have been placed in some plant groups and countries (e.g., *Citrus* in the USA [138]). That germplasm collections may be storing diseased material poses a downstream phytosanitary risk to the collections, species conservation, or future breeding research.

Considering the challenges of sharing material globally, including constrains imposed by the policy environment, plant conservation efforts must include building capacity in the country of origin, notably in biodiverse regions. One example is provided by the Meise Botanic Garden in Belgium, whose scientists have been studying the wild diversity of *Coffea* in Central and West Africa for almost 25 years. In order to ensure the conservation of important diversity in the Democratic Republic of Congo, Meise Botanic Garden has trained a network of local botanists, ensured the rehabilitation of historic field and herbarium collections and supported the collection, conservation and evaluation of newly described plant species in ex situ collections [139]. The biological constraints of the species and/or seed itself can mean that transporting germplasm may be less favourable than storing it locally. Maintaining or developing networks within countries, including with the agricultural sector that often has cryostorage facilities, will enhance the conservation potential for seeds that are short-lived and/or desiccation sensitive.

4. Plant Conservation in the Wider Context

4.1. Integrated Conservation

It has been said that there is no technological reason that any plant need go extinct [26]; however, it is also recognised that seed accessions have a limited shelf life, and that the size and usefulness of accessions will decline over time as they are used for activities such as research, reintroductions, or periodic viability monitoring. Neither seed banks nor botanic gardens, even with support from cryopreservation, are a surrogate for functioning ecosystems, and there will always be a need to preserve plant diversity in situ. Integrated conservation, where botanic gardens and seed banks help ensure plant diversity conservation through outreach and engaging with in situ work, is vital. The following case studies highlight how the knowledge, skills and collections in botanic gardens and seed banks are helping address some of the global challenges facing humanity by halting biodiversity loss, increasing food security and supporting livelihoods.

4.2. Agrobiodiversity—Contribution to Global Food Security

Concern about the impact of climate change on global food security and biodiversity, coupled with continued population growth, has driven initiatives to diversify our food sources and build crop resilience [140].

The Adapting Agriculture to Climate Change, or Crop Wild Relatives, Project (2011–2021), managed by the Global Crop Diversity Trust and RBG Kew, and funded by the Government of Norway is such a project. Crop wild relatives (CWR), the distant wild 'cousins' of domesticated crops, contain genetic traits that can potentially be harnessed through crop breeding techniques to create climate and pest-resilient crop varieties. The unavailability of these vital CWR plant genetic resources for food and agriculture (PGRFA) to crop breeders due to their insufficient conservation in ex situ collections [141], resulted in this concerted international effort to collect the wild relatives of 29 of the most important food crops from across the globe. Occupying a unique position of experience in the ex situ conservation of wild species seeds, the MSB played a pivotal role as the global duplicate repository, via the use of Standard Material Transfer Agreements (SMTA), for almost 4000 seed accessions of over 240 taxa collected by institutional partners in 21 countries. In addition, by making use of Kew's vast herbarium, seed collecting guides for each CWR project country were produced. Training in seed collecting and conservation was also provided at Kew and in-country to 174 individuals associated with the project.

In addition to safeguarding CWR seed collections in the countries of origin and at the MSB, the final component of the project was to distribute small samples (typically 100 seeds) of each accession to specialist CGIAR-affiliated genebanks around the world for incorporation into research programmes, investigating traits within CWRs that confer resilience to abiotic and biotic stresses. As such, the CWR project has by necessity brought the MSB's expertise of wild species seed conservation and the CGIAR genebanks' focus on improved crop varieties together with the common goals of improving global food security through the sustainable use of plant genetic resources in the face of climate change, improving human nutrition and health and reducing poverty.

Many important crop species cannot, however, be stored in traditional seed banks. The National Tropical Botanical Garden in Hawaii is host to the Breadfruit Institute which includes the largest assemblage of breadfruit cultivars in existence. The Institute is using the knowledge acquired by more than 30 years of conserving and studying breadfruit to plant trees in tropical countries for food and reforestation, provide economic opportunity, and to educate the public about the benefits of growing—and eating—this underutilised crop. More than 300 breadfruit trees are conserved in the living collection and, working with its partners, the Institute is sending micropropagated breadfruit trees to tropical countries worldwide. Since the launch of the initiative in 2009, more than 100,000 breadfruit trees have been sent to 44 countries. The breadfruit collection includes accessions from 34 islands across the Pacific including some cultivars that are now rare or vanishing in their homelands. The geographical scope of the collection also includes accessions from Indonesia, the Philippines, Seychelles and Honduras.

4.3. Supporting Food Security and Livelihoods

Throughout human history, wild edible plants (WEP) have been important to rural communities [142–145], but wild populations of WEP are under threat globally [146]. In the South Caucasus, Kew collaborated with long-standing partners of the MSBP, the National Botanical Garden of Georgia, the Institute of Botany Ilia State University of Georgia, and Nature Heritage NGO of the Republic of Armenia to deliver a 3-year Enhancing Rural Caucasian Livelihoods through Fruit and Nut Conservation project funded by the Darwin Initiative. Due to the strong intrinsic link between plants and people in the Caucasus, this project ensured the conservation of wild species without jeopardising the livelihoods and food culture of the target and surrounding communities.

Through a multidisciplinary project team, consisting of botanic gardens (and their associated seed banks), local NGOs, social scientists, research universities and community leaders, the project covered five main themes: engagement; in situ and ex situ conservation; research; and training.

Two communities were engaged with the project, one in the south of Armenia and one in the north of Georgia. Community-led steering groups oversaw various project activities, maintained a participatory approach throughout the project's lifetime and provided a legacy after the project end. An awareness campaign on the importance of plant conservation and sustainable harvesting led by in-country project members supported by a steering group reached 60% of communities that utilise the harvesting landscape. Through community member project interviews the key WEP harvested were identified together with their use and the level of perceived importance of these products. Thirty native fruit and nut plants were brought into cultivation in three community-run orchards, alleviating overharvesting of wild populations whilst simultaneously providing income for local families. Training on how to propagate and care for the plants was led by local horticulturalists based at the relevant botanic gardens.

Georgian and Armenian conservationists received IUCN Global Red List assessment training, enhancing the capacity for in situ conservation activities in-country. The training and close collaboration with national herbaria enabled the assessment of 20 fruit and nut species that are endemic to the region. A mixture of species used by local communities and their threatened wild relatives were targeted for seed collection and conservation: 193 seed accessions from 119 different fruit and nut species were conserved and duplicated to the MSB.

The project aimed to narrow the knowledge gap of WEP and build research capacity within the target countries through engaging two local MSc students. Collaboratively developed research topics revolved around key edible species that are frequently used locally, but rarely investigated (*Rosa* in Armenia and *Prunus* in Georgia). By utilising the expertise both in-country and at RBG Kew, the students were trained in various aspects of plant science, from seed conservation to genomics.

Through the project, the team began to have a better understanding of the way the local landscape is used by local communities, and subsequently the needs for its continued conservation. An important aspect that was outside the scope of the project was studying the impact of commercially driven collectors originating from outside the local area, who can have a significant impact on overall biodiversity at the landscape level if their activities are not conducted sustainably [147]. To explore this further, the partnership will need to develop sustainability training at a commercial scale (e.g., the FairWild model) and expertise in supply chain valuation. Building cryopreservation capacity, both in-country and at the duplication site, for wild edible species that have seeds that are short-lived (e.g.,

Corylus spp.), intermediate (e.g., *Fagus* spp.) and/or recalcitrant (e.g., *Quercus* spp.) would also greatly benefit long-term conservation of these species [67].

The Kew-led Useful Plants Project was implemented on the ground over two phases (2007–2010; 2011–2015) under the MSBP with the aim to enhance the capacity of local communities to successfully conserve and sustainably use important indigenous plants in Botswana, Kenya, Mali, South Africa, and Mexico [148]. The project brought together Kew scientists and a wide range of collaborators from different disciplines, involving botanists, horticulturalists, agronomists, and foresters, who worked closely with rural communities, local authorities, and schools utilising participatory techniques. A scientific approach was applied throughout the main components of the project: selecting useful plant species, ex situ conservation; propagation and planting; and supporting people's livelihoods. Most of the species were reported to have a medicinal use (878) or to be used as food for humans (615) and materials (427). In Africa, prioritised useful plants included: the iconic multipurpose *Adansonia digitata* (Baobab) and the highly valued *Senegalia senegal* (Gum arabica); food plants *Schinziophyton rautanenii* (Mongongo tree) and *Tylosema esculentum* (Morama bean) used in Botswana; the multipurpose timber tree species *Melia volkensii* in Kenya; and several species of columnar cacti and agaves in Mexico used for their edible flowers and fruits [149]. This project has achieved an impact on plant diversity conservation through seed banking of priority useful plants, with 1271 seed lots banked in-country and 952 duplicated in Kew's MSB. Research on seed germination helped support plant propagation activities, and facilities were set up or improved at the local level for the conservation and propagation of the prioritised species. Training and knowledge in seed conservation and plant propagation were also provided to rural communities while facilities were set up or improved locally. Two hundred and sixty-seven species (76,389 seedlings) were planted in community gardens for direct use, 59 of which (seeds/seedlings/part of plants or their plant products) were promoted for income generation in rural communities through workshops and marketing events. Finally, the project supported local education through the establishment of school gardens and increased knowledge on the conservation and sustainable use of native species and by tutoring undergraduate and postgraduate students. This project highlighted the importance of applying an 'holistic approach' to address the dual objective of biodiversity conservation and contribution to improved livelihoods in the local communities [148].

4.4. Restoration and Reforestation

The important role that botanic gardens can play in supporting ecological restoration through the provision of scientific knowledge and plant material has been well documented [150]. In GardenSearch, 329 institutions are listed as having a plant reintroduction programme, and 226 are involved in restoration ecology research. The Ecological Restoration Alliance of Botanic Gardens includes 49 members. Here we consider five examples which show how the integrated plant science and conservation expertise of botanic gardens and associated seed banks are helping deliver practical, on-the-ground solutions to plant and habitat loss.

The **UK Native Seed Hub** (UKNSH), established in 2011, was conceived as part of RBG Kew's response to an independent review commissioned by the UK Government, of England's wildlife sites and the connections between them, entitled 'Making Space for Nature' [151]. The report proposed a long-term strategy, to 2050 and beyond, for conservation in England based on rebuilding nature, fragmented and degraded due to human activities, at a landscape scale, by creating coherent and resilient ecological networks that would link and expand existing habitat patches with buffer zones, wildlife corridors and areas of active restoration and habitat creation.

A significant constraint to effective conservation and habitat restoration is the limited availability of known-origin, high-quality, genetically diverse seeds and plants [112]. The MSB, through its comprehensive UK seed collections coupled with its scientific and technical expertise, was well placed to address this shortfall. Since 2011, the UKNSH has embraced this role by providing plant materials (seeds and plug plants), applied research and technical assistance to over 63 conservation projects, and has partnered with more than 41 organisations in the UK from both the public and the private sector.

One of the chief aims of the UKNSH from the outset was to support the goals outlined in the Lawton Report (2010) to rebuild nature on a landscape scale, enhancing and strengthening ecological networks [151]. Eight years on from the publication of that report, while evidence of success in terms of landscape scale conservation nationally is unclear and at best mixed [152,153], the UK Government published a broader 25 Year Environment Plan [154], which included a commitment to develop a Nature Recovery Network, launched in 2020, to expand, improve and connect a national network of wildlife-rich places across towns, cities and the countryside. Going forwards the UKNSH, in line with Kew's Manifesto for Change 2021–2030 [155], will continue to support these aims and seek further opportunities to work with the government, landowners and managers, businesses, local communities and conservation organisations.

China is home to 10 percent of the world's total plant diversity, some 30,000 higher plant species [156]. However, there are many threats to China's native flora including rapid socio-economic development, climate change, habitat conversion and unsustainable use of native species. The country is a key region for BGCI's mission to mobilise botanic gardens and engage partners in securing plant diversity for the well-being of people and the planet. BGCI launched its China Programme Office (hosted by the South China Botanical Garden Chinese Academy of Sciences, Guangzhou) in 2008 following BGCI's collaboration in the development of China's Strategy for Plant Conservation (CSPC) [157]. The CSPC highlighted China's botanical wealth and the urgent need for conservation action. BGCI activities focus on collaborative action with botanic gardens and other botanical institutions in China to develop and implement practical conservation initiatives for the country's threatened native flora to address the targets of the CSPC. In the past 13 years, through the Global Trees Campaign (a partnership between BGCI and Fauna and Flora International) BGCI has funded the protection and restoration of more than 70 threatened trees in China.

A summary of the work BGCI has been implementing with Chinese partners over the past 10 years was recently published with more than 20 case-studies on the integrated conservation of rare and threatened woody plants [158]. One such example is *Bretschneidera sinensis* an endangered tree native to China. Through comprehensive field surveys, wild populations were identified, and seed collected for storage in the South China Botanical Garden seed bank. Additional seed was collected for reintroduction activities including the reintroduction of 1000 seedlings in Dongguan Forest Park and Shimen National Forest Park of Guangzhou city and 300 seedlings into Nankunshan Mt. Nature Reserve.

The **Global Tree Seed Bank Project** (GTSBP) is one of Kew's major science-based plant conservation programmes. Funded by the Garfield Weston Foundation it aims to secure (in safe, long-term storage) seeds of at least 3500 tree species from across the world. The Latin America programme of the GTSBP was initiated in 2015 and delivered through two projects focused on useful trees in **Mexico** and the **Dominican Republic**.

Mexico is in the top five countries in terms of floral diversity (species richness), and more than 50% of the plant species are endemic [159]. There are ca. 3000 native tree species that have been characterised for their uses, distribution, conservation status and endemism [160]. However, more than 30% of the tree species are threatened due to deforestation and/or climate change. The Science-Based Conservation of Tree Species in Mexico Project addressed this by implementing an integrated conservation programme for endemic, protected and useful tree species, important for the livelihoods of rural communities. Work was undertaken in collaboration with the Facultad de Estudios Superiores Iztacala de Universidad Nacional Autónoma de México (FESI-UNAM). Around 400 species have been conserved ex situ, in country and as duplicates at the MSB, and technical and scientific information has been produced to support the national reforestation programme led by the National Forestry Commission (CONAFOR), including seed germination and propagation requirements. A key challenge is promoting the use of more native species to diversify local nurseries which rely on the economic benefits obtained by selling and distributing seedlings. A lack of information about the uses and propagation of native trees other than timber species (mostly *Pinus*) means demand is very low. To overcome this challenge, in collaboration with partners (University and NGOs) species profile sheets (in the local language) were produced and distributed, describing their germination and propagation requirements, uses, phenology, distribution and conservation status to promote their utilisation in reforestation programmes.

In the Caribbean region, the vegetation cover in the island of Hispaniola, is under threat from the expansion of agriculture and development for tourism, and the unsustainable logging for charcoal production. Kew has worked with the Jardín Botánico Nacional (JBN) "Dr. Ma. Moscoso" of Santo Domingo through the MSBP since 2007 under an Access and Benefit-Sharing Agreement with the aim to support the conservation and sustainable use of the Caribbean native flora. Through this collaboration, a new seed bank was established in 2017 [161] together with a conservation programme, which included the provision of technical and scientific support, capacity building and seed research for ex situ and in situ conservation. The Hispaniola Island climate is mainly tropical, and as in most tropical regions, the percentage of species with recalcitrant (non-bankable) seeds is higher than in temperate climates. Thus, one of the challenges faced was the identification or prediction of native species whose seeds could not be conserved under traditional seed banking conditions [162]. The joint Garfield Weston funded project Saving threatened forests of Hispaniola focused on protecting the forests on the island, by researching, conserving and propagating the seeds of native useful tree species and supporting reforestation activities. Seeds of 250 tree species have been banked and planting activities have been carried out in degraded areas, seedlings have been donated for the establishment of a new Botanic Garden in Santiago and used for enhancing the vegetation of urban parks and other green spaces in Santo Domingo. The collaboration has helped the JBN to become a key player for the Ministry of Environment and Natural Resources in the Dominican Republic in seed conservation and the provision of plant material and information for restoring and recovering protected as well as urban and semi urban areas.

Under the umbrella of the **Great Green Wall Initiative**, Kew's pilot project (2013–2020) built a restoration model and generated environmental and socio-economic information to support larger-scale restoration projects in similar contexts and conditions in the Sahara and Sahel regions [163]. Under Access and Benefit-Sharing Agreements, this collaborative project was implemented in partnership with national institutions and local communities in the cross-border zone between three countries: Bankass in Mali; Djibo and Dori in Burkina Faso; and Téra in Niger. A participatory approach was used to select native useful plant species adapted to local conditions, and those that are important to the communities' livelihoods [164]. The most environmentally well-adapted and economically relevant species were prioritised and authenticated, and seeds of 84 useful woody and herbaceous species were collected from the wild and stored according to international standards in national seed banks with duplicates at the MSB. Seed accessions were tested and research on seed biology and ecology was carried out at Kew to support the conservation and propagation of species. In collaboration with local communities, seeds of 55 woody and herbaceous species were propagated and planted to restore 2235 ha of degraded land and create sustainable income-generating opportunities for up to 32,000 people. Over 1,000,000 seedlings of the selected species were planted and monitored in around 200 experimental plots [164]. An assisted natural regeneration approach was used for some of the most important species, such as *Guiera senegalensis*, as an alternative to reforestation [165], while the planting of tall bare roots of *Adansonia digitata* (Baobab) was found to have a better chance to resist and survive harsh climatic conditions and the species is highly valued by farmers [166]. Over 100 village technicians have been trained in nursery management, tree planting, forest restoration and establishment of demonstration plots during the duration of the project. The project has generated capacity at national and regional level to develop, plan and implement science-based restoration programmes aimed at reestablishing the natural capital of the vegetation and supplying the resource base for the enhancement of local livelihoods.

5. Future Directions

In summary, both living and seed collections held in botanic gardens and seed banks offer huge long-term opportunities for the conservation of wild plant diversity. Such collections are a key input into agricultural research and the development of improved crop varieties, with plant breeding depending fundamentally on the availability and accessibility of useful genetic resources, such as those found in crop wild relatives. In addition, WEP can provide the basis of new crops. Some examples of the value to humanity of successful long-term maintenance of a wide spectrum of genetic diversity are provided by Bretting [167], and other uses of seed and plant collections are described in this paper.

Beyond their immediate use value, seeds stored in genebanks represent a form of insurance against the loss of genetic material in situ. Furthermore, and in common with other scientific collections, they may well have uses beyond their original purpose. The use of herbarium specimens to investigate the impacts of climate change is well recognised, and it may be that seeds stored for the long term will offer opportunities for future research, answering questions so far not asked and using tools not yet developed.

There are, however, challenges to be addressed. There are gaps in the geographic, taxonomic and genetic coverage of collections, as well as gaps in coverage of useful plants. It is apparent that ex situ collections and conservation expertise continue to sit outside the countries with the greatest plant diversity. It is, however, encouraging to see that this situation is changing, through investment in people and facilities within biodiverse regions, but more needs to be done.

Availability of funding on a scale commensurate with the biodiversity crisis remains an issue. The conservation of plant diversity needs a step change in scale and investment if we are to prevent the predicted extinction of two in five plants, and one in three trees. The shortage of funding means that prioritisation of which plants are conserved must take place—and this can take many forms depending on the organisation undertaking the conservation initiative. For example, botanic gardens and seed banks have long focused on the most threatened species, including endemics, but also on economically important species, including medicinal plants and CWR. This does not, however, mean that those plants not prioritised have no intrinsic value of their own, and often reflects limitations in our knowledge of plant properties, uses and threat status.

The costs associated with long-term storage pose another constraint and mean that choices need to be made on what to conserve and how to do this, and there is limited information available to guide such choices. Detailed knowledge of local, regional, and global patterns of plant diversity are often lacking [168], and as mentioned, seed banks are often not located where the greatest diversity exists. Savings should not, however, be made through identifying and removing duplicates from collections, as holding multiple accessions from the same population through time and from different populations of the same species are of great conservation value. False savings should also be avoided, for example, devaluing material that is used the least. At the same time, duplication between multiple types of ex situ conservation should be monitored and could provide potential for rationalising collections.

How long material is held is a further consideration. While funds are generally used for the acquisition of plant genetic resources, their long-term maintenance and monitoring also needs to be funded, often directly by the institutions holding the collections. The MSB was built to last 500 years, and most species banked will survive 10 s to 100 s of years under storage conditions, but it is questionable whether sufficient funds will be available to support the long-term ex situ conservation of the world's genetic resources [168].

The increasing use of genomic information in breeding may suggest a move towards dematerialisation of collections and a greater focus on storing genomic data only. We would of course argue that there are many more uses of seeds beyond their use in breeding programmes and the ability of scientists to work usefully with DNA alone is still some way in the future. We therefore do not believe that conserving dematerialised DNA will be a useful plant conservation strategy for some time to come. Another challenge is the increasing complexity of issues around the ownership and control of genetic resources as well as the sensitivity of the legal and political situations.

Quality control must continue to be addressed, and improved monitoring systems put in place. While a number of standards and guidelines have been published to guide seed banking within the botanic garden sector, it is not clear to what extent these standards are being followed and how much of the seed held by botanic gardens and other institutes is in fact viable and usable. Within the European plant genetic resources community, a seed bank quality system has been put in place which provides a set of policies, processes and procedures that are to be followed by all members of the European Genebank Integrated System (AEGIS) to assure an appropriate quality of activities. The system requires all members to develop an operational genebank manual based on a common template that documents the operating procedures and standards of the genebank. In 2019, in the framework of the EU-funded project GenRes Bridge, three European genebanks were evaluated as part of a pilot project. The project allowed the genebanks to review each other on the basis of their Genebank Manuals and provided a mentoring system to help the genebanks address quality issues identified during the reviews [169]. This pilot project may provide valuable lessons to guide efforts to improve the quality of botanic garden seed banks.

Two further areas stand out as opportunities to improve ex situ conservation for the future: technological advances and networks. Technological advances in cryobiotechnology are already helping some seed banks conserve existing collections of wild species whose seeds are short-lived under conventional seed banking conditions, as well as species with intermediate or recalcitrant seeds (Figure 1). The process for the latter is, however, research intensive as species require tailored protocols to be developed for them to optimise conservation outcomes. The flora of highly diverse and threatened regions are a priority for conservation, and these tend to fall in the tropics—which have both a geographical gap in existing collections, and a technological gap in relation to the conservation of 'exceptional species', which are more prevalent in these regions. There is a need to increase the availability of cryopreservation options for these species through provision of facilities, or access to existing facilities currently used for different purposes (e.g., crops), and through training. Developing and strengthening networks within tropical regions will greatly help.

We have highlighted the importance of networks for sharing knowledge, data, expertise and access to facilities as well as to raise the profile of plant conservation and leverage funding. It is important to grow and maintain conservation networks, not just within the botanic garden and seed bank sectors, but by incorporating other practitioners and local communities working towards the conservation of a given species, habitat or flora. Such networks can help establish integrated conservation planning from the outset of programmes, and lead to greater impacts for conservation. In addition, greater linkages need to be formed between the conservation and agricultural worlds to ensure that we hold the material needed to create resilient crops of the future and to diversify the food plants grown as crops. Forming networks that interlink these sectors will help to increase the impact of global conservation efforts, avoid unnecessary duplication and prioritise resource allocation, as was largely achieved through the Adapting Agriculture to Climate Change project. The diversification of metacollections, currently focused on 'exceptional species' to a wider array of threatened plant taxa should also be encouraged as this networking approach greatly enhances the value of individual collections and improves conservation outcomes for the species involved.

Botanic gardens and seed banks are well placed to respond to the biodiversity crises, but their conservation efforts need to be massively scaled up and supported by long-term funding. They also need to be coordinated across institutions, sectors (government agencies, universities, NGOs, etc.), geographies, and political boundaries [34]. Technological advances offer hope for the ex situ conservation of all plants—we now need to act on this possibility.

Author Contributions: Conceptualization and writing—original draft preparation, review and editing, all authors; visualization, E.B., D.B., K.O., A.F. and S.P. All authors have read and agreed to the published version of the manuscript.

Funding: This research received no external funding, although the work of BGCI and the MSBP have been generously funded over a number of years by a wide variety of grants and philanthropic donors.

Institutional Review Board Statement: Not applicable.

Informed Consent Statement: Not applicable.

Data Availability Statement: Data sharing not applicable.

Acknowledgments: We would like to thank Johannes Engels and Andreas Ebert for their guidance and helpful comments on this manuscript, as well as three anonymous reviewers. In addition, our thanks to the seed processing team at the MSB, and to all our project collaborators, for their work in the conservation of wild plants. Alice Hudson provided data on recollection rates for collections held at the MSB.

Conflicts of Interest: The authors declare no conflict of interest.

References

1. Díaz, S.; Settele, J.; Brondízio, E.; Ngo, H.T.; Guèze, M.; Agard, J.; Arneth, A.; Balvanera, P.; Brauman, K.; Butchart, S.H.M.; et al. (Eds.) *IPBES Summary for Policymakers of the Global Assessment Report on Biodiversity and Ecosystem Services of the Intergovernmental Science-Policy Platform on Biodiversity and Ecosystem Services*; IPBES Secretariat: Bonn, Germany, 2019; ISBN 978-3-947851-13-3.
2. *World Meteorological Organization State of the Global Climate 2020 (WMO-No. 1264)*; WMO: Geneva, Switzerland, 2021; ISBN 978-92-63-11264-4.
3. Pörtner, H.O.; Scholes, R.J.; Agard, J.; Archer, E.; Arneth, A.; Bai, X.; Barnes, D.; Burrows, M.; Chan, L.; Cheung, W.L.; et al. *Scientific Outcome of the IPBES-IPCC Co-Sponsored Workshop on Biodiversity and Climate Change*; IPBES and IPCC: Bonn, Germany, 2021. [CrossRef]
4. *World Economic Forum The Global Risks Report 2020*; World Economic Forum: Geneva, Switzerland, 2020; ISBN 978-2-940631-24-7.
5. Bar-On, Y.M.; Phillips, R.; Milo, R. The Biomass Distribution on Earth. *Proc. Natl. Acad. Sci. USA* **2018**, *115*, 6506–6511. [CrossRef] [PubMed]
6. Lughadha, E.N.; Bachman, S.P.; Leão, T.C.C.; Forest, F.; Halley, J.M.; Moat, J.; Acedo, C.; Bacon, K.L.; Brewer, R.F.A.; Gâteblé, G.; et al. Extinction Risk and Threats to Plants and Fungi. *Plants People Planet* **2020**, *2*, 389–408. [CrossRef]
7. Dasgupta, P. *The Economics of Biodiversity: The Dasgupta Review*; HM Treasury: London, UK, 2021; ISBN 978-1-911680-29-1.
8. Ceballos, G.; Ehrlich, P.R.; Barnosky, A.D.; García, A.; Pringle, R.M.; Palmer, T.M. Accelerated Modern Human–Induced Species Losses: Entering the Sixth Mass Extinction. *Sci. Adv.* **2015**, *1*, e1400253. [CrossRef] [PubMed]
9. BGCI. PlantSearch. Available online: https://tools.bgci.org/plant_search.php (accessed on 10 September 2021).
10. Merritt, D.J.; Whitehouse, K.J.; Hoyle, G.L.; Crawford, A.; Wood, J.A.; Satyanti, A.; Norton, S.; Errington, G.; Martyn Yenson, A.J. Seed banking: Orthodox seeds. In *Plant Germplasm Conservation in Australia*; Martyn Yenson, A.J., Offord, C.A., Meagher, P.F., Auld, T., Bush, D., Coates, D.J., Commander, L.E., Guja, L.K., Norton, S.L., Makinson, R.O., et al., Eds.; Australian Network for Plant Conservation: Canberra, Australia, 2021; pp. 119–158, ISBN 978-0-9752191-4-0.
11. Pence, V.C. In Vitro Methods and the Challenge of Exceptional Species for Target 8 of the Global Strategy for Plant Conservation. *Ann. Mo. Bot. Gard.* **2013**, *99*, 214–220. [CrossRef]
12. Pence, V.C.; Ballesteros, D.; Walters, C.; Reed, B.M.; Philpott, M.; Dixon, K.W.; Pritchard, H.W.; Culley, T.M.; Vanhove, A.-C. Cryobiotechnologies: Tools for Expanding Long-Term Ex Situ Conservation to All Plant Species. *Biol. Conserv.* **2020**, *250*, 108736. [CrossRef]
13. Wyse, S.V.; Dickie, J.B. Predicting the Global Incidence of Seed Desiccation Sensitivity. *J. Ecol.* **2017**, *105*, 1082–1093. [CrossRef]
14. Smith, R.; Dickie, J.B.; Linington, S.H.; Pritchard, H.W.; Probert, R.J. (Eds.) *Seed Conservation: Turning Science into Practice*; Royal Botanic Gardens: London, UK, 2003; Volume 95, ISBN 1-84246-052-8.
15. Way, M. Techniques and key issues in collecting crop wild relatives. In *Plant Genetic Resources: A Review of Current Research and Future Needs*; Dulloo, E., Ed.; Burleigh Dodds: Cambridge, UK, 2021; pp. 1–30.
16. Ballesteros, D.; Pence, V.C. Fern Conservation: Spore, Gametophyte, and Sporophyte Ex Situ Storage, In Vitro Culture, and Cryopreservation. In *Current Advances in Fern Research*; Fernández, H., Ed.; Springer International Publishing: London, UK, 2018; pp. 227–249, ISBN 978-3-319-75103-0.
17. Volk, G. Chapter 25: Collecting pollen for genetic resources conservation. In *Collecting Plant Genetic Diversity: Technical Guidelines—2011 Update*; Guarino, L., Ramanatha, V., Goldberg, E., Eds.; Bioversity International: Rome, Italy, 2011; ISBN 978-92-9043-922-6.
18. Colville, L.; Pritchard, H.W. Seed Life Span and Food Security. *New Phytol.* **2019**, *224*, 557–562. [CrossRef]

19. Chau, M.M.; Chambers, T.; Weisenberger, L.; Keir, M.; Kroessig, T.I.; Wolkis, D.; Kam, R.; Yoshinaga, A.Y. Seed Freeze Sensitivity and Ex Situ Longevity of 295 Species in the Native Hawaiian Flora. *Am. J. Bot.* **2019**, *106*, 1248–1270. [CrossRef]
20. *FAO Genebank Standards for Plant Genetic Resources for Food and Agriculture*; FAO: Rome, Itlay, 2014; ISBN 978-92-5-108262-1.
21. BGCI. GardenSearch. Available online: https://tools.bgci.org/garden_search.php (accessed on 10 September 2021).
22. Griffith, M.P.; Clase, T.; Toribio, P.; Piñeyro, Y.E.; Jimenez, F.; Gratacos, X.; Sanchez, V.; Meerow, A.; Meyer, A.; Kramer, A.; et al. Can a Botanic Garden Metacollection Better Conserve Wild Plant Diversity? A Case Study Comparing Pooled Collections with an Ideal Sampling Model. *Int. J. Plant Sci.* **2020**, *181*, 485–496. [CrossRef]
23. Griffith, P.M.; Beckman, E.; Callicrate, T.; Clark, J.; Clase, T.; Deans, S.; Dosmann, M.; Fant, J.; Gratacos, X.; Havens, K.; et al. *Toward the Metacollection: Safeguarding Plant Diversity and Coordinating Conservation Collections*; Botanic Gardens Conservation International, US: San Marino, CA, USA, 2019; p. 12.
24. *CBD Global Strategy for Plant Conservation*; Secretariat of the Convention on Biological Diversity Botanic Gardens Conservation International: Richmond, UK, 2002; ISBN 978-0-9539141-5-9.
25. United Nations. The 17 Goals | Sustainable Development. Available online: https://sdgs.un.org/goals (accessed on 27 August 2021).
26. Mounce, R.; Smith, P.; Brockington, S. Ex Situ Conservation of Plant Diversity in the World's Botanic Gardens. *Nat. Plants* **2017**, *3*, 795–802. [CrossRef]
27. CBD. The Nagoya Protocol. Available online: https://www.cbd.int/abs/about/ (accessed on 1 September 2021).
28. Spencer, R.; Cross, R. The Origins of Botanic Gardens and Their Relation to Plant Science, with Special Reference to Horticultural Botany and Cultivated Plant Taxonomy. *Muelleria* **2017**, *35*, 43–93.
29. Heywood, V.H. Plant Conservation in the Anthropocene—Challenges and Future Prospects. *Plant Divers.* **2017**, *39*, 314–330. [CrossRef] [PubMed]
30. Jackson, P.S.W. Experimentation on a Large Scale—An Analysis of the Holdings and Resources of Botanic Gardens. *Bot. Gard. Conserv. News* **1999**, *3*, 27–30.
31. BGCI. ThreatSearch. Available online: https://tools.bgci.org/threat_search.php (accessed on 19 September 2021).
32. O'Donnell, K.; Sharrock, S. The Contribution of Botanic Gardens to Ex Situ Conservation through Seed Banking. *Plant Divers.* **2017**, *39*, 373–378. [CrossRef] [PubMed]
33. Plant List Home—The Plant List. Available online: http://www.theplantlist.org/ (accessed on 21 October 2021).
34. Westwood, M.; Cavender, N.; Meyer, A.; Smith, P. Botanic Garden Solutions to the Plant Extinction Crisis. *Plants People Planet* **2021**, *3*, 22–32. [CrossRef]
35. Allen, W. Plant Blindness. *BioScience* **2003**, *53*, 926. [CrossRef]
36. Hudson, A.; Smith, P.; Gori, B.; Sharrock, S. Botanic Garden Collections—An Under-Utilised Resource. *AJPS* **2021**, *12*, 1436–1444. [CrossRef]
37. Montgomery Botanical Center. Montgomery Botanical Center Home. Available online: https://www.montgomerybotanical.org/ (accessed on 19 September 2021).
38. Griffith, P.; Lewis, C.; Francisco-Ortega, J. Palm Conservation at a Botanic Garden: A Case Study of the Keys Thatch Palm. *Palms* **2011**, *55*, 93–101.
39. Griffith, M.P.; Calonje, M.; Meerow, A.W.; Tut, F.; Kramer, A.T.; Hird, A.; Magellan, T.M.; Husby, C.E. Can a Botanic Garden Cycad Collection Capture the Genetic Diversity in a Wild Population? *Int. J. Plant Sci.* **2015**, *176*, 1–10. [CrossRef]
40. Wood, J.; Ballou, J.D.; Callicrate, T.; Fant, J.B.; Griffith, M.P.; Kramer, A.T.; Lacy, R.C.; Meyer, A.; Sullivan, S.; Traylor-Holzer, K.; et al. Applying the Zoo Model to Conservation of Threatened Exceptional Plant Species. *Conserv. Biol.* **2020**, *34*, 1416–1425. [CrossRef]
41. Fant, J.B.; Havens, K.; Kramer, A.T.; Walsh, S.K.; Callicrate, T.; Lacy, R.C.; Maunder, M.; Meyer, A.H.; Smith, P.P. What to Do When We Can't Bank on Seeds: What Botanic Gardens Can Learn from the Zoo Community about Conserving Plants in Living Collections. *Am. J. Bot.* **2016**, *103*, 1541–1543. [CrossRef] [PubMed]
42. American Public Gardens Association Collections Showcase | American Public Gardens Association. Available online: https://www.publicgardens.org/programs/plant-collections-network/collections-showcase (accessed on 17 September 2021).
43. BGCI. Global Conservation Consortia. Available online: https://www.bgci.org/our-work/plant-conservation/global-conservation-consortia/ (accessed on 17 September 2021).
44. Center for Plant Conservation. National Collection. Available online: https://saveplants.org/national-collection/ (accessed on 17 September 2021).
45. WIEWS. Background | WIEWS—World Information and Early Warning System on Plant Genetic Resources for Food and Agriculture | Food and Agriculture Organization of the United Nations. Available online: http://www.fao.org/wiews/background/en/ (accessed on 17 September 2021).
46. CGIAR. Guardians of Diversity: The Network of Genebanks Helping to Feed the World. Available online: https://www.cgiar.org/news-events/news/guardians-of-diversity-the-network-of-genebanks-helping-to-feed-the-world/ (accessed on 17 September 2021).
47. Hernández Bermejo, J.E.; Herrera Molina, F. REDBAG: The Spanish Network of Genebanks for Wild Plants. *BGjournal* **2005**, *2*, 18–20.

48. Martyn Yenson, A.J.; Offord, C.A.; Meagher, P.F.; Auld, T.; Bush, D.; Coates, D.J.; Commander, L.E.; Guja, L.K.; Norton, S.L.; Makinson, R.O.; et al. (Eds.) *Plant Germplasm Conservation in Australia*; Australian Network for Plant Conservation: Canberra, Australia, 2021; ISBN 978-0-9752191-4-0.
49. Sommerville, K.D.; Errington, G.; Newby, Z.-J.; Liyanage, G.S.; Offord, C.A. Assessing the Storage Potential of Australian Rainforest Seeds: A Decision-Making Key to Aid Rapid Conservation. *Biodivers. Conserv.* **2021**, *30*, 3185–3218. [CrossRef]
50. ENSCONET. *ENSCONET Seed Collecting Manual for Wild Species*; RBG Kew, Universidad Politécnica de Madrid, Eds.; 2009; ISBN 978-84-692-3926-1.
51. ENSCONET. *ENSCONET Curation Protocols and Recommendations*; RBG Kew, Ed.; 2009; ISBN 978-84-692-5964-1.
52. ISTA. International Rules for Seed Testing. Available online: https://www.ingentaconnect.com/content/ista/rules/2021/00002021/00000001# (accessed on 31 August 2021).
53. Martin, A.C. The Comparative Internal Morphology of Seeds. *Am. Midl. Nat.* **1946**, *36*, 513. [CrossRef]
54. Breman, E.; Way, M. Safe for the Future: Seed Conservation Standards Developed for the Millennium Seed Bank Partnership. In Proceedings of the EuroGard VII Congress, Paris, France, 6–10 July 2018; pp. 267–274.
55. Menges, E.; Guerrant, E.; Hamzé, S. Effects of seed collection on the extinction risk of perennial plants. In *Ex Situ Plant Conservation: Supporting Species Survival in the Wild*; Guerrant, E., Havens, K., Maunder, M., Eds.; Island Press: Washington, DC, USA, 2004; pp. 305–324, ISBN 978-1-55963-875-3.
56. Offord, C.A.; Meagher, P.F. (Eds.) *Plant Germplasm Conservation in Australia: Strategies and Guidelines for Developing, Managing and Utilising Ex Situ Collections*; Australian Network for Plant Conservation Inc.: Canberra, Australia, 2009; ISBN 978-0-9752191-1-9.
57. *Center for Plant Conservation CPC Best Plant Conservation Practices to Support Species Survival in the Wild*; Center for Plant Conservation: Escondido, CA, USA, 2019.
58. Hoban, S. New Guidance for Ex Situ Gene Conservation: Sampling Realistic Population Systems and Accounting for Collection Attrition. *Biol. Conserv.* **2019**, *235*, 199–208. [CrossRef]
59. Brown, A.H.D.; Marshall, D.R. A basic sampling strategy: Theory and practice. In *Collecting Plant Genetic Diversity: Technical Guidelines—2011 Update*; Guarino, L., Ramanatha, V., Reid, R., Eds.; CABI: Wallingford, UK, 1995; pp. 75–91, ISBN 978-92-9043-922-6.
60. Kallow, S. *UK National Tree Seed Project Manual*; Royal Botanic Gardens Kew: London, UK, 2014; p. 39.
61. Hoban, S.; Schlarbaum, S. Optimal Sampling of Seeds from Plant Populations for Ex-Situ Conservation of Genetic Biodiversity, Considering Realistic Population Structure. *Biol. Conserv.* **2014**, *177*, 90–99. [CrossRef]
62. Trivedi, C.; Kallow, S. Benefits and challenges for gene conservation: A view from the UK National Tree Seed Project. In *Gene Conservation of tree Species—Banking on the Future. Proceedings of a Workshop*; General Technical Report PNW-GTR-963; U.S. Department of Agriculture, Forest Service, Pacific Northwest Research Station: Portland, OR, USA, 2017; pp. 44–47.
63. Commander, L.E. Florabank Guidelines—Best Practice Guidelines for Native Seed Collection and Use. Available online: https://www.florabank.org.au/guidelines/ (accessed on 17 September 2021).
64. Commander, L.E.; Coates, D.J.; Broadhurst, L.; Offord, C.A.; Makinson, R.O.; Matthes, M. (Eds.) *Guidelines for the Translocation of Threatened Plants in Australia*; Australian Network for Plant Conservation: Canberra, Australia, 2018. [CrossRef]
65. Probert, R.J.; Daws, M.I.; Hay, F.R. Ecological Correlates of Ex Situ Seed Longevity: A Comparative Study on 195 Species. *Ann. Bot.* **2009**, *104*, 57–69. [CrossRef]
66. Davies, R.M.; Hudson, A.R.; Dickie, J.B.; Cook, C.; O'Hara, T.; Trivedi, C. Exploring Seed Longevity of UK Native Trees: Implications for Ex Situ Conservation. *Seed Sci. Res.* **2020**, *30*, 101–111. [CrossRef]
67. Ballesteros, D.; Fanega-Sleziak, N.; Davies, R.M. Cryopreservation of seeds and seed embryos in orthodox-, intermediate-, and recalcitrant-seeded species. In *Cryopreservation and Freeze-Drying Protocols*; Wolkers, W.F., Oldenhof, H., Eds.; Methods in Molecular Biology; Springer US: New York, NY, USA, 2021; pp. 663–682, ISBN 978-1-07-160783-1.
68. RBG Kew. Millennium Seed Bank Seed List—Royal Botanic Gardens, Kew. Available online: http://apps.kew.org/seedlist/ (accessed on 20 September 2021).
69. Brummitt, R.K.; Pando, F.; Hollis, S.; Brummitt, N.A. *World Geographic Scheme for Recording Plant Distributions*, 2nd ed.; Hunt Institute for Botanical Documentation, Carnegie Mellon University (Pittsburgh): Pittsburgh, PA, USA, 2001. Available online: http://rs.tdwg.org/wgsrpd/doc/data/ (accessed on 18 September 2021).
70. Govaerts, R.; Nic Lughadha, E.; Black, N.; Turner, R.; Paton, A. The World Checklist of Vascular Plants, a Continuously Updated Resource for Exploring Global Plant Diversity. *Sci. Data* **2021**, *8*, 215. [CrossRef] [PubMed]
71. Wei, X.; Jiang, M. Meta-Analysis of Genetic Representativeness of Plant Populations under Ex Situ Conservation in Contrast to Wild Source Populations. *Conserv. Biol.* **2020**, *35*, 12–23. [CrossRef] [PubMed]
72. Ulian, T.; Pritchard, H.W.; Casas, A.; Mattana, E.; Castillo-Lorenzo, E.; Way, M.; Aranda, P.D.; Lira, R. Neglected and underutilised species in Mexico. In *Orphan Crops for Sustainable Food and Nutrition Security: Promoting Neglected and Underutilized Species*; Padulosi, S., King, E.D.I.O., Hunter, D., Swaminathan, M.S., Eds.; Routledge: London, UK, 2021; ISBN 978-1-00-304480-2.
73. Hurka, H.; Neuffer, B.; Friesen, N. Plant Genetic Resources in Botanical Gardens. *Acta Hortic.* **2004**, 651. [CrossRef]
74. Everingham, S.E.; Offord, C.A.; Sabot, M.E.B.; Moles, A.T. Time-Traveling Seeds Reveal That Plant Regeneration and Growth Traits Are Responding to Climate Change. *Ecology* **2021**, *102*, e03272. [CrossRef] [PubMed]
75. Rossetto, M.; Yap, J.-Y.S.; Lemmon, J.; Bain, D.; Bragg, J.; Hogbin, P.; Gallagher, R.; Rutherford, S.; Summerell, B.; Wilson, T.C. A Conservation Genomics Workflow to Guide Practical Management Actions. *Glob. Ecol. Conserv.* **2021**, *26*, e01492. [CrossRef]

76. Viler, V.; Offord, C.A. Ex Situ Management Including Seed Orchard Establishment for Native Guava (Rhodomyrtus Psidioides) Affected by Myrtle Rust. Austral. *Plant Conserv.* **2020**, *29*, 17–19.
77. Engels, J.M.M.; Ebert, A.W. A Critical Review of the Current Global Ex Situ Conservation System for Plant Agrobiodiversity. I. History of the Development of the Global System in the Context of the Political/Legal Framework and Its Major Conservation Components. *Plants* **2021**, *10*, 1557. [CrossRef] [PubMed]
78. Guerrant, E.O.; Havens, K.; Maunder, M. *Ex Situ Plant Conservation: Supporting Species Survival in the Wild*; Island Press: Washington, DC, USA, 2004; Volume 3, ISBN 978-1-55963-875-3.
79. Li, D.-Z.; Pritchard, H.W. The Science and Economics of Ex Situ Plant Conservation. *Trends Plant Sci.* **2009**, *14*, 614–621. [CrossRef]
80. Reed, B.M. Implementing Cryogenic Storage of Clonally Propagated Plants. *Cryo Lett.* **2001**, *22*, 97–104.
81. Vollmer, R.; Villagaray, R.; Egusquiza, V.; Espirilla, J.; García, M.; Torres, A.; Rojas, E.; Panta, A.; Barkley, N.A.; Ellis, D. The Potato Cryobank at The International Potato Center (Cip): A Model for Long Term Conservation of Clonal Plant Genetic Resources Collections of the Future. *Cryo Lett.* **2016**, *37*, 318–329.
82. Reed, B.M. (Ed.) *Plant Cryopreservation: A Practical Guide*; Springer: New York, NY, USA, 2008; ISBN 978-0-387-72275-7.
83. Ballesteros, D.; Pritchard, H.W. The Cryobiotechnology of Oaks: An Integration of Approaches for the Long-Term Ex Situ Conservation of Quercus Species. *Forests* **2020**, *11*, 1281. [CrossRef]
84. Pritchard, H.W.; Nadarajan, J.; Ballesteros, D.; Thammasiri, K.; Prasongsom, S.; Malik, S.K.; Chaudhury, R.; Kim, H.-H.; Lin, L.; Li, W.-Q.; et al. Cryobiotechnology of Tropical Seeds—Scale, Scope and Hope. *Acta Hortic.* **2017**, 37–48. [CrossRef]
85. Streczynski, R.; Clark, H.; Whelehan, L.M.; Ang, S.-T.; Hardstaff, L.K.; Funnekotter, B.; Bunn, E.; Offord, C.A.; Sommerville, K.D.; Mancera, R.L.; et al. Current Issues in Plant Cryopreservation and Importance for Ex Situ Conservation of Threatened Australian Native Species. *Aust. J. Bot.* **2019**, *67*, 1–15. [CrossRef]
86. Wyse, S.V.; Dickie, J.B. Taxonomic Affinity, Habitat and Seed Mass Strongly Predict Seed Desiccation Response: A Boosted Regression Trees Analysis Based on 17 539 Species. *Ann. Bot.* **2018**, *121*, 71–83. [CrossRef]
87. Acker, J.P.; Adkins, S.; Alves, A.; Horna, D.; Toll, J. *Feasibility Study for a Safety Back-up Cryopreservation Facility. Independent Expert Report: July 2017*; Bioversity International: Rome, Italy, 2017; ISBN 978-92-9255-073-8.
88. Pence, V.C. Evaluating Costs for the in Vitro Propagation and Preservation of Endangered Plants. *In Vitro Cell. Dev. Biol. Plant* **2011**, *47*, 176–187. [CrossRef]
89. Pritchard, H.W.; Moat, J.F.; Ferraz, J.B.S.; Marks, T.R.; Camargo, J.L.C.; Nadarajan, J.; Ferraz, I.D.K. Innovative Approaches to the Preservation of Forest Trees. *For. Ecol. Manag.* **2014**, *333*, 88–98. [CrossRef]
90. Wade, E.M.; Nadarajan, J.; Yang, X.; Ballesteros, D.; Sun, W.; Pritchard, H.W. Plant Species with Extremely Small Populations (PSESP) in China: A Seed and Spore Biology Perspective. *Plant Divers.* **2016**, *38*, 209–220. [CrossRef] [PubMed]
91. Nadarajan, J.; Benson, E.E.; Xaba, P.; Harding, K.; Lindstrom, A.; Donaldson, J.; Seal, C.E.; Kamoga, D.; Agoo, E.M.G.; Li, N.; et al. Comparative Biology of Cycad Pollen, Seed and Tissue—A Plant Conservation Perspective. *Bot. Rev.* **2018**, *84*, 295–314. [CrossRef] [PubMed]
92. Martyn Yenson, A.J.; Commander, L.E.; Offord, C.A.; Makinson, R.O. Chapter 1 Introduction. In *Plant Germplasm Conservation in Australia*; Martyn Yenson, A.J., Offord, C.A., Meagher, P.F., Auld, T., Bush, D., Coates, D.J., Commander, L.E., Guja, L.K., Norton, S.L., Makinson, R.O., et al., Eds.; Australian Network for Plant Conservation: Canberra, Australia, 2021; pp. 1–21, ISBN 978-0-9752191-4-0.
93. Godfree, R.C.; Knerr, N.; Encinas-Viso, F.; Albrecht, D.; Bush, D.; Christine Cargill, D.; Clements, M.; Gueidan, C.; Guja, L.K.; Harwood, T.; et al. Implications of the 2019–2020 Megafires for the Biogeography and Conservation of Australian Vegetation. *Nat. Commun.* **2021**, *12*, 1023. [CrossRef] [PubMed]
94. Hamilton, K.N.; Offord, C.A.; Cuneo, P.; Deseo, M.A. A Comparative Study of Seed Morphology in Relation to Desiccation Tolerance and Other Physiological Responses in 71 Eastern Australian Rainforest Species. *Plant Species Biol.* **2013**, *28*, 51–62. [CrossRef]
95. Ashmore, S.E.; Martyn, A.; Sommerville, K.D.; Offord, C.A. Chapter 14 Seed Biology. In *Crop Wild Relatives and Climate Change*; Redden, R., Yadav, S., Maxted, N., Dulloo, M.E., Guarino, L., Smith, P., Eds.; John Wiley & Sons Inc.: Hoboken, NJ, USA, 2015; pp. 187–211, ISBN 978-1-118-85433-4.
96. Offord, C.A.; Makinson, R.O.; Guja, L.K.; Auld, T. Chapter 2 Options, major considerations and preparation for plant germplasm conservation. In *Plant Germplasm Conservation in Australia*; Martyn Yenson, A.J., Offord, C.A., Meagher, P.F., Auld, T., Bush, D., Coates, D.J., Commander, L.E., Guja, L.K., Norton, S.L., Makinson, R.O., et al., Eds.; Australian Network for Plant Conservation: Canberra, Australia, 2021; pp. 23–52, ISBN 978-0-9752191-4-0.
97. Ashmore, S.E.; Hamilton, K.N.; Offord, C.A. Conservation Technologies for Safeguarding and Restoring Threatened Flora: Case Studies from Eastern Australia. *In Vitro Cell. Dev. Biol. Plant* **2011**, *47*, 99–109. [CrossRef]
98. Hardstaff, L.K.; Sommerville, K.D.; Funnekotter, B.; Bunn, E.; Offord, C.A.; Mancera, R.L. Cryostorage of Australian Rainforest Species. *Austral. Plant Conserv.* **2020**, *28*, 9–11.
99. Fensham, R.J.; Carnegie, A.J.; Laffineur, B.; Makinson, R.O.; Pegg, G.S.; Wills, J. Imminent Extinction of Australian Myrtaceae by Fungal Disease. *Trends Ecol. Evol.* **2020**, *35*, 554–557. [CrossRef] [PubMed]
100. Sommerville, K.D.; Cuneo, P.; Errington, G.; Makinson, R.O.; Pederson, S.; Phillips, G.; Rollason, A.; Viler, V.; Offord, C.A. Conservation in the Wake of Myrtle Rust—A Case Study on Two Critically Endangered Australian Rainforest Plants. *Pac. Conserv. Biol.* **2020**, *26*, 218. [CrossRef]

101. Pearce, T.R.; Antonelli, A.; Brearley, F.Q.; Couch, C.; Forzza, R.C.; Gonçalves, S.C.; Magassouba, S.; Morim, M.P.; Mueller, G.M.; Lughadha, E.N.; et al. International Collaboration between Collections-Based Institutes for Halting Biodiversity Loss and Unlocking the Useful Properties of Plants and Fungi. *Plants People Planet* **2020**, *2*, 515–534. [CrossRef]
102. Pronatura Veracruz, A.C. *La Red de Viveros de Biodiversidad y Reserva de Semillas Como Una Iniciativa de Largo Plazo Para Diversificar Los Viveros Forestales de México*; Sexto Informe Nacional de México ante el CBD: Mexico City, Mexico, 2020.
103. Mariotti, M.; Magrini, S. *The RIBES Seed-Banks for the Conservation of the Crop Wild Relatives (CWR)*; RIBES: Cassano delle Murge, Italy, 2016; ISBN 978-88-940844-2-9.
104. GENMEDA. GENMEDA—Network of Mediterranean Plant Conservation Centres. Available online: http://www.genmeda.net/ (accessed on 18 October 2021).
105. Rivière, S.; Breman, E.; Kiehn, M.; Carta, A.; Müller, J.V. How to Meet the 2020 GSPC Target 8 in Europe: Priority-Setting for Seed Banking of Native Threatened Plants. *Biodivers. Conserv.* **2018**, *27*, 1873–1890. [CrossRef]
106. *Plant Conservation Alliance (PCA) National Seed Strategy for Rehabilitation and Restoration, 2015–2020*; U.S. Department of the Interior, Bureau of Land Management: Washington, DC, USA, 2021.
107. Sutherland, L.A. Safeguarding Australia's Flora: Through the Australian Seed Bank Partnership. *BGjournal* **2012**, *9*, 32–35.
108. FAO (Ed.) *The State of the World's Forest Genetic Resources*; FAO: Rome, Italy, 2014; ISBN 978-92-5-108402-1.
109. BGCI. Exceptional Species. Available online: https://www.bgci.org/our-work/plant-conservation/exceptional-species/ (accessed on 13 September 2021).
110. Cincinnati Zoo & Botanical Garden. Exceptional Plant Conservation Network. Available online: http://cincinnatizoo.org/conservation/crew/exceptional-plant-conservation-network/ (accessed on 13 September 2021).
111. Liu, U.; Breman, E.; Cossu, T.A.; Kenney, S. The Conservation Value of Germplasm Stored at the Millennium Seed Bank, Royal Botanic Gardens, Kew, UK. *Biodivers. Conserv.* **2018**, *27*, 1347–1386. [CrossRef]
112. Chapman, T.; Miles, S.; Trivedi, C. Capturing, Protecting and Restoring Plant Diversity in the UK: RBG Kew and the Millennium Seed Bank. *Plant Divers.* **2019**, *41*, 124–131. [CrossRef] [PubMed]
113. Genesys. Genesys PGR. Available online: https://www.genesys-pgr.org/ (accessed on 18 October 2021).
114. RBG Kew. Millennium Seed Bank Partnership Data Warehouse—BRAHMS Online. Available online: http://brahmsonline.kew.org/msbp/SeedData/DW (accessed on 1 September 2021).
115. ENSCONET. ENSCOBASE—Online Database of the European Native Seed Conservation Network. Available online: http://enscobase.maich.gr/ (accessed on 1 September 2021).
116. Gardens4Science. Gardens4science—Online Catalog of Collections of Botanical Gardens. Germany. Available online: http://gardens4science.biocase.org/ (accessed on 19 September 2021).
117. Lindenmayer, D.; Scheele, B. Do Not Publish. *Science* **2017**, *356*, 800–801. [CrossRef]
118. Rodriguez Medina, L. A Geopolitics of Bad English. *Tapuya Latin Am. Sci. Technol. Soc.* **2019**, *2*, 1–7. [CrossRef]
119. Schellnack-Kelly, I.S. Decolonising the Archives: Languages as Enablers and Barriers to Accessing Public Archives in South Africa. *Arch. Manuscr.* **2020**, *48*, 291–299. [CrossRef]
120. RBG Kew; Instituto Humboldt Columbia. In-Colombia.Org—Proyecto Plantas y Hongos Útiles de Colombia. Available online: https://in-colombia.org/ (accessed on 2 September 2021).
121. Rojas, T.; Cortés, C.; Noguera, M.; Ulian, T.; Diazgranados, M. *Evaluación del Estado de los Desarrollos Bioeconomicos Colombianos en Plantas y Hongos*; Royal Botanic Gardens, Kew e Instituto de Investigaciones en Recursos Biológicos Alexander von Humboldt, 2020; p. 124.
122. GBIF. Global Biodiversity Information Facility. Available online: https://www.gbif.org/ (accessed on 2 September 2021).
123. The World Bank. Individuals Using the Internet (% of Population): International Telecommunication Union (ITU) World Telecommunication/ICT Indicators Database. Available online: https://data.worldbank.org/indicator/IT.NET.USER.ZS (accessed on 27 August 2021).
124. Dempewolf, H.; Eastwood, R.J.; Guarino, L.; Khoury, C.K.; Müller, J.V.; Toll, J. Adapting Agriculture to Climate Change: A Global Initiative to Collect, Conserve, and Use Crop Wild Relatives. *Agroecol. Sustain. Food Syst.* **2014**, *38*, 369–377. [CrossRef]
125. CBD. The Convention on Biological Diversity. Available online: https://www.cbd.int/convention/ (accessed on 2 September 2021).
126. BGCI. Implementing Access and Benefit Sharing. Available online: https://www.bgci.org/our-work/policy-and-advocacy/access-and-benefit-sharing/implementing-access-and-benefit-sharing/ (accessed on 10 September 2021).
127. Reyes-García, V.; Benyei, P. Indigenous Knowledge for Conservation. *Nat. Sustain.* **2019**, *2*, 657–658. [CrossRef]
128. Akeli, S.; Rimmer, Z.; Kautondokwa, N.; Pereira, F. Decolonising collections and exhibition management. In *Guidelines for German Museums: Care of Collections from Colonial Contexts*; Ahrndt, W., Museumsbund, D., Eds.; German Museums Association: Berlin, Germany, 2021; pp. 129–146.
129. Berg, G.; Raaijmakers, J.M. Saving Seed Microbiomes. *ISME J.* **2018**, *12*, 1167–1170. [CrossRef] [PubMed]
130. Wassermann, B.; Müller, H.; Berg, G. An Apple a Day: Which Bacteria Do We Eat With Organic and Conventional Apples? *Front. Microbiol.* **2019**, *10*, 1629. [CrossRef]
131. Carnegie, A.J.; Lidbetter, J.R.; Walker, J.; Horwood, M.A.; Tesoriero, L.; Glen, M.; Priest, M.J. Uredo Rangelii, a Taxon in the Guava Rust Complex, Newly Recorded on Myrtaceae in Australia. *Austral. Plant Pathol.* **2010**, *39*, 463. [CrossRef]

132. Carnegie, A.J.; Kathuria, A.; Pegg, G.S.; Entwistle, P.; Nagel, M.; Giblin, F.R. Impact of the Invasive Rust Puccinia Psidii (Myrtle Rust) on Native Myrtaceae in Natural Ecosystems in Australia. *Biol. Invasions* **2016**, *18*, 127–144. [CrossRef]
133. Toome-Heller, M.; Ho, W.W.H.; Ganley, R.J.; Elliott, C.E.A.; Quinn, B.; Pearson, H.G.; Alexander, B.J.R. Chasing Myrtle Rust in New Zealand: Host Range and Distribution over the First Year after Invasion. *Austral. Plant Pathol.* **2020**, *49*, 221–230. [CrossRef]
134. World Trade Organisation WTO | Understanding the Sanitary and Phytosanitary Measures Agreement. Available online: https://www.wto.org/english/tratop_e/sps_e/spsund_e.htm (accessed on 26 September 2021).
135. EUR-Lex Access to European Law. Document 32016R2031. Available online: https://eur-lex.europa.eu/eli/reg/2016/2031/oj (accessed on 27 August 2021).
136. Bezuidenhout, I. *Development of an Effective Phytosanitary Report Certification System for South African Compliance with the European Market*; University of Pretoria: Pretoria, South Africa, 2009.
137. Mink, G.I. Pollen and Seed-Transmitted Viruses and Viroids. *Annu. Rev. Phytopathol.* **1993**, *31*, 375–402. [CrossRef]
138. USDA USDA APHIS | Citrus Regulated Articles. Available online: https://www.aphis.usda.gov/aphis/ourfocus/planthealth/plant-pest-and-disease-programs/pests-and-diseases/citrus/citrus-federal-orders/citrus-policy-federal-orders (accessed on 17 September 2021).
139. Piet, S.; Mwanga, I.M.; Kambale, B.; Ntore, S.; Shalukoma, C.; Masumbuko, C.; Ramazani, E.; Lomboto, P.; Ebele, T.; Asimonyio, J.; et al. An Answer to the Coffee Challenge: From Herbarium to Coffee Genetic Resource Collections in the Democratic Republic of Congo. *BGjournal* **2019**, *16*, 20–24.
140. Ulian, T.; Diazgranados, M.; Pironon, S.; Padulosi, S.; Liu, U.; Davies, L.; Howes, M.R.; Borrell, J.S.; Ondo, I.; Pérez-Escobar, O.A.; et al. Unlocking Plant Resources to Support Food Security and Promote Sustainable Agriculture. *Plants People Planet* **2020**, *2*, 421–445. [CrossRef]
141. Vincent, H.; Wiersema, J.; Kell, S.; Fielder, H.; Dobbie, S.; Castañeda-Álvarez, N.P.; Guarino, L.; Eastwood, R.; León, B.; Maxted, N. A Prioritized Crop Wild Relative Inventory to Help Underpin Global Food Security. *Biol. Conserv.* **2013**, *167*, 265–275. [CrossRef]
142. Bharucha, Z.; Pretty, J. The Roles and Values of Wild Foods in Agricultural Systems. *Philos. Trans. R. Soc. Lond. B Biol. Sci.* **2010**, *365*, 2913–2926. [CrossRef] [PubMed]
143. Hunter, D.; Borelli, T.; Beltrame, D.M.O.; Oliveira, C.N.S.; Coradin, L.; Wasike, V.W.; Wasilwa, L.; Mwai, J.; Manjella, A.; Samarasinghe, G.W.L.; et al. The Potential of Neglected and Underutilized Species for Improving Diets and Nutrition. *Planta* **2019**, *250*, 709–729. [CrossRef]
144. Borelli, T.; Hunter, D.; Powell, B.; Ulian, T.; Mattana, E.; Termote, C.; Pawera, L.; Beltrame, D.; Penafiel, D.; Tan, A.; et al. Born to Eat Wild: An Integrated Conservation Approach to Secure Wild Food Plants for Food Security and Nutrition. *Plants* **2020**, *9*, 1299. [CrossRef]
145. Pieroni, A.; Hovsepyan, R.; Manduzai, A.K.; Sõukand, R. Wild Food Plants Traditionally Gathered in Central Armenia: Archaic Ingredients or Future Sustainable Foods? *Environ. Dev. Sustain.* **2021**, *23*, 2358–2381. [CrossRef]
146. Khoury, C.K.; Greene, S.L.; Krishnan, S.; Miller, A.J.; Moreau, T. A Road Map for Conservation, Use, and Public Engagement around North America's Crop Wild Relatives and Wild Utilized Plants. *Crop Sci.* **2019**, *59*, 2302–2307. [CrossRef]
147. Maxwell, S.L.; Fuller, R.A.; Brooks, T.M.; Watson, J.E.M. Biodiversity: The Ravages of Guns, Nets and Bulldozers. *Nature* **2016**, *536*, 143–145. [CrossRef] [PubMed]
148. Ulian, T.; Sacandé, M.; Hudson, A.; Mattana, E. Conservation of Indigenous Plants to Support Community Livelihoods: The MGU—Useful Plants Project. *J. Environ. Plan. Manag.* **2017**, *60*, 668–683. [CrossRef]
149. Ulian, T.; Florès, C.; Lira, R.; Mamatsharaga, A.; Mogotsi, K.K.; Muthoka, P.; Ngwako, S.; Nyamongo, D.O.; Omondi, W.; Sanogo, A.K.; et al. (Eds.) *Wild Plants for a Sustainable Future: 110 Multipurpose Species*; Royal Botanic Gardens, Kew: London, UK, 2019; ISBN 978-1-84246-673-5.
150. Hardwick, K.; Fiedler, P.; Lee, L.; Pavlik, B.; Hobbs, R.; Aronson, J.; Bidartondo, M.; Black, E.; Coates, D.; Daws, M.; et al. The Role of Botanic Gardens in the Science and Practice of Ecological Restoration. *Conserv. Biol.* **2011**, *25*, 265–275. [CrossRef] [PubMed]
151. Lawton, J.; Brotherton, P.; Brown, V.; Elphick, C.; Fitter, A.; Forshaw, J.; Haddow, R.; Hilborne, S.; Leafe, R.; Mace, G.; et al. *Making Space for Nature: A Review of England's Wildlife Sites and Ecological Network*; Report to Defra; UK Governement: London, UK, 2010.
152. Cunningham, C.A.; Thomas, C.D.; Morecroft, M.D.; Crick, H.Q.P.; Beale, C.M. The Effectiveness of the Protected Area Network of Great Britain. *Biol. Conserv.* **2021**, *257*, 109146. [CrossRef]
153. Hunter-Ayad, J.; Hassall, C. An Empirical, Cross-Taxon Evaluation of Landscape-Scale Connectivity. *Biodivers. Conserv.* **2020**, *29*, 1339–1359. [CrossRef]
154. HM Government. *A Green Future: Our 25 Year Plan to Improve the Environment*; UK Government: London, UK, 2018; p. 151.
155. RBG Kew. *Our Manifesto for Change 2021–2030*; RBG Kew: London, UK, 2021; p. 36.
156. Wu, Z.; Raven, P.H. *Flora of China*; Missouri Botanical Garden Press: St. Louis, MO, USA, 1994.
157. China Wild Plant Conservation Association. *China Strategy for Plant Conservation 2021–2030*; China Wild Plant Conservation Association: Beijing, China, 2019.
158. Wen, X. *Integrated Conservation of Rare and Threatened Woody Plants: Practice of and Perspectives from BGCI Programs in China (2010–2020)*; China Forestry Press: Beijing, China, 2020.
159. Villaseñor, J.L. Checklist of the Native Vascular Plants of Mexico. *Rev. Mex. Biodivers.* **2016**, *87*, 559–902. [CrossRef]
160. Tellez, O.; Mattana, E.; Diazgranados, M.; Kühn, N.; Castillo-Lorenzo, E.; Lira, R.; Montes-Leyva, L.; Rodriguez, I.; Flores Ortiz, C.M.; Way, M.; et al. Native Trees of Mexico: Diversity, Distribution, Uses and Conservation. *PeerJ* **2020**, *8*, e9898. [CrossRef]

161. Mattana, E.; Manger, K.R.; Way, M.J.; Ulian, T.; Garcia, R.; Encarnacion, W.; Clase, T.; Peguero, B.; Jimenez, F. A New Seed Bank for Hispaniola to Support the Conservation and Sustainable Use of the Caribbean Native Flora. *Oryx* **2017**, *51*, 394–395. [CrossRef]
162. Mattana, E.; Peguero, B.; Di Sacco, A.; Agramonte, W.; Encarnación Castillo, W.R.; Jiménez, F.; Clase, T.; Pritchard, H.W.; Gómez-Barreiro, P.; Castillo-Lorenzo, E.; et al. Assessing Seed Desiccation Responses of Native Trees in the Caribbean. *New For.* **2020**, *51*, 705–721. [CrossRef]
163. Ceci, P.; Sanogo, S.; Ambrose, E.; Sacande, M.; Sanou, L.; Adda, M.; Ulian, T. Reversing Land Degradation and Desertification in Africa's Drylands. *BGjournal* **2018**, *15*, 19–22.
164. Sacande, M.; Berrahmouni, N. Community Participation and Ecological Criteria for Selecting Species and Restoring Natural Capital with Native Species in the Sahel. *Restor. Ecol.* **2016**, *24*, 479–488. [CrossRef]
165. Kelly, B.A.; Sanogo, S.; Sidibé, S.I.; Castillo Lorenzo, E.; Ceci, P.; Ulian, T. Restoring Vegetation and Degraded Lands by Using Assisted Natural Regeneration Approach (ANRA): Case Study at Bankass in the Centre of Mali, West Africa. *Environ. Dev. Sustain.* **2021**, *23*, 14123–14139. [CrossRef]
166. Kelly, B.A.; Sanogo, S.; Sidibé, S.I.; Ceci, P.; Castillo-Lorenzo, E.; Ulian, T. Survival and Growth of Adansonia Digitata L. Tall Bare-Roots in Farm Fields: A Case Study from Bankass, Mali (West Africa). *Environ. Dev. Sustain.* **2021**. [CrossRef]
167. Bretting, P.K. 2017 Frank Meyer Medal for Plant Genetic Resources Lecture: Stewards of Our Agricultural Future. *Crop Sci.* **2018**, *58*, 2233–2240. [CrossRef]
168. Fu, Y.-B. The Vulnerability of Plant Genetic Resources Conserved Ex Situ. *Crop Sci.* **2017**, *57*, 2314–2328. [CrossRef]
169. GenRes Bridge. European Genebanks Start to Review Each Other. Available online: http://www.genresbridge.eu/about-us/news/news/european-genebanks-start-to-review-each-other/ (accessed on 19 September 2021).

plants

MDPI

Review

Effective Coordination and Governance of PGRFA Conservation and Use at the National Level—The Example of Germany

Frank Begemann *, Imke Thormann, Sarah Sensen and Karina Klein

Federal Office for Agriculture and Food, 53179 Bonn, Germany; imke.thormann@ble.de (I.T.); sarah.sensen@ble.de (S.S.); karina.klein@ble.de (K.K.)
* Correspondence: frank.begemann@ble.de

Abstract: Recognition of the importance of plant genetic resources started in Germany at the end of the 19th century. Plant research and breeding began to develop in the 1920s. Formal structures of public institutions were founded, long-term conservation facilities were established, private breeding initiatives developed. In 1990, the German reunification required an assessment of the existing research and breeding landscape. This milestone allowed a comprehensive overview of the great number of stakeholders, active in the entire range of tasks related to plant genetic resources. The Federal Ministry of Agriculture then developed a conceptual approach for an efficient governance structure and published its concept of a national programme for the conservation and sustainable use of genetic resources for food and agriculture in 2000. It recognized the sharing of decentral responsibilities among the respective public and private actors and governmental levels with dis-tributed mandates and funding. It also led to the establishment of a central information and coordination center for genetic resources, which facilitates the data sharing, communication, and co-operation among stakeholders, supports public awareness and advises the Federal Ministry on national policies and efficient European and global cooperation. It also supports efficient contributions of German stakeholders into European structures and international bodies. An equivalent conceptual approach and governance structure is recommended to be established at European level.

Keywords: plant genetic resources for food and agriculture; conservation and use; national coordination and governance structure; Germany

Citation: Begemann, F.; Thormann, I.; Sensen, S.; Klein, K. Effective Coordination and Governance of PGRFA Conservation and Use at the National Level—The Example of Germany. *Plants* **2021**, *10*, 1869. https://doi.org/10.3390/plants10091869

Academic Editor: Oscar Vicente

Received: 17 August 2021
Accepted: 3 September 2021
Published: 9 September 2021

1. Historic Background until 1945

In Germany, traditional, locally adapted landraces were used in agricultural production and by the mid of the 19th century a few individuals began to select and improve such landraces for better yields. Christian Adolf Leberecht Steiger in Leutewitz (fodder beets 1825), Wilhelm von Borries in Eckendorf (fodder beets 1846), Wilfried Paulsen (potato 1846), Mathias Rabbethge, and others (sugar beets 1862) and Wilhelm Rimpau (rye 1867) were a few of them [1]. The introduction of the "squarehead" wheat from England and its appreciated value for crop production created an inspiring environment and further incentives for early breeding initiatives in Germany. In 1869, the first public quality control station for agricultural, forestry, and horticultural seed was founded at the Royal Academy for Foresters and Farmers in Tharandt. The first director, Friedrich Nobbe, published a "handbook of seed science" in 1876, which became influential for further seed research. Seed testing began and the publication of the test results created a growing demand for improved varieties by farmers and also promoted the further development of breeding companies [1].

The importance of landraces as "plant genetic resources" has been recognized in Germany as early as by the end of the 19th century, when Emanuel Ritter von Proskowetz and Franz Schindler acknowledged the "use value" of traditional landraces for agricultural production in general and plant breeding in particular at the International Agricultural and Forestry Congress in Vienna in 1890 [2].

The Russian scientist Nikolai Ivanovich Vavilov started plant explorations in 1916 and collected landraces worldwide [3]. He described centers of diversity of cultivated plants and presented them in 1927 at the 5th International Genetics Congress in Berlin. Vavilov's work greatly influenced further activities in Germany. At the beginning of the 1930s, the General Assembly of the International Plant Breeders Association in Berlin concluded far-reaching recommendations related to the conservation of landraces of cultivated plants. Representatives of all countries were requested to approach their governments to collect and conserve the diversity of traditional landraces existing in their respective countries. For this purpose, in Germany it was recommended that appropriate farmers or institutes maintain and manage traditional landraces in the original planting areas following the old cropping practices [4].

A new Kaiser-Wilhelm-Institute of Research on Plant Breeding was founded in Müncheberg in 1927, with Erwin Baur (1875–1933) becoming its first director. Although this institute collected germplasm samples in Turkey and some Latin American countries its main concern had not been the long-term aspect of conservation but rather topics such as crop plant evolution and genetics. The institute and Erwin Baur developed many fundamental impulses for further research related to plant genetic resources. Hence, in 1943, another Kaiser-Wilhelm-Institute of Crop Plant Research was founded at the Tuttenhof farm near Vienna with its director Hans Stubbe (1902–1989), a student of Erwin Baur. Stubbe laid the foundation of the institute's future crop plant-related research in a wider interpretation, including the collecting and conservation of global crop diversity and their research in the important botanic disciplines such as systematics, ecology, genetics, biochemistry, biophysics, and physiology [2]. By the end of World War II, the institute's important plant collection was moved to Gatersleben in the Eastern part of Germany [5].

It should be noted that with the increased industrialization of German agriculture at the beginning of the 20th century, the appreciation of traditional German landrace diversity was turned upside down. While important plant explorations were organized into other countries and continents (e.g., Asia, Latin America), the NAZI regime conducted a "plant variety cleaning". This cleaning process was meant to protect the German farmers from inferior seed and resulted in a loss of some three quarters of the varieties of the main crops from the seed market [6].

2. Period of a Divided Germany (1945–1990)

After World War II, Germany was divided into a Western part, the Federal Republic of Germany, and an Eastern part, the German Democratic Republic. Due to the geographical division and the different political systems of the two German parts, the plant genetic resources activities developed separately throughout the various locations.

In Eastern Germany important crop collections were located in Gatersleben and Halle and a fruit collection existed in Dresden-Pillnitz. The collection of the Institute for Plant Breeding in Halle was transferred to Gatersleben between 1945 and 1950 [2]. At the same time, in Western Germany, a part of the collections of the former Kaiser-Wilhelm-Institute of Research on Plant Breeding was moved to the Max-Planck-Institute for Plant Breeding Research at Cologne-Vogelsang. Today's Julius Kühn-Institute for Grapevine Breeding near Siebeldingen in the South-Western part of Germany became the location for the national grapevine collection [5]. Beside these public research institutes, many private breeding companies maintained their own breeding collections.

2.1. Eastern Germany

In Eastern Germany, a plant breeding and seed service infrastructure was established by the Sowjet military administration already in 1946. While throughout the country 63 seed selection and development farms (seed stations) with a total area of 27,729 ha were set up, a central coordination mandate of these farms was taken over by the German

Seed Association (DSG). The DSG was responsible for the coordination, documentation, collection, and distribution of seed and planting material. Seed production plans became obligatory. Already existing small seed initiatives and institutes could still remain independent but had to follow the instructions of the DSG [7].

In addition, crop-oriented breeding research institutes were established or continued such as the Institute Quedlinburg (vegetables), Institute Bernburg (cereals, maize, forages, special crops), Institute Groß Lüsewitz (potatoes), Institute of Crop Plant Research, Gatersleben, Institute of Phytopathology, Aschersleben, Agricultural Faculty, Martin-Luther-University Halle-Wittenberg and the Institute of Breeding Research Müncheberg. Important breeders also continued their breeding activities in Hadmersleben, Gülzow, and Klein Wanzleben.

After the foundation of the German Democratic Republic in 1949, the new German Academy for Agricultural Sciences (DAL) was founded in 1951 with Hans Stubbe becoming the first president. The former DSG breeding research institutes were now integrated in the Academy as DAL-institutes [7].

- Institute Quedlinburg (breeding research, vegetables)
- Institute Bernburg (cereals, maize) and Research Station Hadmersleben (cereals, lupines)
- Institute Klein Wanzleben (beets)
- Institute Groß Lüsewitz (potatoes)
- Institute Gülzow (cereals etc.)
- Institute of Fruit Research Dresden-Pillnitz
- Institute of Forages Paulinenaue

Essential research and breeding partners of these DAL-Institutes (later called AdL-Institutes) were the Institute of Crop Plant Research (Genebank) Gatersleben, the Institute of Phytopathology Aschersleben, the Agricultural Faculty, Martin-Luther-University Halle-Wittenberg, and numerous existing seed stations. Additional nationally owned seed and planting material companies (VVB) were responsible for country-wide seed supply. The "VVB Seed and Planting Material" comprised 110 farms with over 100,000 ha farmland area [7].

The Institute of Crop Plant Research in Gatersleben played a unique role as it was the home of the important genebank collection of plant genetic resources. The institute was placed within the German Academy for Sciences (DAW) in 1948 and, later in 1972, within the scope of the Academy of Sciences (AdW). From 1970 to 1991 this Institute was called Central Institute of Genetics and Crop Plant Research (ZIGuK) before it was renamed in 1992 in Institute of Plant Genetics and Crop Plant Research (IPK). The institute, with its first director Hans Stubbe, conducted numerous collecting and exploration missions and received plant genetic resources from other collection holders. This growing genebank collection of global importance (see Table 1) was and still continues to be intensively used for research on taxonomy, genetics, phylogeny, evolution, and breeding of cultivated plants. The genebank places emphasis and conducts research on the systematics, characterization, evaluation, and documentation of the conserved plant genetic resources as a fundamental service for the research and breeding activities.

Table 1. Size of genebank collections maintained at Gatersleben [2].

Time	Number of Accessions
1945	app. 3500
1950	12,550
1960	20,197
1970	32,489
1980	48,959
1989	65,756

After 1945, numerous breeding or seed companies left the Eastern part of Germany and tried to settle in the Western part of Germany. As an indication, a list of companies provided by Röbbelen [8] gives an idea of this process. Out of 74 listed companies from Eastern Germany, 40 companies lost their private identity due to the land reform and 34 companies were reported to have moved to the West. The number of companies, including the ones existing already in the Western part of Germany, with at least one variety in the official seed list, was reported to be 277 in 1949 and 264 in 1975. Many of the companies were constituted as farmers' cooperatives or other forms of production associations [8].

2.2. Western Germany

While in Eastern Germany the research, breeding, and seed systems were almost entirely based upon public institutions and infrastructure, in Western Germany the systems were continued and built upon private initiatives and companies. In the West, public institutions were active especially in the areas of research, education, and seed quality control. The Western part of Germany was controlled after the Second World War by the USA, France, and Great Britain, leading in 1949 to a constitution as a federal republic of states (*Laender*) with different mandates at the Federal and *Laender* levels. As there was no central coordination yet, the breeder associations were also divided in three parts, the Association of Plant Breeders (VdP) in Hanover, the Bavarian Breeders' Association (VBP) in Munich, and the South-Western Breeders' Association (VSWP) in Stuttgart. In 1962 (VdP and VBP) and 1966 (VSWP) these three associations were merged and formed the new German Plant Breeders' Association (BDP) in Bonn (1970). Throughout the years, many private breeding companies extended their businesses based on a particular crop focus, diversification across different crops, and/or by international cooperation [8].

In Western Germany, public research and education related to plant breeding and plant genetic resources were embedded in leading university institutes, especially at Stuttgart-Hohenheim, Göttingen, Freising-Weihenstephan, Hanover, Giessen, Bonn and Kiel, as well as in regional research institutions. Since the 1950s, the Max-Planck-Institute for Plant Breeding Research at Cologne-Vogelsang also worked on forages and other crops.

In 1965, the possibly unique German Federation for the Promotion of Plant Breeding (GFP) was founded in Hanover to maintain and support private breeding initiatives, to assist in knowledge transfer from latest scientific developments, to facilitate the adoption for technical implementation, to support and enable high economic value-addition and to support international cooperation [9]. As such, the GFP (today GFPi) created a link between public research and private plant breeding.

The growing necessity to also conserve plant genetic resources for future research and breeding in Western Germany was addressed and promoted especially by Hermann Kuckuck, Dieter Bommer, and the GFP. Based on their initiative, in 1970, a new genebank in Western Germany was established by the Federal Ministry of Agriculture at the Institute of Agronomy and Plant Breeding of the Federal Research Centre for Agriculture (FAL) in Braunschweig [4]. The genebank in Braunschweig was developed to provide services especially to the privately structured plant breeding system. This concept was, however, different to the much wider comprehensive plant genetic resources approach of the genebank in Gatersleben.

2.3. West German Recognition of International Developments

During the 1970s, the Federal Republic of Germany commenced to support international activities to promote the conservation and use of plant genetic resources. In 1971, the Consultative Group on International Agricultural Research (CGIAR) was established. A number of the international agricultural research centers of the CGIAR focused on the improvement of crop plants, collected plant genetic resources, and established genebanks to support their improvement programmes.

The Food and Agriculture Organization of the United Nations (FAO) played a key role to support international cooperation and communication related to plant genetic resources activities. In 1974, largely initiated by the FAO, the International Board for Plant Genetic Resources (IBPGR) was created to support the research, collecting, conservation, documentation, evaluation, and use of the genetic diversity of cultivated plants worldwide. It was also active to organize a global network of genebanks holding base collections within and outside the CGIAR [10].

The growing awareness of the importance of plant genetic resources led to an international agreement at the FAO in 1983, the so-called International Undertaking on Plant Genetic Resources [11] and the establishment of a Commission on Plant Genetic Resources [12], which the Federal Ministry of Agriculture was committed to. A central element of the International Undertaking was stated in article 7.1 (a):

> *"there develops an internationally coordinated network of national, regional, and international centers, including an international network of base collections in gene banks, under the auspices or the jurisdiction of FAO, that have assumed the responsibility to hold, for the benefit of the international community and on the principle of unrestricted exchange, base or active collections of the plant genetic resources of particular plant species".*

Inspired by these formal international processes and developments in plant research, the Federal Ministry of Agriculture initiated an assessment of related activities in 1985 [10].

3. German Reunification (1990)—An Assessment of the Plant Genetic Resources System

While the initiation of the assessment of PGR in Western Germany was started in 1985, the results were only published in 1990 [10]. It was based on the definition of plant genetic resources as laid down in the International Undertaking [12], where plant genetic resources meant the reproductive or vegetative propagating material of the following categories of plants:

1. Cultivated varieties (cultivars) in current use and newly developed varieties;
2. Obsolete cultivars;
3. Primitive cultivars (landraces);
4. Wild and weedy species, near relatives of cultivated varieties;
5. Special genetic stocks (including elite and current breeders' lines and mutants).

Hence, in a broad interpretation, the assessment addressed resources in Western Germany, which were [10]:

- Important for breeding of actual and potential crops as well as important from an ecology perspective and for the conservation of the vegetation in Germany;
- Required for plant breeding and land improvement, considering the scope of agriculture, horticulture, pomiculture, forestry and landscape management;
- Important in relation to nature conservation, ecosystems and protection of wild species;
- Maintained by Federal and *Laender* institutions and non-governmental organizations.

The most important institution in Western Germany at the time of reunification was the Institute of Agronomy and Plant Breeding of the Federal Research Centre for Agriculture (FAL) in Braunschweig, holding a large genebank collection. Since 1980, endangered wild species in Germany were also integrated in this collection. Reference samples of the Federal Plant Variety Office (BSA) and related information were handed over to the FAL-genebank for varieties after the expiry of their variety protection period. Evaluation and documentation of the resources were a particular priority of the institute. Annually, 7000–8000 samples were provided to recipients at their requests; about 1/3 of the requests were received from abroad [10].

Apart from the registration of test collections maintained at the Federal Plant Variety Office for the testing period, important collections of agricultural crops were held at *Laender* institutes, universities, Max-Planck institutes, and private breeding companies.

For horticultural crops, many fruit and vegetable collections were identified at the FAL, the BSA, the Federal Research Institute for Horticultural Plant Breeding in Ahrensburg and more than 20 *Laender* and county institutes as well as numerous private companies. This diverse situation was similar for genetic resources of ornamental plants, where botanic gardens, universities, and outdoor museums played a significant role as well. The Federal Institute of Grapevine Breeding in Siebeldingen and six *Laender* institutes maintained the main genetic resources collections for grapevine research and breeding.

The plant genetic resources concept with its assessment [10] also covered forest genetic resources. The Federal Forestry and Wood Research Institute, *Laender* forestry research institutes, forest administration, arboreta and private forest owner were the key actors involved. These activities were coordinated by the Federal—*Laender* Working Group "Conservation of Forest Genetic Resources" [13].

Activities for the conservation of genetic diversity of wild species were part of the nature and landscape protection measures. Federal and *Laender* research institutes, universities, botanic gardens, outdoor museums, nature conservation associations and individuals are engaged in these activities. The main conservation responsibilities, however, were located at the *Laender* level.

The main collection of microbial genetic resources existed at the German Collection of Microorganisms (DSM; today DSMZ) in Braunschweig. The DSM was also the International Depositary Authority under the Budapest Treaty on the International Recognition of the Deposit of Microorganisms for the Purposes of Patent Procedure in Germany. The collection consisted of some 4200 strains of microorganisms [10].

At the time of the German reunification in 1990, the study by Bommer and Beese [10] presented the activities related to plant genetic resources in Western Germany. However, due to the reunification an overall assessment in both parts of Germany was needed. Hence, two additional studies were conducted and are important to mention: the so-called "genebank study" issued by the Federal Ministry for Research and Technology and presented by the Association of Academic Societies of Agricultural, Forestry, Food, Veterinary and Environmental Research (DAF) [14], and the assessment by Begemann and Hammer, published by the Federal Ministry for Food, Agriculture and Forestry (BMELF) [5].

These studies provided a comprehensive picture and revealed a large number of public institutions being involved in a wide range of activities related to plant genetic resources for research and breeding. The two large genebanks at the IPK in Gatersleben and at the FAL in Braunschweig played a leading role. They also maintained long-term base collections within the international network of base collections supported by the IBPGR [15]. In addition, in 1992, the Federal Centre for Breeding Research of Cultivated Plants (BAZ) was established in Quedlinburg. The BAZ integrated the former Institute of Phytopathology in Aschersleben, the Institute of Breeding Research in Quedlinburg, the Institute of Potato Research in Groß Lüsewitz, the Institute of Fruit Research in Dresden-Pillnitz (former AdL-institutes), and the former independent Federal Institute of Grapevine Breeding in Siebeldingen and the Federal Research Institute for Horticultural Plant Breeding in Ahrensburg. An overview of the most important genebank collections at IPK, FAL and BAZ in 1992 are shown in Table 2.

Furthermore, about 100 private breeding and seed companies were operational for the German and, partially, for the international seed markets. A number of non-governmental organizations were engaged in plant genetic resources activities such as on-farm conservation and management. Pioneering examples were the Association for the Conservation of Crop Plant Diversity (VEN) founded in 1986, the Association for the Promotion of Seed Research in Biodynamic Agriculture in 1988, the Association of Pomologists in 1991 or later the Association for the Conservation and Recultivation of Cultivated Plants (VERN) in 1996.

It was evident that there were different levels of governmental mandates at the Federal and *Laender* levels. The main responsibility for nature protection, including the conservation of genetic resources, as well as for academic institutions such as universities rests with the *Laender* governments. The Federal government oversees collaborative activities across the *Laender*, providing a national policy framework, national documentation and monitoring, and providing for the international cooperation.

Table 2. Most important genebank collections in Germany at IPK, FAL, and BAZ in 1992 [5].

Crop Species (Groups)	IPK	FAL [1]	BAZ [2]
Cereals	36.095	29.467	
Legumes	16.850	9.030	
Oil crops and fibres	5.711	3.222	
Beets and potatoes	6.580	6.265	
Fodder crops	11.142	2.797	
Tobacco	463	43	
Other agricultural crops		1.155	
Vegetables	9.962	2.237	5.000
Medicinals and spices	2.570	1.090	
Mutants etc.	2.614	1.814	900
Ornamentals	1.961		380
Fruits	1.988		163
Grapevine			2.027
Total	95.936	57.120	8.470

[1] as of 15th August 1991, [2] use of estimates.

The plant genetic resources system was marked by a rich but scattered research landscape, by a few multi-crop and many crop-oriented genebanks and research institutes. A certain degree of duplication of the respective public research centers, in particular between the two large multi-crop genebanks in Gatersleben and Braunschweig, was highlighted. It was recommended to integrate the collection of the genebank in Braunschweig into the genebank in Gatersleben, which was implemented over a certain period of time to avoid any loss of material or knowledge related to the collection and was concluded in 2004.

4. Overall Coordination and Governance Structure of Genetic Resources for Food and Agriculture

As part of the initial assessment in Western Germany, before the reunification, a plant genetic resources concept was elaborated on how to integrate the multitude of stakeholders, measures, and programmes and better prepare the entire plant genetic resources system for future challenges and opportunities [10].

Background to this so-called "Bommer and Beese" concept [10] was the recognition of the loss of species diversity and genetic erosion on the one hand and, on the other hand, new opportunities arising from recent scientific developments especially in the biological sciences and information technology, which allowed new developments of the potential use of genetic resources.

The proposed plant genetic resources concept covered a wide range of agricultural crops including fruit crops, vegetables, ornamentals, grapevine, forest, and wild species including crop wild relatives. Moreover, microorganisms were considered. It emphasized the promotion of research and conservation efforts and recognized different mandates related to the federal structure of Germany. Core elements of the concept were the following: the Federal Ministry of Agriculture itself, an inter-disciplinary Advisory Board for Plant Genetic Resources, an Information and Coordination Centre for Genetic Resources with a central documentation system, and crop committees for different crops (crop groups). Besides these new bodies, the existing system with the well-functioning conservation, research, and breeding institutions at Federal and *Laender* levels should remain as it was operating by that time [10].

This decentral or distributed system with a central coordination unit was considered advantageous over a combined and centralized system in one large plant genetic resources institution or a completely decentral system of individual institutions. Advantages were seen in securing the necessary continuity of this long-term task, in supporting the interdisciplinary collaboration, in maintaining the plant genetic resources activities as part of a broader institutional research setting, and in using available infrastructure for new conservation measures [10].

A first step to implement the new components of the plant genetic resources concept was the establishment of the proposed Information and Coordination Centre for Genetic Resources (IGR) in April 1991. The IGR was located at the German Centre for Agricultural Documentation and Information (ZADI), an institution under the Federal Ministry of Agriculture. The IGR started and evolved in a stepwise manner. Initially, the main task was to develop a national database for plant genetic resources and support the exchange of data with other national and international databases. As from 1993, with its new director Frank Begemann, the following tasks were added:

- To provide advice to the Federal and *Laender* ministries;
- To support the Federal Ministry of Agriculture
    - To represent Germany in international bodies such as the FAO Commission on Plant Genetic Resources, the European Cooperative Programme for Plant Genetic Resources (ECPGR) and bodies at the European Commission;
    - To prepare the 4th International Technical Conference of FAO for Plant Genetic Resources (1996) in Leipzig;
- To provide the Secretariat to the National Committee that was asked to prepare the German national report for this 4th International Technical Conference of FAO;
- To collect, analyze and disseminate information about national and international conservation measures;
- To undertake public awareness activities;
- To support collaboration between the formal and informal sectors;
- To support national conservation measures in line with international activities;
- To contribute to improved links between conservation and use activities.

The National Committee that was created for the 4th International Conference of the FAO, consisted of representatives of all stakeholders involved such as different ministries, science and research, the private sector, non-governmental organizations, international cooperation agencies, associations, farmers organizations, and nature conservation agencies [16]. This inclusive composition proved to be very useful and was taken up a few years later when a formal national committee for plant genetic resources had to be formed.

It is worth mentioning that the IGR could not support and coordinate activities through a dedicated budget line under its control. It rather facilitated conservation, documentation and sustainable use of plant genetic resources merely through appropriate information and communication means. This turned out to be effective and supported the collaboration between stakeholders within the large and diverse plant genetic resources system in Germany.

Step by step, the usefulness of these information and coordination functions of the IGR in the plant domain was recognized by the Federal Ministry officials in the domains of animal, forest, and aquatic genetic resources.

Based on this experience, and considering international processes such as the broadening of the scope of the FAO Commission on Plant Genetic Resources to the new Commission on Genetic Resources for Food and Agriculture (CGRFA) in 1996, the German Federal Ministry of Food, Agriculture and Forestry published a landmark "Concept for the Conservation and Sustainable Use of Genetic Resources for Food, Agriculture and Forestry" in 2000 [17]. This concept was built on the former West German concept for plant genetic resources [10].

As far as the coordination and governance structure is concerned, in essence, the genetic resources concept of 2000 [17] is still operational today. The main components of this coordination and governance structure with updated names of their functional entities are (Figure 1):

- Federal Ministry of Food and Agriculture (BMEL) with a coordination division related to genetic resources for food and agriculture and additional domain-specific divisions for plant, animal, forest and aquatic genetic resources;
- Agrobiodiversity strategy and national programmes for plant, animal, forest and aquatic genetic resources;
- National expert committees for plant, animal, forest and aquatic genetic resources, consisting of *Laender* authorities and a wide range of experts and stakeholders.
- Scientific Advisory Board for Biodiversity and Genetic Resources;
- Information and Coordination Centre for Biological Diversity (IBV) (successor of the former IGR) at the Federal Office for Agriculture and Food (BLE);
- National inventories for plant, animal, forest and aquatic genetic resources;
- National information platform/website (https://genres.de/en/)

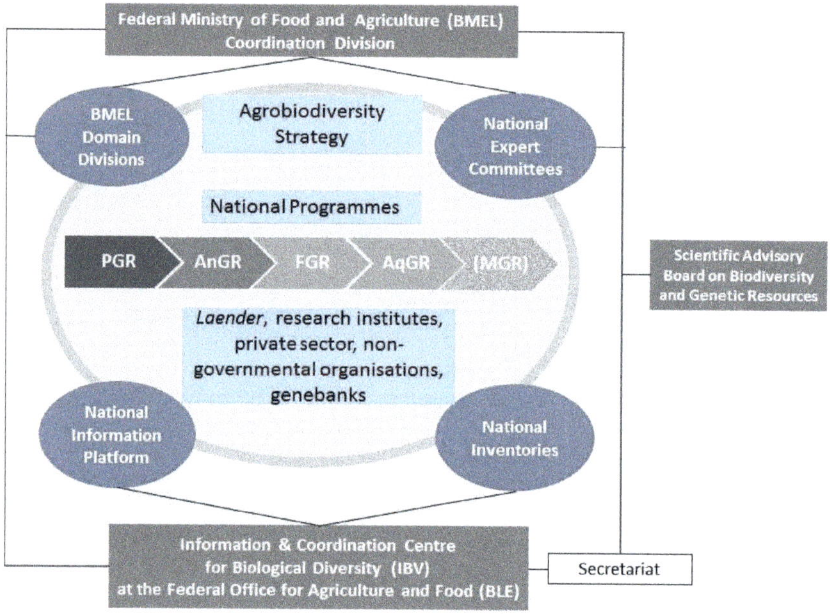

Figure 1. Coordination and governance structure of the National Concept for the Conservation and Sustainable Use of Genetic Resources for Food, Agriculture and Forestry (revised from [17]).

In 2007, the Federal Ministry of Food, Agriculture and Consumer Protection (BMELV) published a so-called "Agrobiodiversity Strategy", in furtherance of the genetic resources concept from 2000 and as a supplement to the National Biodiversity Strategy [18]. The title of the Agrobiodiversity Strategy also sums up the mission statement: "Preserving agrobiodiversity, tapping the potential of agriculture, forestry and fisheries and making sustainable use of it". To achieve this, the strategy has three overarching aims:

- Achieve long-term conservation and broader-based use of genetic resources;
- Achieve sustainable use of agricultural biodiversity while protecting natural ecosystems and threatened species;
- Strengthen international cooperation and a globally coordinated strategy for the management of global resources.

In addition to the overall agrobiodiversity strategy, domain-specific national programmes for genetic resources for plant, animal, forest, and aquatic genetic resources were developed. They are functional instruments to describe the detailed measures to be implemented throughout Germany in each of the domains within a certain time period, to set priorities, to facilitate the monitoring of activities and achievements, and to assist in linking the required stakeholder groups. A potential programme for microbial genetic resources is being considered. The programmes are updated from time to time to remain useful instruments for priority setting and implementation.

The implementation of the programmes is based upon the active involvement of a wide range of stakeholders and constituencies involved in all areas such as the identification, collecting, conservation, documentation, characterization, evaluation, and other uses of genetic resources for research and breeding as well as for direct uses for food and agriculture, horticulture, pomiculture, viticulture, forestry, and fisheries. The stakeholders participate in their respective fields of competence and measures based upon their existing respective responsibilities and budgets.

National expert committees for plant, animal, forest, and aquatic genetic resources, consisting of *Laender* authorities and a wide range of experts and stakeholders are in charge of guiding the implementation of the national programmes.

According to the amended scope of the genetic resources concept and the new agrobiodiversity strategy and in recognition of the usefulness of an information and coordination entity, the former Information and Coordination Centre for Genetic Resources (IGR) was renamed into Information and Coordination Centre for Biological Diversity (IBV) and was integrated in the Federal Office for Agriculture and Food (BLE) in 2005. The current tasks of the IBV relate to agrobiodiversity, in particular to genetic resources for food, agriculture, forestry, and fisheries; they include inter alia:

- To provide advice to the Federal and *Laender* ministries;
- To support the BMEL representing Germany in international bodies such as the FAO Commission on Genetic Resources (CGRFA), the European Cooperative Programme for Plant Genetic Resources (ECPGR), the European Regional Focal Point of Animal Genetic Resources (ERFP) and bodies at the European Commission;
- Provides national coordinator for ECPGR;
- Support of the development and implementation of the national programmes for plant, animal, forest and aquatic genetic resources;
- Secretariat for the national expert committees of the Federal Ministry for Agriculture (BMEL) for plant, animal, forest and aquatic genetic resources, as well as for the Scientific Advisory Board for Biodiversity and Genetic Resources;
- Data collection and documentation of national inventories as well as user-oriented central dissemination of information on occurrences, characteristics and performance of genetic resources for food, agriculture, forestry and fisheries;
- Monitoring and assessment of agrobiodiversity trends in Germany;
- Coordination of conservation activities and assistance to conservation networks;
- Facilitation of national and international support measures and programmes;
- Knowledge transfer and advisory services for political decision makers and other stakeholders;
- Biopatent monitoring and access and benefit-sharing (ABS) issues;
- Public relations and awareness raising.

The Scientific Advisory Board on Biodiversity and Genetic Resources was constituted in 2003. The Board advises the BMEL on general and fundamental issues relating to the conservation and sustainable use of biological diversity at national, EU, and international level. Members of the Board are scientists from different disciplines appointed by the Federal Ministry of Agriculture, the four chairpersons of the national expert committees on plant, animal, forest, and aquatic genetic resources, as well as the director of the IBV. The main topics to be considered by the board are:

- Biological and ecological basics;
- Economic, social and ethical evaluation;
- Development of science and technology, including genetics and breeding;
- Land use, landscaping and rural areas;
- Importance for raw materials, energy, nutrition and health;

- Promotion of strategies and concepts;
- Legal, policy and ethical issues;
- Information and communication, marketing and awareness.

While the BMEL is setting the policy framework, the implementation of the national programmes remains under the responsibility of all stakeholders. Due to the broad scope of the demanding programmes a wide range of actors are involved such as Federal and *Laender* institutes, genebanks, research institutions, fisheries, the private sector and non-governmental organizations, including universities, agricultural and horticultural actors, breeders, farmers and foresters, nature conservation, botanic and zoological gardens.

The IBV is keeping an oversight and facilitates the implementation of the national programmes through information, documentation, communication, and coordination measures. A specific website (https://genres.de/en/) is dedicated to providing the overall information platform of all programmes and stakeholders as well as the national inventories and a newsletter.

5. Coordination and Governance Structure of the German Plant Genetic Resources System

Given the global challenges as agreed by the 2030 Agenda for Sustainable Development with its 17 Sustainable Development Goals (SDG) and the particular importance of climate change, loss of biodiversity, and food security, the essential role of plant genetic resources for food and agriculture is evident. These resources are fundamental for further research and plant breeding to support the diversification of the agriculture and food system and contribute to the climate change adaptation processes.

Based on its historical developments, the German plant genetic resources system (see Figures 2 and 3) is characterized by an effective long-term conservation infrastructure with internationally recognized genebanks and well-qualified plant research institutions. A large number of breeding and seed companies operate in Germany and offer, as of July 2021, 2635 varieties of agricultural species and 640 varieties of vegetable species at the European seed catalogues to farmers. A wide range of non-governmental organizations and individuals are engaged in conservation and management of plant genetic resources on farms or in gardens.

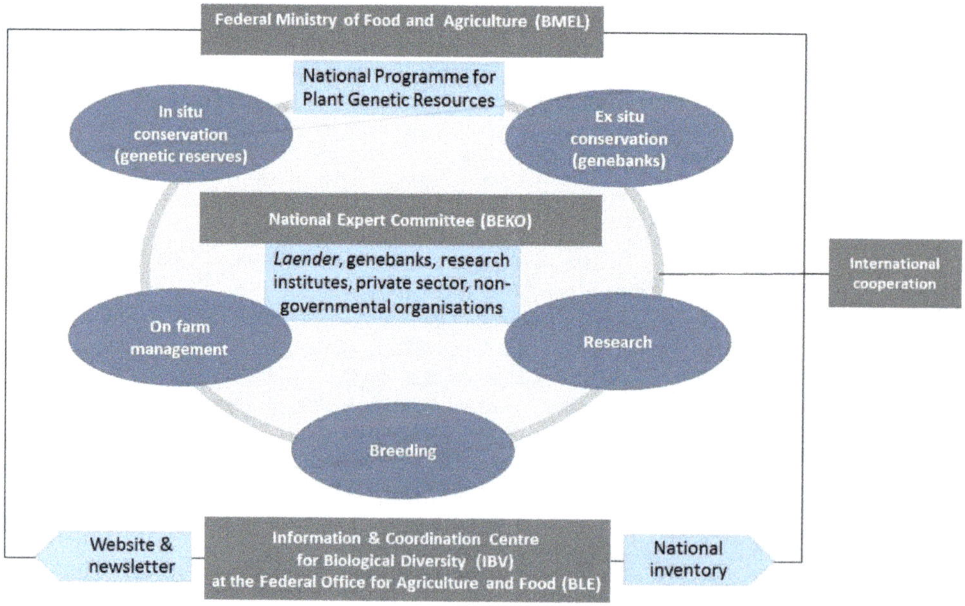

Figure 2. Elements of the German plant genetic resources system.

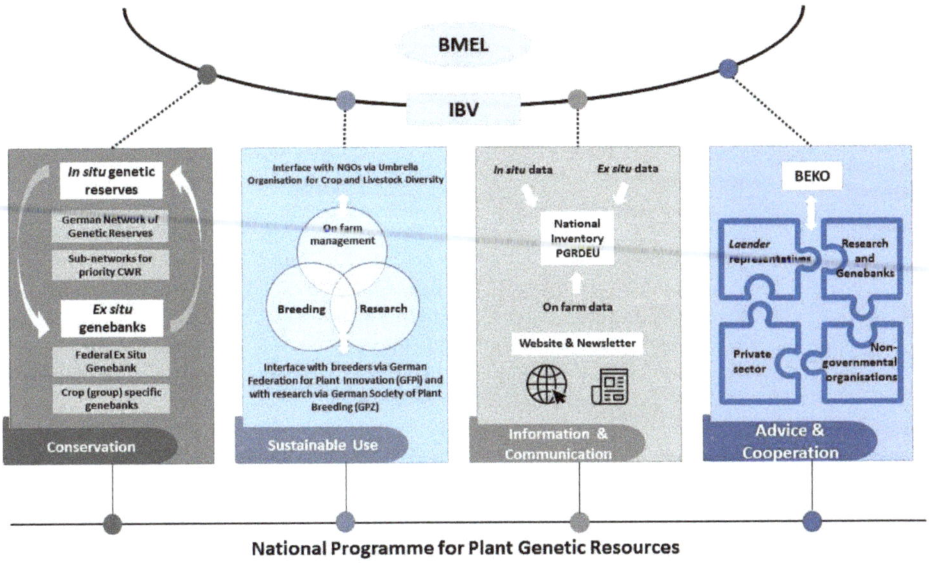

Figure 3. Coordination and governance structure of the German plant genetic resources system.

The National Programme for the Conservation and Sustainable Use of Plant Genetic Resources of Agricultural and Horticultural Crops [19] describes the main objectives and measures to be undertaken at the national level. The first national PGR programme was adopted by the Conference of Agriculture Ministers in 2002. It was fundamentally revised in 2012. This second programme is based on the second Global Plan of Action for Plant Genetic Resources for Food and Agriculture of FAO (GPA2). It describes the political and legislative framework at national, European, and international level. The national priority activities are grouped in accordance with the main elements of GPA2, i.e., specifically address ex situ conservation, in situ conservation including both on farm management and in situ conservation of crop wild relatives (CWR), sustainable use, as well as information and documentation. The next revision and updating cycle of the national programme has already been initiated by the Ministry.

The national expert committee for PGR, called the "Advisory and Coordinating Committee for Agricultural and Horticultural Crops (BEKO)" was established by the BMEL in 2002. It consists of up to 17 members representing Federal and *Laender* authorities, professional associations and organizations from science and research, the private sector, representatives of genebanks, the in situ conservation and on farm management sector, and non-governmental organizations. The terms of office of members are five years. The BEKO has provided reports about the implementation of the national programme for the periods 2008–2014 and 2015 to 2019. Since the beginning of the current term (2020–2024), also the nature protection sector is represented through the Federal Agency for Nature Protection.

5.1. Ex Situ Conservation

In Germany, there are currently six national genebanks (see Table 3). These consist of more than 100 collections hosted and curated by a most varied range of actors at the Federal, *Laender,* and local level, and even by private individuals. Four of these genebanks are in fact decentralized networks that are specialized in the conservation of certain crops, namely the German Genebank for Fruit Crops, the German Genebank for Grapevine, the German Genebank for Ornamentals, and the Genebank for Crop Wild Relatives.

Despite the differences in the species conserved and the actors involved, all four decentralized genebank networks follow the same structure, which is set out in a cooperation agreement. The coordinating organizations of the two larger networks, i.e., the Genebank for Fruit Crops and the Genebank for Ornamentals, are supported by scientific advisory boards.

All genebanks conserve their accessions according to the FAO genebank standards [20]. The IPK genebank is running a quality management system since 2007 and is certified according to ISO 9001:2015. The vast majority of its collection is stored as dry seed at −18 °C. Conservation of vegetatively propagated accessions is facilitated through permanent cultivation in the field or by means of in vitro culture or cryo conservation in liquid Nitrogen.

The IBV, as a higher-level coordinating body, is a partner with defined tasks in all decentralized genebank networks. This includes the integration of the respective genebanks in the national and international processes as well as the integration of the data about the respective genebank holdings into the National Inventory of Plant Genetic Resources (PGRDEU) in Germany.

All plant genetic resources conserved in the German genebanks are distributed under the terms of the International Treaty on Plant Genetic Resources for Food and Agriculture (ITPGRFA) for the purposes of research, breeding, and training with the Standard Material Transfer Agreement (SMTA), or a special Material Transfer Agreement for ornamentals based on the SMTA. The IPK genebank provided in 2019 a total of 20,069 samples to recipients at their requests, under SMTAs; 11,351 samples were requested from abroad. The current genebank holdings are listed in Table 3.

Within the network of IBPGR base collections, the IPK held a global base collection of *Lycopersicon* and *Lupinus*, and the former FAL genebank held global collections of *Avena*, *Beta*, four *Brassica* species and *Sinapis*, as well as a European collection of *Phaseolus* [15]. Germany has placed a significant number of unique accessions within the "virtual" decentralized European collection AEGIS (A European Genebank Integrated System), which is the initiative of the European Cooperative Programme on Plant Genetic Resources (ECPGR) aiming to efficiently conserve and provide access to unique germplasm in Europe through this European Collection. The AEGIS accessions contributed by German genebanks, mostly by IPK, constitute 41% of the European collection as of July 2021. About 65% of the accessions (Table 3), specifically 75% of the IPK collection and accessions held by the CWR and fruit genebanks, are of species included in Annex I of the ITPGRFA and have been notified as part of its Multilateral System.

Table 3. Ex situ conservation of plant genetic resources in Germany (2021).

Ex Situ Conservation	Coordinating Institute	Number of Accessions [1]	Number of Genera
Federal genebank of agricultural and horticultural crop plants	Leibniz-Institute of Plant Genetics and Crop Plant Research (IPK), Gatersleben	150,905	776
German genebank for fruit crops	Julius-Kühn-Institute (JKI), Institute for Breeding Research on fruit crops, Dresden-Pillnitz	5374	7
German genebank for grapevine	Julius-Kühn-Institute (JKI), Institute for Grapevine Breeding Geilweilerhof, Siebeldingen	4224	3
German genebank for ornamentals	Federal Plant Variety Office (BSA), Hanover	16,016	75
German genebank for crop wild relatives	Botanic Garden, University Osnabrueck	4711	178
German genebank for tobacco	NiCoTa GmbH, Rheinstetten	788	1
Total number of accessions		182,018	

[1] Source: PGRDEU, 21 July 2021.

5.2. In Situ Conservation

Attention to in situ conservation of CWR has increased over the past decade. The "German Network of Genetic Reserves" has been established in 2019 as framework for in situ conservation of priority CWR [21]. It consists of networks for specific priority CWR. The CWR networks include genetic reserves harboring populations or plant communities identified based on agreed criteria and managed by coordination units located at agencies or institutions involved in work with PGRFA. The overall network is coordinated by the IBV (Figure 4). The German Network of Genetic Reserves has the following objectives:

- Improvement of priority CWR in situ conservation in their natural habitats, combined with complementary ex situ conservation in genebanks.
- Provision of a framework for coordination, management and integration of CWR into in situ conservation activities and for raising awareness about the importance of CWR conservation.
- Promotion of CWR utilization through documentation and the provision of freely available in situ and ex situ characterization and evaluation data in national and international information systems.
- Supporting the national PGRFA program in international cooperation and the implementation of the CBD, the GPA2, and the International Treaty on PGRFA.
- Supporting the fulfilment of international reporting obligations regarding the implementation of GPA2, the International Treaty, and the State of the World Report on PGRFA.

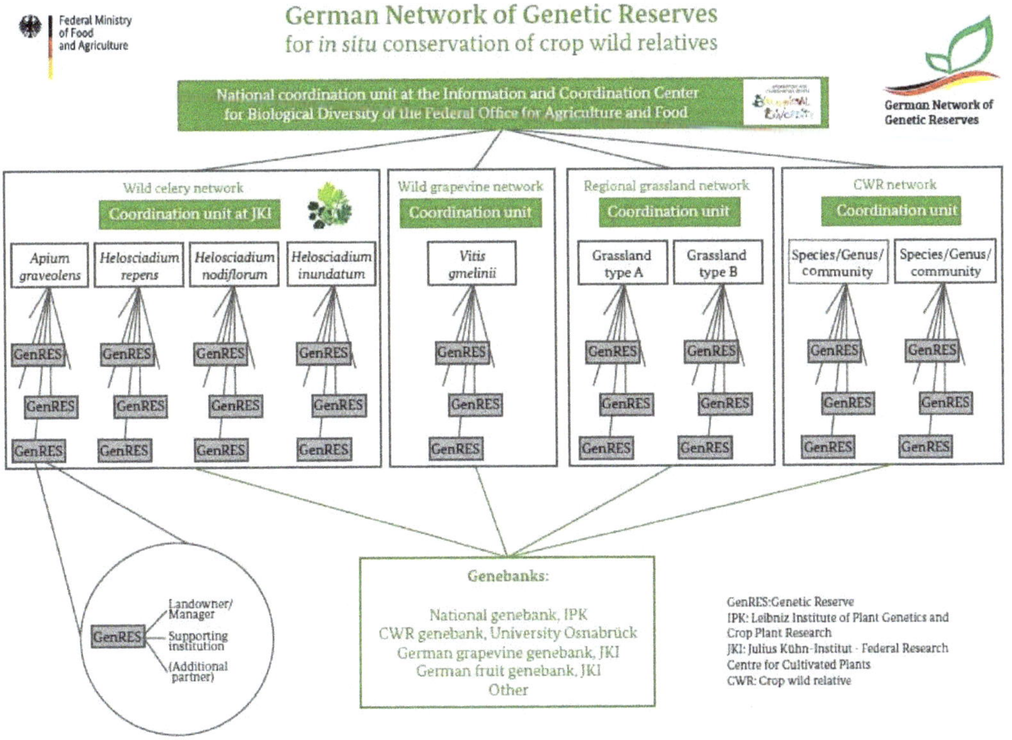

Figure 4. Structure of the German Network of Genetic Reserves.

The first CWR specific network was the wild celery network established in 2019. Currently it is already composed of 17 wild celery genetic reserves; further reserves are in the process of being designated [22]. The coordination unit of the wild celery network is located at the Federal Research Centre for Cultivated Plants (JKI), Institute for Breeding Research on Agricultural Crops, in Quedlinburg.

In situ conservation of wild plant species is a key task of the nature protection sector, while CWR are of particular interest to the agricultural sector given their importance for plant breeding and crop improvement. Hence, their in situ conservation requires collaboration with the nature protection sector, both at the local level, when designating genetic reserves, as well as regional and federal level. This collaboration is currently being established and extended, and has *inter alia* led, as reported above, to the representation of the Federal Agency for Nature Protection in the BEKO.

5.3. On-Farm Sector

In Germany, there are a large number of NGOs, most of which are organized through an umbrella organization for crop and livestock diversity. This umbrella organization is also a member in the BEKO to advice on issues related to on farm conservation and management.

The EU-Regulation on Conservation Varieties adopted in 2009 created the legal prerequisite to permit and market seeds of landraces and varieties of agricultural species and vegetable species, which are relevant for the conservation of genetic resources under facilitated conditions. This was a supportive prerequisite to enable on-farm management of varieties that no longer have seed approval.

The Federal Ministry offers project funding to support on farm conservation and management, through which a large number of projects have been carried out in recent years, including some projects that investigated the recultivation of genebank accessions of old landraces from the IPK genebank.

Within the scope of the EU's European Agricultural Fund for Rural Development (EAFRD) there is also the national cooperative funding instrument of the Federal and *Laender* governments "Improvement of the Agricultural Structure and Coastal Protection" (GAK) to support agriculture and forestry, the development of rural areas and to improve coastal and flood protection. One of the funding measures specifically serves to promote the cultivation and conservation of old landraces/varieties, that are listed on the Red List of endangered local crops in Germany. The *Laender* can offer these funding measures and receive co-financing from the Federal ministry. However, due to the priority setting by the individual *Laender*, the efforts required and the low funding volume so far only one Federal State (*Land*) offers this funding measure for plant genetic resources.

5.4. National Inventory

The National Inventory of Plant Genetic Resources in Germany (PGRDEU) is the central documentation of the ex situ, in situ, and on-farm conserved plant genetic resources in Germany. This includes

- The documentation of the six national genebanks in Germany,
- The data from the "German Network of Genetic Reserves",
- The list of priority CWRs,
- Extensive data on the historically used vegetable varieties and species from the period 1850–1950,
- The red list of endangered indigenous crop landraces/varieties in Germany,
- An inventory of on-farm actors that is currently being developed,
- Variety descriptions of genebank material from cultivation trials.

The national inventory is hosted and managed by the IBV at the BLE. It is regularly updated and, besides being a central resource for national stakeholders, it serves as data source for fulfilling international data reporting obligations to the European Catalogue of plant genetic resources EURISCO managed by the ECPGR and to FAO for SDG Indicator 2.5.1.

5.5. International Cooperation

The European and global collaboration in plant genetic resources for food and agriculture conservation and use is also coordinated by the PGR experts at the IBV. They coordinate the interactions with and the contributions to ECPGR. They advise and represent the Ministry in the collaboration with and sessions of the ITPGRFA and the Intergovernmental Technical Working Group for PGR of the CGRFA and take care of all international reporting obligations to the ITPGRFA, the CGRFA, and GPA. Through IBV's various functions and roles in the BEKO and the CWR genetic reserve and genebank networks it communicates relevant international information and necessary actions to the national stakeholders and vice versa.

6. Conclusions

The coordination and governance structure is functioning well since more than 20 years now. It is a light structure based on information, communication, and coordination elements but without a centralized funding structure. While this structure brought many advantages, still some challenges remain to be addressed. These issues will be elaborated based upon the plant genetic resources domain as follows.

The benefits of the structure—distributed with a central coordination—can be appreciated by the improved national cooperation with a balanced implementation of the national programme across both larger and smaller stakeholders. Additional capacities could be identified and integrated for conservation of plant genetic resources by including very small and also private actors. This is the case, for instance, in the further development of the German genebanks of fruits and ornamentals, where a number of well-qualified and motivated partners are able to contribute to the national endeavor. The comprehensive national programme, the BEKO, as well as the information and communication means allow them to participate in a fair and equitable manner.

The comprehensive representation of a wide range of stakeholders in the BEKO facilitates their engagement and contributions toward the implementation of activities in the national programme. This goes hand in hand with the official acknowledgement and recognition by reporting their valuable activities at national and international levels.

Monitoring, regular revisions, and priority setting of the national programme is facilitated by the BEKO and the supporting activities of the IBV. This approach allows to continuously integrate innovations from science, observations of the private sector, and findings of non-governmental organizations, as well as new developments from political debates and decisions, and international developments.

The capacity to collect information about activities of and contribution from numerous stakeholders, besides the well-known research institutes such as the IPK genebank in Gatersleben, allows enhanced collaboration and facilitates documentation and reporting of the German contributions toward international cooperation. Especially the ECPGR could benefit by coordinated inputs from Germany. At the same time, the active involvement of German members in ECPGR activities facilitates the harmonious implementation of the concepts developed within the ECPGR in Germany.

When developing positions for European and global processes, the German Federal Ministry of Food and Agriculture (BMEL) is benefitting from advice of the BEKO and the IBV and national programme assessments. This is especially the case for FAO processes under the CGRFA and the ITPGRFA, as well as for the CBD processes, including the Nagoya Protocol. An additional advantage of the governance structure is an improved coherence in German positions related to plant genetic resources for food and agriculture at the international level.

Besides such advantages, some challenges of this structure remain to be addressed. In particular, the information flow from *Laender* activities or non-governmental organizations related to in situ conservation and on farm activities to the IBV could be improved. IBV also would benefit from more regular information from research institutes about ongoing projects, especially those funded by third parties (e.g., the EU). Several coordination and conservation activities within the ex situ and in situ conservation networks would benefit from more long-term institutional support to stakeholders carrying out these functions.

While policy coherence could be improved with the new governance structure for policy setting at different national agricultural for a related to plant genetic resources, processes to develop joint positions between the agricultural and environmental sectors at national and international level should still be further enhanced.

A key step in the development of the current coordination and governance structure was the need to thoroughly assess and analyze the national plant genetic resources landscape at the occasion of the German reunification. This assessment has taken into consideration the existing political and administrative structure and the distribution of competencies between the *Laender* and Federal governments. More than 30 years have passed since. Looking back today from within a stable and well-functioning plant genetic resources system, the approach to (only) centralize information and coordination functions in a permanent dedicated unit, while keeping or developing concrete implementation of conservation and use, research and breeding embedded in functioning local, regional, and federal structures or distributed networks have proven to be very sustainable and effective.

It is conducive to establishing long-term collaboration both at national and international levels, having allowed Germany to engage effectively in continued collaboration with all relevant bodies, i.e., ECPGR, FAO, ITPGRFA, and CGRFA, including respective working groups and subsidiary bodies.

This light governance structure provided by a central coordination unit, i.e., the IBV, deserves appropriately staffed offices. The scientists working at the IBV are civil servants, all experts in their respective field of genetic resources, who are entirely dedicated to carrying out the tasks listed in Section 4, for which the IBV is responsible. The financial support provided by the federal government to this permanent information and coordination unit underlines the importance, which the government does attribute to this task. It recognizes the historical developments and achievements and the difficulties faced during World War II and the post-war times. In particular, it values the fundamental importance of plant genetic resources for further research and plant breeding to support the diversification of the agricultural sector and the entire food system and to contribute to climate change adaptation processes and the protection of biological diversity.

It is recommended to establish an equivalent concept and governance structure at European level. Like Germany as a Federal state, with a multitude of stakeholders and significant differences among the *Laender*, and relevant competencies and responsibilities distributed between *Laender* and federal level, similarly Europe presents a highly diverse genetic resources landscape in terms of conservation, management, use, research and breeding, actors and (agro)ecologies, as well as relevant competencies and responsibilities.

Author Contributions: Conceptualization, F.B. and I.T.; formal analysis, F.B., I.T. and S.S.; data curation, I.T. and S.S.; writing—original draft preparation, F.B.; writing—review and editing, F.B., I.T., S.S. and K.K.; visualization, K.K.; supervision, F.B. All authors have read and agreed to the published version of the manuscript.

Funding: This research received no external funding.

Data Availability Statement: Publicly available datasets were analyzed in this study. This data can be found here: https://pgrdeu.genres.de/ex-situ-bestaende/suche-nach-genbanken/?L=0.

Conflicts of Interest: The authors declare no conflict of interest.

References

1. Meinel, A. Die Saatgut- (Saatzucht-) Abteilung der DLG (1886–1993)—Anfänge des landwirtschaftlichen Saatgutwesens. In *Die Entwicklung der Pflanzenzüchtung in Deutschland (1908–2008): 100 Jahre GFP e.V.-Eine Dokumentation*; Röbbelen, G., Ed.; Gesellschaft für Pflanzenzüchtung e.V.: Göttingen, Germany, 2008; Volume 75, pp. 1–17.
2. Lehmann, C.O. Hundert Jahre Sammlung und Nutzung von Landsorten-zur Erinnerung an Emanuel Ritter von Proskowetz und Franz Schindler. In *Gemeinsames Kolloquium "Sicherung und Nutzbarmachung Pflanzengenetischer Ressourcen", 3–5 July 1990, Braunschweig-Gatersleben*; Dambroth, M., Lehmann, C.O., Eds.; Verein zur Erhaltung der Nutzpflanzenvielfalt e.V.: Braunschweig-Gatersleben, Germany, 1990; pp. 10–22.
3. Cohen, B. Nikolai Ivanovich Vavilov: The Explorer and Plant Collector. *Econ. Bot.* **1991**, *45*, 38–46. [CrossRef]
4. Frese, L. Sammlung und Erhaltung der biologischen Vielfalt. In *Die Entwicklung der Pflanzenzüchtung in Deutschland (1908–2008): 100 Jahre GFP e.V.-Eine Dokumentation*; Röbbelen, G., Ed.; Gesellschaft für Pflanzenzüchtung e.V.: Göttingen, Germany, 2008; Volume 75, pp. 510–517.
5. Begemann, F.; Hammer, K. Analyse der Situation pflanzengenetischer Ressourcen in der Bundesrepublik Deutschland nach der Wiedervereinigung—Unter besonderer Berücksichtigung der Genbank in Gatersleben—Sowie konzeptionelle Überlegungen für ein deutsches Gesamtprogramm. In *Schriftenreihe des Bundesministers für Ernährung, Landwirtschaft und Forsten, Reihe A: Angewandte Wissenschaft*; Landwirtschaftsverlag GmbH: Münster-Hiltrup, Germany, 1993; Volume 422.
6. Flitner, M. *Sammler, Räuber und Gelehrte—Die Politischen Interessen An Pflanzengenetischen Ressourcen 1895–1995*; Campus Verlag: Frankfurt, Germany, 1995.
7. Kratzsch, G. Pflanzenzüchtung und Saatgutwesen in der sowjetischen Besatzungszone und der Deutschen Demokratischen Republik (1945–1990). In *Die Entwicklung der Pflanzenzüchtung in Deutschland (1908–2008): 100 Jahre GFP e.V.-Eine Dokumentation*; Röbbelen, G., Ed.; Gesellschaft für Pflanzenzüchtung e.V.: Göttingen, Germany, 2008; Volume 75, pp. 37–52.
8. Röbbelen, G. Pflanzenzüchtung in den westlichen Besatzungszonen und der Bundesrepublik Deutschland (nach 1945). In *Die Entwicklung der Pflanzenzüchtung in Deutschland (1908–2008): 100 Jahre GFP e.V.-Eine Dokumentation*; Röbbelen, G., Ed.; Gesellschaft für Pflanzenzüchtung e.V.: Göttingen, Germany, 2008; Volume 75, pp. 53–78.
9. Bulich, C. Die Gemeinschaft zur Förderung der privaten deutschen landwirtschaftlichen Pflanzenzüchtung e.V., Bonn (seit 1945)—Wissenschaft aus Instituten für Zuchtbetriebe. In *Die Entwicklung der Pflanzenzüchtung in Deutschland (1908–2008): 100 Jahre GFP e.V.-Eine Dokumentation*; Röbbelen, G., Ed.; Gesellschaft für Pflanzenzüchtung e.V.: Göttingen, Germany, 2008; Volume 75, pp. 79–91.
10. Bommer, D.F.R.; Beese, K. *Pflanzengenetische Ressourcen-Ein Konzept zur Erhaltung und Nutzung für die Bundesrepublik Deutschland. Schriftenreihe des Bundesministers für Ernährung, Landwirtschaft und Forsten, Angewandte Wissenschaft*; Landwirtschaftsverlag GmbH: Münster-Hiltrup, Germany, 1990; Volume 388.
11. FAO. *Report of the 22nd Session of the FAO Conference (Resolution 8/83)*; Food and Agriculture Organisation of the United Nations: Rome, Italy, 1983.
12. FAO. *Report of the 22nd Session of the FAO Conference (Resolution 9/83)*; Food and Agriculture Organisation of the United Nations: Rome, Italy, 1983.
13. BLAG. Bund-Länder-Arbeitsgruppe "Erhaltung forstlicher Genressourcen". In *Konzept zur Erhaltung Forstlicher Genressourcen in der Bundesrepublik Deutschland, Großhansdorf*; Forst und Holz: Hannover, Germany, 1989; Volume 44, pp. 379–404.
14. DAF. *Organisationsanalyse zu Pflanzengenetischen Ressourcen für die Forschung im Bereich Landwirtschaftlicher und Gartenbaulicher Kulturpflanzen*; Dachverband Wissenschaftlicher Gesellschaften der Agrar-, Forst-, Ernährungs-, Veterinär- und Umweltforschung e.V. (DAF): Frankfurt am Main, Germany, 1994; ISBN 3-7690-5022-3.
15. Thormann, I.; Engels, J.M.M.; Halewood, M. Are the old International Board for Plant Genetic Resources (IBPGR) base collections available through the Plant Treaty's multilateral system of access and benefit sharing? A review. *Genet. Resour. Crop Evol.* **2019**, *66*, 291–310. [CrossRef]
16. Oetmann, A.; Brockhaus, R.; Begemann, F. Erhaltung und nachhaltige Nutzung pflanzengenetischer Ressourcen—Deutscher Bericht zur Vorbereitung der 4. Internationalen Technischen Konferenz der FAO über pflanzengenetische Ressourcen vom 17.-23 Juni 1996 in Leipzig. In *Schriftenreihe des Bundesministeriums für Ernährung, Landwirtschaft und Forsten, Reihe A: Angewandte Wissenschaft*; Federal Ministry of Food, Agriculture and Forestry: Bonn, Germany, 1995; Volume 441, 178p.
17. Oetmann, A. Genetische Ressourcen für Ernährung und Landwirtschaft—BML-Konzeption zur Erhaltung und nachhaltigen Nutzung genetischer Ressourcen für Ernährung, Landwirtschaft und Forsten. In *Schriftenreihe des Bundesministeriums für Ernährung, Landwirtschaft und Forsten, Reihe A: Angewandte Wissenschaft*; Federal Ministry of Food, Agriculture and Forestry: Bonn, Germany, 2000; Volume 487.
18. BMELV. *Conservation of Agricultural Biodiversity, Development and Sustainable Use, of Its Potentials in Agriculture, Forestry and Fisheries*; Federal Ministry of Food, Agriculture and Consumer Protection: Bonn, Germany, 2007.
19. BMELV. *National Program for the Conservation and Sustainable Use of Plant Genetic Resources of Agricultural and Horticultural Crops*; Federal Ministry of Food, Agriculture and Consumer Protection: Bonn, Germany, 2015.
20. FAO. *Genebank Standards for Plant Genetic Resources for Food and Agriculture*; Food and Agriculture Organization of the United Nations: Rome, Italy, 2013.

21. Thormann, I. The German Network of Genetic Reserves. In *Crop Wild Relative Newsletter*; University of Birmingham: Birmingham, UK, 2020; Issue 12; pp. 23–25. ISSN 1742-3694. (Online). Available online: https://more.bham.ac.uk/farmerspride/wp-content/uploads/sites/19/2020/09/CWR_Newsletter_Issue_12.pdf (accessed on 21 July 2021).
22. Bönisch, M.; Frese, L. Designation of genetic reserves for wild celery species in Germany. In *Crop Wild Relative Newsletter*; University of Birmingham: Birmingham, UK, 2020; Issue 12; pp. 4–7. ISSN 1742-3694. (Online). Available online: https://more.bham.ac.uk/farmerspride/wp-content/uploads/sites/19/2020/09/CWR_Newsletter_Issue_12.pdf (accessed on 21 July 2021).

plants

MDPI

Review

Collaboration between Private and Public Genebanks in Conserving and Using Plant Genetic Resources

Johannes M. M. Engels [1,*], Andreas W. Ebert [2] and Theo van Hintum [3]

1 Independent Researcher, 06062 Citta' della Pieve, PG, Italy
2 Independent Researcher, 73529 Schwäbisch Gmünd, Germany; ebert.andreas6@gmail.com
3 Centre for Genetic Resources, The Netherlands (CGN), Wageningen University & Research, 6700 AA Wageningen, The Netherlands; theo.vanhintum@wur.nl
* Correspondence: j.engels@cgiar.org

Abstract: Among the most important users of plant genetic resources, conserved predominantly in public genebanks around the world, are public and private plant breeders. Through their breeding efforts, they contribute significantly to global, regional, and local food and nutrition security. Plant breeders need genetic diversity to be able to develop competitive new varieties that are adapted to the changing environmental conditions and suit the needs of consumers. To ensure continued and timely access to the genetic resources that contain the required characteristics and traits, plant breeders established working collections with breeding materials and germplasm for the crops they were breeding. However, with the changing and increasingly more restrictive access conditions, triggered by new global legal instruments like the Convention on Biological Diversity/Nagoya Protocol and the International Treaty, plant breeders started to establish their own genebanks at the turn of the 21st century. This paper analyses the conditions that contributed to this situation as well as the historical ways that plant breeders used to acquire the germplasm they needed. Public genebanks played and continue to play a conducive role in providing genetic resources to users, including private-sector plant breeders. However, also the practices of the germplasm curators to collect and distribute germplasm were affected by the new legal framework that had been developed in global fora. It is against this background that the complementarity and collaboration between public and private sector genebanks have been assessed. Whenever possible, vegetable genetic resources and vegetable private breeding companies have been used to analyze and illustrate such collaboration. The authors look at reported successful examples of collaborative efforts and consider opportunities and approaches under which such collaboration can be established and strengthened to ensure the continued availability of the building blocks for food and nutrition security.

Keywords: plant genetic resources; public genebanks; private genebanks; collaboration; conservation; use

Citation: Engels, J.M.M.; Ebert, A.W.; van Hintum, T. Collaboration between Private and Public Genebanks in Conserving and Using Plant Genetic Resources. *Plants* **2024**, *13*, 247. https://doi.org/10.3390/plants13020247

Academic Editors: Yong-Bi Fu and Brad Sherman

Received: 17 September 2023
Revised: 8 January 2024
Accepted: 10 January 2024
Published: 15 January 2024

1. Introduction

1.1. Motivation

Genetic diversity, within and between species, provides the raw material for plant breeders to work with. This diversity has been readily available and without restrictions for a long time, until the 1980s, albeit increasingly threatened by genetic erosion. Plant breeders typically established and maintained working collections of selected materials of a given crop, including their own breeding lines, collected materials, and accessions obtained from public genebanks or (seed) markets elsewhere. With the introduction of a legal framework for access and benefit-sharing (ABS) of this genetic diversity, in fact, of biodiversity at large, access to genetic resources became more restricted or was even, in some cases, completely denied. During the last three decades, access to genetic diversity and benefit-sharing conditions has become more complicated and bureaucratic. Given the fact such access is an essential requirement for any breeder, breeders started to pay more interest in strengthening their own working collections. Eventually, seed companies established their own private genebanks to ensure their in-house breeders have permanent access to the genetic resources they need for developing new varieties.

In parallel to the above developments, many public genebanks, especially in developing countries saw increased budgetary constraints, faced an increasing lack of required expertise and equipment, and thus became increasingly isolated from the global user community [1]. The mentioned legal complexity of providing access to healthy and good quality genetic resources further contributed to a weakening position of part of the public genebanks in offering targeted services to society and consequently resulting in less support for investments in conservation and use.

The above situation has prompted the authors to review the current practices of collecting, conserving, and maintaining germplasm accessions more systematically and critically as well as how these genetic resources are accessed, in particular by the private breeders and their breeding companies, as their role in conservation is steadily increasing, albeit restricted largely for their own use. Therefore, identifying the bottlenecks and constraints in the current global conservation and use system that impact the collaboration between private and public genebanks is the first step towards resolving them. In this review paper, we will assess germplasm flows, procedures, traditions, and other (legal) processes that might include such constraints, bottlenecks, and possibly other negative reasons with respect to the collaboration, followed by presenting possible solutions. This paper intends to contribute currently missing information to the political debate on revising global legal instruments and national/local conservation practices in order to strengthen the effectiveness and efficiency of germplasm conservation and use efforts.

1.2. Why Focus on Vegetables?

Breeding crops is a multidisciplinary activity that strongly varies in methodology from one crop to another. The way that genetic resources are being used in the breeding process, therefore, also varies significantly between crops or groups of crops. The latter depends, among others, on the 'development state' of the crop or group of crops, including to which extent they have been researched. To avoid unnecessary complexity in our assessment of the situation of access, it was deemed easier to restrict the overall range of crops. A somewhat arbitrary decision was made to focus this paper on vegetable crops, whenever possible, meaningful, and applicable. Only a relatively small number of known public genebanks had a strong collaboration history of working with the private breeding sector. Also, only a few curators from private genebanks and only a few private plant breeders had expressed preparedness to share relevant information for this paper. Consequently, this paper has a rather strong focus on a limited number of countries and genebanks (see also Section 2). When no specific information on vegetable genetic resources was available, 'general information' was used with the assumption that this would also apply to vegetable crops.

Accessing new genetic diversity is essential in the breeding process of any vegetable crop and there is, in general, a broad range of genetic diversity available from public and private genebanks for these breeding activities. However, many 'minor' vegetable crops are underrepresented in genebanks, which is typical for the neglected and underutilized species—NUS. Among the major vegetable crops, only the Brassica complex (Brassica et al.), carrots (Daucus), and eggplant (Solanum) form part of Annex I crops of the International Treaty on PGRFA, while all the minor ones are not included. Furthermore, in particular, the minor vegetables are, in general as well as in breeding terms, less 'advanced' compared to the major staple crops as many of the minor vegetable crops have hardly undergone any significant breeding. This means that we have a vast spectrum of vegetable crops with respect to their improvement status and thus, the vegetables at large represent the entire spectrum of their dependency on genetic diversity for the breeding process in creating new, more nutritious, and better-adapted varieties.

Whereas vegetable genetic resources are only a smaller proportion of the total range of PGRFA, the findings on the collaboration between public and private genebanks of vegetable genetic resources are representative of all PGRFA and all genebanks, and very suitable for illustrating the points that are being made. This same line of thought also applies to the limited number of genebanks and countries that provided information, the findings can be extrapolated to other genebanks as well, without too many restrictions.

Some general considerations

While attempting to strengthen the collaboration between public genebanks and breeding companies, one must consider that such collaboration should benefit both sides. As already mentioned, the 'business model' of breeding companies is to breed new varieties that address problems experienced by growers and meet the expectations of consumers. Newly bred varieties need to be competitive and possess environment-friendly properties, thus being sustainable and 'marketable', generating financial benefits to the companies, besides other benefits to the society at large. To do so, the companies need access to plant genetic resources.

Public genebanks aim at effective and efficient long-term conservation of defined national and/or economically important crop genepools and to make these genetic resources available to users. They often also have the mandate to contribute to the conservation of the bio-cultural heritage of their country and region. Genebanks tries to make the conserved accessions available in the most user-friendly form, along with related information. Furthermore, they should assist in the acquisition and research of genetic resources of importance to their national agricultural economy. Many of the (national) public genebanks are well-positioned to facilitate access and benefit-sharing arrangements. Fulfilling all these requirements will allow them to significantly contribute to national and global food and nutritional security and a more sustainable form of agriculture.

2. Materials and Methods

Considering the (political) 'sensitivity' of this subject and the hesitation from the breeding companies to openly engage in assessing and critically reviewing the current practices concerning the conservation of genetic resources in their companies, it was decided to base this review on the information that had been received from a limited number of private-sector breeders and company genebanks, including Rijk Zwaan, the Netherlands, and East-West Seed International Limited, Thailand as well as from the literature. The inclusion of data from the Dutch national genebank CGN is important as it has a significant collection of different vegetable crops is one of the most advanced national genebanks in the world with respect to genebank and germplasm management practices, and has gathered significant experiences in collaborating with breeding companies in the Netherlands and beyond. Another national genebank with a very broad portfolio of field and vegetable crops and their related wild relatives is the Leibniz Institute of Plant Genetics and Crop Plant Research (IPK), Gatersleben, Germany. It is among the most important national genebanks that distribute germplasm worldwide and is, similar to CGN, one of the most advanced genebanks. The only global vegetable genebank is the World Vegetable Center (WorldVeg), with headquarters located in Taiwan. It is the 5th biggest international public genebank with a unique focus on vegetables. It is closely associated with the CGIAR genebank platform and has established significant collaboration with private and public vegetable breeders and companies. The long and extensive experience of the authors in operating public genebanks, including the conservation and facilitation of the use of a wide diversity of (vegetable) crops and their wild relatives, their experience in collaborating with the private genebank/breeding sector as well as with the framing of the global, regional and some national legal frameworks and applying it on a routine basis, provide a good foundation for this review and to identify and present a number of opportunities that would facilitate and strengthen the collaboration between two complementary activities, i.e., conservation and use.

3. A Historical Overview of Accessing and Managing Germplasm by the Private Sector

Since the first steps of domesticating plant species were made some 10,000 years ago in the Fertile Crescent, and even earlier when gathering wild plants, human selection started to play an essential role in shaping the foundation for agriculture. Farmers began to select species and genotypes within species for cultivating 'crops'. This process lasted for millennia and continues even today. It was not until after the rediscovery of Gregor Mendel's principles of inheritance, published in 1865 [2], that scientific breeding based on crossing and selection started at the beginning of the 20th century. Plant breeding as a commercial undertaking is much older, possibly dating back to the start of breeding efforts by de Vilmorin in France in 1743 [3]. Ample genetic diversity is the foundation for selecting a suitable population or the best genotypes within a given species for cultivation, or parents for making crosses and thus to produce varieties that stack desirable characteristics. During the first part of the 20th century, plant introduction centres were established for single crops, like potatoes, or multiple crops [4]. To ensure the continued availability of sufficient genetic diversity to select from, breeders typically establish working collections of genotypes of a given crop that would correspond with their breeding objectives [5].

At a later stage, targeted collecting of genetic diversity of a given crop, in the form of landraces, traditional farmers' varieties, or related wild relatives of such crops, was initiated by plant breeders and scientists, and the first genebanks were established, mainly for research purposes, first in the USA (Harry Harlan) and Russia (Nikolai Vavilov), later in Germany (Hans Stubbe) and in other European countries [4]. However, with the occurrence and increase of the so-called genetic erosion, especially following the large-scale introduction of scientifically bred varieties with a limited genetic base [6], the foundation of crop improvement and plant breeding started to become threatened, and thus, the earlier priority of public genebanks to provide genetic diversity to users changed to a more conservation orientated objective. Consequently, efforts were undertaken to more systematically collect and conserve genetic resources, at the beginning especially landraces of crops, by and in genebanks worldwide.

Until the early 1990s, access to the genetic resources available in situ or conserved in the predominantly public genebanks was free and unrestricted. This access paradigm was embedded in the International Undertaking that FAO had concluded as a voluntary international agreement with their member countries in 1983 [7]. Thus, breeders had ready access to genetic diversity conserved in genebanks and/or available in farmers' fields and natural populations. Generally, this on-farm and in situ germplasm could be collected without major restrictions. As these materials were typically added to or shared with the public genebank collections in the respective countries, and as genebanks exchanged germplasm among them to respond to requests of users/breeders, only limited access problems existed. Mostly public breeders organized collecting missions, frequently together with scientists from universities and/or genebank staff and added collected material to their working collections. Through collaboration with research institutes and universities, private breeders had easy access to these collections and pre-bred materials. For instance, in the Netherlands, public research institutes had a strong focus on the production of pre-bred materials that could be made readily available to private and public breeders [8]. From all these sources, including from public genebanks and research institutes worldwide, germplasm materials and advanced breeding lines have been incorporated into the working collections of private breeders.

Private breeders, in general, prefer to use commercially successful 'elite' varieties of a given crop as parents in their breeding activity and 'only' resort to genetic resources, in particular, landraces and crop wild relatives, when they need to enlarge the genetic diversity pool and/or want to include specific traits in the 'elite variety' they plan to produce [9]. In particular, the use of crop wild relatives in breeding programmes generally requires lengthy and costly back-cross programmes to obtain genotypes/varieties that no longer possess unwanted characteristics from the wild species. In this context, it should be noted that genomic tools such as marker-assisted selection allow a much more effective removal of linkage drag. Furthermore, genome-wide association mapping and comparative genomics allow the identification of marker-trait associations and the prediction of associated candidate genes. This information can then be used for the efficient incorporation of desired traits through marker-assisted breeding (see, for example, Puranik et al. [10]). It should be stated that the above arguments apply to vegetable genetic resources and PGRFA at large.

To give breeders protection for their efforts, a system of intellectual property protection of released varieties, so-called plant breeders' rights, was set up in the second half of the 20th century. The Union for the Protection of Plant Varieties or UPOV is the most widely applied 'system', which gives breeders for up to 20 years the monopoly of reproducing and selling the protected variety. Such protected varieties can still be freely used by competing breeders for further breeding activities under the so-called 'breeders' exemption'. In this context, also the US 1930 Plant Patent Act should be mentioned as it aimed to protect newly released varieties through a patent. Through the 'harmonization' process of the International Undertaking and the CBD the principle of 'common heritage of humankind' was eliminated and replaced by the notion of 'national sovereignty' of states over their natural resources [7]. This process of increased application of legal protection of varieties, especially after the application of patents on genes, traits, and even varieties, resulted in more restrictive accessibility of genetic resources, especially from 'the Global South' and was further fueled by the establishment of the Convention on Biological Diversity (CBD) in 1992, including the recognition of 'national sovereignty' of states over the biological resources in their territories and in 2014 by its related Nagoya Protocol. Subsequently, the FAO had to counter the access and benefit-sharing challenges those bilateral negotiations under the CBD posed to the exchange of plant genetic resources for food and agriculture. This process resulted 2001 in the establishment of the International Treaty on Plant Genetic Resources for Food and Agriculture (hereinafter called ITPGRFA or International Treaty), which entered into force in 2004 [11]. In harmony with the CBD, it made specific arrangements for PGRFA, notably a multilateral system for access and benefit-sharing [4].

The CGIAR genebanks contribute(d) significantly to the collecting, acquisition, and long-term conservation and use of PGRFA, maintained in the public domain for the entire global community, especially regarding staple crops. The developments over the past 30 years with respect to the acquisition and distribution of germplasm by the CGIAR centres were analysed to demonstrate the current situation with respect to ABS by Halewood et al. [12]. A highly political environment was observed, such as countries' unwillingness to share their materials. Restrictive national laws and policies were the reasons most often cited by the CGIAR centres' genebank managers for decreased rates of acquisition of additional materials to conserve in and distribute from their genebanks. Furthermore, the following more specific scenarios that impacted negatively on the acquisition of germplasm by the CGIAR centres were reported: a combination of different elements, including ABS aspects, that have built mistrust on geopolitical levels; intentions or promises to share materials by (technical-level) partners that have been thwarted by political arguments; insecurity on the part of national partners, in particular in the Global South, because of unclear lines of authority and fear of being accused of 'selling out' the country's patrimony; insecurities on the part of the centres to even request for materials, given the vagaries of national procedures and the possibilities of backlashes and a considerable degree of uncertainty throughout the entire national and international system [12].

In a more recent study regarding rates of acquisition and distribution of germplasm materials by the eleven CGIAR genebanks, Halewood et al. [13] observed an increasing geopolitical polarization over access and benefit-sharing arrangements, as well as the unwillingness of several International Treaty member states to share germplasm through the multilateral system as they feel that they do not get sufficient recognition for their germplasm maintained and shared by/with the centres. Furthermore, developing country contracting parties are dissatisfied with the fact that only three payments have been made to the Plant Treaty's Benefit-Sharing Fund by commercial users of materials from the multilateral system, two back in 2016 [14] and one in 2018 [13]. It may be assumed that all this is true for vegetable collections as well. Although benefit-sharing arrangements are not the main focus of this paper, examples of actual or perceived benefit-sharing arrangements are included in Section 5.5.

As a result of the decreasing collecting and access to PGRFA, breeders support joint collecting missions with public genebanks, implemented within the provisions of the existing legal framework. Collecting is predominantly done in centres of diversity and has a focus on landraces and crops of wild relatives. Breeders also increasingly established their own working collections, particularly through the acquisition of germplasm accessions from national and international genebanks, as well as pre-bred materials from research institutes and commercial varieties worldwide. Gradually, these breeding collections increased in size. With the introduction of more complex and restricted access regulations by countries and genebanks, many companies decided to start their own genebanks to maintain this in-house germplasm for the long term, assuring continued access to the basic material of their breeding activities. This will be further elaborated in Section 5.

The traditional way of (private) breeders to acquire the needed germplasm materials either from or with the assistance of a national public genebank or by joining collecting missions to centres of diversity, has become more complicated and bureaucratic. This also applies to the sharing of pre-bred materials among national plant research institutes and private breeders. At the same time, and as a response to the general trend, we observe increasing numbers of consortia, including partners from public genebanks and private breeding companies that facilitate the sharing of selected accessions among the participating institutes.

4. Why Do Breeding Companies Establish Genebanks?

Considering the developments concerning access to plant genetic resources, an increasing number of breeding companies decided to expand their conservation efforts by consciously adding genetic diversity to the existing breeding collections of crops that are part of their breeding activities and thus, establishing their own private genebank for long-term conservation. The first mention of breeding companies to have established their own genebanks for some crop species was made by Kate and Laird [15]. However, this development process of establishing genebanks by companies started much earlier but was possibly not published or reported [9]. In this section, we elaborate on the reasons why this happened. The content presented is based mainly on the information provided, through personal communication, with a managing director and a genebank manager, i.e., Kees Reinink of Rijk Zwaan [3] and Marilyn Belarmino of East-West Seed International Limited [16], respectively as well as on the knowledge and experience of the three authors.

The main reasons to establish a genebank by breeding companies have been grouped into two parts: (1). Cost and efficiency considerations; and (2). Securing future access to needed genetic diversity.

4.1. Cost and Efficiency Considerations

Breeding companies were already managing large working/breeding collections, sometimes of several hundred thousand samples, in the case of larger seed companies. Therefore, adding 'a few ten thousand' accessions acquired by purpose from different sources to increase the diversity would neither create a significant additional cost nor significantly increase the workload. Furthermore, having your own genebank with a good representation and documentation of the genepool(s) concerned contributes to considerable time and cost savings by not being forced to acquire or collect the required genetic diversity each time when needed.

When starting a screening process, it is advantageous to have the entire range of genetic diversity of a given crop already 'in-house'. Especially having sufficient seed quantity of the accessions to be screened saves considerable time and avoids the need to request new/additional germplasm materials, usually requiring extensive correspondence and time. When participating in research or breeding projects, or regional/global consortia, the sharing of germplasm is typically a pre-condition, and this requirement can be more easily met when having an own genebank.

Many national and institutional genebanks in less developed countries are operating on limited budgets and this frequently results in the distribution of poor-quality germplasm samples that often are not yet sufficiently characterized. Furthermore, the amount of seeds/plant propagules per sample provided by public genebanks is always limited, requiring a seed or tissue multiplication step before screening can be initiated, adding to cost and loss of time. Another, albeit less frequent comment related to the quality of the distributed germplasm is that the genetic composition of individual accessions might not meet the expectations of the recipients. For many of the evaluation activities, breeders need uniform germplasm samples. Especially when the requested material is used for molecular activities, the availability of accessions consisting of single seed descents would greatly facilitate their 'instant use'. However, many of the traditional genebank accessions are, heterogeneous, and therefore require an additional step to obtain uniform samples for screening or evaluation activities. Related to this, public genebanks are frequently unable to respond, in a targeted manner, to trait-specific requests made by breeders for germplasm accessions and/or might not have the supporting characterization/evaluation data of the provided accessions at hand.

The phytosanitary status of germplasm accessions is an issue of increasing concern. Materials from public genebanks often contain seed-transmitted infectious pathogens. This might be caused by the fact that many genebanks are not able to keep their collections disease-free, because of high costs or a lack of adequate and up-to-date seed health testing facilities and expertise [17]. Once the germplasm material has been received by a private company genebank, the accessions/samples must be tested and cleaned, often a requirement of the national phytosanitary authorities, and thereafter maintained clean.

4.2. Securing Present and Future Access to Needed Genetic Diversity

Possibly the most critical reason for having an own company genebank is to ensure that good quality and securely conserved genetic resources are readily available to the breeders to support pre-breeding, breeding, and research programmes, now and in the future [18]. Over the past years, it has become increasingly more challenging to acquire germplasm from national or local genebanks. The most important reason for this is that most genebanks have a policy to only provide a limited number of accessions per request and/or year and requester. Another reason is that several public national and many institutional genebanks that frequently manage crop-specific collections, lack a functional information management system and thus are not able to effectively deal with germplasm requests. A very different reason is that used material transfer agreements sometimes include a requirement that the obtained germplasm should be destroyed after a single use or a demand for high royalties upon the release of a commercial variety.

The current complexity of policy and legal instruments regulating ABS arrangements for germplasm is possibly the most critical reason for breeding companies to establish and operate their own genebanks. In addition to current bureaucratic access restrictions, there is also a great level of uncertainty around the ongoing political debate concerning future ABS regulations triggered by the diversity in national legislation to address the requirements of the Nagoya Protocol [5]. This high degree of uncertainty about future germplasm access has motivated breeding companies to establish their own genebanks and to acquire a wide range of potentially useful genetic resources of company-specific target crops from public genebanks or through collecting missions. It should be noted that since the late 1980s discussions ensued about the application of property rights over newly bred varieties, typically protected through plant breeders' rights and increasingly with patents on plant traits and the underlying genes, but also varieties). As many developing countries and NGOs have opposed these developments, this issue has certainly contributed to critical views in the private sector and consequently, becoming more restrictive in providing easy access to genetic resources.

Some more specific points related to international, regional, and national ABS legislations and regulations, that have been raised in this respect include:

- Seed companies decided to maintain all accessions that had been obtained in the past from third parties in-house for long-term conservation to prevent having to do all 'the burdensome paperwork again and again'. Over the past years, it has become more difficult to import germplasm due to burdensome pre-shipping requirements, partly also due to the increased risk of transboundary spread of pathogens and insect pests [19];
- Typically, companies carefully study the conditions included in contracts and SMTAs from public genebanks and other sources that must be signed before accepting genetic material. If the conditions are unacceptable, especially when demanding unrealistic benefit-sharing requirements, such material will not be included in the company's breeding pool. In this context, the Nagoya Protocol should be mentioned as an important 'trigger' for these developments;
- An important legal aspect of the current benefit-sharing arrangements for a breeding company is that they oppose everlasting obligations. These are often included in ABS rules and require that the company that does the introgression of an interesting trait from a genebank accession into breeding material has to pay as long as this trait is present in one of its varieties, whereas all competitors, who just take the trait from a released variety, based on the breeders' exemption, do not have to pay ABS. This rule thus puts the company that does the largest breeding effort in a disadvantageous position.

The expectation that breeding companies share evaluation information on the acquired accessions with the providing genebanks is for many companies difficult to accept. They prefer not to share their internal evaluation results or, when participating in research or other consortia, agreements often include embargo periods before screening results are publicly shared. Thus, having your own germplasm accessions from the 'private' genebank circumvents such requirements.

In summary, the reasons for breeding companies to establish their own genebank are manifold. In many instances, they are directly related to the fear of having to spend increasingly more time and money to obtain sufficient and good quality accessions along with the relevant information from genebanks around the world, or, in some cases, no longer being able to access those resources at all.

5. Practices of Breeding Companies to Acquire Germplasm

As already stated before, breeders have their own working collections for each crop they work on. These collections are typically dynamic, and they reflect the actual breeding objectives and priorities for a given crop. Many collections also include traits in response to current and future demands from their direct customers, farmers, and the market for new varieties. Also, useful breeding materials from the own programme, released varieties from competitors, as well as research materials, can be found in these working collections.

For a correct interpretation of the current global situation with respect to access and benefit sharing, it is important to review how private breeders have traditionally obtained the genetic resources for their breeding programme and, more recently, for the conservation and use of their own genebanks. The most common way has been through formal acquisition from public national or international (e.g., the CGIAR and WorldVeg) genebanks, as will be presented below. Another important way was/is through participating in collecting missions within the country where they are based and/or abroad. The latter is usually through collaborative projects with the (national public) genebank and/or with public or university research teams. It is typically also through these arrangements that possible benefits are shared. Furthermore, the acquisition of pre-bred materials has been and still is a relevant source of diversity to the private sector breeding and genebank activities.

Today, public-private partnership (PPP) research consortia and networks are also sources of germplasm materials shared among and by partners. Unfortunately, it is impossible to quantify these different ways of germplasm acquisitions by private genebanks. Details on the acquisition sources and the common practices of the private sector to obtain the genetic resources they need are elaborated further below.

Regarding the acquisition of germplasm from genebanks, especially national genebanks, a few aspects should be mentioned as background information. Genebanks have included the distribution of germplasm materials they conserve to users as one of their routine operations and responsibilities. In general, public genebanks are well connected to the various research and breeding institutions of the agricultural sector in their country and abroad. Recently, Mekonnen and Spielman [20] correlated historical trends in genebank acquisitions and changes in germplasm exchange over time, with changes in national and global policy environments for seven crops (sorghum, cowpea, pearl millet, beans, maize, rice, and wheat) that are essential for food security in developing countries. Based on these results, the authors concluded that a sharp decline in genebank acquisitions was observed in 1993 when the CBD entered into force and that country's membership in the CBD is closely associated with reductions in the flow of genetic resources. Furthermore, the Nagoya Protocol may affect global PGRFA flows in a potentially negative and unintended manner. In contrast, ITPGRFA membership is likely to moderate the adverse effects of the CBD and the Nagoya Protocol [5].

Since the implementation of ABS regulations, the distribution of requested materials is done under material transfer agreements. In the case of species listed in Annex I of the International Treaty, often the standard material transfer agreement (SMTA) is used [21], and an increasing number of countries are using the SMTA also for germplasm materials not listed in Annex I, e.g., the Netherlands, Germany, and most Nordic countries [5]. The following paragraphs will provide examples of the distribution of predominantly vegetable genetic resources by the Netherlands, Germany as well as the international genebank of WorldVeg.

5.1. Germplasm Distribution by Selected National Genebanks

In the ten-year period 2013–2022, CGN distributed 46,819 samples or an average of 4620 samples per year, of which 22,996 (=49%) samples had been sent to private, predominantly breeding companies. The latter included 11,634 (=25%) samples that were sent to companies outside the Netherlands. Most of this material concerns vegetable seeds of landraces and crop wild relative species (Figure 1).

CGN has the policy that all requests for more than 50 accessions need to be questioned by its curator of the corresponding crop genepool; usually, this process results either in a reduction of the number of accessions being shared in cases where a better selection of accessions or traits can be made, or in an agreement about sharing the results in cases that large scale screenings are done. The screening results are only made public after a negotiable embargo period of about three years. The fluctuations in the number of distributed samples over the years can largely be explained by incidental large-scale screening projects, resulting in an inflation of the number of distributed samples.

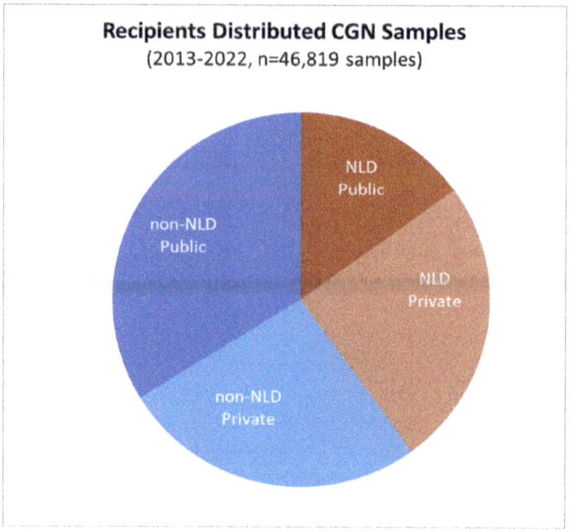

Figure 1. Germplasm distribution from the Dutch national genebank CGN from 2013–2022 to recipients worldwide. 'NLD' stands for the Netherlands, 'Public' refers to public institutions including universities, research organizations and genebanks, and 'Private' to (mainly) breeding companies. To illustrate the development with respect to the distribution of genebank accessions/samples to users for all crops and species, distribution data have been obtained from CGN since its establishment in 1986. The number of annually distributed samples and answered requests are presented in Figure 2. CGN makes its material available to any user who wants to use it for research, breeding, or education, and distributes its material all over the world. All distributions are done under the SMTA. Sometimes the phytosanitary or other import requirements make shipment difficult or impossible.

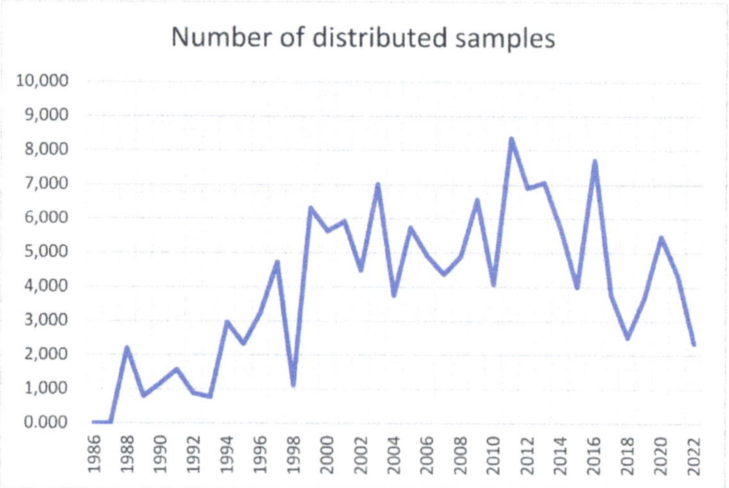

Figure 2. Number of annually distributed germplasm samples by CGN since its establishment in 1986.

As illustrated in Figure 3 and based on data received from Weise and Oppermann [22], IPK distributed during the period 2006–2022 a total of 142,157 vegetable germplasm samples which is 34.0% of the total of 418,197 samples distributed of all crops and species during the same period. It should be noted that the data on vegetables are based on an internal IPK crop classification (55 crops or crop groups, of which 20 were regarded as vegetables) that required in some instances arbitrary decisions to include or not include a given category as a vegetable. Two examples are the inclusion in this study of *Phaseolus vulgaris* and *Solanum tuberosum* as vegetables. Whereas 14.5% of the total distributed 418,197 germplasm samples were sent to German (4.4%) and foreign plant breeders (10.1%), of the 142,157 samples of vegetable germplasm 5.6% were sent to German plant breeders and 11.4% to breeders outside Germany. The 87,214 samples of vegetable germplasm distributed to all German recipients (i.e., 61.4% of the total of 142,157 samples) was more than 1.5 times higher compared with the total samples distributed to all recipients outside Germany, i.e., 54,943 samples or 38.6%).

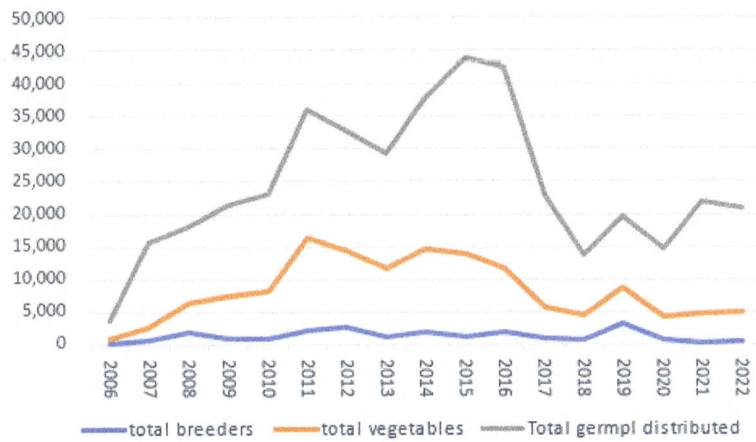

Figure 3. Numbers of distributed samples of vegetable crops by IPK to plant breeders only (blue line), to all users (orange), and total samples of all conserved crops and species distributed per year (light blue), from 2006–2022.

The distribution figures of the Dutch and German national genebanks of all germplasm accessions maintained are impressive and seem to have reached a maximum in 2011 in the Netherlands (almost 8500 samples) and in 2015 in Germany (almost 45,000 samples). The distribution rate to the private sector, predominantly plant breeders, is roughly the same for Dutch and foreign breeders. In Germany, the distribution of all germplasm to foreign breeders is slightly lower (43.9%) than the distribution to German breeders (i.e., 56.1%). The percentage of samples distributed to breeders from all the germplasm distributed during the period 2006–2022 by IPK is 14.5%, with a variation from 9.4% in 2015 to 23.9% in 2019.

Day-Rubenstein et al. [23] reported that about 5% of the 162,673 germplasm samples distributed by the United States National Plant Germplasm System from 1995 to 1999 went to commercial companies outside the United States.

It is interesting to note that the distribution data stems largely from genebanks and countries that have very limited restrictions on the distribution of public germplasm. Unfortunately, several other 'big genebanks', e.g., the Indian, Chinese, and Russian national genebanks, have a far more restrictive policy regarding the international distribution of publicly conserved germplasm.

5.2. Germplasm Distribution by CGIAR and WorldVeg Genebanks

Regarding germplasm acquisition from genebanks by the private sector, one of the very few information sources is the State of the World reports that FAO produces and publishes about every decennium. These reports are based on country reports produced by the formally designated focal institution (usually the national genebank) that gathers the information nationwide. For this reason, the information in the reports is predominantly related to the public genebanks [1,24,25]. The majority of germplasm accessions used in public and private plant breeding were sourced from national genebanks (more than half), followed by CGIAR genebanks and international and regional networks (about one-third), local genebanks, public institutions from developed and developing countries, and the private sector [1]. During the period 2017–2019, the following CGIAR centres were reported to have distributed germplasm to the commercial (mostly breeding) sector: Alliance Bioversity and CIAT, CIMMYT, CIP, ICARDA, ICRISAT, IITA, and IRRI. Seven percent of the approximately 200,000 samples were sent to the commercial sector [13].

Regarding the type of germplasm, some data are available for the period from 1996 to 2006, during which the international agricultural research centres distributed a significant amount of germplasm to the private sector; 51.7% of the distributed accessions were landraces, 36.0% breeding lines, 7.1% crop wild relatives and 5.1% improved varieties [25]. The latter coincides roughly with the data reported by Halewood et al. [13] for the period 2017–2019, i.e., 50% landraces, 24% breeding materials, 13% crop wild relatives, and 6% improved varieties.

More detailed but somewhat dated information from WorldVeg illustrates a more detailed picture of the distribution of vegetable genetic resources. This international nonprofit institute for vegetable research and development actively exchanges genetic resources and related information with national programs, regional organizations, and the private sector. As recent distribution data are not accessible, we refer here to data published earlier, giving a reliable overview of germplasm distribution until about 2014. Until this period, the WorldVeg genebank distributed approximately 6000–7000 accessions and breeding lines each year for crop improvement programs and related research worldwide. Although no detailed figures are available, the ratio of 'pure genebank accessions' to 'improved breeding lines' is approximately 20 to 80% for the vegetable crops for which WorldVeg operates crop-specific breeding programmes. In contrast, this ratio is roughly the reverse for minor vegetables and indigenous/traditional crops without a WorldVeg breeding program. Based on 2012 data, the largest number of annually distributed seed samples is shared in-house with breeders and other scientists, e.g., virologists, entomologists, and molecular biologists (37%), followed by national agricultural research and extension systems (26%), breeding companies (22%), universities (10%), non-governmental organizations (3%), and others (2%) around the world [26]. Based on 2012 data, the efforts of WorldVeg breeders, focusing on several major vegetable crops as well as on traditional crops of more local/regional importance, such as amaranth, African eggplant, okra, roselle, Corchorus, and bitter gourd [27], led to the accumulated release of more than 466 improved vegetable varieties, developed with/using WorldVeg germplasm and breeding lines, around the world [28].

Chilli pepper (*Capsicum*) is the most widely distributed crop by the WorldVeg genebank, followed by tomato. During the period from 2001 to 2012, 29,980 germplasm materials were distributed, comprising 6008 genebank accessions (20%) and 23,972 advanced breeding lines (80%), developed by WorldVeg breeders [29]. The top ten recipient countries of Capsicum germplasm during that 12-year period were India (4671; 15.6%), the Republic of Korea (2710; 9.0%), Thailand (2484, 8.3%), China (2416, 8.1%), USA (2211, 7.4%), Vietnam (1418; 4.7%), Taiwan (1154; 3.8%), Indonesia (1064; 3.5%), the Netherlands (717; 2.4%), and Tanzania (509; 1.7%). The National Agricultural Research and Extension Systems (NARES) received the largest germplasm share (13,672 samples; 57.0%), followed by breeding companies (9741; 32,5%), universities (4558; 15.2%), private individuals (1611; 5.4%), and non-governmental organizations (NGOs) (398; 1.3%).

An analysis of tomato distribution data, the second most widely distributed vegetable crop by the WorldVeg genebank after *Capsicum*, was undertaken during the period from 2001 to 2013 [30]. During that 13-year year period, a total of 27,438 germplasm samples and breeding materials were distributed from headquarters to 138 countries worldwide. Resembling the Capsicum distribution data, the majority of the distributed seed samples (22,258; 81%) were improved lines developed by WorldVeg breeders, while 5180 seed samples (19%) were genebank accessions.

It is quite interesting to note that the top three tomato-producing countries in the world (China, India, and the US) were among the top four tomato germplasm recipient countries of the WorldVeg genebank—an indication of the relevance of WorldVeg tomato germplasm/advanced breeding materials for the top global producer countries. Most countries preferred to receive advanced lines that had been developed by WorldVeg breeders, except for Japan (due to a high demand by local breeding companies to address particular local demands) and Pakistan. For these two countries, the share of genebank accessions clearly dominated and reached 81.8% and 60.6%, respectively [30]. The Netherlands also had a relatively high share of genebank accessions of 41.5%. This is an indication that those countries have strong public and private tomato breeding programmes, capable of exploiting the full potential of the genetic variability of germplasm accessions. The share received by the different categories of users is similar to *Capsicum* distribution data. Government organizations (10,601 seed samples; 39%), breeding companies (8882 samples; 32%), and universities (5873 samples; 21%) were the top three recipient categories of WorldVeg tomato germplasm.

Even though the above distribution figures are somewhat dated, these data are still a good indication of the distribution of WorldVeg advanced breeding materials and genebank accessions to the private sector. It can be assumed that the distribution pattern remained similar until 2019/2020, with COVID-19 most likely causing significant disruption in the flow of germplasm and seed supply chains, especially in the Asia-Pacific region but also in Africa [31,32]. Whereas no data on the distribution of minor vegetable crops by WorldVeg are available, it should be noted that significant germplasm distribution of minor crops, in general, is related to breeding efforts by WorldVeg breeders, either based in Taiwan or Africa. A quote from WorldVeg's [33] strategic breeding plan describes the crop-specific breeding approach to achieve impact: *"It is important to recognize that impacts were achieved through very different pathways and partnerships depending on crop and location. There are contrasting impact pathways between (open-pollinated) varieties and hybrids as well as between countries with developed and underdeveloped seed systems. For instance, WorldVeg breeding lines of tomato and chilli pepper made a large impact through private sector pathways but made very little impact through public sector pathways. In contrast, mungbean breeding lines created tremendous impact through public sector pathways and negligible impact through private sector pathways. This shows that the Center needs to be strategic in how to tailor its breeding products to achieve impact at scale"*. The above distribution figures from WorldVeg beg the question about benefit-sharing. One important dimension of sharing benefits with the 'countries of origin' of the collected germplasm is the facilitation of the distribution of newly bred varieties to farmers and local plant breeders, particularly in developing countries that provide the original germplasm sources. Another dimension is the sharing of information from the private plant breeders back to the genebank and breeders at WorldVeg and thus, at least indirectly, contributing to the sharing of benefits. However, it should be noted that feedback from private breeders is difficult to obtain, whereas public plant breeders tend to be more willing to provide feedback information on the performance of shared germplasm and breeding lines.

The analyses of germplasm distribution data from the national genebanks of the Netherlands and Germany show a significant and steady flow of germplasm from the public genebanks to breeders and breeding companies. Whereas in the case of CGN, 49% of the distributed samples between 2013 and 2022 (annual average 4620 per year) went equally to breeding companies in the Netherlands and abroad, in Germany, 14.5% and an average of 3573 samples per year of all crops/species distributed went to breeders, both public and private, of which about 1.5 times more to breeders abroad. For vegetable germplasm, these figures are 17.0% to plant breeders with an average of 1418 samples per year. The distribution by the CGIAR genebanks fluctuated from year to year and was slightly over 4200 distributed samples in 2019 to the private sector. WorldVeg distributed between 6000 and 7000 samples of germplasm accessions and breeding materials annually.

In 2012, 22% of the distributed samples went to breeding companies. It should be noted that in all analysed cases, the distribution reached a maximum between 2011 and 2016.

5.3. Germplasm Collecting

Regarding collecting missions implemented or supported by breeding companies during the last twenty years or so, only very limited information is available. In general, collecting missions have not been organized and undertaken by breeding companies, in some cases the latter supported such efforts financially. Most information on joint collecting efforts has been obtained from the Netherlands. Collecting missions have been undertaken by CGN, with the financial support of the breeding companies, and by some breeding companies themselves, especially in countries situated within the centres of diversity of a given crop, since the early 1900s. The focus was mainly on local landraces, only during the second half of the last century crop wild relatives were gradually included.

Since its establishment, CGN has been actively collecting PGR material. An overview of all collecting missions it organised is given at https://missions.cgn.wur.nl/; accessed on 23 August 2023 [34]. Details on collecting missions implemented over the past ten years with support from private vegetable breeding companies are presented in Box 1.

Box 1. Collecting missions organized by CGN with the support from vegetable breeding companies over the past ten years.

The Centre for Genetic Resources, The Netherlands has been actively collecting PGR material since its establishment. An overview of all collecting missions it organised is given at [33]. The most recent missions, for which detailed information is available, include:

2013 Armenia and Azerbaijan	115 wild populations of 7 *Lactuca* species
2015 Uzbekistan and Kyrgyzstan	190 *Daucus* wild populations and 22 carrot landraces
2017 Uzbekistan	50 melon landraces
2017 Jordan	51 *Lactuca aculeata* wild populations, 1 *Lactuca serriola*, 1 *Lactuca saligna*, 1 *Lactuca undulata* and 1 *Lactuca orientalis* population
2019 Uzbekistan	21 *Lactuca altaica* wild populations, 28 *L. serriola* populations and 13 mixed populations.

The above joint collecting efforts between a public genebank and private companies stem from one country and are certainly not representative of the global picture. However, it can be noted that the Dutch National Genebank is a good example of how collaboration with breeding companies can contribute to successful conservation efforts, both in the countries where the collected (through training and the deposit of half of the collected samples in the national genebanks) as well as in The Netherlands. However, as can be observed from the above box, almost all collecting missions are conducted in Central Asia and on a limited number of crops. Due to the impact of the CBD and the Nagoya Protocol, no missions have been possible to collect for instance germplasm of the major vegetable crops in Latin America and some countries in Asia.

5.4. Germplasm Exchange through PPPs and Other Research Consortia

As mentioned before, research consortia as public-private partnerships have been established since approximately 2000, frequently under the coordination and execution of a public institute together with variable private sector entities. It should, however, be noted that also during earlier periods, public-private partnerships were established, for instance, to 'hunt' for rare plant materials during the 18th and 19th centuries. In activities that focused on plant breeding, germplasm materials were either resources from public genebanks only or partners were expected to share some of their germplasm. As part of a European Cooperative Programme for Plant Genetic Resources (ECPGR) initiated project on PPP activities, a small database on projects has been established, including details on germplasm source acquisition [35]. PPPs are regarded as a possible approach to address market failure in the field of technology innovation when the public and the private sectors are not able to carry out the required R&D activities on their own. In recent years, many PPPs in plant breeding have been established. Among those is a PPP initiative of the Nordic Council of Ministers for pre-breeding activities in the Nordic countries. The PPP initiative was proposed in 2010, and the first call for proposals was launched in 2012 [36]. The PPP is based on pooled public funding, project-based participation of interested plant breeding companies, engagement of state-of-the-art research facilities for the respective projects, and an equal share of funding from public and private sources. Given the success of the initiated pre-breeding projects in apple, barley, perennial ryegrass, and plant phenotyping, NordGen and the governments of the Nordic countries have decided to continue funding this PPP.

Another successful PPP was established in 2012 in Southeast Asia by the International Potato Center (CIP), HZPC B.V. (a private Dutch potato seed company), and Syngenta Foundation for Sustainable Agriculture (SFSA) [37]. Whereas potatoes are usually regarded as a field crop, Drewnowski and Rehm [38] provided data that justify potatoes to be treated as a vegetable as well. This PPP aims at the collaborative breeding of five tropically adapted potato varieties with high and stable yields, thus enhancing the food security and family income of resource-poor farmers in Southeast Asia. Within a short period of only four years from the first crossings in 2016, five clones have already been identified for variety release in Vietnam. During the next phase, this successful PPP project aims at the development of processing varieties with multiple resistance against biotic stresses.

Phenotyping germplasm collections is laborious and costly, but international research initiatives and public-private partnerships have been established to mitigate this hurdle. For example, under the Horizon2020-funded G2P-SOL project ("Linking genetic resources, genomes and phenotypes of Solanaceous crops"), a collaboration of 19 institutions across Europe, Turkey, Israel, Peru, and Taiwan, global core collections of tomato, potato, pepper, and eggplant, have been generated, genotyped and phenotyped. The results of this endeavour are made publicly accessible [39].

The ECPGR coordinates the European Evaluation Network (EVA), involving genebanks, research institutes and private sector breeding companies and aiming to generate standardized evaluation data (both phenotypic and genotypic data) through participatory plant breeding actions [40]. Every partner contributes according to their expertise and capacity and especially the breeding companies implement field evaluations. The project evaluates accessions of wheat/barley, maize, carrot, lettuce, and pepper in crop-specific networks that currently bring together 29 participating genebanks from 21 countries, 49 breeding companies from 14 countries, and 34 research institutes from 20 countries for a 5-year period, from 2019 to 2024. It aims to establish a self-sustaining long-term project and to continue the evaluation of genebank accessions, also of additional crops. The evaluation activities would enable the participating breeding companies to observe and characterize new diversity that could potentially be of significant interest to them and would give them a few years of a leading edge over their competitors.

The International Lettuce Genomics Consortia (ILGC; [41] is another example of an efficient platform for the exchange of lettuce genetic diversity among breeders and scientists, worldwide. This consortium, led by UC Davis, USA, in which many countries participate and through which countries like the USA are actively distributing germplasm samples from their genebanks, including vegetable crops like lettuce.

Summarizing the different ways through which breeding companies acquire their genetic resources, it seems that the 'traditional ways' of collecting and acquiring from public genebanks are still ongoing. However, it is difficult to quantify at a global level the number of accessions and samples that are being obtained this way. A number of examples and cases are being presented, whenever possible, including data, to illustrate some common practices, including distribution figures from the Dutch and German national genebanks, from the WorldVeg genebank, including through PPP activities coordinated by them, and other examples of PPP projects from European countries as well as from the CGIAR centres. Through PPP initiatives, germplasm is exchanged in a very targeted manner. CGN provided details on joint collecting activities with private primarily Dutch breeding companies. They encountered limitations with the type of germplasm material available for use, in particular as many geographic areas are not accessible. Analysing the distribution data from CGN, IPK and WorldVeg, one could conclude that there is an ongoing flow of germplasm from the public genebanks to the breeding companies; in particular, the data from CGN show this clearly and those from WorldVeg illustrate how international agricultural research centres operate. At the same time, it should be noted that the information on the supply of germplasm materials to the private sector breeders is relatively scarce and limited, although some excellent examples of successful exchanges exist.

5.5. Benefit-Sharing Arrangements by the Private Sector

Whereas benefit-sharing aspects are not the main focus of this paper it was felt necessary to provide examples of concrete and/or perceived benefits shared with the countries of origin or directly with the farming communities that are regarded as the custodians of the genetic diversity and contributions made by the private sector to the global conservation efforts. It should be noted that especially the benefit-sharing with farmers and farming communities has become more complex and complicated due to the decision during the development process of the International Treaty to leave arrangements for Farmers' Rights to the discretion of countries and not as a global responsibility. A second point that should be made in this context is the more recent and ongoing debate on digital sequence information (DSI), biological data associated with, or derived from, genetic resources such as nucleotide sequences and epigenetic, protein, and metabolite data. The benefit-sharing framework for DSI is currently being developed, based on a decision made by the Conference of the Parties of the CBD [42].

The following examples of benefit-sharing by the private sector have been mentioned or published and they demonstrate the different approaches and ideas that underly these arrangements:

- Support of Dutch companies to public national and local genebanks in building up, maintaining and regenerating collections as well as supporting collecting missions, thus contributing to long-term conservation;
- The establishment and operation of a regional (Afrisem) breeding programme by Rijk Zwaan and East-West Seeds in Tanzania allow contributions to the production and consumption of vegetables in Africa [43];
- Three global companies, as well as East-West Seed, reported collaborating with local partners to provide access to specific genetic material or biotechnology traits [43];
- Regional and national companies (e.g., East African Seed, Kenya Seed Company and Seed Co) work with partners in their country of origin, and partner with multiple local seedbanks and global research institutes by supporting genebanks and providing company genetic resources. Some national companies in East Africa also donate their germplasm to public research partners [43];
- Forty-four seed companies offer increasingly more extension services, including technical guidance and training to smallholder farmers in 47 countries on three continents [44];
- KWS reported the support of public genebanks in Peru and Ethiopia, and East-West Seed their support to genebanks in Indonesia and Thailand [43];
- Seven companies reported providing financial and technical support to the public (local/national) genebanks and four companies reported having given access to their own genetic resources [43].

These are just examples, and one may argue that these examples show that the extent of benefit-sharing is limited and incidental. Therefore, it is important to stress that the main (perceived) contribution of the breeding industry to farmers and growers is the added value that is comprised of the release of new varieties that they develop, combined with professional growing advice, that helps farmers achieve a better income. Through the open access system in plant breeding, i.e., the breeder's exemption, these improved varieties are available to anybody for further breeding, including to farmers. Typically, such varieties possess new genetic diversity that allows better adaptation, increases yield and improves the nutritional value, even if the providers of the germplasm materials used in the breeding efforts are not necessarily the same as those that grow the new varieties.

6. Examples of Current Collaboration between Public Genebanks and Private Sector Breeders

Since many public genebanks have developed from germplasm working collections that had been established by predominantly public breeders, it could be expected that collaboration between the two is obvious and intense. However, with the growing importance of and attention to the conservation of threatened plant genetic resources, among others triggered by the leadership and coordinating role of the FAO and, to a lesser extent by the establishment of the Convention on Biological Diversity in 1992, countries had increasingly created national PGRFA programmes and built public genebanks. These developments resulted in more attention in the private sector to become attentive to genetic resources and to safeguard their own growing collections [3]. In addition, possibly stimulated by the global public and critical debate on ownership over PGRFA and on access and benefit sharing issues, the collaboration between the public research and conservation programmes, particularly in the main centres of crop origin and diversity, and breeding companies started to become more constrained. However, there are still some very good and convincing examples of a close collaboration between both sectors on the conservation and use of PGRFA and these are summarized below.

6.1. *Centre for Genetic Resources (CGN)*

Since its establishment, CGN has had a fruitful collaboration with the vivid breeding industry in the Netherlands. This was also built on the existing close collaboration between the pre-breeding programmes of the Dutch public breeding research institutes and the private breeding companies that had evolved over many decennia. CGN involves breeding companies (not only but predominantly Dutch) in many of its activities. The breeding companies advise CGN regarding technical issues, including the composition of the collections [45,46], and assist with the regeneration of the CGN accessions, as an in-kind contribution. Jointly with the companies, CGN also organises large-scale screening experiments of germplasm, among others in search of disease-resistant traits [47]. In these initiatives, the companies advise on what traits need to be traced and identified and on the respective screening protocol [48]. CGN distributes the germplasm accessions to the participating companies who screen them and send the results back to CGN, which combines the data, does a quality check (every accession is sent to two companies) and sends the combined data back to the respective participating companies. After an embargo period of usually three years, CGN makes all data publicly available through its online accessible database. The data are analysed and jointly published in a scientific paper (for example van Treuren et al. [48]). Besides the advisory role, the regeneration and joint phenotyping, companies are also involved in prioritising and funding collecting trips, including the benefit-sharing component, and other acquisition activities. In that context, they also provide material of their own varieties for inclusion in the CGN collection when these varieties are no longer on the market. Overall, CGN and the collaborating breeding companies have an intense and very positive collaboration, contributing to both conservation and use. Participation is, in principle, open for any company to join, initiatives are generally organised by CGN via Plantum, the Dutch association for plant breeders and young-plant growers.

6.2. *World Vegetable Center (WorldVeg)*

WorldVeg maintains the world's largest vegetable genebank with 65,152 accessions encompassing germplasm of 133 genera and 330 species from 155 countries, including some of the world's largest vegetable crop genepool collections held by a single institution, such as chilli pepper, tomato, and eggplant, as well as about 12,000 accessions of indigenous vegetables [49]. Due to a major regeneration backlog of primarily cross-pollinated vegetables, WorldVeg concluded agreements with private-sector companies to rescue original accessions in order to make them available to users worldwide. Companies willing to support WorldVeg in this endeavour include, among others Enza Zaden in the Netherlands for the rescue and multiplication of *Cucurbita moschata* (pumpkin) and *Momordica charantia* (bitter gourd) germplasm and Rijk Zwaan for the rescue and multiplication of *Citrullus lanatus* (watermelon) germplasm. Sakata Seed Corporation, Japan assisted the WorldVeg genebank with the screening of *Brassica* accessions for resistance to *Albugo macrospora* (white rust). The screening data are shared by WorldVeg online with the public after an embargo period of two years.

To accelerate the development and dissemination of elite vegetable crop materials, WorldVeg entered breeding consortia in Asia and Africa. The Asia and Pacific Seed Association (APSA)/World Vegetable Center Vegetable Breeding Consortium was founded in 2017 with 19 members and expanded to 51 in 2023 [50]. Participating companies obtain privileged early access to newly developed lines, for which they pay an annual fee, and are invited to an annual workshop to visit and evaluate field trials and interact with WorldVeg breeders. Feedback obtained from 34 vegetable seed companies in Asia that are part of this APSA/WorldVeg consortium showed that close to 90 commercial varieties of tomato, pepper, pumpkin, and bitter gourd that are currently sold in Asia contain pre-bred germplasm developed by WorldVeg [26]. APSA/WorldVeg consortium members sold 24.7 tons of seeds of these varieties in Asia in 2020. This quantity of seed is sufficient to plant vegetables on 171,000 hectares, benefitting close to half a million smallholder farmers in that region.

Given the positive response from the APSA/WorldVeg breeding consortium, a similar consortium was established in 2018 under the umbrella of the African Seed Trade Association (AFSTA). It is known as the Africa Vegetable Breeding Consortium (AVBC), which counted nine seed company members in 2019 and expanded to 23 members in 2021 [27,51]. AVBC membership grants early access to pre-bred material developed by WorldVeg breeders. During the 2021 AVBC workshop held in Arusha, Tanzania, private seed companies associated with AVBC, were able to evaluate 40 advanced breeding lines of African eggplant, amaranth, mungbean, peppers, pumpkin, and tomato [27].

6.3. East-West Seed International

The Genetic Resource Management section of East-West Seed International has its headquarters in Thailand and provides in-kind support to the national and other domestic genebanks in the Philippines and Indonesia. It also collaborates with CGN in regenerating germplasm materials [16]. Assistance is provided through regeneration support of accessions with low viability or low seed number, thus ensuring that these accessions are preserved for future generations.

In summary, this section describes the ongoing positive routine cooperation between the Dutch national genebank CGN and private sector breeding companies, largely from the Netherlands and coordinated by Plantum. The active engagement of several private companies in several routine operations of the genebank at WorldVeg as well as the active participation of breeding companies in breeding consortia in Asia and Africa using pre-bred materials from WorldVeg as parent lines demonstrate the advantages of such cooperation. The collaborative arrangements between (inter)national public genebanks and vegetable breeding companies, often coordinated by the respective seed associations, contribute significantly to germplasm collecting, conservation, documentation, and their sustainable use.

7. Opportunities for and Advantages of Collaboration between Breeding Companies and Public Genebanks

As already addressed above, the need to create genebank collections in breeding companies has been prompted by, among others, the decreasing access to 'public PGR'. It can be observed, certainly in the Netherlands, that the willingness of the private sector to support public activities that contribute to increasing access to 'public PGR' is strong. There are various opportunities for further collaboration between the breeding companies and the public genebanks to create a win-win situation.

The provision of requested germplasm to breeding companies by genebanks serves as an obvious opportunity to explore, agree, and implement collaborative activities. Regarding the already mentioned germplasm exploration and collecting missions, it seems logical and advantageous to both, if the genebank would use its contacts and experience in planning and implementing collecting missions and, thus, potentially facilitate access to countries and regions that the breeding company otherwise would not have. Furthermore, it seems logical that the breeding companies would participate in the costs and if the collecting mission focuses strongly on the priority crops set by the breeding companies, the funding could be substantial. Such collaboration would also include the sharing of benefits with the countries in which germplasm is being collected.

Building on the specialized and deep knowledge of breeders of the crop(s) or crop genepool(s) they are focused on, it can be expected that the genebank could take advantage when structuring the collection into trait-orientated subsets, core collections, or other priorities and thus, increase the value of the conserved germplasm materials and increase their usability.

An obvious subsequent activity for the breeders to support public genebanks could be the multiplication of the collected materials of the crops of interest to the breeding company as they will have the knowledge and infrastructure to produce high-quality and healthy germplasm for subsequent long-term storage in the public as well as the in-house genebank. Such collaborations could also include germplasm materials already conserved in the public genebank, which require urgent regeneration. For example, from January 2012 to December 2019, more than 2100 accessions from the CGN genebank were regenerated and/or multiplied by private-sector seed companies. Most regenerations of CGN material are done by breeding companies, usually in various company locations in the Netherlands, and sometimes in other countries such as Spain or Morocco [1].

Another comparable opportunity is the joint molecular and/or phenotypic characterization of genebank accessions. Also, in this case all parties involved should benefit. Therefore, it can be attractive in those cases where the characterization is necessary for all parties in the consortium and implementing the characterization separately would only increase the costs. Implementing it jointly, coordinated by the genebank, and applying an embargo on the results for a couple of years after which the genebank can make the data public, can be a construction that would benefit all. CGN applies this approach in evaluating its collections for disease resistance as the participating breeding companies would all need to screen the germplasm for resistance anyway. It is the companies who decide the trait to be searched for and the method applied. It seems to be fair to conclude that the above-mentioned collaborative efforts would be mutually advantageous as the strengths of both partners are being combined. Such collaborations would also benefit other users of germplasm accessions in the country and worldwide as the knowledge on individual accessions will be steadily increased, and the quality of the germplasm samples will get to a higher standard.

Some public genebanks can offer specific expertise and knowledge on technologies such as cryopreservation, information technologies, seed science and molecular technologies that could be made available as part of a collaborative partnership with breeding companies. However, such opportunities are rare, as the breeding companies are generally better equipped. The area in which public genebanks do have a comparative advantage is their knowledge of where to obtain specific germplasm materials and/or related information and which genebanks might be accessible. Public genebanks might also be able to use their international reputation and credibility to facilitate access to germplasm, for instance as a contact point in organizing international collecting missions and/or exchanging germplasm.

Strengthened collaboration between public and private genebanks could significantly contribute to achieving a better complementarity between the strengths and interests of the two sectors, especially with respect to technology transfer, sharing of knowledge and expertise, sharing costs of joint activities, achieving more trust among the two sectors as a basis for more widely accepted legal and policy decisions. While considering such opportunities, one should clearly keep the motivations of both 'sides' in mind. The private sector is certainly willing to act in a 'responsible way' with respect to the genetic resources they obtained and possess, but they will rarely give up these strategic resources. At the same time, the companies need to ensure to have continued access to 'new genetic diversity' for the crops they breed. Thus, here lies possibly the most obvious area and common ground for both, the public genebanks in conserving genetic diversity and making it available for current and future use and the breeding companies in breeding new varieties on a commercial basis and thus directly contributing to food and nutrition security. The latter is also of greatest interest to the global, national, and local human societies, irrespective of whether 'under development' or 'developed' as long as just and transparent benefits at large are shared.

Given the above, the basic question is, whether we want to regard plant genetic resources as a public or a private resource. In this context, we equate the 'common heritage' principle of plant genetic resources with 'a public good', and this concept started changing during the last quarter of the past century. Although the 'private good' option seems tendency-wise to be decided, the authors believe that the arguments in favour of being a public good are strong and worth (re)considering. The breeding companies would certainly prefer that these genetic resources become a public resource (again), a 'heritage of humankind' style as initially treated by the International Undertaking. However, since the establishment of the CBD with the important notion that countries have the sovereign right over the genetic resources present in their territories and, consequently, regarding these resources as their property, private companies also need to treat genetic resources as a 'commodity'. Considering this logic, we are sure that companies like to support developments that make the PGR more public again, albeit except for their own PGR. One must be aware and accept that the business of breeding companies is to breed, and that access to PGR is an essential prerequisite. Consequently, if society doesn't give proper and facilitated access to genetic diversity, it could seriously hamper the breeding progress made by breeding companies. Therefore, private-sector companies must ensure that adequate genetic diversity is available to their breeders, enabling them to fulfil their mission to contribute to food and nutrition security.

8. Approaches to Facilitate Further Public-Private Collaboration

In this section, we identify and assess options for more public-private collaboration based on the discussions with breeders and based on the authors' collective experience with hands-on public genebank and germplasm management practices over almost 50 years, on all continents of the world:

a. Participation of private sector representatives in national genebank advisory committees. This seems to be a logical and important step that strengthens collaborations and thus, contributes to the sharing of responsibilities of joint interest. This would allow the voice of the private sector to be heard during the planning and implementation of the national genebank's activities, allow for identifying and implementing complementary activities on 'both sides', formalize the collaboration, make it more visible and thus facilitating a better coordination and more efficient conservation and use at the national and maybe at the regional or even international level. Such a 'formalization' of the cooperation would undoubtedly increase the trust in each other and could eventually also facilitate discussions on contributions of the private sector to the implementation of the national PGRFA conservation strategy [52,53];

b. Concluding formal, possibly long-term agreements between the public national genebank and breeding companies [53]. This will make the collaboration transparent and thus, facilitate/enhance the acceptability of the arrangement by the society at large, including politicians and thus, results in increased collaboration nationally, regionally, and globally. Furthermore, this will enable better planning of activities by all parties involved;

c. Multilateral initiatives. These are in general the preferred format for collaboration [53], as they will increase the acceptability of the collaboration, ensure more sustainability, and combine a broader array of strengths and capacities to increase the sustainable use as well as the long-term conservation of the PGRFA that are part of such collaboration. One specific advantage is that such initiatives can involve regional partners that might share more common objectives. An excellent example of such an initiative is ECPGR which instigates and coordinates projects in facilitating the use of defined PGRFA through evaluation projects;

d. A specific aspect of the previous point is the cooperation in broader consortia, e.g., EU-funded projects focusing on research activities, (re)sequencing genebank accessions; and joint participation in the ECPGR coordinated EVA project [40]. In general, this type of cooperation facilitates the generation of non-monetary benefits, such as the exchange of germplasm and information, access to and transfer of technology, and capacity building at a large(r) scale [5];

e. Formalization of collaboration. An important aspect is the approval and support by the respective government(s) of the collaboration between the public genebank and the breeding company/ies [53] and thus, to strengthen also the sustainability of the governmental support to the public genebank This can be achieved by demonstrating the increased usefulness of the genebank for the society at large and the acquisition of additional funding for the genebank, for instance for collecting, characterization and evaluation activities enabled by the collaboration with the private sector;

f. Provision of mutual services. For example, CGN organizes collecting missions; provides access to conserved germplasm materials and associated information; advises companies on use and legal (ABS) aspects and, where applicable, assists in the sharing of benefits. Dutch breeders, through their association with Plantum, provide policy advice to CGN through its crop advisory committees; provide technical inputs such as materials and knowledge through established working groups; support collecting missions, including co-funding and multiplication of collected materials; provide in-kind inputs, such as regenerating and evaluating accessions, morphological description as well as trait evaluation;

g. Public genebanks and breeding companies should jointly look for opportunities to collaborate closely in convincing the society at large, including policymakers, that continued and unrestricted access to genetic resources will be the most efficient way to contribute to food and nutrition security. Furthermore, such collaboration will also generate ample benefits for all partners in the food value chain, to be shared with all these stakeholders in a just and transparent manner while recognizing the sovereign rights of states over their genetic resources and adhering to and/or achieving less bureaucratic and more user-friendly ABS regulations.

9. Conclusions and Recommendations

Whereas this paper has a focus on vegetable genetic resources and the sources of information are limited to a few countries and genebanks, the authors are convinced that the findings do apply to PGRFA at large, to their use in breeding activities in general and to public and private genebanks in many other countries and regions of the world.

Breeding companies started to establish their own genebanks at the beginning of the 1990s, based on their traditional working collections and later expanded these collections in a targeted way by acquiring wider genetic diversity of the crop genepools of their interest. It is argued that this development needs to be recognized in the planning and implementation of global and national conservation efforts and that cooperation between the public and private sectors is advisable to facilitate more sustainable, efficient, and effective collaborative efforts with regard to the long-term conservation and use of PGRFA.

Accepting the fact that access to PGRFA is becoming more restrictive and recognizing the need of breeders to have continued access to more and new genetic diversity, it will be indispensable to reform the current ABS arrangements as recently reviewed by Ebert et al. [5]. Such reform should duly consider the contributions the private sector makes to global conservation efforts, to the sharing of benefits (or lack thereof) as well as the importance of a continued flow of germplasm. Such a reform will be necessary to ensure the continued creation of new crop varieties that allow agriculture to cope with climate change and other constraints, thus contributing to more sustainable agriculture and global food and nutrition security.

The prevailing perception that the private sector is not contributing to the cost of long-term conservation of PGRFA undertaken by the public sector is not correct but makes it more difficult for countries to share their genetic resources freely and to arrive at an effective, inclusive, and rational oversight of any global PGRFA initiative and the global conservation and use system. Better and more targeted information, as outlined in this paper, on the role and contributions of the private sector to public conservation efforts is indispensable to change this perception.

The most fundamental reasons for the private sector to establish their own genebanks lay in the perceived shortcomings of the present legal framework, especially caused by the Nagoya Protocol, both for the 'donors' of the germplasm (i.e., predominantly the Global South) as well as the users (i.e., mainly private breeding companies in the North as well as in the Global South). These shortcomings are predominantly caused by the existing ABS arrangements, for which significant differences exist between expectations in the 'Global South' and the actual shared benefits by the 'North' to the biodiverse-rich countries, including the local custodians of this diversity. In this paper, we have taken the stance that the cooperation between private and public genebanks takes place within the existing legal framework of the 'global system', especially regarding access and benefit-sharing aspects, particularly those of the International Treaty and its MLS. This is certainly the perspective that private breeding companies took while providing inputs to the paper in assessing reasons for establishing genebanks and in seeking improvements in collaboration with public genebanks. Consequently, the focus of this paper is on access to genetic resources, as the restrictions were the main reason for the private sector to establish genebanks and less on the benefit-sharing dimension. Nevertheless, and where possible, concrete (or perceived) examples of benefit-sharing contributions by the private sector have been presented in this paper. However, these examples might be seen as 'fragmented' and possibly 'opportunistic', but they give a more realistic and positive picture than what is usually experienced. It can be concluded that, despite these encouraging examples, much stronger incentives and a more facilitating legal ABS framework are needed to ensure that the (perceived) shortcomings can be resolved and thus, the collaboration between private and public genebanks can be further strengthened. The current 'divide' of countries with respect to their willingness to share PGRFA with others can only be overcome through improved communication making it clear to all that sharing of germplasm will be indispensable to achieve increased and long-lasting food and nutrition security globally. More transparent, less bureaucratic, and more efficient benefit-sharing arrangements are required to make this happen, at all levels.

It is argued that closer collaboration between private plant breeding companies and public sector genebanks in routine genebank operations, at the global, regional, and national levels, will benefit all, especially by strengthening the link between conservation and use [54]. This will lead to more efficient and targeted use of conserved genetic resources, to more cost-efficient conservation operations and thus, to a more sustainable agriculture.

Increased trust, possibly achieved through better communication and accepting each other's 'business models', will facilitate such cooperation.

It has become clear that an intensified cooperation between private plant breeding and public genetic resources conservation also requires a critical assessment of routine genebank operations, a more effective germplasm and information management, including improved ways of accession distribution that also facilitates the molecular use of the materials by the recipients.

Author Contributions: J.M.M.E., A.W.E. and T.v.H. contributed equally to the article's conceptualisation, writing, reviewing, and editing. All authors have read and agreed to the published version of the manuscript.

Funding: This research received no external funding.

Data Availability Statement: No new data were created or analyzed in this study. Data sharing does not apply to this article.

Acknowledgments: The extensive contribution of information, including facts and figures, from Kees Reinink, Managing Director Rijk Zwaan Breeding BV, De Lier, The Netherlands; Marilyn Belarmino and her successor Katherine Panergayo, Genetic Resource Manager, East-West Seed International Limited, Thailand; Orlando de Ponti, Wageningen, the Netherlands; Niels Louwaars, Plantum, Gouda, The Netherlands; Rob van Treuren, CGN, Wageningen, the Netherlands; Stephan Weise and Markus Oppermann, IPK, Gatersleben, Germany for providing the distribution data. Marlene Diekmann's critical review of the manuscript is acknowledged herewith.

Conflicts of Interest: The authors declare no conflict of interest.

References

1. Engels, J.M.M.; Città della Pieve, PG, Italy. Personal Observation, 2023.
2. Miko, I. Gregor Mendel and the principles of inheritance. *Nat. Educ.* **2008**, *1*, 134. Available online: https://www.nature.com/scitable/topicpage/gregor-mendel-and-the-principles-of-inheritance-593/ (accessed on 15 April 2023).
3. Reinink, K.; Rijk Zwaan, De Lier, The Netherlands. Personal Communication, 2022.
4. Engels, J.M.M.; Ebert, A.W. A Critical Review of the Current Global Ex Situ Conservation System for Plant Agrobiodiversity. I. History of the Development of the Global System in the Context of the Political/Legal Framework and Its Major Conservation Components. *Plants* **2021**, *10*, 1557. [CrossRef]
5. Ebert, A.W.; Engels, J.M.M.; Schafleitner, R.; van Hintum, T.; Mwila, G. Critical Review of the Increasing Complexity of Access and Benefit-Sharing Policies of Genetic Resources for Genebank Curators and Plant Breeders–A Public and Private Sector Perspective. *Plants* **2023**, *12*, 2992. [CrossRef] [PubMed]
6. Khoury, C.K.; Brush, S.; Costich, D.E.; Curry, H.A.; de Haan, S.; Engels, J.M.M.; Guarino, L.; Hoban, S.; Mercer, K.L.; Miller, A.J.; et al. Crop genetic erosion: Understanding and responding to loss of crop diversity. *New Phytol.* **2022**, *233*, 84–118. [CrossRef] [PubMed]
7. GRAIN. International Undertaking on Plant Genetic Resources: The Final Stretch. Available online: https://grain.org/en/article/90-international-undertaking-on-plant-genetic-resources-the-final-stretch (accessed on 12 September 2023).
8. IVT. *Jaarverslag 1951–1952*; Instituut voor de Veredeling van Tuinbouwgewassen: Wageningen, The Netherlands, 1952; p. 151.
9. De Ponti, O.; Wageningen, The Netherlands. Personal Communication, 2023.
10. Puranik, S.; Sahu, P.P.; Beynon, S.; Srivastava, R.K.; Sehgal, D.; Ojulong, H.; Yadav, R. Genome-wide association mapping and comparative genomics identifies genomic regions governing grain nutritional traits in finger millet (*Eleusine coracana* L. Gaertn.). *Plants People Planet* **2020**, *2*, 649–662. [CrossRef]
11. FAO. International Treaty on Plant Genetic Resources for Food and Agriculture. Available online: https://www.fao.org/plant-treaty/en/ (accessed on 12 September 2023).
12. Halewood, M.; Sood, R.; Hamilton, R.S.; Amri, A.; Van den Houwe, I.; Roux, N.; Dumet, D.; Hanson, J.; Hari, D.; Tay, D.; et al. Changing rates of acquisition of plant genetic resources by international gene banks: Setting the scene to monitor an impact of the International Treaty. In *Crop Genetic Resources as a Global Commons. Challenges in International Law and Governance*; Halewood, M., López Noriega, I., Louafi, S., Eds.; Routledge: Oxon, UK, 2012; pp. 99–132. Available online: https://www.researchgate.net/publication/270548102_Crop_Genetic_Resources_as_a_Global_Commons_Challenges_in_International_Law_and_Governance (accessed on 12 September 2023).
13. Halewood, M.; Jamora, N.; Noriega, I.L.; Anglin, N.L.; Wenzl, P.; Payne, T.; Ndjiondjop, M.-N.; Guarino, L.; Kumar, P.L.; Yazbek, M.; et al. Germplasm Acquisition and Distribution by CGIAR Genebanks. *Plants* **2020**, *9*, 1296. [CrossRef]
14. The Access to Seeds Foundation. Access to Seeds Index Report 2016. Available online: https://www.accesstoseeds.org/media/publications/ (accessed on 22 November 2023).

15. Kate, T.K.; Laird, S.A. *The Commercial Use of Biodiversity. Access to Genetic Resources and Benefit Sharing*, 1st ed.; Earthscan Publications Ltd.: London, UK, 1999; p. 418.
16. Belarmino, M.; Panergayo, K.; East West Seed International Limited, Nonthaburi, Thailand. Personal Communication, 2023.
17. Kumar, P.L.; Cuervo, M.; Kreuze, J.F.; Muller, G.; Kulkarni, G.; Kumari, S.G.; Massart, S.; Mezzalama, M.; Alakonya, A.; Muchugi, A.; et al. Phytosanitary Interventions for Safe Global Germplasm Exchange and the Prevention of Transboundary Pest Spread: The Role of CGIAR Germplasm Health Units. *Plants* **2021**, *10*, 328. [CrossRef]
18. Byrne, P.F.; Volk, G.M.; Gardner, C.; Gore, M.A.; Simon, P.W.; Smith, S. Sustaining the Future of Plant Breeding: The Critical Role of the USDA-ARS National Plant Germplasm System. *Crop. Sci.* **2018**, *58*, 451–468. [CrossRef]
19. European Commission, Food Safety. Plant Health Rules. Available online: https://food.ec.europa.eu/plants/plant-health-and-biosecurity/legislation/plant-health-rules_en (accessed on 12 September 2023).
20. Mekonnen, D.K.; Spielman, D.J. Changing patterns in genebank acquisitions of crop genetic materials: An analysis of global policy drivers and potential consequences. *Food Policy* **2021**, *105*, 102161. [CrossRef]
21. International Treaty on Plant Genetic Resources for Food and Agriculture. The Multilateral System. Available online: https://www.fao.org/plant-treaty/areas-of-work/the-multilateral-system/smta/en/ (accessed on 23 August 2023).
22. Weise, S.; Oppermann, M.; IPK, Gatersleben, Germany. Personal Communication, 2023.
23. Rubenstein, K.D.; Smale, M.; Widrlechner, M.P. Demand for genetic resources and the US national plant germplasm system. *Crop. Sci.* **2006**, *46*, 1021–1031. [CrossRef]
24. FAO. The State of the World's Plant Genetic Resources for Food and Agriculture. Rome, Italy. p. 511. 1998. Available online: https://www.fao.org/3/w7324e/w7324e.pdf (accessed on 23 August 2023).
25. FAO. *The Second Report on the State of the World's Plant Genetic Resources for Food and Agriculture*; Commission on Genetic Resources for Food and Agriculture of the FAO: Rome, Italy, 2010; Available online: https://www.fao.org/3/i1500e/i1500e.pdf (accessed on 23 August 2023).
26. Schreinemachers, P.; Ebert, A.W.; Wu, M.-H. Costing the ex situ conservation of plant genetic resources at AVRDC–The World Vegetable Center. *Genet. Resour. Crop Evol.* **2014**, *61*, 757–773. [CrossRef]
27. World Vegetable Center. *Annual Report 2021*; World Vegetable Center: Shanhua, Taiwan, 2022; 83p.
28. Mecozzi, M.; Ebert, A.W. *At This Bank for Vegetables, Withdrawals Produce Interest*; DFID Case Study on AVRDC Genebank; AVRDC—The World Vegetable Center: Tainan, Taiwan, 2012; p. 7.
29. Lin, S.; Chou, Y.; Shieh, H.; Ebert, A.W.; Kumar, S.; Mavlyanova, R.; Rouamba, A.; Tenkouano, A.; Afari-Sefa, V.; Gniffke, P.A. Pepper (*Capsicum* spp.) germplasm dissemination by AVRDC–The World Vegetable Center: An overview and introspection. *Chron. Hortic.* **2013**, *53*, 21–27.
30. Ebert, A.W.; Chou, Y.-Y. The tomato collection maintained by AVRDC—The World Vegetable Center: Composition, germplasm dissemination and use in breeding. *Acta Hortic.* **2015**, *1101*, 169–176. [CrossRef]
31. Deuss, A.; Gaspar, C.; Bruins, M. *The Impact of the COVID-19 Pandemic on Global and Asian Seed Supply Chains*; OECD Food, Agriculture and Fisheries Papers 2021, No. 168; OECD Publishing: Paris, France, 2021. [CrossRef]
32. Kimani, D.; Adokorach, J.; Munkombwe, G.; Mushita, A.; Otieno, G.A.; Recha, T.; Subedi, A.; Vernooy, R. *Strengthening Community Seed Banks in East and Southern Africa in Times of COVID-19*; Alliance of Bioversity and CIAT: Rome, Italy, 2021; p. 6.
33. World Vegetable Center. *International Vegetable Breeding: A Strategy to Create Development Impact at Scale*; World Vegetable Center: Shanhua, Taiwan, 2019; 31p.
34. CGN. Collecting Missions. Available online: https://missions.cgn.wur.nl/ (accessed on 23 August 2023).
35. ECPGR. Resources. PPP-Knowledge-Base. Available online: https://www.ecpgr.cgiar.org/resources/private-public-partnerships-ppp/ppp-knowledge-base/details?pppId=16 (accessed on 12 September 2023).
36. Nilsson, A.; von Bothmer, R.; Johanesson, T.; Nybom, H.; Bendevis, M.A.; Bengtsson, T.; Rognli, O.A. *Promoting Nordic Plant Breeding for the Future—PPP Public Private Partnership for Pre-Breeding*; Nordic Council of Ministers: Copenhagen, Denmark, 2016. [CrossRef]
37. Sharma, N.; Sanders, W.; Graveland, R.; Schoper, J.; Campos, H. A public–private partnership to speed up potato breeding. *Open Agric.* **2020**, *5*, 826–833. [CrossRef]
38. Drewnowski, A.; Rehm, C.D. Vegetable Cost Metrics Show That Potatoes and Beans Provide Most Nutrients Per Penny. *PLoS ONE* **2013**, *8*, e63277. [CrossRef]
39. G2P-SOL Gateway. Linking Genetic Resources, Genomes and Phenotypes of Solanaceous Crops. Available online: http://www.g2p-sol.eu/G2P-SOL-gateway.html (accessed on 23 August 2023).
40. EVA. European Evaluation Network. Available online: https://www.ecpgr.cgiar.org/eva (accessed on 23 August 2023).
41. Lettuce Genome Resource, UC Davis Genome Center. Available online: https://lgr.genomecenter.ucdavis.edu/ (accessed on 23 August 2023).
42. Halewood, M.; Bagley, M.A.; Wyss, M.; Scholz, A.H. New benefit-sharing principles for digital sequence information. Benefit sharing should be decoupled from access. *Science* **2023**, *382*, 520–522. [CrossRef]
43. The Access to Seeds Foundation. Access to Seeds Index 2019. Synthesis Report. Available online: https://www.accesstoseeds.org/media/publications/ (accessed on 22 November 2023).
44. The World Benchmarking Alliance. Access to Seeds Index 2021. Synthesis Report. Available online: https://www.worldbenchmarkingalliance.org/publication/access-to-seeds-index/about/ (accessed on 22 November 2023).

45. van Treuren, R.; van Hintum, T.J.L.; van de Wiel, C.C.M. Marker-assisted optimization of an expert-based strategy for the acquisition of modern lettuce varieties to improve a genebank collection. *Genet. Resour. Crop Evol.* **2008**, *55*, 319–330. [CrossRef]
46. van Treuren, R.; van Hintum, T.J.L. Comparison of anonymous and targeted molecular markers for the estimation of genetic diversity in ex situ conserved Lactuca. *Theor. Appl. Genet.* **2009**, *119*, 1265–1279. [CrossRef]
47. van Treuren, R.; Arend, A.J.M.; Schut, J.W. Distribution of downy mildew (Bremia lactucae Regel) resistances in a genebank collection of lettuce and its wild relatives. *Plant Genet. Resour. Charact. Util.* **2011**, *11*, 12–25.
48. van Treuren, R.; de Groot, L.; Hisoriev, H.; Khassanov, F.; Farzaliyev, V.; Melyan, G.; Gabrielyan, I.; van Soest, L.; Tulmans, C.; Courand, D.; et al. Acquisition and regeneration of Spinacia turkestanica Iljin and S. tetrandra Steven ex M. Bieb. to improve a spinach gene bank collection. *Genet. Resour. Crop. Evol.* **2020**, *57*, 625–639. [CrossRef]
49. World Vegetable Center. WorldVeg Genebank. Available online: https://avrdc.org/our-work/managing-germplasm/ (accessed on 1 June 2023).
50. Schafleitner, R.; World Vegetable Center, Shanhua, Taiwan. Personal Communication, 2023.
51. Ochieng, J.; Schreinemachers, P.; Ogada, M.; Dinssa, F.F.; Barnos, W.; Mndiga, H. Adoption of improved amaranth varieties and good agricultural practices in East Africa. *Land Use Policy* **2019**, *83*, 187–194. [CrossRef]
52. Spillane, C.; Engels, J.; Fassil, H.; Withers, L.; Cooper, D. *Strengthening National Programmes for Plant Genetic Resources for Food and Agri-Culture: Planning and Coordination*; Issues in Genetic Resources No. 8; International Plant Genetic Resources Institute: Rome, Italy, 1999; Available online: https://www.researchgate.net/publication/252626773_ (accessed on 23 August 2023), ISBN 92-9043-411-2.
53. van Treuren, R. Collaboration between a Vegetable Genebank and the Private Sector. In *International Expert Panel on Fruit and Vegetable Genetic Diversity: The Status and Challenges of Conservation, Exchange and Use*; International Treaty on PGRFA: Rome, Italy, 2021; Available online: https://www.fao.org/plant-treaty/overview/partnerships/international-panel/en/ (accessed on 23 August 2023).
54. Engels, J.M.M.; Ebert, A.W. A Critical Review of the Current Global Ex Situ Conservation System for Plant Agrobiodiversity. II. Strengths and Weaknesses of the Current System and Recommendations for Its Improvement. *Plants* **2021**, *10*, 1904. [CrossRef] [PubMed]

plants

Opinion

AEGIS, the Virtual European Genebank: Why It Is Such a Good Idea, Why It Is Not Working and How It Could Be Improved

Theo van Hintum [1,*], Johannes M. M. Engels [2] and Lorenzo Maggioni [2,3]

1 Centre for Genetic Resources, The Netherlands (CGN), Wageningen University & Research (WUR), P.O. Box 16, 6700AA Wageningen, The Netherlands
2 Alliance of Bioversity International and CIAT, Via di S. Domenico, 1-00153 Rome, Italy; j.engels@cgiar.org (J.M.M.E.); l.maggioni@cgiar.org (L.M.)
3 European Cooperative Programme for Plant Genetic Resources (ECPGR), Via di S. Domenico, 1-00153 Rome, Italy
* Correspondence: theo.vanhintum@wur.nl

Abstract: Europe is very active in terms of conserving plant genetic resources, with hundreds of genebanks and thousands of dedicated people involved. However, the resulting infrastructure is, along with being very expensive, far from efficient and not very reliable. In this opinion paper, the authors describe how this situation arose, and why the European Cooperative Programme for Plant Genetic Resources (ECPGR), the collaborative umbrella organization of the European countries involved, has not been able to improve this situation so far significantly. The principles of the decentralized virtual genebank (AEGIS) are described, and an analysis is made of the reasons for its lack of success. Possible changes for making AEGIS a success, or at least steps in the right direction, are proposed. These changes center around the creation of a system of certified genebanks with proper quality management, guaranteeing the long-term conservation of, and immediate access to the plant genetic resources conserved in it.

Keywords: PGRFA; ECPGR; virtual European genebank; AEGIS; EURISCO; genebank quality management

Citation: Hintum, T.v.; Engels, J.M.M.; Maggioni, L. AEGIS, the Virtual European Genebank: Why It Is Such a Good Idea, Why It Is Not Working and How It Could Be Improved. *Plants* **2021**, *10*, 2165. https://doi.org/10.3390/plants10102165

Academic Editor: Calvin O. Qualset

Received: 6 September 2021
Accepted: 8 October 2021
Published: 12 October 2021

1. Introduction

Plant Genetic Resources for Food and Agriculture (PGRFA) need to be conserved and made accessible for use in crop research and plant breeding. There are various reasons for this, the most prominent being the high rate of genetic erosion of PGRFA, i.e., without conservation, the genetic resources will disappear quickly [1]. Without these genetic resources, the world will not be able to feed its fast-growing population, since crops can no longer be adapted to the drastic, changing environments (biotic and abiotic factors).

The awareness of the importance of the conservation of, and access to PGRFA started with collecting activities and the research of pioneers such as, most prominently, N.I. Vavilov and H.V. Harlan in the first half of the 20th century. The first dedicated genebanks were established in Germany, the USA, and at the large international agriculture institutes of the CGIAR during the second half of the 20th century, and soon after, many crop research institutes upgraded their working collections into genebanks [2]. Many countries wanted to have their own national genebank or a network of genebanks located predominately in research institutes that specialized in specific crop groups. Some of the resulting genebanks are no more than poorly maintained working collections, whereas others are professionally managed long-term collections, stored in well-equipped genebanks, meeting all international genebank standards and operating on the basis of a proper quality management system.

EURISCO, the database that gathers, processes, stores and makes available information about all European genebank collections, identifies at present some 400 institutes maintaining PGRFA collections in Europe (*sensu lato*, including 43 countries; for an overview of the data included in EURISCO, see Table 1). The resulting genebank landscape is very heterogeneous, since it has grown spontaneously, without any prior plan or coordination. It could be described as chaotic, involving hundreds of institutes, thousands of people and costing huge amounts of money. But despite all the money being spent, it is not clear at all if these genebanks and other PGRFA-related institutes are adequately functioning, i.e., if they are properly maintaining the PGRFA that should be conserved for the long term and made available to users. Furthermore, it is clear that this 'system', based on national and institutional centers, was never designed and set up to serve collaborative conservation objectives aimed at increasing efficiency and cost-effectiveness in such a fragmented landscape as Europe. Each European genebank formulated its own goals and has been reaching these goals, to varying extents, while being largely independently from related institutions and certainly from those in other countries. However, now that this European landscape has existed for nearly 50 years, the common goal has become more prominent, as have the doubts about efficacy and efficiency.

Table 1. Salient statistical features of information in EURISCO.

	Number	Percentage of Acc's in EURISCO
Number of accessions	2,056,983	
part of MLS:	430,597	20.90%
part of AEGIS:	65,286	3.20%
with a DOI:	228,078	1.10%
Number of institutes:	401	
Number of countries:	43	
Number of genera:	6725	
Number of species:	45,179	

Source: [3].

Although it is clear that much valuable germplasm material is properly conserved in the well-functioning genebanks, it is also clear that many genebank operations are not effective at all, and certainly from an overarching European or global perspective. For example, (1) much material is duplicated in many collections, while other important material for a given crop or species might be missing in all the genebanks [4]; (2) access to the conserved materials for users, if there is access at all, is often restricted to a small group, consisting of colleagues in the institute, partners of a project or members of a restricted network. Most importantly, (3) the quality of the conservation methodologies and of the conserved material is often very low.

The change in the genetic resources paradigm—caused by the Convention on Biological Diversity (CBD) that was agreed upon in 1992—from genetic resources being a 'heritage of mankind' to a resource 'under national sovereignty' did not improve the situation [5]. Access rules dictated by national governments often did not exist (certainly not before the entrance into force of the CBD in 1993) or became stricter and/or more complex thereafter. Since then, some genebanks have needed permission from national authorities to distribute material abroad for each distribution, and collecting missions of foreign countries have required complex permission procedures and conditions, if such permissions are granted at all [6]. As a result, the cross-border exchange of PGRFA has been severely and increasingly hindered, with obvious consequences for the effectiveness of collaboration among genebanks [7].

Luckily for Europe, there is a well-established umbrella organization of European countries, the European Cooperative Programme for Plant Genetic Resources (ECPGR), which has aimed at the coordination of PGRFA conservation and use activities since the early 1980s [8]. ECPGR is one of the regional PGRFA networks that were considered by the International Board for Plant Genetic Resources (IBPGR) to be part of the global conservation system of the Food and Agricultural Organization of the United Nations (FAO) in the late 1970s [2]. It organizes various collaborative activities, largely through its twenty Crop Working Groups and three Thematic Working Groups, which have resulted in valuable outputs such as the earlier mentioned EURISCO, many crop descriptor lists and crop-specific quality standards, and provides a platform for the formulation of EC-funded collaborative project proposals, and the organization of training workshops, etc. However, it has limited funds (contributed by the member countries) and does not have sufficient political leverage to directly enforce a rationalization of the conservation system at the regional level. The question 'How can we make conservation of PGRFA in Europe more efficient?' has been asked occasionally (e.g., [9]), but the answers have never translated into a restructuring of the existing landscape, since national and institutional interests have prevailed over the possibility to take decisions from a regional interest point of view.

At the global level, other important developments have influenced the political thinking with respect to the management of PGRFA. Besides the already mentioned CBD, the FAO launched the Global Plan of Action (GPA) for the Conservation and Sustainable Utilization of Plant Genetic Resources for Food and Agriculture in 1996 [10]. The GPA, updated in 2011 with the adoption of the second GPA [11], called for a more rational conservation system based on better planning and more collaboration and coordination, while allowing individual countries to maintain their sovereign rights over PGRFA. However, this had very little impact on the actual collaboration among countries, if any. Additionally, the establishment of the International Treaty on Plant Genetic Resources for Food and Agriculture (International Treaty) and its Multilateral System (MLS) for the access to and benefit-sharing of PGRFA in 2004 had little, or possibly even negative, impact on actual collaboration in Europe [7].

With regards to the above-described background, members of the ECPGR Steering Committee strongly felt that something had to fundamentally change, and subsequently formulated an idea that potentially could improve the existing situation: AEGIS, 'A European Genebank Integrated System' ([12], see also Box 1).

2. The Concept of AEGIS

Obviously, if Europe were to start from scratch and could operate as a unity, for instance, a system similar to that of the USDA National Plant Germplasm System [13] could be set up, i.e., including some central and some sub-regionally specialized facilities, with proper quality management and a clear policy regarding access. However, despite the potentially much larger cost-effectiveness of such an optimized and partly centralized system, its establishment is not conceivable, owing to the lack of political unity among the countries of the region, and the lack of strategic support from the European Union. Therefore, AEGIS had to be formulated in such a way that it could be implemented by the current actors, with the consequence that changes would have as little impact on the current activities as possible. ECPGR provided the political and administrative framework needed for the initiative, thus taking advantage of the existing common legal framework established by the International Treaty, as described above.

The concept of AEGIS was formulated in 2004 and resulted in a Policy Guide on AEGIS, endorsed by the ECPGR Steering Committee [14]. AEGIS aims at establishing a European Collection that is maintained in a decentralized virtual genebank consisting of various genebanks scattered over Europe and adhering to the AEGIS concept and principles. The European Collection consists of the collective set of accessions that have been identified and proposed by each country according to agreed criteria. Accessions of the European Collection are maintained for the long-term based on agreed quality standards in the decentralized virtual European genebank, i.e., the various participating genebanks, and are freely available under the terms and conditions set out in the International Treaty on Plant Genetic Resources for Food and Agriculture. A mechanism for the selection of the material to be included in this European Collection was created and eventually simplified [15], and the quality standards for maintenance [16] were defined.

The AEGIS concept was expected to result in a decentralized collection of unique and well-maintained accessions, assuring that the material in the European Collection would be safely conserved for the long term and freely accessible from the virtual European genebank. It was expected to give genebanks the option to drop the responsibility of maintaining accessions that were already well conserved in the European Collection by an identified colleague genebank. The benefits of AEGIS perceived at the time of its establishment are summarized in Table 2.

Table 2. Perceived public benefits of AEGIS.

(a)	Improved collaboration among European countries.
(b)	Cost-efficient conservation activities within and among European genebanks.
(c)	Reduced redundancy in European collections.
(d)	Improvement of quality standards for conservation, information management and the facilitation of the use of conserved germplasm across Europe.
(e)	More effective and better quality regeneration.
(f)	Facilitated access to all the germplasm included in AEGIS.
(g)	Improved security of germplasm through standardized commitments and safety duplication.
(h)	Improved linkages between ex situ and in situ conservation as well as linkages with users.
(i)	Improved sharing of knowledge and information.

Source: [14].

On 23 July 2009, upon signature of the Memorandum of Understanding by the tenth country eligible for membership, AEGIS entered into force. It took until 12 December 2011 for the first accessions to be included in AEGIS. However, retrospectively, there were a few flaws in the concept, and as a result it failed to become a full success in the first decade of its existence.

3. The Status of AEGIS

Currently, the European Collection consists of 65,267 accessions maintained by 46 genebanks. An overview of the current holdings in AEGIS is included in Table 3. This material was largely included because some individual genebanks submitted all the material in their collection that they considered as original material from their respective countries (i.e., collected by, or bred in the country), resulting in five genebanks submitting over three quarters of the total AEGIS accessions. Most of the additionally included accessions were part of joint crop-based and ECPGR-funded projects that required the inclusion of material in AEGIS. As a result, the content of AEGIS, about three percent of the accessions that are documented in EURISCO, is very much clustered around a few genebanks and around a few crops. This would be fine if all 65 thousand accessions in the European Collection were well-managed in accordance with the agreed criteria and met the AEGIS-agreed quality standards. However, this is not clear, and certainly not assured, as there is no operating auditing system.

Because of the slow growth of the European Collection, the ECPGR Steering Committee held an AEGIS review meeting in 2018 in Spain [17]. Its conclusions indicated that AEGIS could become more effective by (1) creating a network of ertified genebanks (with a sui generis certification system and definition of the standards); (2) creating a capacity building system of genebanks that want to become certified; (3) raising sufficient funding to support coordination, monitoring and capacity-building activities.

Table 3. Overview of accessions in AEGIS.

Total number of AEGIS accessions (included in the European Collection)	65,267	
Number of countries with AEGIS accessions	19	Germany 26,725 acc.; Italy 16,336 acc.; Netherlands 5841 acc.; Switzerland 5611 acc.; Nordic Countries 4785 acc.; 14 other countries 5969 acc.

Table 3. *Cont.*

Number of genebanks with AEGIS accessions	46	
Number of AEGIS genera	366	*Hordeum* 15,667 acc.; *Triticum* 11,129 acc.; *Zea* 5699 acc.; *Lolium* 2727 acc.; *Solanum* 2449 acc.; *Pisum* 2341 acc.; *Brassica* 2289 acc.; *Festuca* 2256 acc.; other 358 genera 20,710 acc.

Source: [18].

4. Critical Assessment of the Current Functioning of AEGIS

Considering the foregoing, and including the procedural steps that have been agreed upon by the AEGIS members, one should ask the question why AEGIS, which is such a good idea, has not (yet) become a success? Three elements can be distinguished: (1) too little material has been included for AEGIS to become 'really operational' and thus, to obtain benefits; (2) the quality of the genebanks' operations has not been assured and is possibly too low in many cases; (3) trust in the continued availability of the accessions has not been assured, despite the formal inclusion of this aspect in the MOU and in the AEGIS Associate Membership Agreement. Let us, then, have a close and critical look at these elements.

4.1. Too Little Material Has Been Included in the European Collection

Why would a country/genebank include material in the European Collection? The public benefits of inclusion were spelled out in Table 2, and some high quality genebanks considered these sufficiently convincing arguments and designated accessions for inclusion in AEGIS. However, most countries and genebanks thought they were not or, at least, have not concluded an agreement with ECPGR or designated materials.

Obviously, there are several reasons to be reluctant to submit material. First of all, a country must commit itself (through the MOU) to conserve the designated accessions for the long term and to make them available to users. Furthermore, the genebank commits itself (through the Associate Membership Agreement) to meeting the quality standards (although no auditing system is implemented yet) to conserve the material 'in perpetuity' and provide complete access to the material under the SMTA. Thus, noting these obligations, and consequences such as making long-term commitments, accepting additional workloads and possibly requiring more funds to meet the requirements, one could ask 'Why should they include materials? It only makes their lives difficult!'.

The reasons for joining AEGIS from a genebank point of view are supposed to be related to the advantages generated by membership of a cooperative framework: recognition of quality and the possibility of benefiting from common resources, such as the auditing system, capacity-building opportunities and sharing responsibilities among each other. Apparently, these advantages have not been sufficiently convincing. The label of being an 'AEGIS genebank' has not generated sufficient status yet, and access to the common resources has not materialized yet, e.g., participation in ECPGR-organized capacity-building activities is not dependent on AEGIS membership.

Furthermore, to reduce avoidable redundancy in the European Collection, the procedures for including material, i.e., individual accessions, were initially very complicated. However, these have been simplified, and the only remaining requirement is that the accession is 'original', meaning that it should be either unique, from a European perspective, to the holding genebank and/or that it originated through collecting or breeding from the country where the genebank is located.

Possibly, a reluctance to include material in the European Collection also exists at the national level. The AEGIS concept aims at generating a stable commitment from each country to conserve accessions of the European Collection for the long term. Countries appear very cautious, suggesting that their budgets for long-term conservation have not been secured and indicating that they cannot commit to the obligations formulated in the MOU [19].

The limited participation can also be interpreted as an indication that the perceived benefits of AEGIS for policymakers have not been fully appreciated. These benefits range from the assured compliance with International Treaty/the Nagoya Protocol for all the accessions included in AEGIS, to the fact that AEGIS offers a mechanism to optimize the use of resources by avoiding redundancy and assuring quality. In addition, AEGIS could strengthen the position of the European region in international fora, by offering an example of efficiency and commitment.

The fact that the above-mentioned benefits have not been able to facilitate the rapid growth and implementation of the AEGIS concept probably reflects a more general situation whereby the advantages of cooperating as a region rather than in isolation do not spontaneously emerge. This is especially understandable when considering the short-sighted political attitudes of many countries or segments of society within and outside the EU, which have been increasingly promoting nationalistic approaches for political agendas in the past ten years or so. This situation is somewhat surprising, as one would expect that, in postwar Europe, the extension of the European Union to include 27 countries, the peculiarities of PGRFA and the greatly varying state of development with respect to the conservation and sustainable use of PGRFA among European countries would call for intensified collaboration and coordination. In particular, the sharing of conservation and facilitating the use of PGRFA responsibilities among countries was thought to be a 'given' and as such, was included as a motivating factor for joining and operating AEGIS. However, this assumption seems to have some flaws, as the actual preparedness to share responsibilities among countries remains rather limited.

A few changes could be considered to improve this situation. First, a genebank should have a clear incentive to submit material to the European collection. This incentive is currently lacking, as the designation of accessions does not lower the operational costs of a genebank and does not bring other evident advantages to the genebank; to date, no duplicate accessions have been reported to have been eliminated from a genebank on the basis of AEGIS (one heard argument is that reduced numbers of accessions in a genebank could lead to a reduction in the institutional budget). Possibly, the assumption that individual genebanks would be eager to eliminate redundancies with other genebanks was incorrect. Moving from the 'local genebank' level to the national level, the argument might hold that conservation costs can be reduced, but this appears to be a non-incentive to genebanks without strong political support from the national government.

To establish an incentivizing framework, there might be a need to distinguish among genebanks that are able and willing to follow the agreed FAO Genebank Standards [20] and those that are not. In fact, one could argue that many of the 400 European genebanks or collections/repositories do not properly respond to a working definition of 'genebank', i.e., an institution that meets the requirements for well-managed and effectively operated collections. To distinguish these from 'real genebanks', an AEGIS certification system could be set up, establishing a European circle of AEGIS-certified genebanks to be supported at national and regional levels. Institutions that want to become AEGIS-certified genebanks, but do not meet the requirements yet, would need to be supported by ECPGR and other donors (e.g., the respective governments and the European Union) to reach this goal by capacity building, staff exchanges, support for setting up the required facilities, etc. In fact, it can be argued that the authorities concerned should demand from the genebanks they fund that they should become AEGIS certified and thus become amenable for eventual funding from European Union sources. ECPGR could obviously play an important task in setting up such a system, by creating the certification system and promoting the actual certification of genebanks through its Steering Committee. At the same time, more efforts are possibly required by ECPGR to make a strong case at the European level for such targeted funding. For instance, ECPGR is involved in the EU-funded Genres Bridge project, which has proposed a European Genetic Resources Strategy along the lines described here [21].

Once a critical mass of AEGIS-certified genebanks, i.e., the 'real genebanks', are identified, the inclusion of materials in the European Collection should become much more straightforward. Everything in a real genebank can be included, irrespective of its originality or other criteria. Subsequently, duplicated accessions in the European Collection would be able to be removed from the genebanks involved, as they would be available elsewhere in the European Collection, as indicated in EURISCO.

4.2. The Quality of Genebank Operations Has Not Been Assured

According to the AEGIS concept, if accessions are included in the European Collection, these accessions have to be properly managed, to such an extent that another genebank maintaining the same material can stop doing so. Obviously, proper genebank management requires assurances of the quality of operations of the participating genebanks. This implies that the genebanks (1) operate a quality management system; (2) meet the agreed quality standards; (3) are audited regularly. Obviously, these points would fit perfectly in an AEGIS certification system, as described under the previous point.

AEGIS has already made a good start in developing the standards by designing a quality management system called AQUAS [16], based on the FAO Genebank Standards [20]. The standard operating procedures proposed in these standards are based on realistic but sound quality levels, assuring the proper conservation of, and full access to the material in the European Collection. An outline of the required auditing system has also already been formulated [22], consisting of record keeping, reporting and monitoring steps. However, this has never been implemented, because the low number of AEGIS accessions has not yet justified the launching of a fully fledged auditing system. Another, possibly more important reason for not implementing the auditing system is the fact that the participating genebanks have been reluctant to introduce a monitoring system that could create an unwanted bureaucratic and reporting burden for the genebanks, as well as which could interfere with national or institutional management routines and decisions. Obviously, these considerations are completely contrary to current quality management concepts, which require the proper and transparent documentation of procedures, standardization where possible, and the monitoring of processes with appropriate performance indicators.

As a possible remedial action, to increase the transparency of European genebanks and to boost awareness of the importance of quality management, a genebank peer review process was created and successfully tested in 2019. It involves the documentation of the genebank processes (using AQUAS formats), mutual visits of experts to the genebanks involved, who provide frank and clear comments in a fully transparent way to each other, and reports about these visits [23].

4.3. The Continuity of the Availability of Accessions Has Not Been Assured

Genebanks can not and will not rely on each other if they cannot be sure about the continuity of the collections. History has shown, for instance, that genebanks can disappear, institutional policies can change, crop priorities within the national or institutional context are dynamic and evolve, and that national authorities can decide that germplasm can no longer leave the country without their consent. This obviously makes creating a collaborative system like AEGIS very difficult, and these are important points to be included in the formal agreements.

The MOU between ECPGR and the countries includes precautionary measures with respect to the withdrawal of accessions from the European Collection; a 12 months' notice by the holding genebank is stipulated [19]. In case an associated genebank withdraws, a 12 months' notice is also required, and in case a country wants to terminate the MOU, a 12 months' notice is also required. These terms should give colleague genebanks the opportunity to request and receive the accessions to be withdrawn from the European Collection and include these in another AEGIS-certified genebank. However, the enforcement of the MOU has its limits, and recent history has shown that if a country doesn't want to provide access to information or material, there is very little one can do to get access, irrespective of signed MOUs.

To overcome the risk of losing access to material from the European Collection, there could be an easy solution: if a genebank currently includes accessions that are conserved as seeds in the European Collection, it guarantees the availability of these materials to the rest of the world, and those accessions also have to be safety backed-up in another European genebank (or at the Svalbard Global Seed Vault). This is usually done in a 'black-box' arrangement, i.e., the holding genebank sends a sample of each accession to a colleague genebank that stores it under optimal conditions; this deposit will not affect any property or other rights on the material; the duplicated materials will remain in sealed containers; the terms and conditions governing the deposit will be agreed on between the two genebanks involved; the receiving genebank will not take any actions to further transfer the material other than back to the originator of the duplicated accessions or in accordance with the depositing genebank's instructions, i.e., the material can only be retrieved from the genebank acting as the back-up location by the genebank which sent it there ([19], art. 1-iv). No one else has access. For PGR accessions that are not conserved as seed, alternative solutions are sought, but on the same principles. This safety backup 'system' could easily be modified to serve as an instrument to guarantee continuity of access. This would be guaranteed if the safety duplication is done under the provision that the accessions in the back-up genebank can be used for inclusion in another genebank collection in the undesirable case that the original holding genebank could no longer provide access to these accessions. This simple change would assure that the material stays within the AEGIS system and remains available even if an institute or country is not able to comply with the availability clauses of the MOU anymore, or even decides to withdraw its membership. The material could simply be reintroduced into the European Collection by another AEGIS-certified genebank. The rationale for proposing this change lies in the recognition that AEGIS-designated materials are treated as part of the MLS and, therefore, after their first exchange with an SMTA, they can be indefinitely transferred under the same conditions to any user. Therefore, it does not make sense to maintain the obligation to return the material only to the original depositor such as in most 'black-box' arrangements for safety duplication.

5. Future of AEGIS

A reasonable response to the points above would be "Dream on!", and, indeed, the proposed solutions to the points above will not be easy to implement. A genebank certification system might be very difficult to set up, as some countries will fear that they would never be able to meet the standards and thus prefer to avoid confrontation with the reality that some genebanks do not live up to the level of quality that is expected from public goods institutions operating in the global arena. Furthermore, establishing and running an auditing system requires an adequate budget (setting up a quality management system could cost up to 10% of an annual budget, and operating it up to 5% of the annual budget, but could be much less) and such funds are currently lacking. Therefore, setting up an auditing system without sufficient support or proper funding will be very difficult. Changing the safety back-up system from the current black-box construction to an emergency-access system might be considered undesirable, since it could discourage some genebanks from backing up their material for fear of losing control.

Nevertheless, some suggested steps can be made to improve the situation, paving the way for an easier and more comprehensive implementation of AEGIS and its European Collection. The European genebank community clearly has the desire to professionalize, to move from the first generation of genebank managers to the second. The already mentioned quality management system AQUAS provides valuable tools, such as the Genebank Manual [24], allowing genebanks to describe their current procedures, which is a first step towards proper quality management and a great way to improve transparency. Implemented with the hope of increasing transparency and of moving towards an auditing system, the peer reviews appeared successful, and several genebanks volunteered to participate. The first experiences with this approach have been very positive [23].

Furthermore, other initiatives to improve transparency and create a clearer picture about the quality of the European genebanks can and should be taken. For example, checking the availability and quality of the material in the European Collection is easy; one can simply request the material from the genebank and check its quality. To avoid wasting materials in this process, such requests should be done together with users that actually will use the material. A first attempt to create such a 'system' was undertaken by the Centre for Genetic Resources, The Netherlands (CGN), in 2019, when this genebank asked its users to draft a list of materials from other European genebanks that they would like to receive for their use. CGN requested these materials from these genebanks with full transparency about the context of these requests. Most of the requested material was not received. However, the COVID-19 crisis could be the main explanation for the (temporary?) lack of access that was experienced. The ECPGR Executive Committee was informed accordingly about this experiment, and it was favorable to extending it in scope, in terms of crops, genebanks and breeding community [25].

The above-mentioned steps can improve the situation in the European genebank community and reduce its rather uncoordinated and unharmonized aspects. However, they remain small steps, to be taken slowly. Apparently, national and institutional interests are currently still larger than concerns about an efficient PGRFA conservation system in Europe. All we can do is to continue trying to make small steps in the right direction. The initiatives developed in the framework of the EU-funded Genres Bridge project might have a very positive effect, provided that funds will become available. In particular, the European Genetic Resources Strategy, recently drafted by the GenRes Bridge partners, calls for the establishment of a coherent policy framework for genetic resources in Europe, facilitating and promoting genetic resources conservation, documentation and sustainable use at both national and European levels [21]. It also calls for the further development of a European infrastructure for ex situ and in situ PGRFA conservation and sustainable use. This infrastructure should include, inter alia, the decentralized/virtual European Genebank consisting of certified genebanks, building on the AEGIS experience and principles.

6. Conclusions

The world is not a perfect place, and the European PGRFA activities are no exception. The chaotic, casually grown European landscape of PGRFA actors and activities, although it does a lot of good, is far from ideal. The optimal solution is not feasible, given the current funding and decision-making mechanisms. ECPGR made an excellent attempt to improve the situation with AEGIS; however, this has not really worked so far: the impact on efficacy and efficiency has been very limited.

The possibility of reducing costs by reducing redundancy appears to have very limited appeal to most countries or genebanks. The concept of 'national sovereignty' over PGRFA, promoted by the CBD, appears to have a stronger appeal than the advantages of making PGRFA a common good in Europe. Therefore, implementing AEGIS will remain a significant challenge, as long as genebanks are funded by national authorities. Having additional EU regional funding would make the challenge of creating an effective and efficient European genebank infrastructure much easier.

With additional funding, a system of AEGIS-certified genebanks could be set up, in which the quality and continuity of conservation of, and access to PGRFA could be guaranteed. Joining this system would be attractive as it would certify that a genebank is a 'real genebank', with reliable conservation, access and legal protocols. Regional funding aimed at ex situ PGRFA management should concentrate on these certified genebanks, helping them to make the large steps that are needed to function optimally in a rapidly changing, increasingly -omics, research and breeding oriented environment. Obviously, a capacity-building program supporting genebanks that want to reach certification should be a prominent part of this vision. In the meanwhile, attempts to make small steps towards the goals of AEGIS are ongoing.

A decentralized, 'virtual' European genebank is a valid model, but apparently very hard to implement without proper regional and political visions, funding and decision making.

Author Contributions: Writing—original draft preparation, T.v.H., J.M.M.E. and L.M.; writing—review and editing, T.v.H., J.M.M.E. and L.M. All authors have read and agreed to the published version of the manuscript.

Funding: The contribution of T.v.H. was supported by the "European collaboration" project (WOT-03 Genetic Resources) by the Dutch Ministry of Agriculture, Nature and Food Quality. The opinions expressed are those of the authors, not of the funding bodies.

Institutional Review Board Statement: Not applicable.

Informed Consent Statement: Not applicable.

Acknowledgments: The authors acknowledge the inputs made by Stephan Weise, EURISCO Coordinator.

Conflicts of Interest: The authors declare no conflict of interest.

Abbreviations

Abbreviations and Acronyms

AEGIS	A European Genebank Integrated System
AQUAS	AEGIS Quality System
CBD	Convention on Biological Diversity
CGIAR	originally: Consultative Group on International Agricultural Research, now usedas acronym-name
CGN	Centre for Genetic Resources, The Netherlands
ECPGR	European Cooperative Programme for Plant Genetic Resources
EURISCO	European Search Catalogue for Plant Genetic Resources
FAO	Food and Agriculture Organization of the United Nations
GPA	Global Plan of Action for Plant Genetic Resources for Food and Agriculture
IBPGR	International Board for Plant Genetic Resources
Treaty	International Treaty on Plant Genetic Resources for Food and Agriculture (FAO)
MLS	Multilateral System of Access and Benefit-sharing
MOU	Memorandum of Understanding
PGRFA	Plant Genetic Resources for Food and Agriculture
SMTA	Standard Material Transfer Agreement

References

1. Khoury, C.; Brush, S.; Costich, D.; Curry, H.; de Haan, S.; Engels, J.; Guarino, L.; Hoban, S.; Mercer, K.; Miller, A.; et al. Crop genetic erosion: Understanding and responding to loss of crop diversity. *New Phytol.* in press. **2021**. [CrossRef] [PubMed]
2. Engels, J.M.M.; Ebert, A.W. A Critical Review of the Current Global Ex Situ Conservation System for Plant Agrobiodiversity. I. History of the Development of the Global System in the Context of the Political/Legal Framework and Its Major Conservation Components. *Plants* **2021**, *10*, 1557. [CrossRef]
3. EURISCO Catalogue. Available online: https://eurisco.ipk-gatersleben.de/ (accessed on 13 August 2021).
4. Van Treuren, R.; Coquin, P.; Lohwasser, U. Genetic resources collections of leafy vegetables (lettuce, spinach, chicory, artichoke, asparagus, lamb's lettuce, rhubarb and rocket salad): Composition and gaps. *Genet. Resour. Crop Evol.* **2012**, *59*, 981–997. [CrossRef]
5. Cooper, E.; Engels, J.M.M.; Frison, E.A. *A Multilateral System for Plant Genetic Resources: Imperatives, Achievements and Challenges*; Volume 2; Issues in Genetic Resources; International Plant Genetic Resources Institute (IPGRI): Rome, Italy, 1994.
6. Maggioni, L.; López Noriega, I.; Lapeña, I.; Holubec, V.; Engels, J.M.M. Collecting plant genetic resources in Europe: A survey of legal requirements and practical experiences. In *Implementing the Nagoya Protocol. Comparing Access and Benefit-Sharing Regimes in Europe*; Koninklijke Brill nv: Leiden, The Netherlands, 2015; pp. 327–362. ISBN 9789004293212. [CrossRef]
7. Brink, M.; van Hintum, T.J.L. Genebank Operation in the Arena of Access and Benefit-Sharing Policies. *Front. Plant Sci.* **2020**, *10*, 1712. [CrossRef] [PubMed]

8. UNDP/IBPGR; The ECP/GR. *An Introduction to the European Cooperative Programme for the Conservation and Exchange of Crop Genetic Resources*; UNDP/IBPGR: Rome, Italy, 1984; 20p.
9. Gass, T.; Begemann, F. International efforts to sustain ex situ collections: Options for a closer cooperation in Europe. In *Implementation of the Global Plan of Action in Europe—Conservation and Sustainable Utilization of Plant Genetic Resources for Food and Agriculture, Proceedings of the European Symposium, Braunschweig, Germany, 30 June–3 July 1998*; Gass, T., Frese, L., Begemann, F., Lipman, E., Eds.; ECPGR: Rome, Italy, 1999; pp. 109–115.
10. FAO. *Global Plan of Action for the Conservation and Sustainable Utilization of Plant Genetic Resources for Food and Agriculture and the Leipzig Declaration*; FAO: Rome, Italy, 1996; 63p.
11. FAO. *Second Global Plan of Action for Plant Genetic Resources for Food and Agriculture*; FAO: Rome, Italy, 2011; 96p.
12. AEGIS. A European Genebank Integrated System. Available online: https://www.ecpgr.cgiar.org/aegis (accessed on 1 October 2021).
13. USDA. National Plant Germplasm System. Available online: https://data.nal.usda.gov/dataset/national-plant-germplasm-system (accessed on 14 August 2021).
14. ECPGR. *A Strategic Framework for the Implementation of a European Genebank Integrated System (AEGIS). A Policy Guide*; ECPGR: Rome, Italy, 2009; 32p.
15. ECPGR. Revised Simplified Procedure for the Selection and Flagging of Accessions for the European Collection. Available online: https://www.ecpgr.cgiar.org/fileadmin/templates/ecpgr.org/upload/AEGIS/FOR_WEB_FINAL/Selection_and_flagging_procedure__final_.pdf (accessed on 19 June 2021).
16. ECPGR. A European Genebank Integrated System. *AQUAS: Quality System for AEGIS*. Available online: https://www.ecpgr.cgiar.org/aegis/aquas/overview (accessed on 14 August 2021).
17. ECPGR. A European Genebank Integrated System. *AEGIS Workshop 2018. Assessing Current Practices and Procedures to Strengthen AEGIS*. Available online: https://www.ecpgr.cgiar.org/aegis/aegis-workshop-2018 (accessed on 14 August 2021).
18. ECPGR. A European Genebank Integrated System, European Collection, European Accessions. Available online: https://www.ecpgr.cgiar.org/aegis/european-collection/european-accessions (accessed on 14 August 2021).
19. ECPGR. Memorandum of Understanding for the Establishment of A European Genebank Integrated System (AEGIS). Available online: https://www.ecpgr.cgiar.org/fileadmin/templates/ecpgr.org/upload/AEGIS/FOR_WEB_FINAL/AEGIS_MOU_final130309_original.pdf (accessed on 25 August 2021).
20. FAO. *Genebank Standards for Plant Genetic Resources for Food and Agriculture*; FAO: Rome, Italy, 2013; 181p.
21. GenRes Bridge European Genetic Resources Strategy. Available online: http://www.genresbridge.eu/resources/european-strategy/ (accessed on 25 August 2021).
22. Engels, J.; Maggioni, L. *Record Keeping, Reporting and Monitoring of the European Collection*; ECPGR: Rome, Italy, 2016; 4p.
23. ECPGR. A European Genebank Integrated System. *Genebank Capacity Building Peer Visits*. Available online: https://www.ecpgr.cgiar.org/aegis/aquas/peer-visits (accessed on 25 August 2021).
24. ECPGR. A European Genebank Integrated System. *Genebank Manuals*. Available online: https://www.ecpgr.cgiar.org/aegis/aquas/genebank-manuals (accessed on 14 August 2021).
25. ECPGR. Minutes of the 15th ECPGR Executive Committee Meeting. Available online: https://www.ecpgr.cgiar.org/fileadmin/templates/ecpgr.org/upload/EXCO/Fifteenth_meeting/Minutes_Exco15-_final_19_03_2020_rev_12_01_21.pdf (accessed on 25 August 2021).

plants

MDPI

Review

Envisaging an Effective Global Long-Term Agrobiodiversity Conservation System That Promotes and Facilitates Use

Charlotte Lusty [1,*], Ruaraidh Sackville Hamilton [1], Luigi Guarino [1], Chris Richards [2], Nelissa Jamora [1] and Geoffrey Hawtin [3]

1 Global Crop Diversity Trust, Platz der Vereinten Nationen 7, 53113 Bonn, Germany; ruaraidh.sh@gmail.com (R.S.H.); luigi.guarino@croptrust.org (L.G.); nelissa.jamora@croptrust.org (N.J.)
2 USDA National Laboratory for Genetic Resources Preservation, 1111 South Mason Street, Colorado State University Campus, Fort Collins, CO 80521, USA; chris.richards@ars.usda.gov
3 Alliance of Bioversity International and the International Center for Tropical Agriculture, Via di San Domenico, 1, 00153 Rome, Italy; geoffhawtin@hotmail.com
* Correspondence: charlotte.lusty@croptrust.org

Abstract: Genebanks were established out of a recognised need not just to provide genetic variation to support breeding objectives but to prevent crop diversity from being lost entirely for future users. Such conservation objectives may have led, over the past few decades, to a gradually diminishing connection between genebanks and current users of diversity. While there continues to be large-scale distribution of germplasm from genebanks to recipients worldwide, relatively little is known or published about the detailed trends in the demand for genebank materials. Meanwhile, the rapid expansion of the applications and uses of modern genomic technologies and approaches is, undoubtedly, having a transformational impact on breeding, research and the demand for certain genetic resources and associated data. These trends will require genebanks to be responsive and to adapt. They also provide important opportunities for genebanks to reorganize and become more efficient individually and as a community. Ultimately, future challenges and opportunities are likely to drive more demand for genetic diversity and provide an important basis for genebanks to gear up.

Keywords: genebanks; CGIAR; plant genetic resources; conservation; breeding; genomic research

Citation: Lusty, C.; Sackville Hamilton, R.; Guarino, L.; Richards, C.; Jamora, N.; Hawtin, G. Envisaging an Effective Global Long-Term Agrobiodiversity Conservation System That Promotes and Facilitates Use. *Plants* **2021**, *10*, 2764. https://doi.org/10.3390/plants10122764

Academic Editor: Shri Mohan Jain

Received: 14 November 2021
Accepted: 4 December 2021
Published: 14 December 2021

Publisher's Note: MDPI stays neutral with regard to jurisdictional claims in published maps and institutional affiliations.

1. Introduction

When Nikolai Vavilov began to collect seeds from around the world in the 1920s and 1930s, he was a pioneer in what has now come to be known as genebanking. While he and most other early crop plant collectors primarily sought landraces and traditional varieties to use as a source of novel traits in their breeding programmes, as they continued to travel and collect, it became increasingly apparent to them that many landraces were disappearing from farmers' fields [1]. According to Mooney [2], on a collecting expedition to Turkey in the 1940s, the plant collector Jack Harlan "encountered virtually thousands of flax varieties. When he returned 20 years later only one variety remained—and this was imported from Argentina".

Genetic erosion was especially rapid following the Green Revolution of the 1960s and 1970s, when scientists from the International Rice Research Institute (IRRI) and the International Centre for the improvement of Maize and Wheat (CIMMYT) bred high yielding varieties of rice and wheat, which were then widely disseminated [3]. While contributing enormously to easing the food shortages of the time, the rapid spread of these varieties resulted in the replacement of many indigenous landraces, a process that continues today. Observing this led many early plant collectors and breeders to recognize the importance of creating crop collections not only as a ready source of new alleles for genetic improvement, but also to conserve crop variation for the future, as an insurance against its loss in the wild and in farmers' fields. To meet this dual need, specialized facilities for seed processing, storage and distribution were created, known as genebanks.

As the Green Revolution was taking off, new international agricultural research centres, modelled on IRRI and CIMMYT, were being established around the world and were funded collectively by a consortium of donors called the Consultative Group on International Agricultural Research (CGIAR). Those centres that were concerned with the genetic improvement of crops began to develop germplasm collections, focussed initially on providing genetic variation for the immediate needs of their breeding programmes but increasingly, over time, had a long-term conservation objective as well.

The chickpea, lentil and faba bean collections of the International Centre for Agricultural Research in the Dry Areas (ICARDA) provide one such example. They were originally assembled in Lebanon and Syria by the food legume breeders themselves to ensure that sufficient genetic variation was available to initiate their crop improvement programmes. Large collections were built up both through extensive field collecting in West Asia and North Africa, starting in 1975, and through acquisition, especially from the collections that had been put together by the USDA/USAID-funded Regional Pulse Improvement Project in India and Iran in the 1960s. Over time, as the breeders identified ever more accessions likely to provide alleles for traits of interest to their breeding objectives, so they increasingly came to spend more of their time working with this material and less time looking for additional genetic variation. However, ICARDA recognized that the collections that had been built up were not only of immediate use but had a long-term value for both the Centre's own breeding programmes as well as those of their partners. Thus, in 1983, ICARDA established a Genetic Resources Unit (GRU) to develop and run a genebank [4]. At this point, objectives began to diverge somewhat, with conservation becoming the main priority of the GRU and developing genetically improved varieties the priority of the breeding programmes. While the GRU still aimed to serve the needs of the genetic improvement programmes, in practice, the breeders made fewer forays into the collections as they came to concentrate more on their own 'elite' genepools. The GRU thus took on a life of its own, with a leadership separate from that of the breeding programmes.

A somewhat similar situation occurred at the International Rice Research Institute (IRRI) in the Philippines, where the international rice collection was first assembled in the early 1960s to support the Centre's breeding work. However, in the late 1960s, the geneticist T.T. Chang (one of the members of the team that created the semi-dwarf variety IR8, the key rice variety of the Green Revolution) had a disagreement with the breeders over what to do with material from the breeding programme that was not of immediate interest. As a result, he established the International Rice Germplasm Centre, now called the T.T. Chang Genetic Resources Centre (GRC), as a separate entity within IRRI to conserve germplasm samples that could be of longer-term value. At the same time, there was growing concern in the Philippines and elsewhere over the replacement of local rice landraces and farmers' varieties by IR8. Recognizing this, in the 1970s, T.T. Chang began a large programme of collecting from farmers, with the primary aim of protecting against genetic erosion. The fact that the target was to conserve rather than immediately support genetic improvement contributed to a growing divide between the breeding programme and the GRC.

Since then, the breeder-genebank relationship at IRRI, as in many other CGIAR Centres, has fluctuated widely and for a variety of reasons. In the 1990s, fears over the potential impact of the Convention on Biological Diversity (CBD) resulted in reduced internal distribution of material from the genebank to the breeders as concerns grew that IRRI's breeding lines might become subject to CBD restrictions. Later, with the adoption of the Nagoya Protocol in 2010, uncertainty over how to implement its rules on access and benefit sharing (ABS) resulted in the breeders being reluctant to deposit any of their breeding lines in the GRC. In addition, independently of political concerns, some breeders are averse to 'polluting' their breeding populations with 'inferior' germplasm from the genebank due to linkage drag.

On the positive side, a very large proportion of the IRRI collection has been screened by breeders, geneticists, and physiologists for a wide range of characteristics. In many cases, this has resulted in the identification of an allele or alleles that have enabled breeders to overcome particular bottlenecks in meeting their specific genetic improvement objectives; notable examples include "scuba rice" containing a submergence tolerance gene from a traditional Indian variety, and numerous disease resistance genes identified in and transferred from the wild relatives of rice [5,6].

The slightly ambivalent relationship between breeders and genebanks is not exclusive to the CGIAR but is widely observed in countries around the world. Greater global recognition of the social and cultural relevance of landraces and farmers varieties has given germplasm collections an importance beyond just serving as a resource for plant genetic improvement. This is particularly true at the local and national level, where the conservation of indigenous germplasm has acquired a political dimension that is reflected in the often-acrimonious debates on access and benefit sharing in various international fora, including the Nagoya Protocol and the International Treaty on Plant Genetic Resources for Food and Agriculture (Plant Treaty) [7]. On the one hand, the heightened interest in plant genetic resources has served to underline the importance of conservation, but on the other hand, a growing recognition of its potential value has resulted in greater restrictions on the ability of breeders to access material held within the collections [8].

Given this backdrop, we explore in this article how advances in genomics will further affect the relationship between the genebank and breeders, changing the way genebanks may be used in the future and creating opportunities for collections to be curated differently to maximize both their current usefulness and their efficiency in the long term. Firstly, however, we briefly describe what we know about the use of CGIAR genebanks.

2. Current Use of Material in CGIAR Genebanks

Systematic documentation of the distribution of plant genetic resources for food and agriculture (PGRFA) began in 2007 when the Governing Body of the Plant Treaty determined that providers of PGRFA must report to the Governing Body on the material they provide under the multilateral system of access and benefit-sharing (MLS) [9]. However, details of these reports are confidential, and only aggregated statistics are publicly available [10]. Indeed, general information on germplasm distributions from genebanks is restricted to a basic set of parameters, such as number of requests, number of accessions and samples distributed, and countries receiving germplasm. Deficiencies in data on germplasm flows for informing policy-relevant analysis and guidance was recently highlighted by Mekonnen and Spielman (2021). Private-sector recipients frequently do not want the details of their germplasm requests made known. Furthermore, the Standard Material Transfer Agreement (SMTA), which is issued as part of the transaction between providers and recipients upon the transfer of materials that are included in the MLS, expressly requires the provider to undertake that the material is provided "without the need to track individual accessions" [11].

Such principles and practices aiming to facilitate international germplasm movement and exchange have worked somewhat as a disincentive for genebanks to document and analyse information about user demand and potential future needs for genebank materials, resulting in much less being known about the current or potential deployment of diversity from genebanks than might be expected from a typical service provider.

A voluntary mechanism for uniquely identifying individual samples of PGRFA, which incorporates the possibility to track the movement and use of individual genebank materials and their derivatives, has recently been established by the Plant Treaty Secretariat with unique digital object identifiers (DOIs). At the time of writing, DOIs have been registered for 1,181,758 samples of PGRFA [12] by 2820 registrants [13], including all CGIAR genebanks and some national genebanks, but DOIs have yet to be adopted on a wide scale by genebanks or by CGIAR breeders and researchers and other users of genebanks. Once fully adopted, DOIs have the potential, at least, to allow the tracing of the use of specific accessions in published research and in germplasm exchanges and released materials [14].

The Global Crop Diversity Trust (Crop Trust) also records the numbers of samples and accessions distributed from international genebanks receiving long-term funding (including nine CGIAR genebanks). Although these efforts do not provide the kind of market intelligence that may help genebank managers and staff cater to trends or manage the collections more rationally, some patterns may be discerned.

In 2020, CGIAR breeding programs and genebanks accounted for 89% of the germplasm distributed in the MLS [15]. During the 8-year period between 2012 and 2019, a total of 853,808 germplasm samples were distributed from CGIAR genebanks, of which just over half were requested by CGIAR scientists and 47% were shipped outside CGIAR to an average of nearly 2000 requesters annually (Table 1).

Table 1. Germplasm distribution from CGIAR genebanks between 2012–2019, showing the proportion of rice and wheat compared to the other 27 mandate crop species that are available from CGIAR and the proportion of samples shipped outside CGIAR to users making requests.

Column1	Total	Annual Average	Rice & Wheat	% Rice & Wheat	Other Crops	% Other Crops
Total number of samples distributed	853,808	106,726	460,976		392,832	
% of total distributed			**54%**		**46%**	
Number of samples distributed internally within CGIAR	452,966	56,621	301,942	**66%**	151,024	**38%**
% of total internally distributed	53%		**65.5%**		**38%**	
Number of samples distributed to users outside CGIAR	400,842	50,105	159,034	**34%**	241,808	**62%**
% of total externally distributed	**47%**		**34.5%**		**62%**	

The only genebank system to distribute more is the US Department of Agriculture National Plant Germplasm System, which distributes about 250,000 samples yearly, of which 25% are distributed internationally [16]. Trends in germplasm distribution from CGIAR over the past four decades have been volatile, though with a gradual upward trend [15]. Germplasm-related projects (e.g., large-sale sequencing/genotyping of rice, wheat and maize) are responsible for some of the recent peaks in demand.

Rice is, by some margin, the most distributed of the 29 crop species (which include banana and plantain, Bambara groundnut, barley, beans, cassava, chickpea, cowpea, faba bean, temperate and tropical forages, fruit and multi-purpose trees, grasspea, groundnut, lentil, maize, various underutilized legumes, pea, pearl millet, pigeon pea, potato, rice, small millets, sorghum, soybean, sweetpotato, wheat, yam, and Andean roots and tubers) conserved by CGIAR genebanks, accounting for 36% of total CGIAR germplasm distributions between 2012 and 2019 (Figure 1).

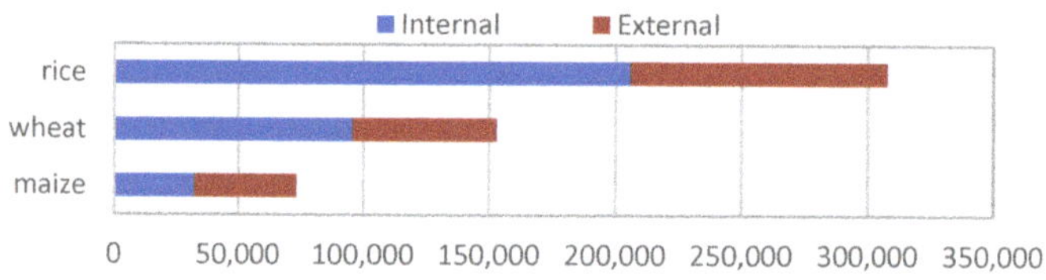

Figure 1. Distribution of germplasm samples of rice, wheat and maize shipped from CGIAR genebanks between 1980 and 2019 to users who are internal or external to CGIAR.

Rice and wheat together made up more than half (54%) the germplasm distributions from CGIAR genebanks between 2012 and 2019, with the majority (65.5%) of rice and wheat samples going to CGIAR breeders and researchers (Table 1). Thus, CGIAR wheat and rice breeding and research accounted for more than one-third (35%) of the overall germplasm distribution from CGIAR genebanks. Future research efforts by CGIAR on these two crops will, no doubt, continue to have a significant influence on CGIAR genebank use.

However, it is possible to imagine that future use of CGIAR genebanks may change, given the growing range of crops of interest in agricultural research and development. The data on the distribution of germplasm of CGIAR's mandate crops other than wheat and rice hint at the potential for such a change. Over the same 8-year period, the germplasm of mandate crops other than rice and wheat have been predominantly distributed to external users (62%) rather than to CGIAR breeders (Table 1, Figure 1).

Between 2017 and 2019 (when data are available), the external demand came mainly from the public sector: 81% was from national agriculture research institutes, universities and advanced research institutes, 12% were farmers, NGOs and individuals and 7% were commercial sector users (Figure 2).

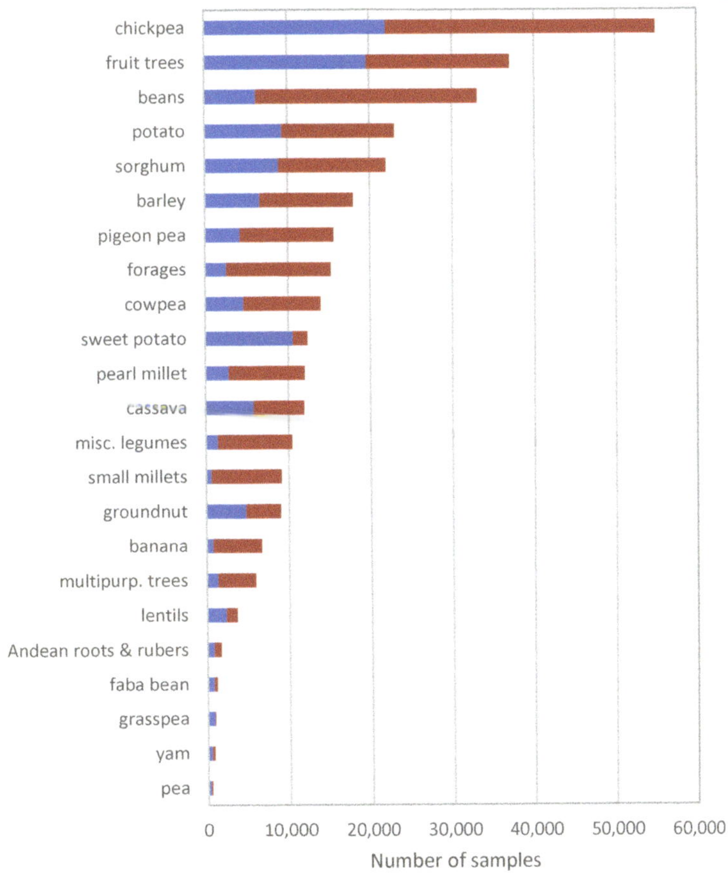

Figure 2. Distribution of germplasm samples of crops (exc. rice, wheat and maize) shipped from CGIAR genebanks between 1980 and 2019 to users who are internal or external to CGIAR.

Germplasm samples were shipped to every region and sub-region of the world in response to requests for a diverse range of crop species (e.g., 20 species were shipped to both Africa and Asia). Asia received the most germplasm samples (42%), followed by Africa (23%) and the Americas (19%) (Figures 3 and 4). All regions show similar proportions of germplasm going to different user categories, although most of the samples requested by farmers and NGOs were shipped to Africa.

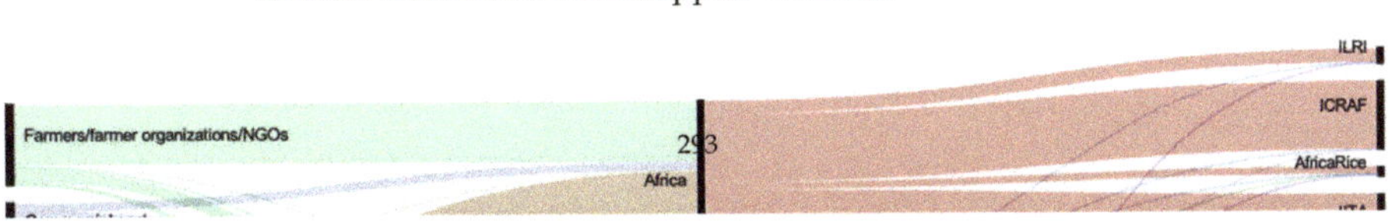

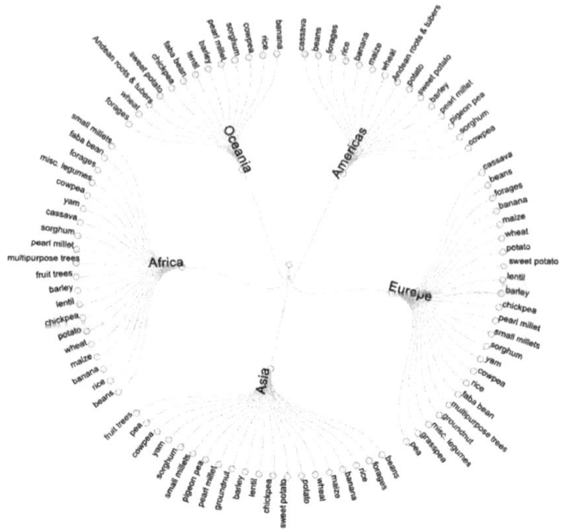

Figure 4. Crop species distributed by CGIAR genebanks to different geographical regions between 2017 and 2019.

At a country level, distribution figures are skewed towards countries that host CGIAR genebanks (Figures 5–8).

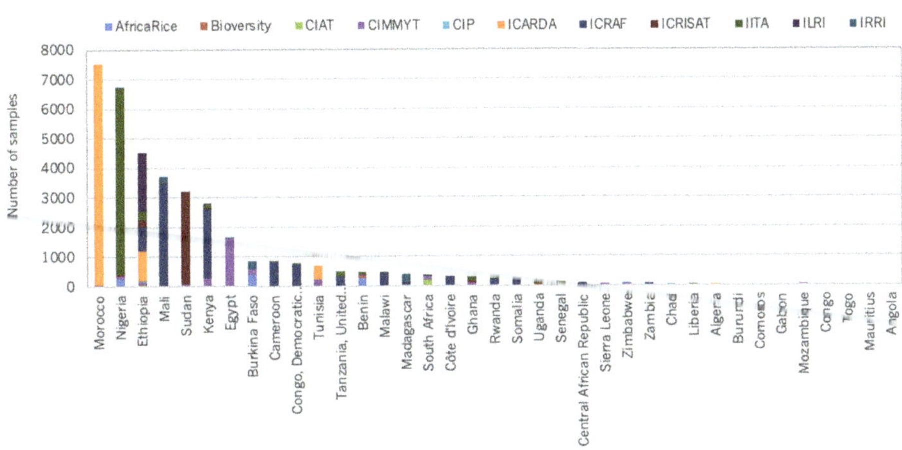

Figure 5. Number of germplasm samples distributed by CGIAR genebanks to external users in Sub-Saharan Africa.

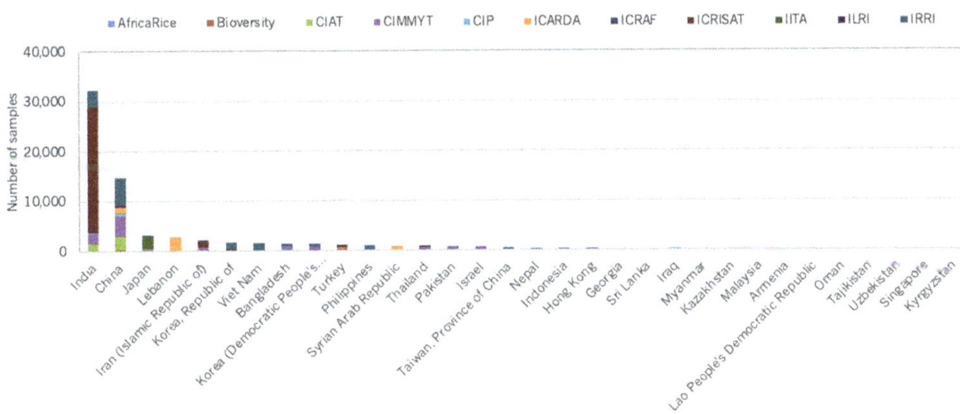

Figure 6. Number of germplasm samples distributed by CGIAR genebanks to external users in Asia.

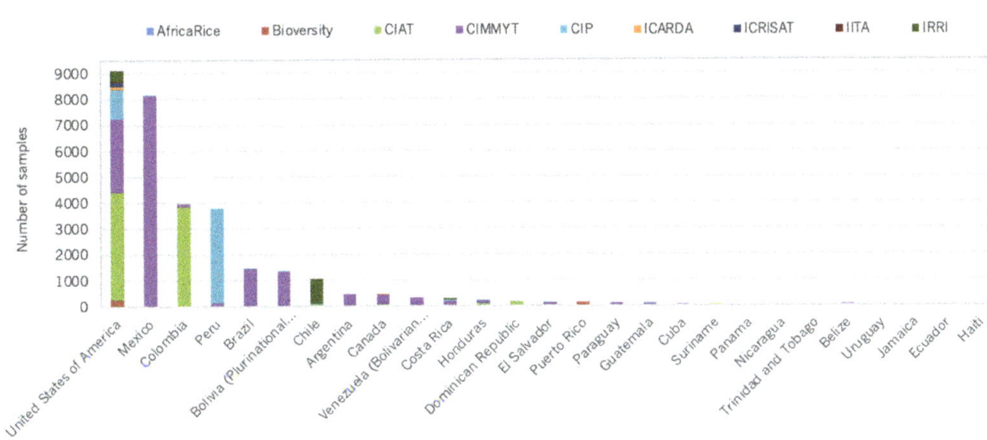

Figure 7. Number of germplasm samples distributed by CGIAR genebanks to external users in the Americas.

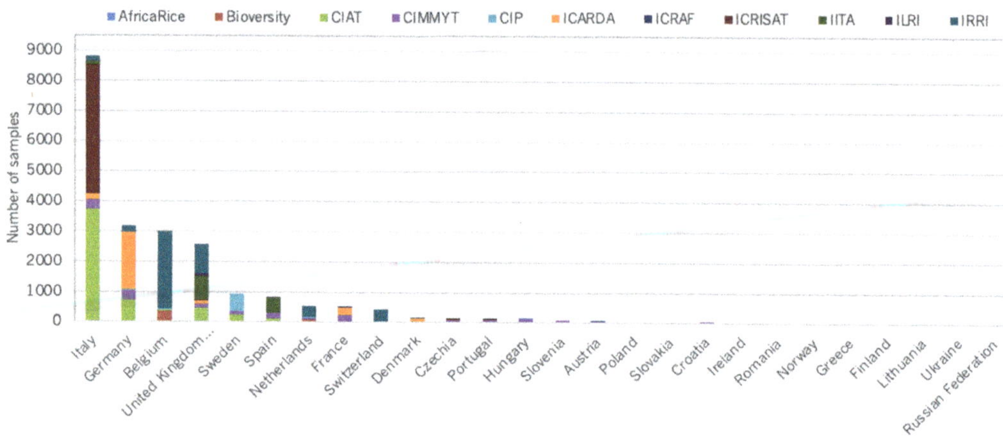

Figure 8. Number of germplasm samples distributed by CGIAR genebanks to external users in Europe.

In Africa, the top three recipient countries are Morocco (now hosting ICARDA), Nigeria (IITA) and Ethiopia (ILRI), and, combined, they receive more germplasm than the rest of the continent put together. Are other African countries accessing crop genetic resources from their national genebanks or genebanks other than CGIAR? The Tropical Agricultural Research and Higher Education Center (CATIE), the Centre for Pacific Crops and Trees (CePaCT), International Center for Biosaline Agriculture (ICBA) and the World Vegetable Centre have international genebanks and also provide germplasm globally, but mostly of different crop species than CGIAR. Such patterns pose many questions about germplasm distribution and demand that, for now, remain unanswered.

Given the unpredictability of future challenges and opportunities, the multitude of ways agriculture producers and consumers may respond to them, and the increasing capacity to generate knowledge to facilitate the use of genetic diversity, these patterns could suggest that there is a significant latent demand in many middle- and low-income countries for germplasm and services from CGIAR and possibly other genebanks. Without gathering more detailed data from users or potential users concerning the deployment and need of germplasm from genebanks, and analysing trends as a routine, it is difficult to improve our understanding of user demand or potential demand and respond appropriately to it. However, there are major advances in science that are impacting and will continue to impact the use of genebanks, and these, too, should have a major influence on how germplasm samples are delivered and conserved. We will now turn the discussion to these points.

3. Advances in Genomics and Their Influence on Breeding and the Role and Structure of Genebanks

3.1. Advances in Breeding

Modern breeders typically work with a limited diversity of painstakingly chosen potential parents, enabling reliable progress that does not risk breaking up the superior combinations of genes in elite breeding material. Crosses between more genetically distant parents bring the potential for much greater stepwise progress, and become essential when breeding objectives change (such as when a new disease appears) or when an existing breeding programme stagnates for lack of diversity to work with (The amount of additive genetic variation for a trait under selection in a breeding population (σ_a) is one of the determinants of the annual rate of genetic gain $G = (\sigma_a ir)/L$, where i = selection intensity, r = selection accuracy and L = number of years per cycle [17]). However, such "wide crosses" bring the risk (or even the near certainty in crops such as maize or in interspecific crosses between crops and their wild relatives) of breaking up desirable combinations of genes [18].

Fortunately, advances in genomics are creating new opportunities for exploring and utilizing crop diversity [19–22]. Breeders can now choose parents and select progeny based directly on genotype rather than phenotype, which can be much faster and cheaper. Selections based only on phenotype are often challenging because of the low heritability and polygenic nature of the desired traits, their high dependence on the environment in which they are assessed, and the genetic background of the genes involved. In contrast, for single-gene traits for which there are good markers (Ideally markers are within the gene controlling the trait, but they do not have to be. Good markers are often close to (and therefore genetically linked with), but not part of, the gene controlling the trait. The further the marker is from the gene, the less tightly it is linked, and therefore the more dependent on the specific materials being bred.) for the desired functional genetic variants, the required genotypes can be selected with high certainty at the seedling stage.

In addition, through gene editing it is now possible to add, delete or change a single gene in a genome. Where a causal relationship has been established between a single gene and a desired trait, this makes it possible for breeders to add a high-value gene (e.g., the sub1 gene responsible for "scuba" rice, see above) into a high-value genome in a single step, enabling large improvements without the risk of breaking up desirable gene combinations. In conjunction with synthetic biology, it will eventually become possible for breeders to even edit the gene without accessing physical material in genebanks. However, gene editing is an effective breeding tool only after research to determine the sequence and function of the "best" functional genetic variant for any given objective. This research will rely on continued access to physical genetic resources for the foreseeable future. Additionally, gene editing typically addresses only one gene at a time or about 0.002% of the genes in the genome (although techniques are being introduced to edit several genes simultaneously [23]), but it is much faster and more precise than conventional backcrossing. For traits with complex or uncertain genetic control, genomic selection may be used [24]. The prerequisites in this case are an effective, intelligent algorithm for the selection of variants across the whole genome and a good training population for the algorithm to learn from. Genomic prediction also requires high-throughput phenotyping to arrive at, and to validate, the algorithms [25].

Advances thus continue to bring gains in efficiency and effectiveness to the slow-but-sure approach of modern breeding, but there remains a glaring need to explore the much greater potential of recombining widely different genomes—an area where genebanks can uniquely contribute. The client base for genebank material is increasingly shifting from breeders towards upstream researchers. There is a rapidly emerging client base for "digital genebanks", i.e., for comprehensive online searchable repositories of information on genetic resources, as users increasingly require access to digital information associated with accessions. It is clear that genebanks will need to evolve, not only to improve how they work and catch up with the advanced state of breeding (particularly for the most advanced crops, although under-utilized and less intensively bred crops will follow), but also to accommodate a changing role [26–29].

3.2. Advances in the Role of Genebanks

Given the advances in breeding and genomics described above, we can envisage a not-too-distant time when every gene or haplotype (including coding, non-coding and regulatory regions) within the crop genepool being conserved will be catalogued and searchable, along with every existing potentially functional variant of each of these. The development of a comprehensive catalogue of the functionally significant genetic variants of each accession can thus become a feasible target for the ideal genebank of the future. Many of those variants will have their phenotypic effect either predicted or empirically demonstrated in at least one environment, genetic background and epigenetic status, or at least imputed from their homology to other known sequences. Whole genome sequences help reveal functional variants, including structural variations such as inversions and deletions that are hard to identify and map using conventional methods but may have significant impact on phenotypes. Pangenomics analyses enable the discovery of such variations that cannot be seen with genotyping. Even with as many as a million genomes per crop for 20 crops, with around 25,000–75,000 genes per crop genome, the data in the catalogue might require only about 20 terabytes of storage capacity (Very approximately and subject to revision: a million genomes per crop for 20 crops gives 20 million genomes. Multiplying 20 million genomes by 50,000 genes per genome gives a trillion records. Each record would be a pointer to an entry in a dictionary of gene variants: at, say, 20 bytes per record, that is 20 terabytes. The dictionary itself would be a fraction of that size at about 5 gigabytes (20 crops * 50,000 genes * approximately 5000 bytes per gene based on a full sequence for the most common variant and differences for the other variants). This is tiny relative to modern "big data" applications and readily tractable. It would be a game-changing contrast to relying solely on the never-ending treadmill of phenotyping: a digital genebank that provides material and information that meets users' needs with a precision that is currently unachievable.

Given the rate of progress to date, including automated algorithms for genome annotation, it should be possible to build an initial, reasonably comprehensive, multi-crop catalogue of functional genetic variants within 20 years. However, the catalogue would need to be progressively refined continuously after that.

In the meantime, to explore diversity and to develop the catalogue, a range of options needs to be built up to stratify collections for easier research and use. Many genebanks have already identified traditional core or mini-core subsets intended simply to make the task of phenotyping large collections more manageable [30]. Alternatively, accessions have been selected based on specific user-defined criteria (usually combinations of passport, phenotypic and genetic data), including using machine learning software such as the Focused Identification of Germplasm Strategy (FIGS) developed by ICARDA to create subsets that are more likely to contain adaptive traits that users want [31,32]. One of the reasons for higher distribution figures from genebanks for some crops, such as rice from IRRI, over the past decade has been the increased demand for subsets of accessions that have been sequenced [33–37]. The reason is that this enables users to conduct their own genome-wide association studies, which is an increasingly important first step in understanding the genetic control of a trait: a single sequenced subset can be used to support gene discovery for multiple traits. Hence, a short-term objective for genebanks should be to replicate this for all crops by sequencing the genomes of well-chosen core collections of all their crops.

In addition, genebanks should be invested in becoming more proactive in designing and creating novel genetic resources in support of breeders and researchers. Importantly, they must complement rather than duplicate breeders' own trait-discovery or "pre-breeding" work, and hence must undertake such efforts in consultation and collaboration with breeders. Breeders' pre-breeding initiatives are typically trait-specific, focusing on introgressing high-value traits from "undesirable" genomes into elite breeding lines. Genebanks may play a complementary role by "pyramiding" multiple known high-value traits into easily useable material [38]. They could also take a more exploratory or trait-agnostic approach, combining divergent genomes that have never previously been crossed with the aim of exposing large amounts of novel phenotypic diversity by creating radically different genomes, supporting rapid response to change. A range of possible crossing designs already exist, such as MAGIC (Multi-parent Advanced Generation Inter-Cross) and NAM (Nested Association Mapping); their exploratory value can be maximised by using genomic information to select the parents. Genebanks have a particularly complementary role to breeders in exploring the variation available in crop wild relatives by crossing them with elite material to tease out hidden characteristics and developing combinations that may eventually be more attractive to the breeder to work with [39].

For the purposes of gene discovery, it is important to phenotype the exact same genome (i.e., the precise individual) that has been sequenced or genotyped. This may require the sequenced genome to be managed, conserved and distributed separately from the accession from which the genome was taken. However, if many accessions are conserved in their original form and also in the form of pure lines, the size of the collection would at least double. Clearly management decisions will need to take this into account and genebanks will need to adapt to conserve such genetic stocks on a short-term basis and provide them to breeders and researchers.

3.3. Advances in the Structure of Genebanks

The ability to ensure delivery of materials more precisely corresponding to the demand of users will not only speed up crop improvement but vastly increase the return on investment in genebanks, and it may even lead to a change in genebank funding models. Today, contrary to normal practice for other services, whereby users pay for the services provided, genebanks are effectively paid to provide genetic resources to users by governments and donors. The justification for such public spending is compelling: users need access to genebanks to broaden the diversity of materials they use for agricultural development to everyone's benefit, but, as the genebank cannot know which accessions will actually help any given user, users will understandably not pay for such services. Once genebanks start delivering well-targeted materials that meet users' needs, however, the more usual "user pays" funding model may work for genebanks as well, subject to the provisions of the Plant Treaty.

If users were to pay for more precisely attuned services, genebanks would need to learn how to place a value on the resources they conserve and provide. Resource economists have established ways of conceptualizing different categories of value for economic research: use, non-use and option values [40]. In the future, new tools could help to quantify the value of germplasm appropriately. Advanced algorithms, based on genome-wide selection, may be used to explore the likely consequences of combining different genomes. The result would be a purpose-specific "current value", or "use value", for each accession, i.e., the extent to which that accession could enable a breeder or researcher to meet their known current needs. These values would be highly dynamic, increasing as accessions are found to contain genes needed by the breeders, and decreasing as those genes enter the breeders' own genepools. They would be used to select the most appropriate materials for specific current users. They could also be used more proactively to guide the creation and management of a large, dynamic set of user-oriented accessions, pre-bred by the genebanks (or others) and designed to meet current needs of researchers and breeders as effectively as possible.

However, such a focus on current value must not detract from the role of genebanks in long-term sustainability. Other measures of accession value must also be introduced to ensure an effective long-term agrobiodiversity conservation system. Unique diversity that differs genotypically from varieties in current use will have an "option value": even if this diversity has no value for today's food production, conserving it keeps open options for responding to future challenges as they emerge. The option value of an accession will be a function of the number of functional genetic variants (including epigenetic factors, structural variants and transposable elements) that are present in the accession but that are either not known or at risk of extinction outside the genebank. Such materials could include originally collected materials, heterogeneous accessions and populations. These accessions need to be conserved in a way that efficiently keeps their unique genes available for future use without needing to invest in their current use.

In addition, objects whose very existence is prized have a "non-use value". This concept applies, for example, to "heirloom varieties" that may be considered part of the heritage of a particular country or culture or community and may be at risk from changing conditions, practices and priorities if not conserved in a genebank. This should not be taken to imply that heirloom varieties do not have a use value. On the contrary, their use within certain cultures may be vital for those cultures. It just means they have a value beyond their use value, as implied by the very term "heirloom", one of the central underlying themes of non-use value. The importance of some of these types of material will clearly reside in the variety as a whole; that is, in the entire genome, rather than in specific rare genes or gene combinations.

It is important to recognize that these different values are independent concepts and are not mutually exclusive. A heritage variety may contain functional variants that breeders do not have in their collections but that would help meet their objectives and may also contain other unique functional variants with unknown value. An accession of such a variety would have high use value, high non-use value, and high option value. A well-researched and used accession that has functional variants that are already well represented in other accessions may have much less option value and, therefore, be a much lower priority to conserve long term. Purified lines are a clear example.

Beyond uncovering the genetic mechanisms underlying agronomic traits, genomic data can provide detailed analysis of the population genetic history shaping diversity in situ and the mode and tempo of selection during domestication of crop plants [41,42], as well as the long-term effects of keeping genetic diversity ex situ in a genebank. Most accessions in crop collections do not represent uniform sets of genotypes, but rather heterogeneous populations of genotypes, reflecting the mutational and migration effects that are captured at the moment at which the sample is collected. Accessions of crops' wild relatives have inbreeding and levels of differentiation that reflect the sampling effects and logistical constraints of the collector, as well as the inherent breeding system, life history, and ecogeographical range of the species [43]. Any effort to estimate or put a value on this diversity, therefore, must take a population genetic approach to sampling and prioritization.

Curation for long-term conservation and for current use will diverge: the original, heterogeneous accessions may be a cost-effective way of conserving genes and populations long term but may have less value for current use, while the reverse is true for sequenced, purified lines. The future genebank system may be viewed as structured collections with varying levels of intra-accession diversity, different conservation objectives, and varied precision in characterization data. Wild and landrace accessions may be conserved to represent diverse populations with wide-ranging characteristics. Population genetic approaches will be best used to evaluate the diversity of these accessions. Improved accessions and genetic stocks will have increased uniformity with increasingly precise characterization data. The different levels of diversity complement each other. The more purified accessions can serve as a starting point to dissect the genetic architecture of agronomic traits and query the more diverse accessions to find useful allelic variants at key loci [44].

Parameters revealing diversity and differentiation, relatedness and admixture within the collection, and analyses that seek to understand the population genomics of domestication history will be of critical interest. In addition, rounds of regeneration subject the diversity (of individual genotypes or individual haplotypes) to sampling variation, resulting in genetic change in accessions that are not initially uniform. Because this process is driven by sampling (the larger the sample or effective population size, the lower the drift), a genebank can try and maintain genetic integrity through large regeneration populations or through extending generation time intervals between regenerations by ensuring storage conditions are optimized for the long term [45].

Deriving the current use value of accessions is not yet achievable and will, of course, depend on having access to that digital genebank of functional variants. Even with the future digital genebank, details of how to assign a current use value to each accession will depend on various factors, including progress in the development of genetic algorithms, experience in the extent to which genomic predictions must be supported by direct phenotypic observations, and evolution of the ways in which genebanks monitor the changing needs of users. Before providing a practical way of managing germplasm collections, much further work will be needed to develop methodologies to quantify current value. What is important to recognise, however, is that once users are able to more effectively select the materials they need from genebanks, there will be a need to re-structure collections in ways to accommodate fast access to accessions and research-ready materials with a relatively high turnover compared to conventional genebank collections. A well-established collection would be expected to contain a relatively small and stable set of accessions with high option value and high non-use value. It would contain a larger and more dynamic set of accessions with high current use value, changing as users' needs change. New accessions would only be added to the genebank's collection where assessment of their potential value demonstrates that they add significantly to the collection's overall value. The technology already exists to obtain a genome sequence in the field in real time [46]; this would be used to sequence a sample and determine if the sample should be added to the collection or discarded, based on its complementarity to the existing collection.

Deriving the future use value of accessions faces different challenges. Whilst not requiring phenotypic information, it does require consideration of within-accession heterogeneity. However, methods to study the genomics of variable accessions are limited. Attempting to discover the full set of functional variants within one accession by genotyping every individual is not a viable option. This will need to be taken into account in developing methods for handling within-accession heterogeneity, rare functional variants, and their contribution to the future use value of an accession.

A picture starts to emerge whereby the distinction between long-term conservation and immediate use in genebanks will become more pronounced and functional. Up until now, general practice and guidelines have divided genebanks into base collections made up of relatively small samples of the most original seeds held in long-term storage conditions (at a temperature of -18 ± 3 °C and relative humidity of $15 \pm 3\%$) and larger, more dynamic, so-called "active" collections that hold larger samples for distribution and use under refrigeration (at 5–10 °C and relative humidity of $15 \pm 3\%$) [28]. There are exceptions; some genebanks (e.g., CIAT) hold their collections entirely in long-term storage conditions in batches destined for different purposes (e.g., long-term storage, viability monitoring, repatriation, distribution and safety duplication). However, in most cases there is good economic sense in making the distinction between samples that can be left relatively undisturbed for long-term conservation and those for immediate use, since long-term storage at -20 °C is slightly more expensive to run and cannot be staffed in the same way as medium-term storage at 5 °C because of the working conditions. It does not necessarily follow that materials that are held in long-term storage conditions are bound to be conserved for the long term. However, in practice whatever is in long-term storage tends to be challenging to discard and, thereby, becomes a long-term obligation to conserve.

In future, with in depth analysis of the genetic composition of collections, samples of high option value accessions may be prioritized for long term conservation. By contrast, for immediate use, smaller, more dynamic collections of breeder-ready resources, populations, trait subsets, phenotyped core collections, purified lines and other high current value materials that are the subject of active research and phenotyping will be maintained in various derived forms (i.e., not necessarily in the form in which they were collected) (Figure 9).

Given this vision, it is not a giant leap to suggest that genebanks will have significant opportunities to concentrate conservation activities and to specialise. Specialist genebanks already exist; ICBA is developing a specialist collection for salt tolerance, and different CGIAR genebanks focus on tropical agriculture, drylands, semi-arid regions and specific crops. In the future sketched out above, a limited number of genebanks would need to focus on, and specialise in, the long-term conservation of particular crop types: orthodox seeded crops, clonal crops, fruits and vegetables, and wild species, for example. Specialisation would facilitate deep innovations in crop and germplasm management protocols to improve quality, increase reliability, enlarge capacity, and reduce costs. If there are highly repetitive tasks, there are possibilities to automate them for consistently high quality, high throughput, and low cost. Automated processes and better materials management also introduce the possibility of tighter control through remote management, allowing genebank curators to control processes no matter where they are—an advantage that has shown its relevance during pandemic lockdowns. The composition of staff would change, with new technical expertise required for machine and process maintenance, and there would also be a shift in staff balance, with a higher proportion dedicated to information processing and management.

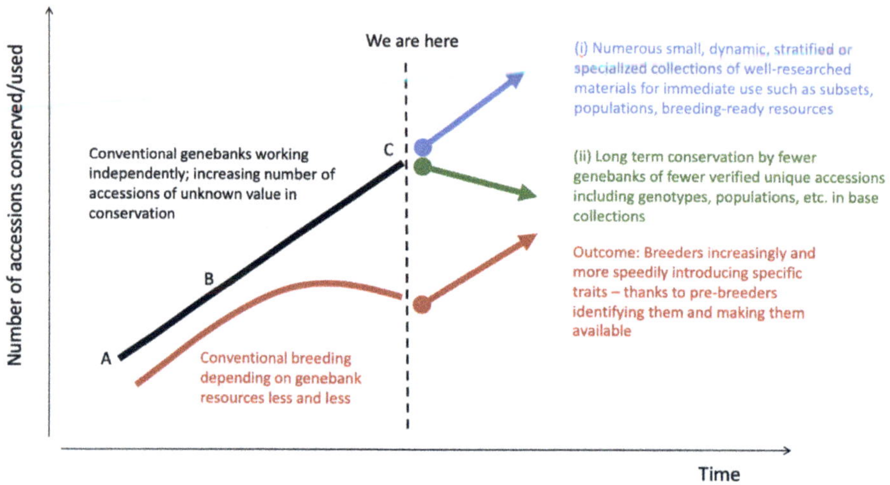

Figure 9. The evolution of genebanks over time—(**A**) originating as a breeding resource in the 1950–60s, (**B**) transforming into the subject of long-term conservation as well as a breeding resource, (**C**) evolving into (i) specialised, dynamic collections for breeders and other users and (ii) base collections for long-term conservation.

IRRI has recently piloted the automation of rice seed sorting. In collaboration with the private sector, the genebank has developed bespoke robotic seed imaging machinery that can be trained for each individual accession to sort high quality seed for storage [47]. Some preliminary attempts have been made to automate various other operations in large genebanks, for example planting, phenotyping, harvesting, viability testing, packing, labelling, and the storage and retrieval of materials from a seed store [48]. The most widely adopted and successful advance, so far, has been the introduction of bar-coded or QR-coded labels for inventory management and for tracking samples through workflows. Generally speaking, it is considerably more challenging to automate genebanks managing multiple crops with diverse, heterogeneous accessions. Only concentration into fewer, larger collections, enabling higher throughput, will tilt the balance towards more automation being appropriate and effective.

4. From Vision to Reality

Comparing today's germplasm distribution data with the vision that we have described above reveals what appears to be an abyss or, perhaps less dramatically, a mismatch between theory and reality. It is important, firstly, to note that distribution statistics will never accurately reflect the use of genebank materials and data or their impact. Distribution is merely the first step in use, not the end result. Nevertheless, a more accurate method of gathering and monitoring germplasm distribution data will be essential in informing the directions to be pursued by genebanks and the institutes and donors that support them. We need basic but detailed and consistent data on every genebank request: the type of material requested, when, by whom, for what purpose, and under what ABS conditions.

However, feedback from users about the use and performance of distributed germplasm samples would also be highly desirable, though it presents significant legal and technical obstacles. The SMTA prohibits providers from requiring such feedback from users of genebank accessions (but allows it for breeding lines that can be categorised as "PGRFA under development"). Hence, traditional attempts to promote feedback will never be particularly effective. On the other hand, the SMTA obliges users to provide such feedback through the Global Information System (GLIS). This potentially opens the door to an effective system, although the GLIS currently has only a rudimentary mechanism to receive feedback. This mechanism relies on GLIS DOIs being used to identify the material and on those DOIs being used by breeders in publications and in online datasets. The only legal and operational mechanism currently available to obtain feedback on the use and performance of distributed germplasm samples is DOIs. Everyone who believes crop diversity and the genebanks that conserve it are important should promote the use of DOIs by germplasm users.

Distribution data hints that there is a wide range of current users, but potentially many more users who may want to request a wider range of genebank materials and crops. CGIAR and other genebanks should not only be gathering and curating more and better data on existing requests, but actively scanning and assessing potential and future demand by better characterising and understanding their users, their users' capacity and their germplasm needs. The reasons behind the apparent geographic patterns of distribution should be better understood. This means a more proactive effort to engage users, follow up after requests, analyse demand patterns and identify potential users; carrying out survey work and promotional work; and collaborating closely with activities to gather market intelligence to determine breeding priorities.

Genebanks will need to meet the needs of users both in terms of genetic resources and associated data more accurately and efficiently than they do today, including those of a burgeoning community of upstream researchers needing both material and in-depth genomic information. As well as responding to requests, genebank activities need to proactively explore hidden traits in collections and develop breeding-ready subsets and resources more closely matching analysed needs. Interacting more closely with the user community, and in particular those involved in pre-breeding, will be crucial to ensure that genebanks conserve the right genetic resources in the right way and closely match resources to the priorities. It is important to stress that these activities need to be funded and should not take the place of important ongoing conservation work.

Although only acquisition and curation have been discussed here, all other genebank processes must become more dynamic as well. Procedures for managing materials, information and processes must be streamlined to maximize efficiency, maintaining consistently high and demonstrable quality while reducing costs in a system that matches throughput capacity to demand. CGIAR recently endorsed a policy framework for the strategic curation of collections under its management, involving the establishment of different curation categories, including the option to "partially curate" or "archive" accessions, formalizing a practice that many genebanks have had in place for years that allows them to adapt the usual sequence of genebank processes for specific accessions where appropriate.

As CGIAR evolves under the current One CGIAR reform, its genebanks will continue to play a pivotal role in a global system for the conservation and use of genetic resources and have an opportunity to contribute to fulfilling the vision outlined above in a number of ways, including the following:

- Providing facilities for the effective management of long-term conservation of an increasing number of crops, and collaborating with others, including the Svalbard Global Seed Vault, in this process. Through enhanced collaboration, consolidation, and division of labour, possibly also involving the private sector, it should be possible to significantly increase the efficiency and effectiveness of long-term conservation;
- Developing novel diversity, e.g., through wide and inter-specific crossing, and creating value-added subsets of materials for breeders, e.g., for genome-wide association studies. Again, there should be scope here for enhanced partnership with private companies;

- Developing methods for assigning current and future values to accessions and for using such values for decision making with respect to curating conserved materials and promoting use, taking into account within-accession heterogeneity;
- Working with national, regional and international partners to develop a system of distribution hubs so as to more efficiently and effectively provide germplasm to those that need it around the world. This is likely to involve the maintenance of dispersed active collections linked to facilities for ensuring the health status of distributed germplasm [49];
- Large-scale sequencing of accessions of all mandated (and, in time, other) crops and making this information available in conformity with applicable ABS regulations;
- Providing input to the future development of international policies, rules and regulations regarding the conservation and use of plant genetic resources, including the equitable sharing of benefits arising from such use;
- Promoting the use of GLIS DOIs by all users and providers as the globally unique public identifier for germplasm samples;
- Providing the training needed within CGIAR and partner institutions and securing adequate financial and other resources to enable this vision of the future to become a reality.

5. Conclusions

This is an extraordinary time of change throughout the world. Climate change and the consequent increase in extreme weather events is already having a significant impact; biodiversity is disappearing despite massive efforts to conserve it; Covid-19 has resulted in a huge increase in human misery and slowed down large sectors of the economy; and the increasing polarization of society and political views is threatening long-established governance systems. To exacerbate things, the UN has estimated that the world's population will grow by almost 1.9 billion people between 2021 and 2050. Plant genetic resources for food and agriculture have an important role to play in addressing many challenges by the development of higher yielding, more nutritious and resilient crops that can help increase rural incomes and avert malnutrition and social unrest, and of crops and cropping systems that require less land and fewer external inputs, or that release less greenhouse gasses.

As this article has attempted to show, future plant breeders are likely to require more, not less genetic diversity than at present, but in a different form and accompanied by larger amounts of reliable data. It is probable that demand will continue to grow for genebank materials that can be used in gene discovery and for the identification of functional variants, shifting the client base toward more upstream scientific researchers. If genebanks are to remain relevant, it will be important that they are able to adapt and cater to new demands. This has important implications for the types of material they maintain and the form in which it, and the associated data, are made available. At the same time, new conservation technologies, policies and institutional arrangements offer ways to improve the efficiency and effectiveness of conservation to the long-term benefit of all.

Author Contributions: Conceptualization, R.S.H.; C.L.; G.H. and L.G.; data curation, N.J.; writing, C.L.; R.S.H.; G.H.; C.R. All authors have read and agreed to the published version of the manuscript.

Funding: This paper results from a genebank review process under the CGIAR Genebank Platform funded by CGIAR donors and the Global Crop Diversity Trust.

Data Availability Statement: Data on germplasm distribution from CGIAR genebanks are available at www.genebanks.org/resources/genebanks-in-numbers/distribution/ and from the genebanks online reporting tool made available upon request to nelissa.jamora@croptrust.org.

Acknowledgments: This paper is based upon an unpublished paper, Sackville Hamilton, R.; *Scope and roles of the CGIAR genebanks: 2030 Vision*, available at www.genebanks.org/wp-content/uploads/2020/09/Session-2_CGIAR-genebanks-2030-vision.pdf last accessed on 26 November 2021.

Conflicts of Interest: The authors declare no conflict of interest.

References

1. Engels, J.M.M.; Ebert, A.W. A critical review of the current global ex situ conservation system for plant agrobiodiversity. I. History of the development of the global system in the context of the political/legal framework and its major conservation components. *Plants* **2021**, *10*, 1557. [CrossRef]
2. Mooney, P.R. *Seeds of the Earth: A Private or Public Resource?* InterPares: Ottawa, ON, Canada, 1979; 125p.
3. Gollin, D.; Hansen, C.W.; Wingender, A.M. Two blades of grass: The impact of the green revolution. *J. Polit. Econ.* **2021**, *129*, 2344–2384. [CrossRef]
4. Nour Mohammed, A. *ICARDA25: A Promise of Hope*; International Center for Agricultural Research in the Dry Areas: Aleppo, Syria, 2002; 165p.
5. Mackill, D.J.; Ismail, A.M.; Singh, U.S.; Labios, R.V.; Paris, T.R. Chapter Six—Development and Rapid Adoption of Submergence-Tolerant (Sub1) Rice Varieties. In *Advances in Agronomy*; Sparks, D.L., Ed.; Academic Press: Cambridge, MA, USA, 2012; Volume 115, pp. 299–352. ISBN 9780123942760. [CrossRef]
6. Brar, D.S.; Khush, G.S. Wild Relatives of Rice: A Valuable Genetic Resource for Genomics and Breeding Research. In *The Wild Oryza Genomes. Compendium of Plant Genomes*; Mondal, T., Henry, R., Eds.; Springer: Cham, Switzerland, 2018. [CrossRef]
7. Halewood, M.; Andrieux, E.; Crisson, L.; Rwihaniza Gapusi, J.; Wasswa Mulumba, J.; Kouablan Koffi, E.; Yangzome Dorji, T.; Raj Bhatta, M.; Balma, D. Implementing 'mutually supportive' access and benefit sharing mechanisms under the Plant Treaty, Convention on Biological Diversity, and Nagoya Protocol. *J. Law Environ. Dev.* **2013**, *9*, 68–96.
8. Mekonnen, D.K.; Spielman, D.J. Changing patterns in genebank acquisitions of crop genetic materials: An analysis of global policy drivers and potential consequences. *Food Policy* **2021**, *105*, 102161. [CrossRef]
9. System-Wide Genetic Resources Programme (SGRP). *A de Facto Certificate of Source: The Standard Material Transfer Agreement under the International Treaty*; Bioversity: Rome, Italy, 2007.
10. International Treaty on Plant Genetic Resources for Food and Agriculture. Germplasm Flow. Available online: https://mls.planttreaty.org/itt/index.php?r=stats/pubStats (accessed on 1 November 2021).
11. International Treaty on Plant Genetic Resources for Food and Agriculture. Standard Material Transfer Agreement. Article 5a. Available online: http://www.fao.org/3/bc083e/bc083e.pdf (accessed on 1 November 2021).
12. International Treaty on Plant Genetic Resources for Food and Agriculture. DOI Module. Available online: https://ssl.fao.org/glis/site/doiindex (accessed on 1 November 2021).
13. International Treaty on Plant Genetic Resources for Food and Agriculture. DOI Registrants. Available online: https://ssl.fao.org/glis/user/search-registrants (accessed on 1 November 2021).
14. Alercia, A.; López, F.M.; Sackville Hamilton, N.R.; Marsella, M. *Digital Object Identifiers for Food Crops—Descriptors and Guidelines of the Global Information System*; FAO: Rome, Italy, 2018.
15. Halewood, M.; Jamora, N.; Lopez Noriega, I.; Anglin, N.L.; Wenzl, P.; Payne, T.; Njiondjop, M.; Guarino, L.; Kumar, P.L.; Yazbek, M.; et al. Germplasm acquisition and distribution by CGIAR genebanks. *Plants* **2020**, *9*, 1296. [CrossRef] [PubMed]
16. Bretting, P. The US national plant germplasm system in an era of shifting international norms for germplasm exchange. *Acta Hortic.* **2007**, *760*, 55–60. [CrossRef]
17. Cobb, J.N.; Juma, R.U.; Biswas, P.S.; Arbelaez, J.D.; Rutkoski, J.; Atlin, G.; Hagen, T.; Quinn, M.; Ng, E.H. Enhancing the rate of genetic gain in public-sector plant breeding programs: Lessons from the breeder's equation. *Theor. Appl. Genet.* **2019**, *132*, 627–645. [CrossRef]
18. Wang, C.; Hu, S.; Gardner, C.; Lübberstedt, T. Emerging Avenues for Utilization of Exotic Germplasm. *Trends Plant Sci.* **2017**, *22*, 624–637, ISSN 1360-1385. [CrossRef]
19. Kilian, B.; Graner, A. NGS technologies for analyzing germplasm diversity in genebanks. *Brief Funct. Genom.* **2012**, *11*, 38–50. [CrossRef]
20. Van Treuren, R.; van Hintum, T.J.L. Next-generation genebanking: Plant genetic resources management and utilization in the sequencing era. *Plant Genet. Res.* **2014**, *12*, 298–307. [CrossRef]
21. Wambugu, P.W.; Ndjiondjop, M.N.; Henry, R.J. Role of genomics in promoting the utilization of plant genetic resources in genebanks. *Brief Funct. Genom.* **2018**, *17*, 198–206. [CrossRef] [PubMed]
22. Holliday, J.A.; Hallerman, E.M.; Haak, D.C. Genotyping and Sequencing Technologies in Population Genetics and Genomics. In *Population Genomics: Concepts, Approaches and Applications*; Rajora, O.P., Ed.; Springer Nature Switzerland AG: Berlin, Germany, 2019; pp. 83–126.
23. Campa, C.C.; Weisbach, N.R.; Santinha, A.J. Multiplexed genome engineering by Cas12a and CRISPR arrays encoded on single transcripts. *Nat. Methods* **2019**, *16*, 887–893. [CrossRef]
24. Bhat, J.A.; Ali, S.; Salgotra, R.K.; Mir, Z.A.; Dutta, S.; Jadon, V.; Tyagi, A.; Mushtaq, M.; Jain, N.; Singh, P.K.; et al. Genomic Selection in the Era of Next Generation Sequencing for Complex Traits in Plant Breeding. *Front. Genet.* **2016**, *7*, 221. [CrossRef] [PubMed]
25. Akdemir, D.; Isidro-Sánchez, J. Design of training populations for selective phenotyping in genomic prediction. *Sci. Rep.* **2019**, *9*, 1–15. [CrossRef] [PubMed]
26. Poland, J.A.; Rife, T.W. Genotyping-by-Sequencing for Plant Breeding and Genetics. *Plant Genome* **2012**. [CrossRef]
27. Mascher, M.; Schreiber, M.; Scholz, U.; Graner, A.; Reif, J.C.; Stein, N. Genebank genomics bridges the gap between the conservation of crop diversity and plant breeding. *Nat. Genet.* **2019**, *51*, 1076–1081. [CrossRef]

28. Langridge, P.; Waugh, R. Harnessing the potential of germplasm collections. *Nat. Genet.* **2019**, *51*, 200–201. [CrossRef]
29. Belzile, F.; Abed, A.; Torkamaneh, D. Time for a paradigm shift in the use of plant genetic resources. *Genome* **2020**, *63*, 189–194. [CrossRef] [PubMed]
30. Engels, J.M.M.; Ebert, A.W. A critical review of the current global ex situ conservation system for plant agrobiodiversity. II Strengths and weaknesses of the current system and recommendations for its improvement. *Plants* **2021**, *10*, 1904. [CrossRef] [PubMed]
31. Bari, A.; Street, K.; Mackay, M.; Endresen, D.T.F.; De Pauw, E.; Amri, A. Focussed identification of germplasm strategy (FIGS) detects wheat stem rust resistance linked to environmental variables. *Genet. Resourc. Crop Evol.* **2012**, *59*, 1465–1481. [CrossRef]
32. Reeves, P.A.; Tetreault, H.M.; Richards, C.M. Bioinformatic Extraction of Functional Genetic Diversity from Heterogeneous Germplasm Collections for Crop Improvement. *Agronomy* **2020**, *10*, 593. [CrossRef]
33. Gouesnard, B.; Bataillon, T.M.; Decoux, G.; Rozale, C.; Schoen, D.J.; David, J.L. MSTRAT: An algorithm for building germ plasm core collections by maximizing allelic or phenotypic richness. *J. Hered.* **2001**, *92*, 93–94. [CrossRef] [PubMed]
34. Reeves, P.A.; Richards, C.M. Capturing haplotypes in germplasm core collections using bioinformatics. *Genet. Resour. Crop Evol.* **2017**, *64*, 1821–1828. [CrossRef]
35. Thachuk, C.; Crossa, J.; Franco, J.; Dreisigacker, S.; Warburton, M.; Davenport, G.F. Core Hunter: An algorithm for sampling genetic resources based on multiple genetic measures. *BMC Bioinform.* **2009**, *10*, 243. [CrossRef] [PubMed]
36. Upadhyaya, H.D.; Ortiz, R. A mini core subset for capturing diversity and promoting utilization of chickpea genetic resources in crop improvement. *Theor. Appl. Genet.* **2001**, *102*, 1292–1298. [CrossRef]
37. Van Hintum, T.J.L. Hierarchical approaches to the analysis of genetic diversity in crop plants. In *Core Collections of Plant Genetic Resources*; Hodgkin, T., Brown, A.H.D., van Hintum, T.J.L., Morales, E.A.V., Eds.; John Wiley and Sons: Chichester, UK, 1995.
38. Mundt, C.C. Pyramiding for Resistance Durability: Theory and Practice. *Phytopathology* **2018**, *108*, 792–802. [CrossRef]
39. Prohens, J.; Gramazio, P.; Plazas, M.; Dempewolf, H.; Kilian, B.; Diez, M.J.; Fita, A.; Herraiz, F.J.; Rodriguez-Burruezo, A.; Soler, S.; et al. Introgressiomics: A new approach for using crop wild relatives in breeding for adaptation to climate change. *Euphytica* **2017**, *213*, 158. [CrossRef]
40. Smale, M. *A Sourcebook: Economics of Conserving Plant Genetic Resources in Genebanks*; Genebank Impacts Working Paper No. 1; CGIAR Genebank Platform and the Crop Trust: Bonn, Germany, 2019.
41. Ross-Ibarra, J.; Morrell, P.L.; Gaut, B.S. Plant domestication, a unique opportunity to identify the genetic basis of adaptation. *Proc. Natl. Acad. Sci. USA* **2007**, *104* (Suppl. 1), 8641–8648. [CrossRef]
42. Purugganan, M.D. Evolutionary Insights into the Nature of Plant Domestication. *Curr. Biol.* **2019**, *29*, R705–R714. [CrossRef]
43. Hoban, S.; Volk, G.; Routson, K.J.; Walters, C.; Richards, C. Sampling Wild Species to Conserve Genetic Diversity. In *North American Crop Wild Relatives*; Greene, S., Williams, K., Khoury, C., Kantar, M., Marek, L., Eds.; Springer: Cham, Switzerland, 2018; Volume 1. [CrossRef]
44. Kumari, R.; Kumar, P.; Sharma, V.K.; Kumar, H. Allele mining for crop improvement. *Int. J. Pure Appl. Biosci.* **2018**, *6*, 1456–1465. [CrossRef]
45. Richards, C.M.; Lockwood, D.R.; Volk, G.M.; Walters, C. Modeling Demographics and Genetic Diversity in Ex Situ Collections during Seed Storage and Regeneration. *Crop. Sci.* **2010**, *50*, 2440–2447. [CrossRef]
46. Parker, J.; Helmstetter, A.J.; Devey, D.; Wilkinson, T.; Papadopulos, A.S. Field-based species identification of closely-related plants using real-time nanopore sequencing. *Sci. Rep.* **2017**, *7*, 8345. [CrossRef] [PubMed]
47. McCartney, E. Keeping up with the Demand for Rice Seeds: The Automation of Seed Sorting at the International Rice Genebank. Available online: https://www.genebanks.org/news-activities/news/seed-sorting/ (accessed on 1 November 2021).
48. Hay, F.R. New technologies to improve the ex situ conservation of plant genetic resources. In *Plant Genetic Resources: A Review of Current Research and Future Needs*; Dulloo, M.E., Ed.; Burleigh Dodds Science Publishing: Cambridge, UK, 2021.
49. CGIAR-Crop Trust System Level Review of Genebank Costs and Operations. Report October 2020. Available online: https://www.genebanks.org/wp-content/uploads/2020/11/GCO-Report_261020.pdf (accessed on 1 November 2021).

plants MDPI

Review

Critical Review of the Increasing Complexity of Access and Benefit-Sharing Policies of Genetic Resources for Genebank Curators and Plant Breeders–A Public and Private Sector Perspective

Andreas W. Ebert [1,*], Johannes M. M. Engels [2], Roland Schafleitner [3], Theo van Hintum [4] and Godfrey Mwila [5]

1 Independent Researcher, 73529 Schwäbisch Gmünd, Germany
2 Independent Researcher, Voc. Podere Sansano 5, 06062 Citta' della Pieve (PG), Italy; j.engels@cgiar.org
3 World Vegetable Center, 60 Yi-Min Liao, Shanhua, Tainan 74151, Taiwan; roland.schafleitner@worldveg.org
4 Centre for Genetic Resources, the Netherlands (CGN), Wageningen University & Research, 6700 AA Wageningen, The Netherlands; theo.vanhintum@wur.nl
5 Executive Secretary of the Zambia Seed Trade Association (ZASTA), Sulmach Buldings, Tiyende Pamodzi, Off Nangweya, Lusaka, Zambia; godfrey.mwila@gmail.com
* Correspondence: ebert.andreas6@gmail.com

Abstract: Plant breeders develop competitive, high-yielding, resistant crop varieties that can cope with the challenges of biotic stresses and tolerate abiotic stresses, resulting in nutritious food for consumers worldwide. To achieve this, plant breeders need continuous and easy access to plant genetic resources (PGR) for trait screening, to generate new diversity that can be built into newly improved varieties. International agreements such as the Convention on Biological Diversity (CBD), the International Treaty on Plant Genetic Resources for Food and Agriculture (ITPGRFA) and the Nagoya Protocol recognised the sovereign rights of countries over their genetic resources. Under the CBD/Nagoya Protocol, countries are free to establish specific national legislations regulating germplasm access and benefit-sharing to be negotiated bilaterally. Consequently, access to PGR became increasingly restricted and cumbersome, resulting in a decrease in germplasm exchange. The ITPGRFA attempted to ease this situation by establishing a globally harmonised multilateral system (MLS). Unfortunately, the MLS is (still) restricted to a limited number of food and forage crops, with very few vegetable crops. Easy and continuous access to genetic diversity combined with equitable and fair sharing of derived benefits is a prerequisite to breeding new varieties. Facilitated access contributes to sustainable crop production and food and nutrition security; therefore, access to and, consequently, use of PGRFA needs to be improved. Thus, the authors recommend, among others, expanding the scope of the ITPGRFA to include all PGRFA and making them and all related information accessible under a Standard Material Transfer Agreement (SMTA) combined, if necessary, with a subscription system or a seed sales tax. Such a transparent, functional and efficient system would erase legal uncertainties and minimise transaction costs for conservers, curators and users of genetic resources, thus aiding plant breeders to fulfil their mission.

Keywords: vegetable genetic resources; global germplasm conservation and use systems; plant breeding; access and benefit-sharing; digital sequence information; international treaty for plant genetic resources; Convention on Biological Diversity; Nagoya Protocol

Citation: Ebert, A.W.; Engels, J.M.M.; Schafleitner, R.; Hintum, T.v.; Mwila, G. Critical Review of the Increasing Complexity of Access and Benefit-Sharing Policies of Genetic Resources for Genebank Curators and Plant Breeders–A Public and Private Sector Perspective. *Plants* **2023**, *12*, 2992. https://doi.org/10.3390/plants12162992

Academic Editor: Othmane Merah

Received: 16 July 2023
Revised: 9 August 2023
Accepted: 17 August 2023
Published: 19 August 2023

1. Introduction

Free access to and exchange of germplasm have been the foundation for all plant domestication and improvement efforts since the start of sedentary farming. Through most of human history, access has been constrained by physical distance and limited knowledge, not by an unwillingness to share or legal instruments. Until the signing of the Convention on Biological Diversity (CBD) in 1992–1993 (Table 1), germplasm was considered a common heritage of humankind to be preserved and to be freely available for use, for the benefit of present and future generations as per the International Undertaking (IU) established by the FAO Commission on PGR in 1983 [1–3]. Plant breeders obtained the required germplasm for their crop-improvement efforts from a wide variety of existing commercial varieties, public and private genebanks, public and private collecting missions, working collections maintained at research institutions and private companies, and from farmers' fields and stores.

Table 1. Main (legal) instruments regarding access and benefit-sharing of PGRFA and some of their main features.

Legal Instrument	Year of Entering into Force	Hosting Organisation and Location of Secretariat	Year of Termination	Main Legal Principles/ Aspects	PGRFA Coverage	Number of Parties
International Undertaking (IU) *	1983	FAO, Rome, Italy	2004?	Voluntary agreement; common heritage principle	All plant species for food and agriculture	n.a.
CBD	1993	UNEP, Montreal, Canada	ongoing	National sovereignty; PIC; agreed terms for use and benefit-sharing; Cartagena Protocol (biosafety)	All plant species (focus on wild); information	196 contracting parties (31 July 2023)
ITPGRFA	2004	FAO, Rome, Italy	ongoing	MLS, SMTA	All PGRFA and information; Materials in MLS: Annex I plus Art. 15 collections (i.e., CGIAR), plus voluntarily added materials	150 contracting parties (1 January 2023)
Nagoya Protocol	2014	UNEP, Nagoya, Japan	ongoing	ABS; Clearing House Mechanism	All crops and species that do not fall under ITPGRFA	140 contracting parties (31 July 2023)

* The IU was a voluntary agreement, not a body of international law.

Plant breeding is a long and tedious process and requires a lot of investment. Vegetable seed companies use up to 30% of their turnover for research and development. With the aim of encouraging continuous development of new plant varieties for the benefit of society at large, plant breeders' rights (PBR) were introduced through the creation of plant variety protection and internationally harmonised through the International Union for the Protection of New Varieties of Plants (UPOV) Convention, adopted in Paris in 1961 and revised in 1972, 1978 and 1991 [4]. Article 15 of the UPOV Convention provides a compulsory breeders' exemption to the exclusive right [5], allowing everyone to freely use any protected variety for further breeding and commercialising the new ones without any obligation to the original PBR holder as long as the newly developed product is sufficiently different from the protected variety. This provision constitutes an essential and principal element towards ensuring continued access of plant breeders worldwide to elite privately owned germplasm as parental material [6].

With the advent of biotechnological innovations during the 1980s, some countries allowed certain inventions to be protected through patents. The patenting of biotechnological inventions can be traced back to 1980 when the Supreme Court of the United States decided that a genetically modified organism, in that specific case a bacterium, is patentable [7]. Thereafter, several proprietary products were released in plant sciences, such as traits/genes and genetically engineered varieties.

Irregular access and use of genetic resources and related traditional knowledge of countries, indigenous peoples and local communities without their consent and the patenting of derived or associated information for further commodification is understood as biopiracy [8]. Cases of biopiracy and the perception in the Global South that the breeding industry in the Global North was earning money based on the genetic resources collected in the Global South without sharing due benefits were major reasons why the continuous free availability and accessibility of genetic resources as foreseen under the IU was no longer considered an acceptable paradigm [9]. This led to the development of new global legal frameworks (see Section 3).

Intergovernmental negotiations with the aim of protecting and conserving biological resources, making them available under the assumption of sharing benefits derived from their use on agreed terms, led to the adoption of several international agreements, such as the CBD in 1992 and the subsequent so-called Nagoya Protocol in 2010, that advise countries on how to implement Access and Benefit-sharing (ABS) regulations in their national legislations, and the International Treaty on Plant Genetic Resources for Food and Agriculture (ITPGRFA) in 2001 (Table 1). Section 3 of this paper will deal with these international agreements in more detail. Both the CBD and the ITPGRFA had, amongst others, the objective of facilitating access to PGRFA [2]). However, due to low compliance and complicated, rather bureaucratic implementation procedures, especially under the bilateral regulations of the Nagoya Protocol, it did not have the desired effect.

Screening programs for desirable traits in crop breeding require access to a large quantity of plant genetic resources from different sources and countries. At the end of the screening process, only a few breeding lines will be used in generating the final, new commercial variety. Negotiating and securing PIC and MAT for each germplasm source via bilateral agreements is a complex and time-consuming process [10]. Differing national and even local ABS laws and regulations create a significant entry barrier and represent a major challenge for seed companies to establish a relevant collection of starting materials for breeding. Bilateral ABS contracts under the Nagoya Protocol require extensive tracking and tracing of every germplasm transfer and subsequent use and movement worldwide. Tracking systems are meant to provide the link between access and use by following the international movements of genetic resources, from original provision until the inclusion in a commercial product, either a new plant variety or other inventions, which could be patentable [11]. Commercial plant breeding programmes, in general, have multiple breeding cycles running in parallel, often in different countries with different climatic conditions and seasons, involving exchanges of breeding material between countries. In this process, ABS tracking requirements create significant complexity in the breeding workflow [10]. For the aforementioned reasons, access to germplasm is often limited under the Nagoya Protocol [10,12,13]. The unresolved regulation of access to digital sequence information (DSI) associated with germplasm accessions complicates things even further [14], resulting in a further decline in the use of genetic resources for crop improvement.

In order to cope with the ever-increasing threats from biotic and abiotic stresses, exacerbated by climate change, the recommendation to consume more biodiverse food to counter the increase in diet-related diseases, and the need to feed a still-growing global population with healthy diets, plant breeders do need continuous access to cultivated and non-cultivated (crop wild relatives-CWR) genetic diversity for trait screening and to generate new diversity with the aim of developing competitive, high-yielding, and nutritious new varieties for the farming community. Loss of crop diversity and erosion of genetic diversity due to a variety of reasons is of general concern, requiring continued efforts to mitigate further loss by safeguarding crop diversity ex situ [15]. However, the decision process to obtain collecting permits and to acquire germplasm in compliance with recently developed legal requirements of international rules and regulations is complex and cumbersome and creates uncertainties for genebank curators and plant breeders alike [10,12,13]. Moreover, the complex/bureaucratic accessibility of genetic resources becomes an additional criterion for deciding whether or not to use genetic resources regulated by the CBD and the ITPGRFA. Although the International Treaty established standard rules and procedures (i.e., the SMTA) for accessing PGRFA, it does not prevent member countries from implementing related legislation, for example, concerning the inclusion of specific PGRFA in the MLS [13]. Such legislation might differ from country to country.

By its very nature, the ABS procedures under the CBD/Nagoya Protocol differ from country to country, are often unclear and highly bureaucratic, and are still evolving. In India, for example, the National Biodiversity Authority and the State Biodiversity Boards are required to consult with the respective local Biodiversity Management Committee (out of approx. 270,000 existing in the country) regarding the access conditions expected by the conservers and holders of biological resources and associated traditional knowledge. Once this internal consultation process is completed, foreign users need to "negotiate" the contractual ABS clauses with national authorities [10]. Public breeders and breeders from small and medium-sized enterprises do not usually have the necessary expertise and resources to navigate such complex arrangements and, therefore, often prefer to stay away from such complexity.

Mekonnen and Spielman [16] correlated historical trends in genebank acquisitions and changes in germplasm exchange over time, with changes in the international policy environment for seven crops that are essential for food security in developing countries. Based on these results, the authors concluded that a country's membership in the CBD is closely associated with reductions in the flow of genetic resources and that the Nagoya Protocol may affect global PGRFA flows in a potentially negative and unintended manner. In contrast, ITPGRFA membership is likely to moderate the negative effects of the CBD and the Nagoya Protocol [16].

Nutritionists and health sector specialists are increasingly highlighting the role of vegetables, fruit and nuts for their potential in combating the triple burden of malnutrition (undernutrition, hidden hunger and overnutrition) [17]. Unfortunately, facilitated access to vegetable genetic resources under the less cumbersome multilateral agreement of the ITPGRFA is rather limited, as the majority of vegetable crops are not included in the Annex I list of the MLS and thus, automatically fall under the Nagoya Protocol obligations. However, genetic diversity is needed to develop new resilient varieties with multiple resistances against ever-increasing biotic stresses and tolerance to abiotic stresses that are exacerbated by climate change. Therefore, the authors of this review emphasise in particular the case of vegetable genetic resources due to their unique role in nutrition security. Vegetable breeders from the public and private sectors face considerable difficulties in accessing and using the required genetic diversity for breeding elite, nutrient-dense and resilient vegetable crop varieties. The vegetable breeding sector deals with a wide range of species and an enormous diversity of diseases and insect pests and is, therefore, perhaps even more reliant on germplasm from genebanks than breeders dealing with other horticultural and agronomic crops. Nevertheless, the challenges and legal uncertainties in accessing and using germplasm and related information for breeding discussed in this paper apply to all PGRFA, and most references cited are not restricted to vegetable crops.

This paper highlights the importance of genetic diversity and plant breeding for sustainable agricultural production and food and nutrition security. In this context, the authors focus on the increasing complexity of access and benefit-sharing policies and their implications for crop germplasm collecting and conservation, and access to and utilisation of the conserved crop genetic diversity by plant breeders from the public and private sectors. Several options for addressing current constraints regarding ABS of PGRFA are discussed. It is essential to develop a more satisfying and functional global germplasm conservation and use system to halt further genetic erosion of threatened and endangered PGRFA and preserve it for use by current and future generations of breeders, farmers and consumers, and society as a whole.

2. The Importance of Genetic Diversity and Plant Breeding for Agricultural Production and Food and Nutrition Security

Since the transition from hunting–gathering to sedentary farming, producing enough food for a growing population has always been a significant challenge. The origins of agriculture can be traced back to about 12,000 years ago, when wheat and barley domestication and cultivation started in the Fertile Crescent in the Near East [18], and a 'crop package' spread from there into Europe, Asia and Africa several thousand years later. Climate change and population growth are considered to have major impacts on sedentary farming. Today, population growth and greater per capita purchasing power, coupled with higher meat, dairy and egg consumption, and the use of agricultural crops for biofuel production are considered to be major driving forces for the continuously growing global demand for food, fibre and fuel crops until 2050 and beyond [19,20]. However, the increasing human population, scarcity of fertile land for the expansion of cropping areas, the negative impact of agriculture on the environment and the increasing threats from climate change mean that further increases in food production must primarily be based on yield enhancement and productivity growth. This can be achieved through continuous plant breeding efforts and sustainable intensification of crop production practises on existing croplands, on which current crop yields are well below the yield potential [21].

In many parts of the world, plant breeding has contributed considerably to increased productivity, apart from increased use of agricultural inputs such as irrigation water, chemical fertilisers and pesticides. This led to stable markets, lower food prices and reduced price volatility [22,23], among others, evidenced by the 'Green Revolution' [24]. Studies conducted by Noleppa and Cartsburg [23] indicated that plant breeding has contributed, on average for all major arable crops grown in the European Union (EU), a yield increase of about 67% since the turn of the millennium. This translates into an average yield enhancement of 1.16% per annum for the major crops. These values are higher than the individual crop yield gains reported by Evenson and Gollin [25] from 1960 to 2000. The development of high-yielding varieties with multiple disease resistances and enhanced water- and nutrient-use efficiency also has considerable societal and environmental benefits, reducing pesticide- and fertiliser-induced hazards and greenhouse gas emissions, apart from avoiding the further expansion of agricultural land [23]. In terms of production volume, similar observations have also been made for tomatoes, the globally dominant vegetable crop, and alfalfa, a globally important forage crop [26].

Breeding and agricultural intensification efforts led to a significant availability of food, which, in turn, contributed to a notable decline in the number of people suffering from chronic hunger. However, after years of steady decline, the trend in world hunger reverted in 2015 and remained relatively constant until 2019 (618.4 million undernourished; 8%). From 2019 to 2020, the prevalence of undernourished people rose sharply, from 8.0 to 9.3%, and to 9.8% in 2021, meaning that approximately 767.9 million people were affected by hunger in 2021 [27]. Current projections indicate that close to 670 million people, or about 8% of the global population, will still face chronic hunger in 2030, approximately the same proportion of the population as in 2015 when the Zero Hunger target of the 2030 Agenda for Sustainable Development was launched by the United Nations [28]. In 2021, 425 million people in Asia, 278 in Africa and 56.5 in Latin America and the Caribbean were suffering from hunger. All in all, around 2.3 billion people (nearly one-third of the world population) were moderately or severely food insecure in 2021 and suffered from chronic micronutrient deficiencies [27,29].

Promoting the production and consumption of vegetables (and fruit) is a valid approach to alleviating 'hidden hunger' and enhancing nutrition security, especially in the case of diets that are dominated by high-energy foods with low levels of micronutrients [30]. This requires significant efforts in crop breeding for sustainable intensification and adaptation to changing climates. During a recent 10-year period (2008–2018), there was indeed a significant increase (24%) in global commercial vegetable production, mainly attributable to production increases in Africa (32%) and Asia (28.3%) [31].

According to the Food and Agriculture Organisation, global vegetable and fruit production in 2020 was estimated to be around 1128 and 887 million metric tons, respectively [32], which would result, in theory, in vegetable and fruit availability of almost 700 g per person per day, assuming 8 billion consumers. This amount is well above the 400 g of fruit and vegetables recommended for daily consumption by the World Health Organisation (WHO) [33] but does not reflect the much lower edible portions of the harvested produce and considerable losses along the value chain. By 2015, only 55% of the global population had an average fruit and vegetable availability above WHO's minimum intake target (400 g), while people in Sub-Saharan Africa, on average, only have access to about 200 g of fruits and vegetables per day [34].

Crop domestication and improvement were based on intentional, ongoing selection for traits that improved the quality and palatability of plant organs for human consumption, facilitated crop cultivation and harvesting (e.g., suitable for mechanical harvesting and non-shattering seeds), enhanced yield and productivity, resistance against pests and diseases and tolerance to a variety of environmental stresses [35,36]. Professional plant breeding basically started with the re-discovery of the laws of inheritance by Gregor Mendel, first published in 1866 in the Proceedings of the Natural History Society of Brno, 157 years ago [37]. Many scientists consider Mendel the father of modern genetics. Various methods are used in plant breeding [38,39]. They can be based on the visual selection of plants with desired variants occurring in nature or within traditional varieties. Often, new genetic diversity is introduced into breeding populations by intercrossing selected elite plants with desired traits that complement each other or by introgression of desired traits/genes from CWR into an advanced breeding line. Modern marker-assisted precision breeding is based on monitored recombination of specific genes with the help of molecular tools that systematically track within-genome variation.

The choice of the breeding method being applied is often crop-specific, determined by the mode of reproduction and the breeding objectives [39]. In the commercial breeding of vegetable crops, the production of hybrids is steadily increasing, as it allows the exploitation of heterosis and facilitates the multiple stacking of desired traits. Careful pollination control is required to ensure efficient hybrid production. Depending on the crop, technologies that inhibit pollen production in mother plants may include manual or mechanical emasculation and genetically controlled systems, such as male sterility [40]. Once desired traits have been fixed in a new variety, and genetic uniformity, yield stability and local adaptation have been verified, seed production and commercialisation of the new variety commence.

3. Current and Evolving Policies and Procedures for Germplasm Collecting, Conservation, Exchange, Use and Related Benefit-Sharing—CBD, International Treaty, Nagoya Protocol, DSI Debate

The intergovernmental negotiations that led to the signing of the CBD in 1992–1993 made member states agree to conserve the biological resources existing in their respective territories, make them available and share benefits deriving from their use on agreed terms. These terms were further specified and formalised in the Nagoya Protocol, which entered into force in October 2014 [41]. With the objective of harmonising the existing access and benefit-sharing (ABS) regulations for plant genetic resources for food and agriculture (PGRFA), as established by the FAO Commission for Genetic Resources for Food and Agriculture (known as International Undertaking), with those established under the CBD, the International Treaty for PGRFA (ITPGRFA) was negotiated by the member states of FAO, adopted in 2001 and came into force in 2004. As part of the ITPGRFA, the Multilateral System (MLS) addresses facilitated access to PGRFA for specific uses, such as conservation, research, breeding and training for food and agriculture. The MLS provides a transparent, multilateral access and benefit-sharing mechanism for both providers and users of PGRFA through the signing of a Standard Material Transfer Agreement (SMTA), thus reducing transaction costs. However, it only covers a limited number of so-called Annex I crops [42] but includes the collections maintained by the CGIAR and other international agricultural research centres that fall under Art. 15 (see below). For access to germplasm of other crops and plant species not covered by the MLS and Art. 15 of the ITPGRFA, the ABS regulations of the CBD, as specified under the Nagoya Protocol, are applicable.

Navigating and complying with these international treaties and their specific ABS regulations, especially the Nagoya Protocol, is complex. There is apparent ambiguity about the scope and application of many provisions of the international agreements regulating access to PGRFA and related benefit-sharing schedules. Access to vegetable genetic resources is severely restricted, especially since most vegetable crops are not included in the list of Annex I crops of the ITPGRFA. Access to genetic sequence information is still being debated and is subject to change.

3.1. *The Convention on Biological Diversity*

Prior to the intergovernmental agreement of the CBD, plant genetic resources were recognised as a common heritage of humankind [1,2] and could be, as stipulated in the IU established by FAO in 1983, freely collected, integrated into and accessed from genebanks, and shared with other countries and public and private genebanks [3,43]. This approach changed completely with the CBD entering into force in December 1993, which had as its main objectives the conservation of biological diversity, the sustainable use of its components and the fair and equitable sharing of the benefits arising from the utilisation of genetic resources [44]. Under the CBD, states can use their national sovereignty to exercise control over the genetic resources within their national territories. According to Article 2 of the CBD, genetic resources "mean genetic material of actual or potential value", and genetic material is defined as "any material of plant, animal, microbial or other origin containing functional units of heredity" [45]. Thus, the CBD covers a broad range of wild and cultivated biodiversity occurring and originating in the 196 member countries and any current and potential use of this diversity [13]. Human genetic resources and genetic resources occurring beyond national jurisdictions are excluded from CBD coverage.

Access to genetic resources is granted under prior informed consent (PIC) and mutually agreed terms (MAT) by the party owning and providing those resources (Article 15 of the CBD) [46]. The CBD does not define or describe terms of ABS of genetic resources but expects the contracting parties to develop and implement national laws to that effect (Articles 6 and 15), based on guidelines provided in the Nagoya Protocol.

3.2. *The Nagoya Protocol*

The Nagoya Protocol on Access to Genetic Resources and the Fair and Equitable Sharing of Benefits Arising from their Utilisation is a supplementary agreement to the CBD [41]. It was adopted in October 2010 in Nagoya, Japan, and entered into force in October 2014. The Nagoya Protocol focuses on the third objective of the CBD, the fair and equitable sharing of benefits arising from the utilisation of genetic resources. It aims to create greater legal certainty and transparency for providers and users of genetic resources by establishing predictable conditions for access to genetic resources and helping to ensure benefit-sharing when genetic resources leave the provider country. The Nagoya Protocol has a membership of 140 Parties (140 ratifications and 92 signatories) (https://www.cbd.int/abs/nagoya-protocol/signatories/; accessed on 31 July 2023). Contracting parties of the Nagoya Protocol are required to implement and adhere to measures regarding access to genetic resources, benefit-sharing and compliance. Provider countries must set out clear conditions for granting PIC and establishing MAT, including access to traditional knowledge associated with genetic resources that are held by indigenous communities, while user countries are required to monitor use and compliance with the MAT.

The Nagoya Protocol refers to several tools and mechanisms that assist with implementing its objective. Among those are the establishment of national focal points (NFPs) and competent national authorities (CNAs) that serve as contact points for information, grant access and cooperate on issues of compliance [41]. Other provisions include an Access and Benefit-sharing Clearing-House (ABSCH) to share relevant and updated information on national regulatory ABS requirements and contact details of NFPs and CNAs [47]. Unfortunately, the information available on the ABSCH website is incomplete and, for a number of countries, not up to date. It does not provide all the details required to access PGR in a particular country. Addresses of NFPs are often not indicated, the CNAs are not clearly defined and clear ABS instructions are often missing. Moreover, law texts are often only available in the local language.

The text of the Nagoya Protocol also provides an Annex with a list of possible monetary and non-monetary benefits that can satisfy the Convention's requirement for benefit sharing (https://www.cbd.int/abs/text/articles/?sec=abs-37; accessed on 31 July 2023). The non-monetary benefits listed under the Nagoya Protocol are numerous and include, among others, sharing of know-how, collaboration, cooperation and contribution to scientific research and development programmes. The value of those non-monetary benefits is often overlooked in the current heated discussions, which mostly centre around monetary benefits. The social, economic and ecological benefits and impacts that accrue to countries through the sharing and use of improved and well-adapted crop germplasm and associated technologies might far exceed the levels of monetary benefits that could be generated under the benefit-sharing arrangements of the CBD and the ITPGRFA [13,48].

3.3. *The International Treaty for Plant Genetic Resources for Food and Agriculture*

The International Undertaking (IU) on Plant Genetic Resources, a voluntary agreement established by the FAO Commission on Genetic Resources for Food and Agriculture in 1983 to promote international collecting, conservation, exchange and use of plant genetic resources for breeding and scientific purposes [2], conflicted with the ABS policy of the CBD, allowing each contracting party to exercise sovereign rights over natural resources. As a consequence, the IU became obsolete and the FAO started its revision process in 1994. The negotiations of the FAO member states were concluded in November 2001 with the adoption of the International Treaty, which came into force in June 2004 [49]. The ITPGRFA is in full compliance with the CBD. In contrast to the bilateral negotiations required under the CBD/Nagoya Protocol, the ITPGRFA establishes, through its MLS, a globally harmonised system to provide farmers, plant breeders and the scientific community with transparent and facilitated access to plant genetic resources. The International Treaty also ensures recipients share the benefits of using these genetic resources. As of 1 January 2023, the ITPGRFA has 150 Contracting Parties, including the European Union as a member organisation (https://www.fao.org/plant-treaty/countries/membership/en/; accessed on 31 July 2023).

An essential provision of the ITPGRFA is the multilateral system (MLS) that facilitates access and benefit-sharing of PGR of crops specified on a defined list, known as Annex I, consisting of 35 food and 29 forage crops. From the contracting parties, all genetic resources that are in the public domain and under governmental control and management should be included in the MLS and as such made freely available (or at minimal transaction cost) for the purposes of conservation, research, breeding and training under a Standard Material Transfer Agreement (SMTA), instead of the use of prior informed consent and mutually agreed terms, on a country-by-country or case-by-case basis, as prescribed by the CBD. However, if the PGRFA materials in the MLS are intended for other purposes only, such as biofuels, pharmaceuticals or other industrial uses, they are excluded from the scope of the International Treaty's MLS [13]. The germplasm recipients shall not claim any intellectual property or other rights that limit access to these resources or their genetic parts or components in the form received from the MLS [49].

As outlined in Article 15 of the ITPGRFA, the MLS also includes Annex I PGRFA materials that are maintained by genebanks of the CGIAR or other regional or international agricultural research centres that have signed agreements with the Governing Body of the International Treaty [42]. These agreements with the Governing Body stipulate that the international centres will also make non-Annex I plant materials held in their collections, under their 1994 "in trust" agreements with the FAO, available to users under the same SMTA terms as foreseen for the exchange of Annex I germplasm [13]. As of August 2018, the genebanks of the CGIAR centres conserved and made more than 730,000 accessions of crop, tree and forage germplasm available under the MLS [48].

The European Cooperative Programme for Plant Genetic Resources (ECPGR) stated in its 2016 report under Article 32–ECPGR and the Treaty/Nagoya Protocol: "It is recommended that all ECPGR member countries (36 as of June 2023 for the current phase (2019–2023)) as appropriate and in line with national legislation, use the SMTA for distribution of both Annex I and non-Annex I PGRFA accessions, independently of whether material is conserved in ex situ collections or held in situ" [50]. Meanwhile, a few countries, such as Germany and the Netherlands as well as NordGen, acting for the Nordic countries (Denmark, Finland, Iceland, Norway, Sweden and the autonomous territories of the Faroe Islands and Greenland) [51], have already adopted this recommendation. The Nordic countries signed the so-called Kalmar Declaration, which recommended facilitated access to PGRFA in general. Greenland is the only Nordic country that introduced "classic" ABS legislation with requirements for PIC and MAT for access to genetic resources or related TK, while Norway has enacted an authorisation for such legislation concerning wild genetic resources. For access to TK held by indigenous peoples and local communities (ILCs), Norway has enacted legislation that requires PIC for the use of TK associated with genetic resources from ILCs. Finland has enacted a PIC requirement for using TK held by the Sámi people, and Sweden is establishing formal consultation with the Sámi Parliament of Sweden on matters dealing with TK [51].

The large collection of plant genetic resources of both Annex I and non-Annex I material of the NordGen genebank has been placed in the Nordic public domain, making all accessions available through the International Treaty's SMTA. It is interesting to note that from 2018 to 2020, NordGen had to deal with a considerable increase (89%) in the number of ordered seed samples, primarily requested by Nordic and other European countries [51]. This could be attributed to the increased complexity of obtaining germplasm from elsewhere. It is also important to state that all accessions (including non-Annex I) designated by ECPGR member countries to the European collection AEGIS (A European Genebank Integrated System) are distributed to users under the terms of the SMTA (https://www.ecpgr.cgiar.org/aegis/about-aegis/aegis-and-the-treaty; accessed on 31 July 2023).

According to Article 13 of the ITPGRFA, facilitated access to germplasm itself is considered to be an important benefit. Other non-monetary benefits include the exchange of information, access to and transfer of technology and capacity building [42]. In practice, this happens to some extent in public–private partnerships between genebanks and the private breeding sector through multiplication and/or evaluation of selected accessions. The SMTA of the ITPGRFA also foresees monetary payments by germplasm recipients under Article 6.7, if material received under an SMTA is used to create PGRFA materials that are not freely available for research and breeding by others (i.e., 0.77% of the sales of those PGRFA or, according to Article 6.11, an alternative payment of 0.5% of all sales of PGRFA belonging to the same Annex I species, to an international benefit-sharing fund (www.fao.org/plant-treaty/areas-of-work/benefit-sharing-fund; accessed on 31 July 2023)). This fund is used to support the conservation and sustainable utilisation of PGRFA, especially among local traditional farming communities.

Payments under SMTAs take time to materialise, as product development with the received germplasm and testing the final product (variety) containing this germplasm is time-consuming. In addition, it is important to consider that most breeders may not seek Intellectual Property protection through patent application for new varieties, hence not limiting the free use of the final product (variety) for further breeding and thus not resulting in mandatory payments. In 2013, the Governing Body of the ITPGRFA established an intergovernmental Open-ended Working Group to negotiate and recommend measures to enhance the functioning of the MLS under the ITPGRFA for consideration by the Governing Body. Various stakeholder groups (seed industry, CGIAR, farmer organisations and civil society) contributed to this process [52]. A group of seed companies proposed a 'subscription-only model', which was meant to replace one-time payments for receiving and using a specific genetic resource in case the final product is not freely available for further breeding. Companies would instead contribute annually a percentage of their annual seed sales for selected Annex I crops as listed under the ITPGRFA for an initial 10-year period, thus ensuring an ongoing and consistent contribution to the ITPGRFA's Benefit-sharing Fund [53] from the very moment of signing the subscription agreement for one or more selected Annex I crops. Forty-one members of the International Seed Federation (ISF) indicated that, if the subscription covers all crops, a single rate of 0.01% on sales of Annex I crops would be acceptable [54]. For those companies that would not participate in the subscription system, the continuation of a single access mechanism with payments based on the use of accessed genetic resources as reflected in current Articles 6.7 and 6.8 of the SMTA [54] should remain, next to the subscription model. ISF indicated that its members would be ready to make payments based on Articles 6.7 and 6.8, as long as the Article 6.8 option was manyfold lower than the payment under Article 6.7.

The intergovernmental Open-ended Working Group considered this idea and proposed a subscription system linking payment obligations to *access* rather than *use* and commercialisation, whereby users pay a fee for being granted access to all or a selected group of PGRFA within the MLS. Such a subscription system would start from the date of signing the subscription agreement (prior to shipping any materials) and thereby would eliminate the need to track single germplasm samples obtained under the MLS down to the final commercialisation of products [52]. However, no agreement has been reached up to now because of the difference in opinion on payment rates, DSI and the expansion of Annex I to all PGRFA. The negotiations were stopped during the eighth Session of the Governing Body of the ITPGRFA (GB 8) in November 2019.

During GB9 held in September 2022 in New Delhi, India, delegates agreed to re-establish the Ad Hoc Open-ended Working Group to Enhance the Functioning of the MLS to finalise the process of adopting a series of options/measures by GB 11 in 2026 [55]. The envisaged measures aim to (i) increase the monetary and non-monetary benefits arising from the MLS for all providers and users; (ii) ensure a sustainable and predictable long-term increase in user-based income to the Benefit-Sharing Fund (BSF); (iii) expand the MLS to include all PGRFA and improve their availability; (iv) make the MLS more dynamic and responsive to new developments and innovations, and to create legal certainty, administrative simplicity, and transparency for all participants.

3.4. The Digital Sequence Information Debate

Rapid advances in genomics and the open access to digital sequence information (DSI), available to all through public sequence databases, have contributed considerably to the recent progress in plant and life sciences, including health, conservation and use of (agro)biodiversity [14]. Plant genomics is evolving rapidly; single reference genomes for plant species are now complemented by multiple sequences and pangenomes, depicting the diversity within crop species or genera.

The term DSI still lacks a precise and generally accepted definition [56–58]. It has been invented as a place-holder term by negotiators covering under a narrow interpretation genetic/genomic sequence data only, while under a wide interpretation, biological data, such as passport, characterisation and evaluation data, ecological adaptation and traditional knowledge may be included [59]. The increasing role and importance of DSI, the speed by which it is developed and analysed, potentially allowing the circumvention of the very use of material genetic resources, contributes to an uneasy feeling among several countries, who consider DSI as a threat to the exertion of their sovereign rights over genetic resources and associated information [59,60]. Despite the rapid progress in genomics and gene editing, DSI is not expected to replace physical access to PRGFA, but the value of DSI for research, breeding and variety development is obvious.

The International Nucleotide Sequence Database Collaboration (INSDC) is the central foundation for global sequence information. It connects over 1700 scientific databases and platforms [61]. The INSDC provides the free core infrastructure for DSI deposition, preservation, and global dissemination as part of a scientific collaboration between the European Molecular Biology Laboratory (EMBL), an intergovernmental organisation with more than 80 independent research groups, the U.S. National Center for Biotechnology Information (NCBI), which is hosting the collaboration through its GenBank, and the DNA Databank of Japan (DDBJ) [62]. Tracking and tracing the movement of nucleotide sequence data (NSD) in GenBank, the largest public database platform, is challenging [59]. Scientists worldwide search this platform for their work; hence, any financial or administrative burden for accessing NSD will affect all scientists, globally, and limit their ability to undertake individual or collaborative research.

There is consensus among genebank curators that open access to and free use of DSI is essential for facilitating adequate conservation and use of PGRFA [43,63]. However, contracting parties to the international treaties express contradicting views on whether and how access to DSI and benefit-sharing from its utilisation should be regulated. Countries in the Global South consider free-for-all access to sequence information (with or without the associated germplasm) to be mainly beneficial for the biotechnology industry in the Global North, and as such, counterproductive for other countries, their local communities and indigenous people who are the custodians of plant agrobiodiversity and who fear that they will not be able to benefit if access to DSI is not subject to ABS regulations of PIC and MAT under the Nagoya Protocol of the CBD [64,65].

Discussions on the current and future access to digital sequence information are ongoing in several fora, under the International Treaty, CBD, Nagoya Protocol, the multilateral Prepared Influenza Preparedness (PIP) Framework, the Antarctic Treaty (dealing with Antarctic species) and the United Nations Convention on the Law of the Sea (UNCLOS), which is developing a multilateral treaty aiming at establishing best practices that regulate access to marine genetic resources and related sequencing data, while sharing the benefits derived from such access to enhance the conservation and sustainable use of marine biological diversity in areas beyond national jurisdiction [66–68].

Despite the rapid progress in genomics and gene editing, the successful transfer of useful traits across different life forms appears to be rather unlikely in the near future due to the complexity of biological systems [69]. Many traits are multigenic, and gene expression may also depend on epigenetics and other factors; hence, it is not straightforward. Therefore, simply disconnecting the study and use of DNA sequences from the physical germplasm resource conserved in genebanks, from which DNA was extracted, will not suffice to solve the food and nutrition security of humankind. The combination of natural selection and professional plant breeding will still be required for crop variety development and adaptation to local agroecological conditions in the foreseeable future to ensure food and nutrition security.

Scholz et al. [14] showed that a benefit-sharing scheme for DSI modelled according to the Nagoya Protocol would likely totally disrupt plant breeding and genomics research, as sequence datasets are downloaded from INSDC 34 million times per year by 10–15 million unique users, making a bilateral system requiring permissions between an end-user and country of origin prohibitively complex. All countries provide and use DSI for basic and applied research, in both the public and private sectors. For example, DSI from Brazil is used by 111 countries worldwide, while scientists in Brazil use DSI from 153 different countries [70]. In the case of Kenya, DSI from this country is used by 79 countries, while scientists in Kenya use DSI from 83 countries. Alternative methods where DSI remains open access and benefits are shared in a fair and practical manner are mandatory and should be the aim of intergovernmental negotiators [63]. A system which requires tracking, tracing, reporting and monitoring DSI information is entirely unsustainable. Multilateral approaches, which delink access to DSI and benefit-sharing, are the most appropriate for all stakeholders involved.

4. Implications of the Complexity of ABS Regulations for Germplasm and Related Information for Genebank Curators and Public and Private Sector Breeders

4.1. Uncertainties Regarding the Current ABS Regulations

Due to the ongoing genetic erosion of valuable (agro)biodiversity, germplasm collecting and ex situ conservation are of high priority to secure the necessary genetic diversity for developing new crop varieties with resilience to biotic and abiotic stresses and of high nutritional value for the benefit of humankind [15]. Genebank curators and breeders understand the national sovereignty of countries over their genetic resources and are willing to comply with clear and realistic standardised ABS regulations established by the countries in which collecting germplasm material is of high priority. The private sector also encourages clear, transparent and easy-to-implement ABS agreements for the sustainable use of genetic resources [71]. Germplasm users and conservationists are ready to share benefits, especially through capacity development in the countries where the collecting is taking place. However, many germplasm collectors struggle with the complexity and lack of clarity of how this sovereignty is exercised and interpreted by individual countries [12,43,58]. The same is true for many breeders, as clearly described by Michiels et al. [10]. Seed companies need to keep themselves up to date with the complexity created by up to 200 different national ABS frameworks, which keep changing and evolving over time. Moreover, there are further differences at provincial or local levels within a given country. Uncertainties regarding ABS regulations and procedures to secure ABS compliance include the questions on how and with whom ABS can be negotiated bilaterally, who is subject to those conditions, how compliance is monitored and how regulations apply to plant biodiversity beyond the time frame of the current instruments [10,58].

Only 35 food and 29 forage crops are covered by the MLS of the ITPGRFA and can be relatively easily accessed under the terms of the SMTA, provided they are placed under the MLS. Access to other food and particularly vegetable crops, including a wide range of underutilised genera with current or potential value as food plants, is governed instead by CBD/Nagoya Protocol-based diverse national regulations, and each single resource exchange needs to be negotiated on a case-by-case basis [72].

4.2. The Issue of Stacking Obligations, Retroactive Effect, Tracking and Tracing and Related Costs

The complex, and often unclear ABS regulations, implemented with a high degree of variation among countries, sometimes with retroactive effect, and the use of different germplasm sources in the long process of developing new varieties require extensive tracking and tracing and raise associated costs and administrative burdens for genebanks, botanic gardens and public and private breeders alike [10,12]. A typical example of a retroactive effect is the case of a melon variety, accessed by an American seed company from the USDA genebank, which originated in and had been received from India long before the country introduced its Biological Diversity Act in 2002. The seed company derived progenies resistant to Cucurbit Yellow Stunting Disorder Virus (CYSDV) from this melon variety and received a patent on this trait from the European Patent Office (EPO). In 2016, the National Biodiversity Office filed a non-compliance case with the EPO [73]. Even the purchase of commercial seed, with unknown history, for use in further breeding in a country that does not impose ABS obligations for the use of its genetic resources may also result in non-compliance issues with retroactive effect if it has parent material in its pedigree from a country such as India [10]. Therefore, the implementation of ABS laws in one country may result in legal uncertainty at the global level.

Moreover, the development of improved, resilient varieties may stack obligations and costs as it involves the incorporation of PGRFA, traditional knowledge (TK) and/or DSI from several countries. Every cross made by plant breeders stacks the contractual ABS requirements of its parental lines. Once a plant breeder has decided to make a cross, i.e., has 'utilised' a genetic resource in a cross, this resource is part of the genotype of all progenies. A sudden request to stop using a specific genetic resource in an ongoing breeding process due to the entry into force of new ABS regulations or different interpretations of their scope requires discarding all breeding materials that used the specific genetic resource in question.

The complexity and legal uncertainty regarding access to in situ and ex situ genetic resources and traditional knowledge (TK) in provider countries are described in great detail for selected countries by Michiels et al. [10]. The legal uncertainty ensuing from the complex and unclear national ABS regulations and the length of time needed and the costs involved in navigating the ABS regulations and negotiating PIC and MAT will often obstruct rather than facilitate ex situ conservation and associated research by genebanks and research institutions [12] as well as R&D investments by the public and private sector into horticultural innovations [10,74]. The efforts in terms of time and human resources that seed companies need to spend on ABS tracking and compliance issues are often considered disproportionate to the benefits that are generated and shared through bilateral ABS agreements [10]. According to Rabitz [75], costs related to germplasm access and use arise from three distinct sources: (a) transaction costs associated with access (potential administrative barriers, negotiation of bilateral ABS contracts, involvement of lawyers for clarification of uncertainties regarding legal and regulatory requirements); (b) the costs due to benefit-sharing obligations, including the costs of mandatory or voluntary tracking of PGRFA through the value chain; and (c) compliance costs after accessing the genetic resource, including monitoring costs to follow the utilisation of accessed materials throughout the value chain and to provide documentary evidence of utilisation in accordance with the applicable ABS laws and regulations. The described costs are considerable and might serve as a disincentive to access new genetic diversity. Therefore, tracking and monitoring mechanisms should be low-cost and should exclude standard genetic or biochemical analysis of individual samples since the cost of such analyses might exceed the expected benefit-sharing levels [11]. Rather than risking a company's reputation by trying to comply with unclear national ABS regulations, which could potentially result in a non-compliance case, companies might prefer to stay away from accessing such genetic resources. Complex ABS rules may also serve as a disincentive for the development of public and private breeding programs for more regional, less profitable crops, thus even threatening progress towards CBD objectives and the United Nations' SDGs.

The mentioned uncertainties also adversely affect collaboration among genebanks as well as the collaboration between genebanks and breeding companies. Genebank curators will be more hesitant to rationalise their own collections by reducing duplication with other genebanks since they cannot be sure of access to other collections in the future [43]. Such uncertainties are forcing countries to stockpile plant genetic resources to ensure future access to genetic diversity for their own research organisations and plant breeders, resulting in redundancies and further stress on the already limited capacity of the PGRFA community. Similarly, private-sector breeding companies feel compelled to stockpile and conserve for the long term the currently available PGRFA in working collections and breeding lines in already existing or yet-to-be-established private genebanks.

4.3. The Added Complexity Due to DSI Inclusion

The current stalemate regarding the inclusion of DSI into existing ABS mechanisms or the creation of separate ones for DSI only is of major concern. Genebank curators, conservationists and plant breeders agree that access to and use of DSI is essential for adequate conservation, sustainable use of plant genetic resources [43] and the development of elite crop varieties. A major headache for genebank curators is the unresolved definition of DSI, as genebanks share germplasm with associated accession-level information [59]. Should countries pre-emptively include DSI in their national ABS legislation before standard access mechanisms have been agreed upon in international fora, genebanks may no longer wish to conserve and exchange material from those countries due to the increased complexity of germplasm handling and distribution and consequent compliance issues. Similarly, the botanical gardens community also fears that individual countries may implement disparate regulations regarding access to DSI as has been the case for physical access to genetic resources [57].

Dozens of countries have already adopted legislation on DSI, often obliging users to obtain PIC and/or MAT to work with DSI [59]. Among those are six African countries (Kenya, Malawi, Mozambique, Namibia, South Africa, and Uganda) that have already put DSI domestic measures of a legal, administrative or policy nature in place [76]. Cameroon has passed a new instrument on ABS according to the ABS Clearing House Mechanism of the CBD. This instrument provides that the 'use of genetic information' is considered an activity relating to the use of genetic resources, and as such is subject to PIC and MAT. Those examples may serve as proxies for potential future restrictions on many, or all, species of plants, thus hindering research and innovation and being counterproductive to the efforts of continuously improving and diversifying our food systems [77].

4.4. The Consequence of Overly Complex ABS Regulations

There is evidence that non-strategic national ABS regulations have threatened or even impeded access to genetic resources for non-commercial research [12,78,79]. In Argentina and Brazil, for example, newly implemented national ABS regulations prevented domestic research organisations from studying local biodiversity, even if such research did not involve international partners or the export of local genetic resources [80–82]. Unclear national ABS legislations and a high level of bureaucratisation with concomitant high transaction costs and compliance risks are also a disincentive for commercial exploration of biodiversity [83,84].

The abundance of diverse ABS regulations established by individual countries, based on the Nagoya Protocol, combined with legal uncertainties regarding their interpretation and implementation may hamper conservation and the exchange of biodiversity. Consequently, food and nutrition security, which are within the scope of international agreements such as the CBD, the ITPGRFA and the Sustainable Development Goals of the United Nations, may also be negatively impacted. Reduced international collaboration as a consequence of overly complex regulations for access to PGRFA and associated information will likely slow down capacity building and technology transfer to less advanced countries, thus deepening global inequalities.

5. Options for Addressing Current Constraints on Access and Benefit-Sharing of Genetic Resources and Related Information, at the Policy Level

5.1. Expanding the MLS of the ITPGRFA to Cover All PGRFA, Combined with a Subscription System

Most plant breeders would prefer the open-source option of the heritage of humankind principle as established under the IU. However, this is no longer a viable option. The MLS of the ITPGRFA significantly reduces the burden of tracking and tracing and associated costs, avoids bilateral contracts and is, therefore, an option widely supported by public and private-sector breeders. In the current situation, the most straightforward ABS option for PGRFA would be an expansion of the MLS of the ITPGRFA to cover all genetic resources for food and agriculture, amongst others including all vegetable crops and species. This proposal is in line with the current efforts of the Open-ended Working Group to Enhance the Functioning of the MLS (see Section 3.3; [55]). If an expanded MLS covering all PGRFA is then based on a subscription system linking payment obligations to access rather than use or commercialisation, there would be no need to track single germplasm samples obtained under the MLS down to the final commercialisation of a new variety, plus its use by further breeding with this new variety, over and over again. Alternatively, if users do not want to adopt the subscription system, commercial utilisation of germplasm obtained under the MLS could be determined with the help of intellectual property tools, aided by using digital object identifiers (DOIs) [85]. Confirmed commercial utilisation of genetic resources would then trigger monetary benefit-sharing payments [56].

5.2. MLS of the ITPGRFA Covering all PGRFA, Combined with a Seed Sale Tax or Levy Option at the National Level

A seed sales tax or levy paid by contracting parties of the ITPGRFA would allow easy access to PGRFA under an expanded MLS covering all PGRFA, without the need for tracking and tracing measures to follow the movement of PGRFA up to the final product: a newly released variety. For 12 consecutive years, Norway has paid annual contributions to the benefit-sharing fund of the International Treaty, equivalent to 0.1% of the value of annual seed and plant material sales in the agricultural sector in Norway [86]. The CGIAR is recommending that contracting parties make annual payments to the Plant Treaty's benefit-sharing fund based on seed sales within their jurisdictions (similar to the Norway levy), using a fixed royalty rate that corresponds to the value of access to, and use of, both PGRFA and DSI. Contracting parties would then have the option to recoup a portion of that levy payment from commercial users in their jurisdictions [63]. Other countries could also follow the example of Germany, the Netherlands and the Nordic countries and share all PGRFA, whether they are within or outside of the MLS of the ITPGRFA, under the SMTA.

5.3. Harmonised, Multilateral ABS Regulations under the CBD/Nagoya Protocol

Harmonisation of ABS regulations under the CBD/Nagoya Protocol at the global level, in a similar manner as under the MLS of the ITPGRFA, would genuinely facilitate the work of all curators of collections at genebanks and botanic gardens and users in the public and private sectors alike [10]. The European Union's ABS Regulation No. 511/2014 (https://eur-lex.europa.eu/legal-content/EN/TXT/PDF/?uri=CELEX:32014R0511; accessed on 31 July 2023) is a first step in that direction. However, this directive defines only the compliance measures that must be followed by all 27 member states of the EU to secure ABS compliance under the Nagoya Protocol, while ABS regulations still follow the principle of national sovereignty and can vary significantly from country to country.

5.4. Harmonised National ABS Regulations and Compliance Measures

There is an urgent need to improve the current bilateral ABS systems. To effectively guide germplasm users trying to comply with all the different national ABS rules and regulations under the Nagoya Protocol, it would be conducive if the country profile on the ABS Clearing House website contained factual, concise information in English on how to access PGRFA in a given provider country and with whom to negotiate PIC and MAT, also including TK. All national legislation should be published on the ABS Clearing House website uniformly, including the scope and jurisdiction of the applicable ABS regulations and measures. A summary of the steps and obligations to follow should also be included.

5.5. Options for the Contentious DSI Issue

The term DSI is still used as a placeholder term and lacks a generally accepted definition. For most scientists, genebank curators and breeders, a narrow interpretation restricting DSI to sequence information only would be preferred, as other information associated with germplasm (passport, characterisation, and evaluation data) could be shared with users together with the exchange of germplasm as is still common practice. In case of a wider definition of DSI, Lawson et al. [66] proposed two options consisting of (i) a risk framework matrix for valuing information as part of the ABS transaction by attributing an estimated value to a particular kind of information; alternatively, (ii) a charge, tax or levy that would externalise the costs so that information would remain available and accessible to all, thus benefitting the global scientific community. Under the matrix option, passport data on accessions would be considered of low value, available without restrictions (public domain data), while descriptive (phenotypic) data would be treated as restricted public access data, and sequence data would have an embargo period until the results obtained need to be reported to the germplasm/DNA provider. Under the tax or levy model, the party accessing the resources would need to pay a tax (called 'Partnership Contribution' under the PIP Framework) or it would be levied on contracting parties, similar to the Norway seed sales tax under the MLS of the International Treaty. The tax or levy option is appealing, as it avoids the high transaction costs required to negotiate the value of information in every single transaction and allows the scientific community to disclose and share the generated information freely [66].

Given the complexity of the DSI issue and the diversity of stakeholders involved, Scholz et al. [61] proposed the creation of a public–private partnership (PPP) to govern the implementation of any future policy framework around DSI and offered five policy options for the sharing of monetary benefits.

5.6. Summary and Concluding Remarks of Section 5

Curators of plant genetic resources and public and private sector breeders hope for a transparent, functional and expanded multilateral system under the International Treaty covering all PGRFA, thereby erasing all legal uncertainties and minimising transaction costs for conservers and users of genetic resources and DSI. The authors of this paper strongly support a single, multilateral access mechanism for both PGRFA and DSI, if necessary combined with a subscription system as currently being negotiated under the MLS of the ITPGRFA or with a national tax or levy, similar to the Norwegian seed sales tax. If current and future international, regional, national and bilateral collaborative efforts would be guided by a focus on the promotion of inclusive innovation and enhanced equity in research, utilisation, and commercialisation of (agro)biodiversity and broader public and social benefits from the outcomes of science, instead of a predominant focus on immediate monetary benefits, greater benefits for all could be expected over time.

6. Recommendations and Concluding Remarks

Breeding improved varieties is a continuous and even cyclic effort that is essential for enhancing food and nutrition security. Crop improvement depends on access to biodiversity to source new genetic variation for breeding. Fair and non-bureaucratic rules to access and use germplasm in breeding is therefore a predisposition for food and nutrition security. Providers and users of plant genetic resources need clear information on the conditions under which the germplasm material can be accessed and used for research and breeding. It has to be clear whether the ITPGRFA, the CBD/Nagoya Protocol or any other ABS tool applies. Furthermore, adjustments to the current texts of these legal instruments are clearly needed to ensure legal certainty and strengthen access to genetic resources. Extending the list of Annex I crops of the ITPGRFA to include all PGRFA, as well as related organisms like pathogens and pests, would greatly benefit the use of new germplasm in breeding and lead to the creation of improved varieties that can cope with climate change challenges and will contribute to more sustainable forms of agriculture. Identification and documentation of the flow of benefits from the use of plant genetic resources to the different stakeholders could contribute to a better understanding of the value of plant genetic resources and related research on this material for humankind. Such a move might reduce current tension between germplasm providers and users, and eventually lead to more transparent and easy-to-follow access provisions. Crop diversity can only benefit humanity if it is not only conserved but also used.

Germplasm conserved in genebanks is most useful when it is distributed together with relevant information. Clarity on the scope of biodiversity data subject to ABS is essential for any future progress. High-throughput approaches have greatly improved genotypic and phenotypic data collection from genebank accessions. Such information can be used to strengthen germplasm management, elucidate questions regarding the taxonomy of accessions, assist in germplasm exchange through diagnostic tools for the detection of viruses and other pathogens, as well as for selecting plant genetic resources and specific traits for research and breeding. Such information can also assist in determining gaps in existing collections and help fine-tune the objectives of new collecting missions. These data could also be used to train artificial intelligence (AI) tools for a wide range of purposes, including ecophysiological crop modelling and identifying germplasm material adapted to climate change.

The outcome of the debate on the nature of DSI and the conditions for access and its use will be critical to actually using the data generated for plant genetic resources in research and breeding. Bilateral provider–user interactions for the use of DSI may be far too complex for regulating the DSI information flow. DSI policies should acknowledge the importance of using DSI across low-, middle- and high-income countries and strive to preserve open access to this crucial common good [14]. Non-monetary benefits that help bridge the scientific and technological gaps in developing countries should also be considered, as these stimulate international public–private partnerships and collaborations [81]. Such non-monetary benefits should include capacity building and technology transfer.

Curators of plant genetic resources in genebanks and botanical gardens as well as public and private sector breeders would benefit from a transparent, functional and efficient multilateral system under the International Treaty covering all PGRFA, thereby erasing all legal uncertainties and minimising transaction costs for conservers and users of genetic resources and DSI. Similarly, multilateral or fully open systems for exchanging biodiversity data are preferred by the wider scientific community [58]. The decision by Germany, the Netherlands and the Nordic countries to share all PGRFA under the ITPGRFA's SMTA is an encouraging example.

Author Contributions: All authors contributed to the article's conceptualisation, writing, reviewing, and editing. All authors have read and agreed to the published version of the manuscript.

Funding: This research received no external funding.

Data Availability Statement: No new data were created or analyzed in this study. Data sharing is not applicable to this article.

Acknowledgments: The authors wish to express their gratitude to Anke van den Hurk and Orlando de Ponti for their encouragement, input, critical review, and comments on a previous version of this manuscript. The critical review of the original version of this manuscript and the valuable recommendations made by three anonymous reviewers are acknowledged as they helped to improve the quality of this manuscript.

Conflicts of Interest: The authors declare no conflict of interest. As this review did not receive external funding, the writing of the manuscript and the decision to publish it is the sole responsibility of the authors.

References

1. Kho, T.Y. Intellectual property rights and the North-South contest for agricultural germplasm: A historical perspective. *Agrar. South J. Political Econ.* **2012**, *1*, 255–290. [CrossRef]
2. Engels, J.M.M.; Ebert, A.W. A Critical Review of the Current Global Ex Situ Conservation System for Plant Agrobiodiversity. I. History of the Development of the Global System in the Context of the Political/Legal Framework and Its Major Conservation Components. *Plants* **2021**, *10*, 1557. [CrossRef] [PubMed]
3. Rose, G.L. The International Undertaking on Plant Genetic Resources for Food and Agriculture: Will the Paper Be Worth the Trees? Available online: https://ro.uow.edu.au/cgi/viewcontent.cgi?article=1033&context=lawpapers (accessed on 21 June 2023).
4. International Union for the Protection of New Varieties of Plants (UPOV a). Overview of UPOV. Available online: https://www.upov.int/edocs/pubdocs/en/upov_pub_437.pdf (accessed on 13 May 2023).
5. UPOV, b. Guidance for the Preparation of Laws Based on the 1991 Act of the UPOV Convention. Available online: https://www.upov.int/edocs/infdocs/en/upov_inf_6.pdf (accessed on 13 May 2023).
6. Euroseeds. Benefits of the "Breeding Exemption" to Society. Position, Brussels, February 2012; Last Update: July 2019. Available online: https://euroseeds.eu/app/uploads/2019/07/12.0100-Euroseeds-Position-Benefits-breeding-exemption-1.pdf (accessed on 12 May 2023).
7. Feindt, P.H. The politics of biopatents in food and agriculture, 1950–2010: Value conflict, competing paradigms and contested institutionalisation in multi-level governance. *Policy Soc.* **2012**, *31*, 281–293. [CrossRef]
8. Wynberg, R. Biopiracy: Crying wolf or a lever for equity and conservation? *Res. Policy* **2023**, *52*, 104674. [CrossRef]
9. Fredriksson, M. Dilemmas of protection: Decolonising the regulation of genetic resources as cultural heritage. *Int. J. Herit. Stud.* **2021**, *27*, 720–733. [CrossRef]

10. Michiels, F.; Feiter, U.; Paquin-Jaloux, S.; Jungmann, D.; Braun, A.; Sayoc, M.A.P.; Armengol, R.; Wyss, M.; David, B. Facing the harsh reality of access and benefit sharing (ABS) legislation: An industry perspective. *Sustainability* **2022**, *14*, 277. [CrossRef]
11. Eaton, D.; Visser, B. Transaction costs of tracking and monitoring the Flows of Genetic Resources. In *A Moving Target: Genetic Resources and Options for Tracking and Monitoring their International Flows*; Muller, M.R., Lapeña, I., Eds.; ABS Series No. 3; IUCN: Gland, Switzerland, 2007; pp. 19–45.
12. Morgera, E.; Geelhoed, M. *Consultancy on the Notion of 'Utilisation' in the Nagoya Protocol and the EU ABS Regulation for the Upstream Actors*; University of Edinburgh: Edinburgh, UK, 2016. [CrossRef]
13. Marden, E.; Hamilton, R.S.; Halewood, M.; McCouch, S. International agreements and the plant genetics research community: A guide to practice. *Proc. Natl. Acad. Sci. USA* **2023**, *120*, e2205773119. [CrossRef]
14. Scholz, A.H.; Freitag, J.; Lyal, C.H.; Sara, R.; Cepeda, M.L.; Cancio, I.; Sett, S.; Hufton, A.L.; Abebaw, Y.; Bansal, K.; et al. Multilateral benefit-sharing from digital sequence information will support both science and biodiversity conservation. *Nat. Commun.* **2022**, *13*, 1086. [CrossRef]
15. Khoury, C.K.; Brush, S.; Costich, D.E.; Curry, H.A.; de Haan, S.; Engels, J.M.M.; Guarino, L.; Hoban, S.; Mercer, K.L.; Miller, A.J.; et al. Crop genetic erosion: Understanding and responding to loss of crop diversity. *New Phytol.* **2022**, *233*, 84–118. [CrossRef]
16. Mekonnen, D.K.; Spielman, D.J. Changing patterns in genebank acquisitions of crop genetic materials: An analysis of global policy drivers and potential consequences. *Food Policy* **2021**, *105*, 102161. [CrossRef]
17. Padulosi, S.; Sthapit, B.; Lamers, H.; Kennedy, G.; Hunter, D. Horticultural biodiversity to attain sustainable food and nutrition security. *Acta Hortic.* **2018**, *1205*, 21–34. [CrossRef]
18. Carey, J. Unearthing the origins of agriculture. *Proc. Natl. Acad. Sci. USA* **2023**, *120*, e2304407120. [CrossRef] [PubMed]
19. Valin, H.; Sands, R.D.; Van der Mensbrugghe, D.; Nelson, G.C.; Ahammad, H.; Blanc, E.; Bodirsky, B.; Fujimori, S.; Hasegawa, T.; Havlik, P.; et al. The future of food demand: Understanding differences in global economic models. *Agric. Econ.* **2014**, *45*, 51–67. [CrossRef]
20. Röös, E.; Bajželj, B.; Smith, P.; Patel, M.; Little, D.; Garnett, T. Greedy or needy? Land use and climate impacts of food in 2050 under different livestock futures. *Glob. Environ. Chang.* **2017**, *47*, 1–12. [CrossRef]
21. Tilman, D.; Balzer, C.; Hill, J.; Befort, B.L. Global food demand and the sustainable intensification of agriculture. *Proc. Natl. Acad. Sci. USA* **2011**, *108*, 20260–20264. [CrossRef] [PubMed]
22. Eriksson, D.; Akoroda, M.; Azmach, G.; Labuschagne, M.; Mahungu, N.; Ortiz, R. 2018. Measuring the impact of plant breeding on sub-Saharan African staple crops. *Outlook Agric.* **2018**, *47*, 163–180. [CrossRef]
23. Noleppa, S.; Cartsburg, M. The socio-economic and environmental values of plant breeding in the EU and for selected EU member states. In *An Ex-Post Evaluation and Ex-Ante Assessment Considering the "Farm to Fork" and "Biodiversity" Strategies*; HFFA Research Paper: Berlin, Germany, 2021; 327p. Available online: https://euroseeds.eu/app/uploads/2021/07/HFFA-Research-The-socio-economic-and-environmental-values-of-plant-breeding-in-the-EU.pdf (accessed on 31 July 2023).
24. Qaim, M. Globalisation of Agrifood Systems and Sustainable Nutrition. *Proc. Nutr. Soc.* **2017**, *76*, 12–21. [CrossRef]
25. Evenson, R.E.; Gollin, D. Assessing the impact of the Green Revolution, 1960 to 2000. *Science* **2003**, *300*, 758–762. [CrossRef]
26. Noleppa, S.; Cartsburg, M. The socio-economic and environmental values of plant breeding in the EU and for selected EU member states. In *An Addendum to Prior Research for Tomato and Alfalfa*; HFFA Research Paper: Berlin, Germany, 2021; 71p. Available online: https://euroseeds.eu/app/uploads/2021/07/HFFA-Research-The-socio-economic-and-environmental-values-of-plant-breeding-in-the-EU-addendum-for-tomato-and-alfalfa.pdf (accessed on 31 July 2023).
27. FAO; IFAD; UNICEF; WFP; WHO. The State of Food Security and Nutrition in the World 2022. In *Repurposing Food and Agricultural Policies to Make Healthy Diets More Affordable*; FAO: Rome, Italy, 2022. [CrossRef]
28. United Nations Development Programme (UNDP). Sustainable Development Goals. Available online: https://www.undp.org/content/undp/en/home/sustainable-development-goals.html (accessed on 10 May 2023).
29. Yadav, L.; Maurya, N.K. Fight Hidden Hunger through National Programs and Food Based Approaches. In *Combating Malnutrition through Sustainable Approaches*; Saeed, F., Ahmed, A., Afzaal, M., Eds.; Intech Open: London, UK, 2022; 19p. [CrossRef]
30. Oyebode, O.; Gordon-Dseagu, V.; Walker, A.; Mindell, J.S. Fruit and vegetable consumption and all-cause, cancer and CVD mortality: Analysis of Health Survey for England data. *J. Epidemiol. Commun. Health* **2014**, *68*, 856–862. [CrossRef]
31. Ebert, A.W. The Role of Vegetable Genetic Resources in Nutrition Security and Vegetable Breeding. *Plants* **2020**, *9*, 736. [CrossRef]
32. FAO. Agricultural production statistics. 2000–2020. In *FAOSTAT Analytical Brief Series No. 41*; FAO: Rome, Italy, 2022; 15p. Available online: https://www.fao.org/3/cb9180en/cb9180en.pdf (accessed on 31 July 2023).
33. World Health Organization (WHO). Diet, Nutrition, and the Prevention of Chronic Diseases. In *Report of A Joint WHO/FAO Expert Consultation*; WHO technical report series 916; WHO: Geneva, Switzerland, 2003; 160p, ISBN 92-4-120916-X.
34. Mason-D'Croz, D.; Bogard, J.R.; Sulser, T.B.; Cenacchi, N.; Dunston, S.; Herrero, M.; Wiebe, K. Gaps between fruit and vegetable production, demand, and recommended consumption at global and national levels: An integrated modelling study. *Lancet Planet. Health* **2019**, *3*, e318–e329. [CrossRef] [PubMed]
35. Smýkal, P.; Nelson, M.N.; Berger, J.D.; Von Wettberg, E.J. The Impact of Genetic Changes during Crop Domestication. *Agronomy* **2018**, *8*, 119. [CrossRef]
36. Turner-Hissong, S.D.; Mabry, M.E.; Beissinger, T.M.; Ross-Ibarra, J.; Pires, J.C. Evolutionary insights into plant breeding. *Curr. Opin. Plant Biol.* **2020**, *54*, 93–100. [CrossRef] [PubMed]

37. De Castro, M. Johann Gregor Mendel: Paragon of experimental science. *Mol. Genet. Genom. Med.* **2016**, *4*, 3–8. [CrossRef] [PubMed]
38. Breseghello, F.; Coelho, A.S. Traditional and modern plant breeding methods with examples in rice (*Oryza sativa* L.). *J. Agric. Food Chem.* **2013**, *61*, 8277–8286. [CrossRef]
39. Fu, Y.B. Understanding crop genetic diversity under modern plant breeding. *Theor. Appl. Genet.* **2015**, *128*, 2131–2142. [CrossRef]
40. Colombo, N.; Galmarini, C.R. The use of genetic, manual and chemical methods to control pollination in vegetable hybrid seed production: A review. *Plant Breed.* **2017**, *136*, 287–299. [CrossRef]
41. Convention on Biological Diversity (CBD). The Nagoya Protocol on Access and Benefit-Sharing. Available online: https://www.cbd.int/abs/about/default.shtml/ (accessed on 13 May 2023).
42. FAO. International Treaty on Plant Genetic Resources for Food and Agriculture. Available online: https://www.fao.org/3/i0510e/i0510e.pdf (accessed on 13 May 2023).
43. Brink, M.; Van Hintum, T. Genebank operation in the arena of access and benefit-sharing policies. *Front. Plant Sci.* **2020**, *10*, 1712. [CrossRef]
44. Convention on Biological Diversity (CBD) The Convention on Biological Diversity. Introduction. Available online: https://www.cbd.int/intro/ (accessed on 15 May 2023).
45. Convention on Biological Diversity (CBD) The Convention on Biological Diversity. Article 2. Use of Terms. Available online: https://www.cbd.int/convention/articles/?a=cbd-02 (accessed on 15 May 2023).
46. Convention on Biological Diversity (CBD) The Convention on Biological Diversity. Article 15. Access to Genetic Resources. Available online: https://www.cbd.int/convention/articles/?a=cbd-15 (accessed on 15 May 2023).
47. Convention on Biological Diversity (CBD) Access and Benefit-Sharing Clearing House (ABSCH). Available online: https://absch.cbd.int/en/ (accessed on 18 May 2023).
48. Noriega, I.L.; Halewood, M.; Abberton, M.; Amri, A.; Angarawai, I.I.; Anglin, N.; Blümmel, M.; Bouman, B.; Campos, H.; Costich, D.; et al. CGIAR operations under the plant treaty framework. *Crop Sci.* **2019**, *59*, 819–832. [CrossRef]
49. FAO. International Treaty on Plant Genetic Resources for Food and Agriculture. Available online: https://www.fao.org/plant-treaty/overview/en/ (accessed on 17 May 2023).
50. ECPGR. Report of the Fourteenth Steering Committee Meeting, 31 May–2 June 2016, Višegrad, Bosnia and Herzegovina. European Cooperative Programme for Plant Genetic Resources/Bioversity International, Rome, Italy. 2016. Available online: https://www.gzr.cz/wp-content/uploads/2018/05/Report-of-the-ECPGR-Steering-Committee.pdf (accessed on 31 July 2023).
51. Nordic Genetic Resource Centre (NordGen). *Access and Rights to Genetic Resources: A Nordic Approach (II)*; Nordic Genetic Resource Centre: Alnarp, Sweden, 2023; p. 80. Available online: https://norden.diva-portal.org/smash/get/diva2:1744186/FULLTEXT01.pdf (accessed on 31 July 2023).
52. Manzella, D. The Multilateral System of Access and Benefit-sharing. In Proceedings of the International Workshop on Access and Benefit-Sharing for Genetic Resources for Food and Agriculture, Rome, Italy, 10–12 January 2018; FAO: Rome, Italy, 2018; pp. 7–17. Available online: https://www.fao.org/3/CA0099EN/ca0099en.pdf#page=29 (accessed on 27 May 2023).
53. Access to Seeds Foundation. Access and Benefit Sharing. Access to Seeds Index 2019—Global Seed Companies. Available online: Accesstoseeds.org/index/global-seed-companies/key-findings/access-and-benefit-sharing/ (accessed on 8 June 2023).
54. FAO; ITPGRFA. International Treaty on Plant Genetic Resources for Food and Agriculture. In Proceedings of the Ninth Meeting of the Ad Hoc Open-Ended Working Group to Enhance the Functioning of the Multilateral System. Submissions from Contracting Parties and Stakeholders on Matters to be Discussed at the Ninth Meeting of the Working Group, Rome, Italy, 17–21 June 2019. Available online: https://www.fao.org/3/ca4962en/ca4962en.pdf (accessed on 8 June 2023).
55. Tsioumani, E.; Kantai, T.; Nyingi, W.; Tsioumanis, A. Summary of the Ninth Session of the Governing Body of the International Treaty on Plant Genetic Resources for Food and Agriculture: 19–24 September 2022. *Earth Negot. Bull.* **2022**, *9*, 1–13. Available online: https://enb.iisd.org/sites/default/files/2022-09/enb09782e_0.pdf (accessed on 15 June 2023).
56. Laird, S.; Wynberg, R.; Rourke, M.; Humphries, F.; Muller, M.R.; Lawson, C. Rethink the expansion of access and benefit sharing. *Science* **2020**, *367*, 1200–1202. [CrossRef] [PubMed]
57. Cowell, C.; Paton, A.; Borrell, J.S.; Williams, C.; Wilkin, P.; Antonelli, A.; Baker, W.J.; Buggs, R.; Fay, M.F.; Gargiulo, R.; et al. Uses and benefits of digital sequence information from plant genetic resources: Lessons learnt from botanical collections. *Plants People Planet* **2022**, *4*, 33–43. [CrossRef]
58. von Wettberg, E.; Khoury, C.K. Biodiversity data: The importance of access and the challenges regarding benefit sharing. *Plants People Planet* **2022**, *4*, 2–4. [CrossRef]
59. Rhoden, F.; Scholz, A.H. The international political process around Digital Sequence Information under the Convention on Biological Diversity and the 2018–2020 intersessional period. *Plants People Planet* **2021**, *4*, 51–60. [CrossRef]
60. Gilliot, P.A.; Gorochowski, T.E. Sequencing enabling design and learning in synthetic biology. *Curr. Opin. Chem. Biol.* **2020**, *58*, 54–62. [CrossRef]
61. Scholz, A.; Upneet, H.; Freitag, J.; Cancio, I.; dos Santos Ribeiro, C.; Haringhuizen, G.; Oldham, P.; Saxena, D.; Seitz, C.; Thiele, T.; et al. *Finding Compromise on ABS & DSI in the CBD: Requirements & Policy Ideas from a Scientific Perspective*; Leibniz Institute DSMZ-German Collection of Microorganisms and Cell Cultures GmbH: Braunschweig, Germany, 2020. [CrossRef]
62. Cochrane, G.; Karsch-Mizrachi, I.; Takagi, T. International Nucleotide Sequence Database Collaboration. The international nucleotide sequence database collaboration. *Nucleic Acids Res.* **2016**, *44*, D48–D50. [CrossRef]

63. Hamilton, R.S.; Halewood, M.; López Noriega, I.; Rouard, M.; Tiambo, C.K.; Azevedo, V.; Bhattacharjee, R.; Cuéllar, W.J.; Gisel, A.; Hearne, S.J.; et al. Digital sequence information is changing the way genetic resources are used in agricultural research and development: Implications for new benefit-sharing norms. In *A Discussion Paper from CGIAR for Consideration by Delegates to the 15th Session of the Conference of the Parties to the Convention on Biological Diversity*; CGIAR Initiative on Genebanks: Lima, Peru, 2022; 49p. Available online: https://cgspace.cgiar.org/bitstream/handle/10568/125749/CGIAR%20DSI%20COP%2015%20discussion%20paper_12.05.22.pdf?sequence=4&isAllowed=y (accessed on 24 June 2023).
64. Kharb, D. The legal conundrum over regulation of access and benefit sharing obligations in digital sequence information over genetic resources-assessing Indian position. *J. World Intellect. Prop.* **2021**, *24*, 152–166. [CrossRef]
65. Nawaz, S.; Satterfield, T.; Hagerman, S. From seed to sequence: Dematerialization and the battle to (re) define genetic resources. *Glob. Environ. Chang.* **2021**, *68*, 102260. [CrossRef]
66. Lawson, C.; Humphries, F.; Rourke, M. The future of information under the CBD, Nagoya Protocol, Plant Treaty, and PIP Framework. *J. World Intellect. Prop.* **2019**, *22*, 103–119. [CrossRef]
67. Aubry, S.; Frison, C.; Medaglia, J.C.; Frison, E.; Jaspars, M.; Rabone, M.; Sirakaya, A.; Saxena, D.; Van Zimmeren, E. Bringing access and benefit sharing into the digital age. *Plants People Planet* **2022**, *4*, 5–12. [CrossRef]
68. Rabone, M.; Harden-Davies, H.; Collins, J.E.; Zajderman, S.; Appeltans, W.; Droege, G.; Brandt, A.; Pardo-Lopez, L.; Dahlgren, T.G.; Glover, A.G.; et al. Access to marine genetic resources (MGR): Raising awareness of best-practice through a new agreement for biodiversity beyond national jurisdiction (BBNJ). *Front. Mar. Sci.* **2019**, *6*, 520. [CrossRef]
69. Gollin, D. Conserving genetic resources for agriculture: Economic implications of emerging science. *Food Secur.* **2020**, *12*, 919–927. [CrossRef]
70. DSI Scientific Network. Giving the Scientific Community A Voice on Digital Sequence Information. Available online: https://www.dsiscientificnetwork.org/ (accessed on 10 June 2023).
71. International Chamber of Commerce. Encouraging ABS Agreements and Sustainable Use of Genetic Resources. 2018. Available online: https://iccwbo.org/publication/encouraging-abs-agreements-sustainable-use-genetic-resources/ (accessed on 7 June 2023).
72. Aubry, S. The future of digital sequence information for plant genetic resources for food and agriculture [Policy and Practice Reviews]. *Front. Plant Sci.* **2019**, *10*, 1046. [CrossRef]
73. India. Observation Filed under Article 115 of European Patent Convention Regarding the Use of Indian Biological Resource and Non-Compliance with India's Biological Diversity Act. 2002. Available online: https://www.ip-watch.org/weblog/wp-content/uploads/2016/01/National-Biodiversity-Authority-India.pdf (accessed on 30 July 2023).
74. Normand, V.; Oliva, M.J.; Müller, S.; al-Janabi, S.; Zohren, Y.; Nöske, N.; Zedda, L.; Paulsch, A. The Contribution of Access and Benefit-Sharing (ABS) to the Sustainable Development Goals. Lessons Learned and Best Practices. 2021. Available online: https://absch.cbd.int/api/v2013/documents/22130401-0F0C-688C-1548-C6E4F1E53BE1/attachments/213523/Compendium%20-%20The%20contribution%20of%20ABS%20to%20the%20Sustainable%20Development%20Goals.pdf (accessed on 9 June 2023).
75. Rabitz, F. Access without benefit-sharing: Design, effectiveness and reform of the FAO seed treaty. *Int. J. Commons* **2017**, *11*, 621–640. [CrossRef]
76. Adebola, T.; Manzella, D. Access and Benefit-Sharing and Digital Sequence Information in Africa: A Critical Analysis of Contemporary Concerns in Regional Governance. In *Access and Benefit Sharing of Genetic Resources, Information and Traditional Knowledge*, 1st ed.; Lawson, C., Rourke, M., Humphries, F., Eds.; Routledge: Abingdon-on-Thames, UK, 2022; Available online: https://ssrn.com/abstract=4307587 (accessed on 31 July 2023).
77. Borrell, J.S.; Dodsworth, S.; Forest, F.; Perez-Escobar, O.A.; Lee, M.A.; Mattana, E.; Stevenson, P.C.; Howes, M.J.R.; Pritchard, H.W.; Ballesteros, D.; et al. The climatic challenge: Which plants will people use in the next century? *Environ. Exp. Bot.* **2020**, *170*, 103872. [CrossRef]
78. Neumann, D.; Borisenko, A.V.; Coddington, J.A.; Hauser, L.C.; Butler, R.C.; Casino, A.; Vogel, J.C.; Haszprunar, G.; Giere, P. Global biodiversity research tied up by juridical interpretations of access and benefit-sharing. *Org. Divers. Evol.* **2018**, *18*, 1–12. [CrossRef]
79. Prathapan, K.D.; Pethiyagoda, R.; Bawa, K.S.; Raven, P.H.; Rajan, P.D.; 172 Co-Signatories from 35 Countries. When the cure kills-CBD limits biodiversity research. *Science* **2018**, *360*, 1405–1406. [CrossRef]
80. Silvestri, L.; Sosa, A.; Mc Kay, F.; Diniz Vitorino, M.; Hill, M.; Zachariades, C.; Hight, S.; Weyl, P.; Smith, D.; Djeddour, D.; et al. Implementation of access and benefit sharing measures has consequences for classical biological control of weeds. *BioControl* **2020**, *65*, 125–141. [CrossRef]
81. Silvestri, L.C.; Mason, P.G. Improved access to biological control genetic resources: Navigating through the Convention on Biological Diversity and the Nagoya Protocol. *BioControl* **2023**, *68*, 299–310. [CrossRef]
82. Mc Kay, F.; Sosa, A.; Cabrera Walsh, G. Accessing biological control genetic resources: The South American experience. *BioControl* **2023**, *68*, 261–267. [CrossRef]
83. Silvestri, L. Acceso a recursos genéticos y distribución de beneficios en Colombia: Desafíos del régimen normativo. *Rev. Investig. Y Desarro.* **2016**, *24*, 1–25. [CrossRef]
84. Silvestri, L.C. Conservación de la diversidad genética en el Perú: Desafíos en la implementación del régimen de acceso a recursos genéticos y distribución de beneficios. *Rev. Peru. Biol.* **2016**, *23*, 73–79. [CrossRef]

85. FAO. Digital Object Identifiers for Food Crops. Descriptors and Guidelines of the Global Information System. Available online: https://www.fao.org/documents/card/en/c/I8840EN (accessed on 28 May 2023).
86. FAO. *Norway Contributes to International Treaty*; FAO: Rome, Italy, 2020. Available online: http://www.fao.org/plant-treaty/news/news-detail/en/c/1296008/ (accessed on 20 May 2023).

MDPI
St. Alban-Anlage 66
4052 Basel
Switzerland
www.mdpi.com

Plants Editorial Office
E-mail: plants@mdpi.com
www.mdpi.com/journal/plants

www.ingramcontent.com/pod-product-compliance
Lightning Source LLC
Chambersburg PA
CBHW061611220926
51850CB00005B/477
9783725805891